NURSERY REARING OF NONHUMAN PRIMATES IN THE 21st CENTURY

DEVELOPMENTS IN PRIMATOLOGY: PROGRESS AND PROSPECTS

Series Editor:
Russell H. Tuttle
University of Chicago, Chicago, Illinois

This peer-reviewed book series will meld the facts of organic diversity with the continuity of the evolutionary process. The volumes in this series will exemplify the diversity of theoretical perspectives and methodological approaches currently employed by primatologists and physical anthropologists. Specific coverage includes: primate behavior in natural habitats and captive settings; primate ecology and conservation; functional morphology and developmental biology of primates; primate systematics; genetic and phenotypic differences among living primates; and paleoprimatology.

ALL APES GREAT AND SMALL
VOLUME I: AFRICAN APES
Edited by Birute M.F. Galdikas, Nancy Erickson Briggs, Lori K. Sheeran, Gary L. Shapiro and Jane Goodall

THE GUENONS: DIVERSITY AND ADAPTATION IN AFRICAN MONKEYS
Edited by Mary E. Glenn and Marina Cords

ANIMAL MINDS, HUMAN BODIES
By W.A. Hillix and Duane Rumbaugh

COMPARATIVE VERTEBRATE COGNITION
Edited by Lesley J. Rogers and Gisela Kaplan

ANTHROPOID ORIGINS: NEW VISIONS
Edited by Callum F. Ross and Richard F. Kay

MODERN MORPHOMETRICS IN PHYSICAL ANTHROPOLOGY
Edited by Dennis E. Slice

BEHAVIORAL FLEXIBILITY IN PRIMATES: CAUSES AND CONSEQUENCES
By Clara B. Jones

NURSERY REARING OF NONHUMAN PRIMATES IN THE 21st CENTURY
Edited by Gene P. Sackett, Gerald C. Ruppenthal and Kate Elias

NURSERY REARING OF NONHUMAN PRIMATES IN THE 21st CENTURY

Edited by

Gene P. Sackett
Washington National Primate Research Center
University of Washington
Seattle, WA, USA

Gerald C. Ruppentahal
Washington National Primate Research Center
University of Washington
Seattle, WA, USA

Kate Elias
Washington National Primate Research Center
University of Washington
Seattle, WA, USA

Springer

Gene P. Sackett
Washington National Primate Research Center
University of Washington
Seattle, WA
USA

Gerald C. Ruppentahal
Washington National Primate Research Center
University of Washington
Seattle, WA
USA

Kate Elias
Washington National Primate Research Center
University of Washington
Seattle, WA
USA

Library of Congress Control Number: 2005924427

ISBN-10: 0-387-25632-6
ISBN-13: 978-0387-25632-0

Printed on acid-free paper.

Printed in the United States of America. (BS/EB)

9 8 7 6 5 4 3 2 1

springeronline.com

PREFACE: OUR HISTORICAL NOTE

It was 1970 when Gerry Ruppenthal and Jim Sackett moved from the University of Wisconsin in Madison to the University of Washington in Seattle. In Wisconsin, Gerry had been working for Harry and Margaret Harlow for over a decade and supervised many of their classic studies of the 1960s. Jim was an associate professor in the psychology department and was in his seventh year of primate research in what is now the Harlow Primate Laboratory. Collectively, we were experienced in studying infant, juvenile, and adult rhesus macaques that had been reared under a variety of captive conditions. Many of these monkeys had been reared in the Harlow Primate Laboratory nursery.

We were brought to Seattle by the Regional Primate Research Center (PC) and the Child Development and Mental Retardation Center (CDMRC), now called the Center on Human Development and Disability (CHDD). Gerry was a research scientist with both NIH-funded centers, while Jim was a professor in the psychology department, a PC core staff member, and PI of a CDMRC 3-year new program grant. Harry Harlow generously allowed us to take a great deal of equipment from Wisconsin—cages, rearing units, developmental testing apparatus, and recording devices. If he had paid more attention, we probably would not have gotten away with as much "loot," but the University of Wisconsin and NIH approved it all. With some additional local start-up funds and a really large amount of space, we started a primate nursery and organized a developmental rearing room and some testing rooms. Our equipment and methods were essentially identical to those used for over 15 years in the Wisconsin laboratory.

Our initial goal was to replicate the effects of isolation and peer rearing found in rhesus macaques in the classic Harlow work. We wanted to do this to learn why male rhesus macaques were affected so much more than females by asocial rearing. Because the PC breeding colony consisted of pigtailed macaques (*Macaca nemestrina*), we needed to demonstrate that they responded like rhesus macaques with the same sex and rearing differences before going on to mechanistic studies of possible causes. Not only did we fail to find sex differences following asocial rearing of pigtails, we also failed to find the same devastating effects of isolation rearing on postrearing social and exploratory behavior. This convinced us that gene–environment interactions were to be expected in assessing acute and chronic effects of rearing conditions on behavior, and probably also on physiology. This theme, in one guise or another, will be found through much of this book.

Our nursery is situated adjacent to the University Hospital, just below the human neonatal intensive care unit, and word soon got out that it was possible to study monkey pregnancy, neonates, and infants in a primate nursery located almost next to one's own office and laboratory. This led to requests to use our facilities by a number of medical researchers, especially a group of neonatologists who were studying lung function in premature newborns. With their help, our nursery came to include a primate neonatal intensive care unit. Also, rather than euthanizing newborns that were premature or low birth weight, ill, had life-threatening birth defects, or whose mothers were ill, wounded, or dead, the PC breeding colony managers began sending such at-risk neonates and young infants to our nursery, initiating an "Infant-Save" program that continues to this day.

By 1971 it was obvious that our nursery was a valuable resource for scientists who were interested in prenatal, perinatal, and infancy studies. As both the PC and CDMRC 5-year core grants were being written for renewal in 1972, we convinced the directors of both centers to include a proposal to support a nursery facility, the Infant Primate Research Laboratory (IPRL). In what appears to be a unique relationship among NIH-supported university centers, our proposal was funded by both core grant requests and we have shared this funding ever since.

From our point of view, the IPRL has had two main purposes. The first has been to use primate models to study important human medical and behavioral problems. This has been the major reason for continued grant success in the CDMRC arm of our endeavors. Equally important, we have spent much of our PC-based resources studying primate medicine and husbandry issues related to breeding, pregnancy and fetal development, hand-rearing methods, and methods of assessing growth, physiology, and behavioral development. Our NIH-supported efforts and those of many other researchers and veterinarians led to the then state-of-the-art publication, *Nursery Care of Nonhuman Primates*, edited by Gerry Ruppenthal and Dorothy Reese, published by Plenum in 1979.

Although much of that book is still relevant today, methods of nursery care, methods of testing, and types of experimental and husbandry problems have changed markedly since 1979. This has resulted in new challenges involving nursery rearing of monkeys with many types of naturally occurring and experimentally induced medical and developmental conditions. New challenges have arisen through changes in attitudes toward animal testing and resulting changes in standards of animal care involving concerns for both the physical and psychological well-being of captive primates. One goal of this book is to describe how these challenges have been met over the past 25 years. The other goal is to show how changes in rearing methods have altered for the better the developmental outcomes of nursery rearing, at least in some species and some facilities. We hope that our story, begun in the 1979 book and continued in the current one, will produce a more realistic view of nursery rearing and its effects than that claimed by opponents of nursery rearing on the basis of antiquated methods now used by only a few facilities or individuals.

This book originated in a workshop of the same name, *Nursery Rearing of Nonhuman Primates in the 21st Century*, held in 2002 at the Oklahoma City meeting of the American Society of Primatologists. All of the workshop presenters are represented, in addition to a number of authors recruited to present important topics not covered in the workshop. We are grateful to all our contributors. We had hoped to include either a section or a CD of basic growth and health data for nursery-compared with mother-reared primates that would serve as normative comparison data for current and future research. Unfortunately, we were able to collect such data on only a few species, although they are ones that are frequently nursery reared in current work. Data on health statistics are included in the final section as an example of basic data that can, and probably should, be collected and disseminated for all laboratory and zoo nursery-reared primate species.

Over the decades our work has involved a large number of students, scientists, health workers, technical personnel, and administrators. They are too numerous to list here, but we must thank our earliest University of Washington students, Dick Holm, Sharon Ramey (nee Landesman), and Jon Lewis, who helped us start the IPRL in both concept and fact. Carol Fahrenbruch, Sherry Savage, Colleen Walker-Gelatt, and Gary Bartram provided invaluable effort in developing and implementing our rearing and developmental testing methods. Without the support of our Primate Center directors, especially Orville Smith, and our CDMRC directors, especially Irvin Emanuel and Michael Guralnick, we would have had quite different careers. We are also grateful to the NIH for its continued support from the National Center for Research Resources, grant RR00166, and NICHHD Mental Retardation Branch, grant HD02274.

Gene P. (Jim) Sackett
Gerald C. Ruppenthal
Kate Elias
Seattle and Pittsburgh, 2004

CONTENTS

CONTRIBUTORS

Christian R. Abee, Department of Comparative Medicine, University of South Alabama, Mobile, AL 36688, USA

Kate C. Baker, Tulane National Primate Research Center, Veterinary Medicine, Covington, LA 70433, USA

Allyson J. Bennett, Physiology/Pharmacology and Pediatrics, Wake Forest University School of Medicine, Winston-Salem, NC 27157, USA

Tamara Bettinger, Cleveland Metroparks Zoo, Conservation and Science, Cleveland, OH 44109, USA

Mollie A. Bloomsmith, Zoo Atlanta, Research/TECH Lab, Atlanta, GA 30315

Ann Bode, Southwest Foundation for Biomedical Research San Antonio, TX 78245, USA

Alan G. Brady, Department of Comparative Medicine, University of South Alabama Mobile, AL 36688, USA

Linda Brent, President and Director, Chimp Haven, Inc., Leithville, LA 71047, USA

John P. Capitanio, California National Primate Research Center, University of California, Davis, CA 95616, USA

Arnold S. Chamove, Massey University School of Psychology, Palmerston North, 5331 New Zealand

Maribeth Champoux, Laboratory of Comparative Ethology, National Institutes Animal Center, NIH, National Institute of Child Health and Human Development, Poolesville, MD 20837, USA

Christopher L. Coe, Harlow Primate Laboratory, University of Wisconsin, Madison, WI 53715-1239, USA

Carolyn M. Crockett, Washington National Primate Research Center, University of Washington, Seattle, WA 98195, USA

Arthur E. Davis, Washington National Primate Research Center, University of Washington, Seattle, WA 98195, USA

Laura DelRosso, California National Primate Research Center, University of California, Davis, CA 95616, USA

Debra L. Durham, Young Hall Department of Anthropology, University of California Animal Behavior Graduate Group, Davis, CA 95616, USA

Jo Fritz, Primate Foundation of Arizona, Mesa, AZ 85277, USA

Susan V. Gibson, Department of Comparative Medicine, University of South Alabama, Mobile, AL 36688, USA

James C. Ha, Washington National Primate Research Center, University of Washington, Seattle, WA 98195, USA

Christy Harris, Washington National Primate Research Center, University of Washington, Seattle, WA 98195, USA

Rodney J. R. Ho, Department of Pharmaceutics, University of Washington, Seattle, WA 98195, USA

Kathleen Hoffman, Primate Foundation of Arizona, Mesa, AZ 85277, USA

Susan M. Howell, Alpha Genesis Research, Yemassee, SC 29945, USA

Ludwig Huber, Institute of Zoology, University of Vienna, Vienna A-1090, Austria

M. Kay Izard, Alamogordo Primate Facility, Holloman AFB, NM 88330, USA

Susan P. Lambeth, Department of Veterinary Sciences, University of Texas, M. D. Anderson Cancer Center, Bastrop, TX 78602, USA

Mark L. Laudenslager, Department of Psychiatry, University of Colorado Health Sciences Center, Denver, CO 80220, USA

Jennifer Leigh, Washington National Primate Research Center, University of Washington, Seattle, WA 98195, USA

Gabriele R. Lubach, Harlow Primate Laboratory, University of Wisconsin, Madison, WI 53715, USA

Mary W. Marzke, Department of Anthropology, Arizona State University, Tempe, AZ 85287, USA

William A. Mason, California National Primate Research Center, University of California, Davis, CA 95616, USA

Sally P. Mendoza, California National Primate Research Center, University of California, Davis, CA 95616, USA

Colleen F. Moore, Department of Psychology, Harlow Primate Laboratory, University of Wisconsin, Madison, Madison, WI 53715, USA

James Murphy, Primate Foundation of Arizona, Mesa, AZ 85277, USA

Laura L. Newell-Morris, Department of Anthropology, University of Washington, Seattle, WA 98195, USA

Kelli Niebruegge, Saint Louis Zoo, St. Louis, MO 63110, USA

Matthew Francis Stuart Xavier Novak, Laboratory of Comparative Ethology, National Institutes Animal Center, NIH, National Institute of Child Health and Human Development, Poolesville, MD 20837, USA

Melinda A. Novak, Department of Psychology, University of Massachusetts, Amherst, MA 01003, USA

Peter J. Pierre, Physiology/Pharmacology and Pediatrics, Wake Forest University School of Medicine, Winston-Salem, NC 27157, USA

Ingrid Porton, Saint Louis Zoo, St. Louis, MO 63110, USA

Erika Rainwater, Washington National Primate Research Center, University of Washington, Seattle, WA 98195, USA

James K. Rilling, Department of Anthropology, Emory University, Atlanta, GA 30322, USA

Jeffrey A. Roberts, California National Primate Research Center, University of California, Davis, CA 95616, USA

Peter G. Roma, NIH Animal Center, Poolesville, MD 20837, USA

Stephen R. Ross, Lincoln Park Zoo, Conservation and Science, Chicago, IL 60614, USA

Gerald C. Ruppenthal, Washington National Primate Research Center, University of Washington, Seattle, WA 98195, USA

Gene P. Sackett, Washington National Primate Research Center, University of Washington, Seattle, WA 98195, USA

Mary L. Schneider, Departments of Kinesiology and Psychology, Harlow Primate Laboratory, University of Wisconsin, Madison, WI 53715, USA

Melanie Schwandt, National Institutes Animal Center, NIH/NIAAA, Poolesville, MD 20837, USA

Kelsey Stratton, Washington National Primate Research Center, University of Washington, Seattle, WA 98195, USA

Stephen J. Suomi, Laboratory of Comparative Ethology, National Institute of Child Health and Human Development, Bethesda, MD 20892, USA

Bernhard Voelkl, Institute of Zoology, University of Vienna, Vienna A-1090, Austria

Lawrence E. Williams, Department of Comparative Medicine, University of South Alabama, Mobile, AL 36688-0002, USA

Julie M. Worlein, Washington National Primate Research Center, University of Washington, Seattle, WA 98195, USA

Dennis Young, Department of Anthropology, Arizona State University, Tempe, AZ 85287, USA

INTRODUCTION

As described in the preface, this book originated in a workshop focused on changes in nursery-rearing practices over the past 35 years and the state of the art in the early 21st century. In designing the workshop, we identified four areas of change: (1) new or modified goals of nursery rearing, (2) new concepts concerning these goals, (3) new methods for attaining these goals, and (4) new data concerning the effects of nursery or hand rearing on the health and biobehavioral and social development of infant and juvenile primates.

1. GOALS

There are two primary goals of contemporary nursery rearing. The first involves saving at-risk newborns and young infants that cannot be reared by their mothers. This goal serves the purposes of conservation of endangered species, display in zoos, production of future breeders, and preservation of natural models of human health or behavior problems. Examples of the latter include low-birth-weight or premature neonates and neonates with genetic defects such as trisomic chromosome conditions. The second goal involves nursery rearing and its variations as specific experimental procedures in research studies. Examples of new research goals include the need for biological containment of neonates and infants in viral and other disease research, the production of specific pathogen-free (SPF) colonies by removal of neonates from the mother, and the production of phenotypes for genome manipulation, assisted reproduction technology, and molecular biology studies.

2. CONCEPTS

A major change in concepts concerning nursery rearing is that nursery-reared primates no longer are simply warehoused for future assignment to research projects. Instead, modern methods are aimed at providing psychologically rich environments to foster reasonably normative behavioral and physiological development as compared with some standard such as rearing with mothers in captive environments. The underlying concept is that normal animals make better research subjects. In the extreme view, only normal animals provide valid subjects in most primate research projects, whether these are behavioral or biological in nature. A variant of this view is that some degree of normal development is necessary for producing successful breeders and healthy colony members. A major problem, however, is how to define "normal." Many of the chapters in this book are relevant to this definitional issue.

On the other hand, research over the past 50 years in developmental sciences, ranging from molecular biology to developmental psychology, shows that genes and environment work together to produce a variety of "normal" phenotypes at all levels of study (e.g., Gottlieb, 1998). With respect to the development of nonhuman primates following variations in rearing conditions, gene–environment interactions appear to be the norm rather than the exception. An important example of this interaction in nonhuman primates is seen in work by Suomi and his colleagues at the Laboratory of Comparative Ethology of the National Institute of Child Health and Human Development.

Rhesus macaques have a variation in the serotonin transporter gene regulatory region (5-HTTLPR), with some individuals having a long allele and some having a short allele. Humans have a similar polymorphism in this same region, which appears to be associated with phenotypic variation in levels of anxiety, depression, affective disorders, and aggression. In one study, Bennet *et al.* (2002) compared concentrations of serotonin in the cerebrospinal fluid of monkeys that were reared with their mothers versus monkeys that were reared in a nursery and then housed with agemates. Among nursery-peer monkeys, those with the short allele had a marked reduction in serotonin compared with monkeys that had the long allele. Similarly. in a study of social play and aggression (Barr *et al.*, 2003), nursery-peer monkeys with the long allele displayed more play and less aggression than those with the short allele, whereas mother-reared infants showed no allele-related differences in behavior. Thus, the effects of genotype on both neurochemistry and behavior depended on how the monkeys were reared.

We were not surprised by these results. In two studies we addressed the question of species differences in the effects of social-isolation rearing (Sackett *et al.*, 1976, 1981). Three species of macaque—rhesus (*Macaca mulatta*), pigtailed (*M. nemestrina*), and longtailed (*M. fascicularus*)—were reared in identical environments with no social or sensory contact with other monkeys for 6–7 months from birth. During the rearing period the rhesus macaques developed a typical isolation syndrome: they displayed no play or exploration behaviors and spent most of their time in self-directed and repetitive behaviors. Longtailed macaques displayed more moderate levels of the isolation syndrome, while pigtailed macaques showed much lower levels of isolate behavior and a relatively high level of play. During postrearing social behavior tests, rhesus macaques continued to show mostly isolate syndrome behaviors, with no play or socially initiated activity. Pigtailed macaques had greatly reduced isolate syndrome behavior, and engaged in some positive social behavior and the same high levels of environmental exploration as controls reared with mothers and peer experience. Longtailed macaques also had a great deal of isolate behavior, but engaged in as much positive social behavior as did control animals reared with mothers and peers. Thus, when reared in an identical impoverished environment, genetically different primate species were differentially affected in degree of both deviant and species-typical behaviors. Furthermore, the results showed that the classic "isolation rearing syndrome" of abnormal personal behavior, neophobia, and lack of social behavior was valid only for the rhesus macaque species.

The lesson of studies such as these is that we should expect to see variations in the effects of nursery rearing both within and between species. This variation does not necessarily mean that we have produced abnormal monkeys. Rather, depending on genotype, each species and individual responds to environmental variation with behaviors that are adaptive in that environment. Motherless rearing in a nursery imposes a major challenge on this adaptability during and after the rearing experience. Chapters in this book suggest that some primate species, especially prosimians, may be difficult to rear under motherless conditions, whereas other species appear to thrive when reared under modern husbandry and social conditions (e.g., Sackett *et al.*, 2002). We believe that there are nursery-rearing conditions that can produce adaptable juveniles and adults, capable of reproducing their species, for all normal genotypes of all primate species. Although this ideal has not been accomplished for all species to date, we hope the data presented in this book and the questions these data raise for future research will help us learn how to attain this goal.

3. METHODS

Research over the past 35 years has led to the development of new nursery-rearing methods. So far, these methods have been specific for particular primate species and institutions that rear primates in captivity. In addition, the psychological well-being movement and the Institutional Animal Care Committee (IACC) have introduced new dimensions into the methods for rearing primates in captivity. These dimensions involve both the physical environment and social–behavioral considerations. In this context, rules and regulations regarding rearing methods may bring bureaucracy into conflict with the conditions that actually foster development for meeting the goals of research, breeding and husbandry, conservation, or public display. IACC and related regulations specify factors ranging from cage sizes to protective gear for primate researchers and caregivers. This has greatly increased the costs of nursery rearing and of doing research with nonhuman primates in general. It has also greatly reduced both the quantity and quality of contact between humans and their monkey charges. On the other hand, these influences have greatly increased the quality and quantity of peer social contact and nonsocial environmental enrichment afforded nursery-reared individuals. A number of methods to cope with these influences are illustrated in chapters in this book, but many issues still remain to be solved by future research and technological innovation.

4. DATA

Newer data of particular importance concern variations in physiological and growth systems that differentially affect nursery-reared compared with mother-reared primates. Examples of such data are maturational, immunological, neurochemical, and hormonal effects that impact all aspects of development. Such information is critical for identifying appropriate research subjects in biobehavioral experiments, as well as for understanding deviations from normative developmental patterns. These new data are also of importance for understanding nongenetic intergenerational phenomena that affect the health and behavior of future offspring. Of course, new data identifying conditions that produce healthy and adaptive nursery-reared individuals are equally important. Chapters throughout the first four sections of this book present such data.

5. CHAPTER ORGANIZATION

The chapters in this book represent as wide a range of genera—great apes, macques and baboons, squirrel and marmoset monkeys, and prosimians—as we could identify for which developmental data on nursery-reared animals have been collected systematically under describable conditions. The book is organized in five sections. Section 1 presents a brief history of nursery rearing and some practical and theoretical issues bearing on contemporary nursery-rearing practices. Sections 2 and 3 present methods and outcomes of nursery rearing in prosimian and simian species. Section 4 deals with general veterinary issues, the rearing of high-risk infants, and some contemporary rearing and research methods important for current and future studies of primate development. The fifth section presents some difficult-to-find health data for representative species commonly reared in nursery environments.

Gene P. Sackett
Gerald C. Ruppenthal
Kate Elias

REFERENCES

Barr, C.S.D., Newman, T.K., Becker, M.L., Parker, C.C., Champoux, M., Lesch, K.P., Goldman, D., Suomi, S.J., and Higley, J.D. (2003). The utility of the non-human primate model for studying gene by environment interactions in behavioral research. *Gene, Brain, Behav.* 2:336–340.

Bennett, A.J., Lesch, K.P., Heils, A., Long, J.C., Lorenz, J.G., Shoaf, S.E., Champoux, M., Suomi, S.J., Linnoila, M.V., and Higley, J.D. (2002). Early experience and serotonin transporter gene variation interact to influence primate CNS function. *Mol. Psychiat.* 7:118–122.

Gottlieb, G. (1998). Normally occurring environmental and behavioral influences on gene activity: From central dogma to probabilistic epigenesis. *Psychol. Rev.* (4):792–802.

Sackett, G.P., Holm, R.A., and Ruppenthal, G.C. (1976). Social isolation rearing: Species differences in behavior of macaque monkeys. *Dev. Psychol.* 12:283–288.

Sackett, G.P., Ruppenthal, G.C., Fahrenbruch, C., Holm, R.A., and Greenough, W.A. (1981). Social isolation rearing effects in monkeys vary with genotype. *Dev. Psychobiol.* 17:313–318.

Sackett, G.P., Ruppenthal, G.C., and Davis, A.E. (2002). Survival, growth, health, and reproduction following nursery rearing compared with mother rearing in pigtailed monkeys (*Macaca nemestrina*). *Am. J. Primatol.* 56:165–184.

SECTION ONE

Introduction to Section 1: The History of Nursery Rearing and a Glimpse into the Future

The first two chapters of this section present brief historical summaries of nursery rearing in laboratories and zoos. These are followed by a chapter on federal regulations for the welfare of captive primates, especially as they may influence current and future nursery-rearing practices. Data management has been an important issue for the past 30 years in the international zoo effort to preserve endangered species. Although similar efforts have been made to systematize data management in primate research institutions, success has been slow in coming. Chapter 4 illustrates the need for modern data management and a possible strategy for development of a basic system of data to be collected in primate research, including nursery-rearing situations. Chapter 5 presents a view of contemporary developmental theory concerning the role of the prenatal period in understanding development over the life span. The critical influence of prenatal development on adult health and behavior is a recent concept not only in psychology, but also in epidemiology and medicine. We believe that prenatal development followed by nursery rearing will have a central role in future studies of nonhuman primate genetic, reproductive technology, and behavior development models relevant to human health and behavior at both the molar and molecular levels.

CHAPTER ONE

The Effects of Rearing Experiences: The Early Years

Melinda A. Novak and Gene P. Sackett

1. HISTORICAL PERSPECTIVE

Perhaps the most basic tenet of development is that behavior is subject to the multiple influences of genes and environments. In fact, it is the dynamic, temporal interaction between genetic and environmental factors that yields different developmental outcomes. Few would disagree with this view today. However, in the middle of the 20th century, when experimental rearing paradigms were first developed and studied in rhesus macaques, the view was different. Behavior patterns were construed as being controlled by either genes or the environment. Ethologists studied species-typical behaviors called fixed action patterns—"fixed" because they were thought to be uninfluenced by experience. In a similar manner, psychologists studied learned behavior that was thought not to have any genetic basis. This dichotomous view permeated the science of the time. Some took on the task of "proving" that specific behavior patterns were innate and developing the methodology to do so (Lorenz, 1950); others attempted to show that organisms were essentially molded by their experiences (Skinner, 1975). There were other voices arguing that all behavior was under both genetic and environmental control (Lehrman, 1970; Schneirla, 1957), but this interactionist notion had not yet taken firm hold. It is important to remember that the early studies of rearing in nonhuman primates took place in this context.

One of the pioneers of research on early rearing experience in nonhuman primates was Harry Harlow. He and his wife, Margaret, along with students and colleagues, began a series of studies beginning in the late 1950s that would transform our understanding of social/emotional development in primates. Harlow eagerly challenged commonly held beliefs of the time and went on to conduct a series of elegant studies that placed social behavior on the environmental side of the equation, or so it seemed. Harlow's research was designed to address two general questions, one related to developmental plasticity and the other to reversibility. His research on plasticity was concerned with identifying the necessary ingredients for social development. To this end, he designed different rearing environments and determined how monkeys fared in them (Harlow and Harlow, 1965). As history has shown, not all rearing environments were effective in eliciting normal social behavior. However, Harlow also was interested in determining whether the deleterious consequences of poor or inadequate early rearing environments could be overcome later in development. Thus, his second basic question was concerned with reversibility (Harlow and Suomi, 1971).

One of Harlow's first forays into research on early experience began with a study comparing the relative importance of milk reinforcement versus contact comfort in maintaining attachment between mother and infant rhesus macaques (Harlow and Zimmerman, 1959). To make this comparison, Harlow created four inanimate surrogate mothers, of which two provided contact comfort either with or without milk and two provided a bare wire surface with or without milk. It came as no surprise to Harlow that infants preferred comfort to milk, a finding that made the researchers of human behavior decidedly uncomfortable because it challenged their strongly held assumptions. This study also had implications beyond that of attachment because monkeys reared in this manner behaved strangely as they grew older and did not appear to show normal social development. In fact, both John Bowlby, a child development researcher, and Robert Hinde, a primatologist, separately pointed out the strange behavior to Harlow, thus facilitating an extensive effort to identify the factors that would lead to normal social development in this species or that conversely would produce pathology.

2. EARLY REARING EXPERIENCE PARADIGM: THE STUDY OF PLASTICITY

The basic paradigm developed by Harlow and others, still in use today, had four principal features. First, the rearing environment commenced shortly after birth and often necessitated the separation of mother and infant. Second, many of these infants were hand reared in a nursery environment where they received extensive human contact (Ruppenthal, 1979). In this context, nursery rearing was the beginning stage upon which specific rearing experiences were built. Third, monkeys were typically exposed to these rearing environments for the first 6 months of life, with a range of 3–12 months. And fourth, infants reared in these altered environments were compared directly or indirectly with infants reared with mothers and peers, a rearing condition that appeared to yield normal social behavior. During a 15-year period, Harlow and his colleagues and other scientists traced both the short- and long-term effects of diverse early rearing experiences on social development. As we shall see, the effects could not be predicted simply by the degree of deprivation versus the mother–peer rearing environment.

2.1. Total Isolation Rearing

One of the first manipulations studied was that of rearing monkeys in total isolation from species members. At the time, many scientists believed that social behavior was instinctive and that the way to "prove this fact" was to deprive animals of the necessary releasers that could trigger social expression. Thus, the objective of this research was to determine whether normal social behavior developed in rhesus macaques in the absence of social stimulation from mothers and/or other companions. The monkeys had to be hand reared, and for at least a month or more they received some contact with humans.

Rearing monkeys in total isolation from birth had devastating effects on behavior and social development. Monkeys reared in this manner showed a constellation of characteristics that became known as the "isolation syndrome" (Mason, 1968). These characteristics included abnormal behavior (e.g., clutching, rocking, huddling), heightened fear or aggression, inadequate motor coordination as in the double-foot clasp mount posture, and deficits in communication and in the establishment of normal social relationships. These patterns persisted as the isolates grew older (Mitchell *et al.*, 1966; Mitchell, 1968; Sackett, 1967). Females were generally inadequate or abusive mothers to their first infants (Arling and Harlow, 1967; Harlow *et al.*, 1966), although most of them improved with a second infant (Ruppenthal *et al.*, 1976). Despite significant deficits in social development, isolation-reared monkeys did not appear to differ from controls on a standard learning test battery (Harlow *et al.*, 1969). However, other studies revealed that isolation-reared monkeys adjusted more slowly to reinforcement contingencies, displayed atypical reactions to noxious stimuli (Lichstein and Sackett, 1971), showed impairments in response inhibition (Gluck and Sackett, 1976), and performed poorly on complex tasks such as oddity learning set (Gluck and Harlow, 1971; Gluck *et al.*, 1973). These social and cognitive deficits were not simply a result of sensory deprivation inasmuch as monkeys provided with toys or manipulanda or exposed to pictures or videotapes of monkeys while in isolation did not fare much better in terms of outcome (Sackett *et al.*, 1982).

Total isolation rearing was also associated with a number of physiological anomalies. At a metabolic level, isolation-reared monkeys showed abnormal eating and drinking patterns (Miller *et al.*, 1971). Isolation rearing affected cell-mediated immunity over the lifespan (Lewis *et al.*, 2000), altered dopamine sensitivity (Lewis *et al.*, 1990), and appeared to lead to cellular modifications in certain brain regions including the striatum (Martin *et al.*, 1991).

Thus, rearing monkeys in isolation from mothers and companions produced individuals with severely abnormal behavior and inadequate social behavior combined with substantial cognitive and physiological deficits. However, this picture turned out to be specific to rhesus macaques and did not generalize well to other monkey species even within the same genus, *Macaca*. For example, isolation-reared pigtailed macaques mirrored rhesus macaques with their inadequate social behavior but differed from rhesus macaques in their low levels of abnormal behavior. Cynomolgus (also called longtailed) macaques displayed yet a different profile of effects ranging from adequate levels of social behavior to high levels of abnormal behavior (Sackett *et al.*, 1976). This was the first inkling that genetic factors might play a role in the response of nonhuman primates to isolation. However, in the case of the rhesus macaque, it was clear that these monkeys needed some kind of experience to develop normally. Now the question was focused on what that experience might be.

2.2. Surrogate-Only Rearing

Monkeys reared in this condition differed from isolation-reared monkeys in that they had continuous access to an inanimate, terry-cloth-covered surrogate "mother." Research had shown that infant monkeys developed a strong attachment to such surrogates, using them as a secure base in novel situations and showing strong protest reactions when the inanimate mother was removed. Could the establishment of a strong attachment with an inanimate mother be sufficient to induce normal social development, or would these infants be as deficient as total isolates? Actually, neither proved to be true. Monkeys reared with inanimate surrogate mothers failed to develop appropriate social behavior or display suitable communication skills (Harlow and Harlow, 1962). However, abnormal behavior was altered with the presence of a surrogate mother, and the alterations were linked directly to the features of the surrogate. Self-clasping decreased, presumably because infants directed their clasping response to the surrogate. Indeed, additional modifications to the surrogate also appeared to reduce other forms of abnormal behavior. For example, surrogate mothers that moved back and forth appeared to reduce rocking behavior in infants (Mason and Berkson, 1975). Despite these improvements in abnormal behavior, surrogate-only rearing was not the path to normal social development.

2.3. Partial Isolation Rearing

This rearing environment differed from total isolation in that infant monkeys who had been removed from their mothers at birth and reared in a nursery could nonetheless continuously see, hear, and smell but not physically interact with other nearby monkeys. Could the mere observation of other monkeys elicit normal social development? In this case, the answer appeared to be a resounding no. Monkeys reared in partial isolation developed a syndrome that was similar to, although not quite as severe as, that exhibited by monkeys reared in total isolation (Cross and Harlow, 1965; Suomi *et al.*, 1971). However, the significance of these findings was clouded by the fact that partial isolates typically observed other partial isolates like themselves. Thus, it remained unresolved whether continuous visual and auditory exposure to normal monkeys could elicit some aspects of normative development in the observer.

2.4. Peer-Only Rearing

Monkeys reared in this manner were separated from their mothers at birth and placed in a nursery. After the first month of life, infants were placed in small groups of two to six wherein they had continuous access to other infants of comparable age. Would physical contact with naive infants be sufficient to induce normal social development, or would monkeys need models (older, socially sophisticated monkeys) to acquire species-appropriate social behavior? The answer was very surprising. Monkeys reared with naive infants seemed to readily acquire most forms of social behavior, including all the relevant behaviors associated with play, grooming, sex, and aggression (Chamove, 1973; Chamove *et al.*, 1973). Female infants often showed appropriate maternal behavior as adults, and both males and females were observed to use communication signals in the relevant contexts. These findings led some people to argue that social behavior was actually instinctive in rhesus macaques. However, this view ignored both the prenatal environment and the infant as potential sources of this effect. In fact, Gilbert Gottlieb (1968) had already shown that Peking duck offspring could be agents in their own development with minimal kinds of stimulation. Thus it seemed likely that genetic predispositions and the prenatal environment combined with the presence of warm, moving bodies in the postnatal environment were sufficient to propel rhesus macaque infants along a path that ultimately yielded some kinds of social behavior.

However, infants reared with each other were not normal in all respects. They showed heightened emotionality, excessive clinging behavior, and some mild forms of abnormal behavior. In addition, the emergence of social behavior was developmentally delayed, with monkeys acquiring sophisticated play patterns considerably later than their mother-reared counterparts (Chamove, 1973). The delay was thought to be caused by excessive clinging behavior. Whereas mothers could encourage their infants to move away, peer-reared infants were frequently trapped by the embraces of other infants. This hypothesis gained some credence when the effects of different group sizes were examined. Clinging was more excessive and persisted for a longer time in paired infants than in groups of four infants, presumably because the paired infants had only each other to hold on to (Chamove *et al.*, 1973). Despite the problems described above, monkeys reared with naive infants nonetheless acquired most forms of social behavior.

2.5. Surrogate-Peer Rearing

This rearing condition overcame a basic problem with peer-only rearing, namely, that infants served in dual roles as attachment figure and playmate. To separate these roles, infant monkeys received continuous exposure to inanimate surrogate mothers (the attachment figure) supplemented with brief daily exposure to infants of the same age (the playmates). In essence, this rearing condition attempted to mirror the balance of time young infants spend with mothers and playmates under more natural conditions. The daily duration of peer exposure was limited to no more than 2 hr per day (range across studies was 15 min to 2 hr). A more lengthy daily exposure was thought to increase the risk that infants would transfer their primary attachment from the surrogate to the other infants in the group, a view supported by anecdotal evidence. An important question was whether this rearing condition could yield benefits beyond that seen in the peer-only rearing situation. As it turned out, there were substantial differences between the two rearing conditions. Perhaps most significant was the marked reduction in the rates at which infants clung to one another. Surrogate-peer-reared monkeys directed most of their clasping responses to their inanimate surrogate and allocated much of their peer interaction time to play and exploration. In addition, surrogate-peer-reared monkeys showed little in the way of developmental delays (Hansen, 1966). Minor differences in disturbance behavior between surrogate-peer-reared and mother-peer-reared monkeys all but disappeared after the first few months of life. Surrogate-peer-reared monkeys also displayed mild forms of abnormal behavior (e.g., self-orality) that declined to the level shown by mother-peer-reared monkeys after the first year of life (Hansen, 1966). Subsequent studies demonstrated that surrogate-peer-reared monkeys showed normal patterns of sexual behavior and reproduction (Novak *et al.*, 1992). In addition, studies of a semi-free-ranging group of surrogate-peer-reared monkeys revealed the development of species-typical patterns of social organization ranging from matrilines to the emigration of adolescent males (Novak *et al.*, 1992). Thus, once again it appeared that peer interaction, particularly of the playful variety, was important in the emergence of species-typical social behavior.

2.6. Mother-Only Rearing

Data derived from the surrogate-peer and peer-only rearing condition suggested that peers were crucially important in the development of social behavior in rhesus macaques. What was the role of the mother in fostering social development? Clearly the rearing conditions described above would never have occurred under natural conditions. It was the feasibility of hand-rearing rhesus macaque infants in a nursery that allowed for the systematic exploration of early experience. Could mothers impart all the relevant social behaviors to their infants without the influence of peers? The mother-only rearing condition was designed to explore this question. The results confirmed expectations. Infants reared with mothers developed the full range of species-typical social behavior and showed little, if any, abnormal behavior (Alexander and Harlow, 1965). Intriguingly, however, as juveniles these monkeys tended to be more aggressive to peers than juvenile mother-peer-reared monkeys. This finding has been interpreted to suggest that monkeys learn to moderate their behavioral responses to others through play.

Taken in its context, the work described above provided an intriguing view of the effects of early experience on social development in the rhesus macaque. Of particular significance was the finding that only the most severe forms of deprivation (e.g., isolation rearing) produced major deficits in social behavior. Thus, the prevailing view at the time was that social development in this species was relatively buffered from environmental perturbation. This research also served to highlight the role of infants as potential agents in their own development. None of this information would have emerged without the establishment of an effective nursery-rearing procedure.

3. REVERSIBILITY OF NEGATIVE REARING OUTCOMES

The early pioneering work by Harlow and others demonstrated that monkeys reared in extremely impoverished environments displayed high levels of aberrant behavior and were socially incompetent. Were these effects permanent or could subsequent interventions lead to rehabilitation? In the 1970s, Harlow and colleagues explored this possibility with monkeys that had been reared in total isolation. At the time, a number of theories were invoked to account for the isolation syndrome. Some of these theories predicted that the isolation effects were permanent. This prediction was based either on some neural deterioration of specific brain regions or on the lack of adequate stimulation during a critical period for social development that might have been present in early infancy. Other theories predicted that the syndrome might be reversed with appropriate intervention. In this case, the isolation syndrome was ascribed to the stress of exposure to a significantly more complex environment following isolation and/or to the lack of sufficient learning opportunities. These theories gave rise to different treatment regimens that were designed to rehabilitate isolation-reared monkeys.

3.1. Agemate Therapy

If monkeys need learning opportunities to develop appropriate social behavior, then exposing isolation-reared monkeys to normally reared monkeys of the same age might serve to create an effective stimulus environment. Shortly after removal from isolation, the deprived monkeys were given exposure to normally reared monkeys in a playroom setting. However, this intervention was short-lived because of its unfortunate consequences. The isolation-reared monkeys showed high levels of disturbance behavior that increased over time in conjunction with the high level of aggression that they received from the normal monkeys (Harlow *et al.*, 1964; Rowland, 1964).

3.2. Adaptation Therapy

If monkeys show stress and withdrawal because of their emergence into a much more complex environment, then adaptation to that environment should facilitate social interactions. In this study, isolation-reared monkeys were extensively adapted to the playroom environment prior to any social exposure. Despite this adaptation, isolation-reared monkeys continued to show high levels of abnormal behavior and did not exhibit appropriate social behavior when tested with other isolates or normal agemates (Clark, 1969).

3.3. Attachment Therapy

The goal of this therapy was to determine if the provision of an inanimate surrogate could reverse the consequences of 6 months of total isolation. Monkeys provided with warm surrogates appeared to develop some form of attachment and spent a significant amount of time clinging to the surrogates. Surrogate therapy proved to be effective in reducing some forms of abnormal behavior such as self-clasping and rocking, but failed to reverse most of the social deficits (Suomi, 1973). These findings paralleled those observed with surrogate-only rearing described above.

3.4. Training Therapy

Isolation-reared monkeys typically showed strong "contact aversion" when tested with other monkeys. The aim of this therapy was to establish whether isolation-reared monkeys could be conditioned to contact other monkeys and to determine if this conditioning would lead to the development of more-normal social interactions (Sackett, 1968). Young adult monkeys that had been reared in total isolation for the first 6–9 months of life were the subjects of this research. During the 3- to 4-year period before treatment, the monkeys were housed with peer-reared monkeys and tested for social behavior in a playroom environment. Even though the isolates had continuous exposure to other monkeys, they showed almost no positive social behavior. With the use of an avoidance conditioning procedure, the isolates were then successfully trained to initiate and maintain nonaggressive physical contact with a peer-reared monkey for 30-min periods to avoid a mild electrical shock on the floor of the test cage. Despite the success of this training, the conditioned contact did not translate into improved social interaction. Isolation-reared monkeys continued to show little in the way of positive social behavior in all the settings in which they were observed. It is possible that the use of positive rather than negative reinforcement might have yielded better results, but such studies were not conducted.

3.5. Younger-Monkey Therapy

Previous research had demonstrated that isolates did not fare well in interactions with normally reared agemates. Did this interaction fail because a normally reared monkey was too complex a social stimulus for the isolates? Did it fail because normally reared monkeys had expectations about the social propensities of monkeys their size? These questions could not be readily answered. However, they led the way to a different kind of social therapy in which isolates were given opportunities to interact with much younger surrogate-peer-reared monkeys (therapists). Thus, isolates were matched to peers on the basis not of chronological age but of social sophistication. Could much younger therapist monkeys undo the effects of early total isolation? The outcome was striking. When 6-month isolates were exposed to much younger peers in the second 6 months of their lives, abnormal behavior decreased substantially and social interaction emerged (Suomi and Harlow, 1972; Suomi *et al.*, 1974). Isolation-reared monkeys eventually responded to the gentle contacts of the therapist monkeys with play. A similar outcome was observed when 12-month isolates were exposed to much younger therapist monkeys (Novak and Harlow, 1975). These findings pertained only to infantile patterns of social behavior (e.g., contact and play). In a subsequent follow-up study, older isolation-reared monkeys interacted effectively with surrogate-peer-reared agemates and exhibited more complex forms of social behavior such as grooming (Novak, 1979). However, it was also clear in this follow-up study that younger-monkey therapy had not reversed all the consequences of isolation rearing. Isolates were less likely to engage in some kinds of social activities, and the form of these patterns was more awkward and strained than in more normally reared monkeys. In addition, some kinds of abnormal behavior persisted. Despite these caveats, the research highlighted once again the importance of the environment in overcoming some of the effects of adverse early experiences.

4. FROM THE PAST TO THE PRESENT

Much has transpired in the field of early-experience research in the intervening third of a century. Gone is the old view of nature versus nurture and in its place is the interactionist view in which different genes or gene alleles can selectively modify the impact of environments on behavior. We have thus become more sophisticated in the questions we pose about development. The focus has expanded beyond group data to include comparisons between individuals and with it has come the understanding that two monkeys reared in the same environment might nonetheless respond quite differently to it. New nursery-rearing procedures and other methodologies have allowed us to explore the effects of early experience not just on behavior but on many other biological systems ranging from neurotransmitters to cardiac function. New ways of measuring behavior beyond that of frequencies and durations have allowed us to identify the more subtle effects of rearing environments. Integrating behavior patterns into new constructs (e.g., reconciliation or recruitment of agonistic aid) has underscored the structure and complexity of behavioral response. All of these changes have made the field of early-experience research as exciting as it was 35 years ago.

REFERENCES

Alexander, B.K., and Harlow, H.F. (1965). Social behavior of juvenile rhesus monkeys subjected to different rearing conditions during the first six months of life. *Zool. Jb. Abt. Allgemeine Zool. Physiol. Tiere* 71:489–508.

Arling, G.L., and Harlow, H.F. (1967). Effects of social deprivation on maternal behavior of rhesus monkeys. *J. Comp. Physiol. Psychol.* 64:371–377.

Chamove, A.S. (1973). Rearing infant rhesus together. *Behaviour* 47:48–66.

Chamove, A.S., Rosenblum, L.A., and Harlow, H.F. (1973). Monkeys (Macaca mulatta) raised only with peers. A pilot study. *Anim. Behav.* 21:316–325.

Clark, D.L. (1969). Immediate and delayed effects of early, intermediate, and late social isolation in the rhesus monkey. *Diss. Abstr.* B29:4862.

Cross, H.A., and Harlow, H.F. (1965). Prolonged and progressive effects of partial isolation on the behavior of macaque monkeys. *J. Exp. Res. Personal.* 1:39–49.

Gluck, J.P., and Harlow, H.F. (1971). The effects of deprived and enriched rearing conditions on later learning: A review. In: Jarrard, L.E. (ed.), *Cognitive Processes of Nonhuman Primates.* Academic Press, New York, pp. 103–119.

Gluck, J.P., and Sackett, G.P. (1976). Extinction deficits in socially isolated rhesus monkeys (Macaca mulatta). *Dev. Psychol.* 12:173–174.

Gluck, J.P., Harlow, H.F., and Schiltz, K.A. (1973). Differential effect of early enrichment and deprivation on learning in the rhesus monkey (*Macaca mulatta*). *J. Comp. Physiol. Psychol.* 84:598–604.

Gottlieb, G. (1968). Prenatal behavior of birds. *Q. Rev. Biol.* 43:148–174.

Hansen, E.W. (1966). The development of maternal and infant behavior in the rhesus monkey. *Behaviour* 27:107–149.

Harlow, H.F., and Harlow, M.K. (1962). The effect of rearing conditions on behavior. *Bull. Menninger Clinic* 26:213–224.

Harlow, H.F., and Harlow, M.K. (1965). The effect of rearing conditions on behavior. *Int. J. Psychiatr.* 1:43–51.

Harlow, H.F., and Suomi, S.J. (1971). Social recovery by isolation-reared monkeys. *Proc. Natl. Acad. Sci. USA* 68:1534–1538.

Harlow, H.F., and Zimmermann, R.R. (1959). Affectional responses in the infant monkey. *Science* 130:421–432.

Harlow, H.F., Rowland, G.L., and Griffin, G.A. (1964). The effect of total social deprivation on the development of monkey behavior. *Psychiatr. Res. Rep. Am. Psychiatr. Assoc.* 19:116–135.

Harlow, H.F., Harlow, M.K., Dodsworth, R.O., and Arling, G.L. (1966). Maternal behavior of rhesus monkeys deprived of mothering and peer associations in infancy. *Proc. Am. Philos. Soc.* 110:58–66.

Harlow, H.F., Schlitz, K.A., and Harlow, M.K. (1969). Effects of social isolation on the learning performance of rhesus monkeys. In: Carpenter, C.R. (ed.),

Proceedings of the Second International Congress of Primatology, Vol. 1. Karger, Basel, pp. 178–185.

Lehrman, D.S. (1970). Semantic and conceptual issues in the nature-nurture problem. In: Aronson, L.A., Tobach, E., Lehrman, D.S., and Rosenblatt, J. (eds.), *Development and Evolution of Behavior: Essays in Memory of T.C. Schneirla*. W.H. Freeman, San Francisco, pp. 17–52.

Lewis, M.H., Gluck, J.P., Beauchamp, A.J., Keresztury, M.F., and Mailman, R.B. (1990). Long-term effects of early social isolation in Macaca mulatta: Changes in dopamine receptor function following apomorphine challenge. *Brain Res.* 513:67–73.

Lewis, M.H., Gluck, J.P., Petitto, J.M., Hensley, L.L., and Ozer, H. (2000). Early social deprivation in nonhuman primates: Long-term effects on survival and cell-mediated immunity. *Biol. Psychiatry.* 47:119–126.

Lichstein, L., and Sackett, G.P. (1971). Reactions by differentially raised rhesus monkeys to noxious stimulation. *Dev. Psychobiol.* 4:339–352.

Lorenz, K.Z. (1950). The comparative method in studying innate behavior patterns. *Symp. Exp. Biol.* 4:221–268.

Martin, L.J., Spicer, D.M., Lewis, M.H., Gluck, J.P., and Cork, L.C. (1991). Social deprivation of infant rhesus monkeys alters the chemoarchitecture of the brain. I. Subcortical regions. *J. Neurosci.* 11:3344–3358.

Mason, W.A. (1968). Early social deprivation in nonhuman primates: Implications for human behavior. In: Glass, D.C. (ed.), *Environmental Influences.* Rockefeller University and Russell Sage, New York, pp. 70–100.

Mason, W.A., and Berkson, G. (1975). Effects of maternal mobility on the development of rocking and other behaviors in rhesus monkeys: A study with artificial mothers. *Dev. Psychobiol.* 8:197–211.

Miller, R.E., Caul, W.F., and Mirsky, I.A. (1971). Patterns of eating and drinking in socially isolated rhesus monkeys. *Physiol. Behav.* 7:127–134.

Mitchell, G.D. (1968). Persistent behavior pathology in rhesus monkeys following early social isolation. *Folia Primatol.* 8:132–147.

Mitchell, G.D., Raymond, E.J., Ruppenthal, G.C., and Harlow, H.F. (1966). Long-term effects of total social isolation upon behavior of rhesus monkeys. *Psychol. Rep.* 18:567–580.

Novak, M.A. (1979). Social recovery of monkeys isolated for the first year of life. II. Long-term assessment. *Dev. Psychol.* 15:50–61.

Novak, M.A., and Harlow, H.F. (1975). Social recovery of monkeys isolated for the first year of life. I. Rehabilitation and therapy. *De. Psychol.* 11:453–465.

Novak, M.A., O'Neill, P., and Suomi, S.J. (1992). Adjustments and adaptations to indoor and outdoor environments: Continuity and change in young adult rhesus monkeys. *Am. J. Primatol.* 28:125–138.

Rowland, G.L. (1964). The effects of total social isolation upon learning and social behavior of rhesus monkeys. *Diss. Abstr.* 25:1364–1365.

Ruppenthal, G.C. (ed.) (1979). *Nursery Care of Nonhuman Primates.* Plenum Press, New York.

Ruppenthal, G.C., Arling, G.L., Harlow, H.F., Sackett, G.P., and Suomi, S.J. (1976). A 10-year perspective of motherless-mother monkey behavior. *J. Abnorm. Psychol.* 85:341–349.

Sackett, G.P. (1967). Some persistent effects of different rearing conditions on preadult social behavior of monkeys. *J. Comp. Physiol. Psychol.* 64:363–365.

Sackett, G.P. (1968). Abnormal behavior in laboratory-reared monkeys. In: Fox, M. (ed.), *Abnormal Behavior in Animals.* W.B. Saunders, Philadelphia, 1968.

Sackett, G.P., Holm, R.A., and Ruppenthal, G.C. (1976). Social isolation rearing: Species differences in behavior of macaque monkeys. *Dev. Psychol.* 12:283–288.

Sackett, G.P., Tripp, R., and Grady, S. (1982). Rhesus monkeys reared in isolation with added social, nonsocial and electrical brain stimulation. *Ann. Istitut. Superiore Di Sanita* 18:203–213.

Schneirla, T.C. (1957). Behavioral development and comparative psychology. *Q. Rev. Biol.* 43:283–303.

Skinner, B.F. (1975). The steep and thorny ways to a science of behavior. *Am. Psychol.* 30:42–49.

Suomi, S.J. (1973). Surrogate rehabilitation of monkeys reared in total social isolation. *J. Child Psychol. Psychiatry Allied Disc.* 14:71–77.

Suomi, S.J., and Harlow, H.F. (1972). Social rehabilitation of isolate-reared monkeys. *Dev. Psychol.* 6:487–496.

Suomi, S.J., Harlow, H.F., and Kimball, S.D. (1971). Behavioral effects of prolonged partial social isolation in the rhesus monkey. *Psychol. Rep.* 29:1171–1177.

Suomi, S.J., Harlow, H.F., and Novak, M.A. (1974). Reversal of social deficits produced by isolation rearing in monkeys. *J. Hum. Evol.* 3:527–534.

CHAPTER TWO

The Changing Role of Hand Rearing in Zoo-Based Primate Breeding Programs

Ingrid Porton and Kelli Niebruegge

1. THE HISTORY OF HAND-REARING PRIMATES IN ZOOS

The husbandry of nonhuman primates in zoos and laboratories has improved significantly over the past 50 years, with concomitant improvements in the ability of captive primates to breed and successfully rear their own offspring. Nevertheless, there will always be a need for zoos to hand raise some infant primates. In the 1950s and 1960s nurseries devoted to hand rearing infant mammals became a prominent feature in zoos (Ogden and Kasielke, 2001). Reproductive success among the majority of exotic mammals was comparatively low relative to current conditions, and management strategies designed to increase the probability of infant survival were prioritized. For many species there was a zero risk policy and a mother was given little time to become comfortable and adjust her own behavior to that of the newborn infant. At the first indication of infant distress, managers rushed in to remove the neonate from its mother. Some zoos developed a policy of automatically hand rearing high-profile primates, especially the great apes.

The welfare of high-risk infants was not the only consideration when these nurseries were built. Humans caring for and playing with infants in view of the public was a recipe for an instantly successful exhibit. The popularity of nurseries as a zoo exhibit was evident in the number that were built (Ogden and Kasielke, 2001). Slow to mature and so much like human infants, nonhuman primate infants were particularly attractive exhibits. It was not unusual for zoos to prolong a young primate's stay in the nursery well past the age it could have been introduced to a social group. Indeed, an empty nursery was itself a motivation to remove an infant from its mother. In some cases, primates were purposefully hand reared so they could be tamed and used for shows or educational presentations (Taylor, 1978). That the interbirth interval of primates such as gorillas (*Gorilla gorilla gorilla*) was decreased was seen as an added benefit of hand rearing (Taylor, 1978).

The practice of hand rearing primates caused two problems. First, female primates that were raised by humans had no chance to learn parenting skills from their mothers. And second, zoo managers had no opportunity to see that first-time mothers can develop from awkward, rough, and seemingly dangerous caretakers to proficient parents. Thus, the hand rearing cycle continued.

A review of the early nursery literature from the zoo community reveals the focus of caring for the infant was almost exclusively on the physical aspects of hand rearing such as formulas, weight gain curves, and medical issues (e.g., Frueh, 1968; Kirchshofer *et al.*, 1968; Breznock *et al.*, 1979). Indeed, the primary emphasis when hand rearing an infant primate was to keep it alive. The overriding importance placed on a sterile, sanitary environment often translated into a sterile social environment for the infant. For some youngsters, socialization into an adult group was delayed until the individuals were fully capable of defending themselves, which placed them well into adolescence. Unfortunately, too many of these hand-reared primates exhibited stereotypic behaviors and were socially and/or sexually incompetent.

The zoo literature reveals a shifting attitude during the 1970s and 1980s. Increasingly, published articles drew attention to the social needs of all infants and particularly primates (Ott and Joslin, 1981; Miller, 1982; Collier, 1983). Forward-thinking professionals advocated the investigation of alternative hand-rearing methods that allowed early socialization opportunities with peers and reintegration into adult groups at younger ages. Authors were clearly influenced by the findings of Harlow and colleagues (1971), who experimentally demonstrated the adverse and long-term negative consequences of social isolation upon primates.

Several influential papers helped change the ways in which nonhuman primates are hand reared in zoos. One change was an increased emphasis on encouraging and facilitating maternal rearing. Particularly valuable was the implementation of a 72-hr postpartum observation protocol for apes (Rosenthal, 1989) based on information that human neonates can survive for over 72 hr without nursing. As many ape infants were removed from their mothers precisely because nursing had not been observed in the first day, the protocol provided zoo managers with the confidence to leave infants with nonabusive mothers despite the absence of nursing. This approach was validated by earlier observations that chimpanzee (*Pan troglodytes*) mother/infant pairs sometimes need several days to synchronize their behaviors to facilitate nursing (Rogers and Davenport, 1969). Other methods to keep infant primates with their mothers included maternal skills training programs for apes (Joines, 1977; Keiter and Pichette, 1977; Schildkraut, 1982), distracting new mothers to allow nursing, or tranquilizing mothers to give the infant and dam more time to become accustomed to nursing (Ott and Joslin, 1981).

If infants could not be raised by their mothers, fostering them to an available lactating female was another approach advocated to avoid hand rearing. Examples of this technique were the successful fostering of rejected callitrichid infants in laboratories and the Los Angeles Zoo Marmoset Colony (Collier *et al.*, 1981). However, because zoos typically house fewer individuals per species than, for example, primate laboratories, fostering is more difficult to achieve.

Other important changes in philosophy included the actual methods used to hand rear infant primates. Maple (1980) challenged the idea that extensive human contact with great ape infants was detrimental to their social and psychological development. A concern held by many nursery personnel was that frequent and prolonged interactions with an infant would result in imprinted individuals unable to function appropriately with conspecifics. Maple (1980) suggested that it was better for caretakers to more closely mimic the amount of social contact infant apes receive from their mothers and, if limited staff prevented realization of the goal, he recommended that volunteers be used.

Peer rearing of infants was recommended as one method to facilitate early and safe exposure to conspecifics. Bringing the infant(s) to the adult facility to gain familiarity with the sight, sound, and smell of adults was viewed as a component of the socialization process. Some (e.g., Meyer and Wilcox, 1982) advocated that infants be moved from the nursery into the adult facility and then into the adult group as soon as possible.

Although these approaches were promoted in the early 1980s, it took time for zoo managers to become comfortable with the added risks associated with these methods. For these procedures to become the norm rather than the exception, they first needed to produce a credible record of positive results. A significant development that advanced best-methods practices was the emergence within zoos of cooperatively managed breeding programs such as the American Zoo and Aquarium Association's (AZA) Species Survival Plan (SSP®). These programs manage all the individuals housed in participating institutions as one population. The SSP® population is genetically and demographically managed through analysis of studbook records and issuance of yearly breeding recommendations. In addition, SSP® Committees are charged with attending to behavioral and husbandry issues. Many of the primate SSP® programs (e.g., Gorilla, Chimpanzee, Golden Lion Tamarin, *Leontopithecus rosali*) directly addressed the hand-rearing issue and developed goals to eliminate all unnecessary hand rearing. In the instances when hand rearing was required, the objective became early resocialization of infants with conspecifics (Porton, 1992 and 1997). SSP® programs also coordinated interzoo transfers of infants to establish peer groups or facilitate the infant's integration into a more appropriate foster group.

2. RESOCIALIZATION GOALS AND TECHNIQUES

The emphasis that zoo professionals have placed on developing hand-rearing protocols that aim to produce socially and sexually competent adults is driven by two goals. One is to increasingly provide for the well-being of all the animals housed in AZA zoos. Lacking certain social skills, some hand-reared primates may be peripherilized from their social group, subjected to a higher proportion of aggression, and/or exposed to higher levels of stress or depression. Zoo professionals endeavor to develop hand-rearing protocols that do not produce individuals compromised by abnormal or neurotic behaviors.

The second goal is to reduce the number of individuals that are either nonreproductive or have special housing requirements. Zoos face a space crunch. The number of breeding programs that zoos can manage is limited by space (Earnhardt *et al.*, 2001). A minimum viable population size is calculated for each managed program and that figure is used to determine the total number of programs that can be accommodated in participating zoos. Dysfunctional hand-reared individuals that are reproductively incompetent decrease the ratio of effective population size to actual population size and thereby unnecessarily increase the population size required to reach program goals. These individuals may also need special housing accommodations if they cannot live compatibly and safely in social groups. Transferring such individuals outside of AZA zoos is also limited because more stringent disposition policies prevent sending primates to facilities that do not meet AZA housing and care standards.

To achieve their goal, primate managers have pushed the limits to improve resocialization techniques used in zoos. Although infants removed from their mothers and raised by humans are all termed "hand reared," the amount of time infants are exclusively in human care varies to such a degree that the term has become misleading. In reality, there is a large continuum and in some cases the distinction between mother reared and hand reared is becoming increasingly blurred. New terms should be coined to capture these differences and add clarity to discussions. Several examples can illustrate the point. A Sumatran orangutan (*Pongo pygmaeus abelii*) at the Brookfield Zoo was removed from its mother at 1 week for hand rearing. The Brookfield staff developed a training program for the mother and infant with the purpose of reuniting the pair as soon as the infant could be fed reliably with a bottle. When strong enough, the infant was trained to cling to a wiremesh panel while nursing at the bottle; the mother, meanwhile, was trained to stay at the front of its cage while the infant nursed. Familiarity between mother and infant was maintained through the training sessions, and at 5 months the infant was returned to its mother (Sodaro and Weber, 2000). Although fed by the keepers, the infant was mother reared from that time on.

Black lemur (*Eulemur macaco macaco*) and black-and-white ruffed lemur (*Varecia variegata*) infants that had to be hand reared at the Saint Louis Zoo were fostered into a family group as young as 6 weeks of age. In the case of the black lemurs, the staff took advantage of the foster mother's tolerance of humans and entered the exhibit with the lemurs to continue feeding formula to the reintegrated infants (Knobbe, 1991). Provisioning formula after reintroducing a litter of three 6-week-old ruffed lemurs to their parents and older sibling was solved by building a creep feeder, which the infants readily entered to obtain their designated diet (I. Porton, unpublished observations).

These days, fewer primate infants are being hand reared in isolation from adult conspecifics. Rather, temporary nursery areas are often set up in the same facility in which the adult group is housed. For example, at the Toledo Zoo, a gorilla infant was carried in a sling whenever possible while the keepers worked in the great ape facility. Periodically the infant was shown to the adult gorillas to encourage interest and interaction (Petiniot *et al.*, 1988). Two 3-month-old chimpanzee females from two different zoos were transferred to the Saint Louis Zoo for peer socialization and eventual integration into an adult group. Whereas previous chimpanzees were raised in the Children's Zoo Nursery, these infants were reared by the keepers and a group of docents in the holding cage next to the adults. The infants were thoroughly familiar with their future family and facility well before the physical introduction was carried out (Knobbe and Porton, 2001). Methods for hand rearing Goeldi's monkey (*Callimico goeldii*) at the Brookfield Zoo were modified in the 1990s. As soon as the infant is removed from its mother, it is housed first in an incubator and later in a wire "howdy" cage that sits directly in the parents' cage (Sodaro, 2000).

3. EVALUATION OF HAND-REARED PRIMATES

Despite the number of primates that have been handreared in zoos, there are surprisingly few population level analyses evaluating the outcome of these procedures. Systematic behavioral research comparing the behavior of hand-reared versus parent-reared individuals is rare, as is studbook-based analysis comparing the reproductive rates of hand-reared versus parent-reared primates. Research using studbook records can be hindered by missing data on rearing history, insufficient information on reproductive opportunities, and small sample size. Small sample size also restricts more detailed comparisons between hand-rearing methods and subsequent adult social behavior.

The species for which there is the most information on the consequences of hand rearing is the western lowland gorilla. This is due to the large proportion of gorillas that has been hand reared in zoos, the popularity of gorillas, and the desire by managers to understand the causes of reproductive and parental deficiencies. The first investigation found that mother-reared female gorillas were significantly more likely to copulate and raise their offspring than hand-reared females (Beck and Power, 1988). No statistical difference was evident in the copulatory behavior of hand-reared versus parent-reared males but the sample size was small. A study of the international studbook data revealed similar results for females and also showed that hand-reared males were less successful reproductively than their parent-reared counterparts (Meder, 1993). An updated analysis of the Gorilla SSP® population again confirmed that mother-reared male and female gorillas were more successful reproductively than hand-reared individuals (Ryan *et al.*, 2002). Hand-reared gorillas are more likely than mother-reared gorillas to exhibit solitary play, self-directed behaviors, regurgitation and reingestion, and inappropriate aggressive behavior towards same-age or adult conspecifics (Meder, 1989; Olson and Gold, 1985; Gold, 1992).

In a study of copulatory behavior in zoo-born male chimpanzees, 54% of those that were hand reared with peers showed appropriate copulatory behavior versus just 30% of the males that were raised in isolation from conspecifics (King and Mellen, 1994). Thus, hand rearing of male chimpanzees without access to peers produces adults that are very likely to be sexually incompetent.

In a study of captive golden lion tamarins, parent rearing significantly increased offspring survival rates to both 7 days and 16 months of age (Rettberg-Beck and Ballou, 1988). Of 101 tamarins that were hand reared, 32 lived to breeding age. Of these, 15 were allowed to breed, 13 with a parent-reared mate and two with each other. Eight of the ten males and two of the five females bred. The pair comprised of two hand-reared individuals did not produce young but did copulate. Overall, hand-reared tamarins were less reproductively successful than their parent-reared conspecifics, but the results were not statistically significant. Unexpectedly, infants reared without a sibling were more successful breeders than peer-reared infants (Rettberg-Beck and Ballou, 1988).

In another chapter in this volume, we used studbook and survey data to show that parent-reared black-and-white ruffed lemurs, red ruffed lemurs (*Varecia rubra*), and black lemurs were more successful reproductively than their hand-reared counterparts. Nevertheless, over 60% of the male and female hand-reared ruffed lemurs and female black lemurs were reproductively competent.

None of the above studies focused on the effect of methods used to resocialize hand-reared infants on adult reproductive performance. It is likely, however, that the greater emphasis placed on resocializing primate infants at a younger age has contributed to the improved reproductive success observed in hand-reared zoo primates.

4. SUMMARY

Over the past 50 years, zoo managers have displayed many changes in their attitude towards hand rearing of nonhuman primates and in their choice of methods when hand-rearing is necessary. Today most zoos hand rear primates only when it is absolutely necessary. Resocializing infants at a young age is a priority, and many innovative management strategies to accomplish this goal have proven successful. There are numerous opportunities for research that will help guide further improvements including retrospective research using studbook data. Such research would greatly benefit from better management records that could provide the needed data with which to evaluate resocialization techniques. Research aimed at understanding the impact of hand rearing on social behaviors other than reproduction is also encouraged, for the primary goal of zoo management must be the overall welfare of each animal.

ACKNOWLEDGMENTS

We thank Alice Seyfried for reviewing this manuscript and providing valuable input.

REFERENCES

Beck, B.B., and Power, M.L. (1988). Correlates of sexual and maternal competence in captive gorillas. *Zoo Biol.* 7:339–350.

Breznock, A.W., Porter, S., Harrold, J.B., and Kawakami, T.G. (1979). Hand-rearing infant gibbons. In: Ruppenthal, G.C. (ed.), *Nursery Care of Nonhuman Primates.* Plenum Press, New York, pp. 287–298.

Collier, C.E. (1983). The distorted mirror: Avoiding abnormal behavior in hand reared mammals. *AAZPA 1983 Annual Proceedings.* American Association of Zoological Parks and Aquariums, Wheeling, WV.

Collier, C., Kaida, S., and Brody, J. (1981). Fostering techniques with cotton-top tamarins *Saguinus oedipus oedipus* at Los Angeles Zoo. *Int. Zoo Yb.* 21:224–225.

Earnhardt, J.M., Thompson, S.D., and Marhevsky, E.A. (2001). Interactions of target population size, population parameters, and program management on viability of captive populations. *Zoo Biol.* 20:169–183.

Frueh, F.J. (1968). A captive-born gorilla *Gorilla g. gorilla* at St. Louis Zoo. *Int. Zoo Yb.* 8:128–131.

Gold, K.C. (1992). Nonsocial behavior of captive infant gorillas. *Am. J. Primatol.* 26:65–72.

Harlow, H.K., Harlow, M.K., and Suomi, S.T. (1971). From thought to therapy: Lessons from a primate laboratory. *Am. Sci.* 59:538–549.

Joines, S. (1977). A training programme designed to induce maternal behavior in a multparous female gorilla at the San Diego Wild Animal Park. *Int. Zoo Yb.* 17:185–188.

Keiter, M., and Pichette, P. (1977). Surrogate infant prepares a lowland gorilla for motherhood. *Int. Zoo Yb.* 17:188–189.

King, N.E., and Mellen, J.D. (1994). The effects of early experience on adult copulatory behavior in zoo-born chimpanzees (*Pan troglodytes*). *Zoo Biol.* 13:51–59.

Kirchshofer, R., Weisse, K., Berenz, K., and Klose, H.I. (1968). A preliminary account of the physical and behavioural development during the first 10 weeks of the hand-reared gorilla twins *Gorilla g. gorilla* at Frankfurt Zoo. *Int. Zoo Yb.* 8:121–128.

Knobbe, J. (1991). Early resocialization of hand-reared primates. In: *AAZPA Regional Conference Proceedings.* American Association of Zoological Parks and Aquariums, Wheeling, WV, pp. 763–770.

Knobbe, J., and Porton, I. (2001). You want me to adopt how many kids? Fostering family life to meet Chimpanzee SSP goals. In: *AZA Annual Conference Proceedings* 2001. American Zoo and Aquarium Association, Silver Springs, MD, pp. 29–35.

Maple, T.L. (1980). Breaking the hand-rearing syndrome in captive apes. In: *AAZPA Annual Conference Proceedings.* American Association of Zoological Parks and Aquariums, Wheeling, WV.

Meder, A. (1989). Effects of hand-rearing on the behavioral development of infant and juvenile gorillas (*Gorilla g. gorilla*). *Dev. Psychobiol.* 22:357–376.

Meder, A. (1993). The effect of familiarity, age, dominance and rearing on reproductive success of captive gorillas. In: Kirchshofer, R. (ed.), *International Studbook for the Gorilla.* Frankfurt Zoological Garden, Frankfurt, pp. 227–236.

Meyer, J.R., and Wilcox, C. (1982). The reintroduction of a hand-reared lion-tailed macaque or wanderoo *Macaca silenus. Int. Zoo Yb.* 22:252–255.

Miller, A. (1982). Raising the whole baby: Strategies for the successful rearing of orphaned wild animals. In: *AAZPA 1982 Regional Proceedings.* American Association of Zoological Parks and Aquariums, Wheeling, WV.

Ogden, J., and Kasielke, S. (2001). Nursery. In: Bell, C. (ed.), *Encyclopedia of the World's Zoos,* Vol. I. Fitzroy Dearborn Publishers, pp. 886–890.

Olson, F., and Gold, K.C. (1985). Behavioral differences in mother-reared and hand-reared infant lowland gorillas. In: *AAZPA Annual Conference Proceedings.* American Association of Zoological Parks and Aquariums, Wheeling, WV, pp. 143–152.

Ott, J.E., and Joslin, P. (1981). Alternatives to hand-rearing. In: *AAZPA Regional Conference Proceedings.* American Association of Zoological Parks and Aquariums, Wheeling, WV, pp. 246–251.

Petiniot, C., Gheres, A., and Husband, S. (1988). A road to mother rearing. *Gorilla Gazette* 2(1):1–2.

Porton, I. (1992). Hand-rearing. In: Garland, C., and Fulk, R. (eds.), *Chimpanzee SSP Husbandry Manual.* North Carolina Zoological Society.

Porton, I. (1997). Hand-rearing gorillas. In: Ogden, J., and Wharton, D. (eds.), *Gorilla SSP Husbandry Manual.* Wildlife Conservation Society.

Rettberg-Beck, B., and Ballou, J.D. (1988). Survival and reproduction of hand-reared golden lion tamarins. In: *Golden Lion Tamarin International Studbook.* National Zoological Park, Washington, DC.

Rogers, C.M., and Davenport, R.K. (1969). Effects of restricted rearing on sexual behavior of chimpanzees. *Dev. Psychol.* 1:200–204.

Rosenthal, M. (1989). The management of lowland gorillas (*Gorilla gorilla*) at the Lincoln Park Zoo. In: *AAZPA Regional Conference Proceedings*. American Association of Zoological Parks and Aquariums, Wheeling, WV.

Ryan, S., Thompson, S.D., Roth, A.M., and Gold, K.C. (2002). Effects of hand-rearing on the reproductive success of western lowland gorillas in North America. *Zoo Biol.* 21:389–401.

Schildkraut, D.S. (1982). Labor in vain: The problem of inadequate mothering in captive gorillas. In: *AAZPA 1982 Regional Proceedings*. American Association of Zoological Parks and Aquariums, Wheeling, WV.

Sodaro, V. (2000). A review of hand-reared Goeldi's monkey *Callimico goeldii* at Brookfield Zoo 1977–1997. *Int. Zoo Yb.* 37:360–366.

Sodaro, C., and Weber, B. (2000). Hand-rearing and early resocialization of a Sumatran orang-utan *Pongo pygmaeus abelii* at Brookfield Zoo. *Int. Zoo Yb.* 37:374–380.

Taylor, S.H. (1978). Hand-raising zoo animals: A review. In: *AAZPA Annual Conference Proceedings*. American Association of Zoological Parks and Aquariums, Wheeling, WV.

CHAPTER THREE

Animal Welfare Regulations and Nursery Rearing

Carolyn M. Crockett

The subject of this chapter is U.S. animal welfare regulations as they pertain to infant nonhuman primates, along with possible changes that have been discussed in the past few years. The goal is to summarize administrative considerations and daily operations that are necessary to comply with existing animal welfare regulations, and to prepare for anticipated changes that may impact nursery rearing of nonhuman primates. The intent is not to regurgitate all of the regulations but to provide an overview of those that apply to nonhuman primates in general, and to identify those rules that especially pertain to infants, whether or not they are nursery reared.

1. HISTORY AND OVERVIEW OF REGULATIONS

Animal welfare regulations vary from country to country (Bayne and deGreeve, 2003). In the United States, they originate from the Animal Welfare Act (AWA) that was passed by Congress in 1966 and amended in 1970, 1976, 1985, and 1990. Certain animals that are bred for commercial sale, transported commercially, exhibited in zoos, or used in research are covered by the AWA and must be provided with care and treatment that meet or exceed minimum standards. These standards pertain to veterinary care, housing, sanitation, food and water, environmental temperature, and protection from extreme weather conditions. Some categories of animal, such as purpose-bred rats and mice, birds, reptiles, privately owned pets of any species, and animals in retail pet shops, are not covered by the AWA. The Animal and Plant Health Inspection Service (APHIS) of the U.S. Department of Agriculture (USDA) administers the standards and regulations required by the AWA. Regulated facilities are subject to unannounced inspections by the USDA to evaluate compliance with the regulations. Much of the information cited here is available on the USDA website, www.aphis.usda.gov/ac/publications.html.

The 1985 Amendment to the AWA introduced the concept of psychological well-being of nonhuman primates to the regulatory environment. This amendment led to the current U.S. Animal Welfare Regulations pertaining to captive nonhuman primates held by research facilities, zoos, and dealers (U.S. Department of Agriculture, 1991). The regulations specify minimum cage sizes, based primarily on body weight, and the requirement that "Dealers, exhibitors, and research facilities must develop, document, and follow an appropriate plan for environment enhancement adequate to promote the psychological well-being of nonhuman primates" (U.S. Department of Agriculture, 1991, §3.81). Cage-size regulations fall in the category of "engineering standards" because they are based on straightforward measurements of the physical environment. However, the environmental enhancement regulations imply a "performance standard" because "promotion of psychological well-being" is the stated goal. For details on the distinction between these two types of standard, see *The Guide for the Care and Use of Laboratory Animals* [National Research Council (Institute for Laboratory Animal Research), 1996]. Although examples of elements considered important in an environmental enhancement plan were listed in §3.81, institutions were required to write, implement, and document their own plans, to be tailored for the type of facility and species housed therein. The USDA did not specify strict engineering standards but suggested areas that facilities must address to pass this part of the inspection. By stating that the environmental en-hancement plan must be "in accordance with currently accepted professional standards as cited in appropriate professional journals or reference guides," the USDA seemed to allow for data-driven and empirically based enrichment.

It became apparent over time that some USDA inspectors were unsure of what to look for when enforcing §3.81, leading the USDA to publish a Draft Policy clarifying what it believed was required to meet the law (U.S. Department of Agriculture, 1999a). The supporting evidence came from a report written primarily by USDA veterinarians (U.S. Department of Agriculture, 1999b). The Draft Policy included more engineering standards than the 1991 regulations, and inspired many complaints by the research community during the comment period. An ad hoc committee of the American Society of Primatologists (ASP), with input from the membership and other primate experts, prepared a comprehensive comment on the Draft Policy (American Society of Primatologists ad hoc committee *et al.*, 1999). Upon approval by the ASP Board, the comment was sent to the USDA. The entire text is available on-line (http://www.asp.org/resolutions/usda_response.html). Among other comments, the ASP stated that various aspects of the Draft Policy were very costly and not supported by evidence that the imposed elements would translate into appreciable increments in psychological well-being. The Association of Primate Veterinarians also prepared a response, but it is not available to the public.

The regulated community, especially research facilities that would have to devote substantial increases in their budgets to meet some of the recommendations of the Draft Policy, awaited the Final Policy. In 2002, the proposed USDA policy was withdrawn and, according to Charles Gipson of the USDA, is to be replaced by a "best practices" document (American Society of Primatologists, 2002). The extent to which this document might resemble the Draft Policy is unclear. However, in addition to the numerous comments sent by organizations and individuals, the Office of Management and Budget had substantial concerns related to "funding mandates" (American Society of Primatologists, 2002, p. 5).

In July 2003, the Animal Legal Defense Fund and the Animal Welfare Institute filed suit in U.S. District Court, challenging the USDA's failure to carry out the 1985 AWA Amendment by dropping the Draft Policy (ALDF press release, 7-22-03). At the conference of the American Association for Laboratory Animal Science in October 2003, Dr. Gipson confirmed that the USDA had decided not to go forward with the policy but, because of litigation, is not allowed to discuss the matter. On March 2, 2004, the U.S. District Court in Northern California dismissed the ALDF-AWI lawsuit in its entirety, agreeing with the National Association for Biomedical Research (NABR) and the USDA that the lawsuit lacks merit (Anonymous, 2004). This presumably means that the USDA will go forward in preparing the "best practices" document.

2. CURRENT USDA REGULATIONS PERTAINING TO INFANTS

According to the existing USDA regulations regarding nonhuman primates (U.S. Department of Agriculture, 1991), infants and young juveniles are covered by the same regulations as adults but are to be given special consideration in institutional environmental enhancement plans. "Certain nonhuman primates must be provided special attention regarding enhancement of their environment, based on the needs of the individual species and in accordance with the instructions of the attending veterinarian. Nonhuman primates requiring special attention are the following: (1) Infants and young juveniles; (2) Those that show signs of being in psychological distress through behavior or appearance . . ." [U.S. Department of Agriculture, 1991, §3.81(c)(1)]. Infants that are removed from the mother for nursery rearing might exhibit psychological distress and therefore merit special attention to the enhancement of their environment.

In general, environmental enhancement plans must address two major components, social grouping and environmental enhancement. Social grouping "must include specific provisions to address the social needs of nonhuman primates of species known to exist in social groups in nature" [§3.81(a)]. Animals might be exempted based on compatibility, health, research protocol considerations, and other reasons approved by the Institutional Animal Care and Use Committee (IACUC) or the Attending Veterinarian. Concerning environmental enrichment, "The physical environment in the primary enclosures must be enriched by providing means of expressing noninjurious species-typical activities" [§3.81(b)].

Examples given include objects to manipulate, mirrors, perches, food treats, and foraging opportunities. Details concerning the "special consideration" required for infants and young juveniles in §3.81(c)(1) are not provided. Under current U.S. law, it is up to institutions to include appropriate considerations in their environmental enhancement plans.

Infants are also mentioned in §3.80 (Primary enclosures), which specifies the minimum floor area and enclosure height that must be provided to each nonhuman primate.

> The different species of nonhuman primates are divided into six weight groups for determining minimum space requirements, except that all brachiating species of any weight are grouped together since they require additional space to engage in species-typical behavior. The grouping provided is based upon the typical weight for various species and not on changes associated with obesity, aging, or pregnancy. These conditions will not be considered in determining a nonhuman primate's weight group unless the animal is obviously unable to make normal postural adjustments and movements within the primary enclosure. Different species of prosimians vary in weight and should be grouped with their appropriate weight group. They have not been included in the weight table since different species typically fall into different weight groups. Infants and juveniles of certain species are substantially lower in weight than adults of those species and require the minimum space requirements of lighter weight species, unless the animal is obviously unable to make normal postural adjustments and movements within the primary enclosure. [§3.80(b)(2)(i)]

Thus, for example, it is possible that a USDA inspector might judge that some very active young animals need a larger area than required by weight alone to allow for normal play behaviors.

Infants are mentioned again in §3.80(b)(2)(iv):

> When more than one nonhuman primate is housed in a primary enclosure, the minimum space requirement for the enclosure is the sum of the minimum floor area space required for each individual nonhuman primate in the table in paragraph (b)(2)(i) of this section, and the minimum height requirement for the largest nonhuman primate housed in the enclosure. Provided however, that mothers with infants less than 6 months of age may be maintained together in primary enclosures that meet the floor area space and height requirements of the mother.

This regulation allows for mothers and infants under 6 months to be kept in the minimum cage size legal for the mother, and, as indicated in the previous paragraph, for pregnant females to be housed in cages corresponding to minimum size for their nonpregnant weight. If, after removal from their mothers, infants and small juveniles are housed in peer groups, the legal minimum floor area is based on the sum of each individual's weight group requirement, not the sum of their combined weights. For example, three 1.5-kg ("group 2") and four 0.9-kg ("group 1") primates require a minimum of $(3 \times 3.0) + (4 \times 1.6)$ or 15.4 square feet, not 4.3 square feet, which would be the legal minimum area for a single primate of 8.1 kg ("group 3"), equivalent to the combined weight of the seven hypothetical primates.

A potentially relevant section is §3.80(b)(c):

> Innovative primary enclosures not precisely meeting the floor area and height requirements provided in paragraphs (b)(1) and (b)(2) of this section, but that do provide nonhuman primates with a sufficient volume of space and the opportunity to express species-typical behavior, may be used at research facilities when approved by the Committee, and by dealers and exhibitors when approved by the Administrator.

For example, a tall, enriched-activity cage with a floor area somewhat smaller than the minimum floor area specified in §3.80(b)(2)(iv) might be approved by the IACUC as acceptable housing for groups of infants and juveniles if furnished with extra perches and other structures providing additional surfaces.

Only two other sections single out infants for special attention. Regulations on feeding specify that "infant and juvenile nonhuman primates must be fed as often as necessary in accordance with generally accepted professional and husbandry practices and nutritional standards, based upon the animals' age and condition" [§3.82(b)]. Regulations for enclosures used to transport nonhuman primates permit a mother and her nursing infant [§3.87(d)(1)(i)] or "a compatible pair of juveniles of the same species that have not reached puberty" [§3.87(d)(1)(iii)] to be transported together. Elsewhere in the regulations, infant nonhuman primates are defined as less than 6 months old and juveniles are defined as 6 months to 3 years [§3.80(b)(2)(i)]. Therefore, it appears to be legal to ship two infants or juveniles together in the same enclosure.

The animal welfare regulations do not specifically address nursery rearing or conditions under which infants might be removed from the mother. However, "Individually housed nonhuman primates must be able to see and hear nonhuman primates of their own or compatible species unless the attending veterinarian determines that it would endanger their health, safety, or well-being" [§3.81(a)(3)]. A research facility's IACUC might grant an exemption from visual contact based on a research protocol with convincing scientific justification. Therefore, a single nonhuman primate infant cannot be housed alone in a nursery without approval from the attending veterinarian, or, in the case of a research protocol, the IACUC.

3. USDA DRAFT POLICY AND ASP COMMENTS

The *Draft Policy on Environment Enhancement of Nonhuman Primates* (U.S. Department of Agriculture, 1999a) and relevant criticisms (American Society of Primatologists ad hoc committee *et al.*, 1999) provide examples of possible changes in regulations applying to infants and corresponding comments from an informed group, including regulated primate researchers.

Given the possibility that animal welfare regulations may be modified in the near future, individuals with oversight of nursery-rearing practices must be aware of potential impending changes. Although we cannot predict exactly what changes may be coming, proposed USDA policy identifies some possibilities. An entire section of the Draft Policy was devoted to Social Needs of Infants (U.S. Department of Agriculture, 1999a, p. 38148). Because psychological well-being of nonhuman primate infants depends on "appropriate infant development," the Draft Policy specifies that the "optimal" environment is one in which the infant stays with its mother through weaning in the company of a species-typical group." According to the Draft Policy, the Environmental Enhancement Plan should include a program to "ensure" (note: a stronger word than "promote," as specified in the 1991 regulations) "species-typical sensory, motor, psychological and social development of infants." The plan must also include criteria for removal of infants from parents, and infants should not be removed at an age younger than that approximating infant independence in nature unless "necessary for the health and well-being of the infant or dam." The plan must describe procedures to minimize the distress of separation, methods for hand rearing, and provision of suitable socialization with conspecifics. In short, far more details were expected in environmental enhancement plans under the Draft Policy than specified in the 1991 regulations, and a substantial emphasis is on infants and nursery rearing.

The ASP response agreed that special attention to infants and young juveniles is warranted, and that being reared by the mother in a social group is optimum for normal development (American Society of Primatologists ad hoc committee *et al.*, 1999, section III). However, the ASP suggested that "age of completed weaning" is better defined than "independence" for an appropriate minimum age of separation from the mother. The ASP also recommended that the phrase "or as part of an IACUC-approved protocol" be added to the end of the sentence referring to conditions of separation of infant from dam. Furthermore, the ASP cited several research studies supporting the recommendation that infants separated from the mother not be peer reared on a continuous basis. Continuous peer-rearing of macaque infants has been found to promote clinging and delayed behavioral maturation in infancy, and social behavior deficits in adulthood (Champoux *et al.*, 1999; Ruppenthal *et al.*, 1991).

4. NRC-ILAR VOLUME ON PSYCHOLOGICAL WELL-BEING OF NONHUMAN PRIMATES

The Psychological Well-being of Nonhuman Primates [National Research Council (Institute for Laboratory Animal Research), 1998] presents widely accepted professional standards based on several years of effort by a National Academy of Sciences Committee on Well-being of Nonhuman Primates. The committee consists of veterinarians and PhDs, many of whom have contributed to professional standards directly through research studies. The volume, available on-line at http://www.nap.edu/, presents the committee's consensus on the essential elements of an environmental enhancement plan for the promotion of psychological well-being and includes some sample plans. Criteria to evaluate psychological well-being include the animal's coping ability with respect to social and physical environmental changes, the presence of appropriate species-typical behaviors and a balanced temperament, and the absence of signs of chronic distress and of self-injurious or other maladaptive abnormal behavior. To promote psychological well-being, as assessed by the above criteria, an environmental enhancement plan should provide for the following:

- Appropriate social companionship.
- Opportunities to engage in behavior related to foraging, exploration, and other activities appropriate to the species, age, sex, and condition of the animals.
- Housing that permits suitable postural and locomotor expression.
- Interactions with personnel that are generally positive and not a source of unnecessary stress.
- Freedom from unnecessary pain and distress (p. 16).

Because early rearing without mother or peers can lead to a variety of abnormal and stereotypic behaviors and inadequate coping strategies, the committee concluded that it is especially important to provide appropriate social contact to very young primates. Behavior disorders that might develop as a result of deficient early rearing can be resistant to treatment. It is important for facilities that house nonhuman primates to have records of early rearing conditions for individual animals because idiosyncratic behaviors in adulthood might reflect early experiences rather than the current housing environment.

With respect to nursery rearing, the volume includes a paragraph on hand rearing (Chapter 3) and a section on conditions involving atypical rearing environments (Chapter 4). The committee urges that when an infant must be separated from its mother, "every effort should be made to provide infants and other immature animals with appropriate social stimulation so as to minimize the adverse effects of rearing in socially restricted environments" (p. 49). Separate chapters on prosimians, New World monkeys, Old World monkeys, and apes include recommendations and information relevant to nursery rearing, environmental enrichment, and selection of objects to encourage species-typical manipulation. Concerning the latter, the committee advises that such objects should be beneficial and appropriately sanitizable or replaceable.

The last chapter of the NRC-ILAR document identifies areas of needed research on psychological well-being. Regarding individual development, the committee calls for more research on the types and extent of social and other stimulation necessary for normal development when primates are separated from the mother in infancy or early juvenility.

It has been suggested that the NRC report on *The Psychological Well-being of Nonhuman Primates* [National Research Council (Institute for Laboratory Animal Research), 1998] has been endorsed in lieu of the USDA Draft Policy (American Society of Primatologists, 2002, p. 11). Although the NRC volume summarizes "professional standards," it is not clear how USDA inspectors will use it in lieu of the apparently abandoned Draft Policy. Regardless, this volume has a wealth of information for those involved in writing and implementing environmental enhancement plans.

5. GUIDE FOR THE CARE AND USE OF LABORATORY ANIMALS

The purpose of the *Guide for the Care and Use of Laboratory Animals* is "to assist institutions in caring for and using animals in ways judged to be scientifically, technically, and humanely appropriate" [National Research Council (Institute for Laboratory Animal Research), 1996, p. ix]. Where the *Guide* differs from current animal welfare regulations, as it does in a few matters, USDA inspectors will follow the regulations. However, institutions seeking animal-research funding from the National Institutes of Health or other Public Health Service agencies must comply with the *Guide* and some other policies (Bayne and deGreeve, 2003).

The *Guide* applies to animals used in laboratory research, summarizes institutional policies and responsibilities; animal environment, housing, and management; veterinary medical care; physical plant considerations; and occupational health and safety programs. It specifies that IACUCs must inspect animal facilities at least every 6 months. When nursery rearing of NHPs is part of a research protocol, such a protocol must be approved by the IACUC. The *Guide* lists a number of points that must be addressed in animal care and use protocols, including "unusual housing and husbandry requirements" (p. 10). Elsewhere, the *Guide* indicates that "Whenever it is appropriate, social animals should be housed in pairs or groups, rather than individually, provided that such housing is not contraindicated by the protocol in question and does not pose an undue risk to the animals" (p. 26). This is an essential resource volume for facilities managers and animal research program coordinators.

6. AAALAC INTERNATIONAL

The Association for Assessment and Accreditation of Laboratory Animal Care International (AAALAC International) is a nonprofit organization that provides voluntary peer review of animal care and use programs (Bayne and deGreeve, 2003). AAALAC International conducts site visits every 3 years, relying on regulations, policies, the *Guide*, and other reference resources and position statements to evaluate programs. AAALAC accreditation is very important because it represents quality, promotes scientific validity, demonstrates accountability, and shows a commitment to humane animal care. AAALAC uses the *Guide* [National Research Council (Institute for Laboratory Animal Research), 1996] as the primary reference document for assessing and accrediting animal care and use programs. AAALAC also refers to other specialty publications, such as *The Psychological Well-being of Nonhuman Primates* [National Research Council (Institute for Laboratory Animal Research), 1998], when conducting program evaluations. See www.aaalac.org/resources.htm for a complete list of reference resources used by AAALAC International.

7. THE AMERICAN ZOO AND AQUARIUM ASSOCIATION (AZA)

To belong to the AZA, organizations must be accredited (zoos, aquariums) or certified (wildlife refuges, rehabilitation centers, and related facilities) as described on the AZA website (http://www.aza.org/Accreditation/). Member organizations must meet standards of animal care and husbandry and follow the *American Association of Zoo Veterinarians Guidelines for Zoo and Aquarium Veterinary Medical Programs and Veterinary Hospitals.* The website also has resource information on government regulations (http://www.aza.org/GovAffairs/GovAffairsResources/).

Taxonomic Advisory Groups (TAGs) of the AZA are in the process of writing documents on standards of husbandry, housing, enrichment, transport, behavior management, and other aspects of care specific to particular mammal groups at AZA zoos. Upon approval by the AZA Animal Welfare Committee, these documents will become official AZA recommendations. The primate documents are still in draft form and cannot yet be considered final, but judging from the draft document on Cebids (Andy Baker, New World Primate TAG Chair, personal communication, August 2003), they will be very useful guidelines for zoos and related facilities. Pertinent to the present volume, the Draft Cebid Husbandry Standards document includes a section titled, "Address what, if any, circumstances might warrant hand-rearing and identify acceptable hand-rearing and introduction protocols." If the final AZA primate Husbandry Standards documents become available on the AZA web page, they will be another useful resource to consult in preparing institutional environmental enhancement plans and in managing facilities of many types, including research laboratories and primate nurseries.

8. INTERNATIONAL REGULATIONS

As noted in the introduction, the focus of this chapter was animal welfare regulations in the United States. For an overview of regulations, policies, and legislation on the use of animals in research in Europe, Canada, and other parts of the world, see Bayne and deGreeve (2003). Although they do not mention infant primates specifically, their material is a good resource on the general topic of research animals. A body called the Multilateral Consultation of Parties to the European Convention for the Protection of Vertebrate Animals Used for Experimental and Other Scientific Purposes has drafted guidelines for accommodation and care of animals. Once finalized, the documents may be available on the Council of Europe website (http://www.coe.int/). According to sources at the 2003 AALAS conference, primate guidelines were expected to be completed in 2004, but were not on the website by mid-2005.

9. RECOMMENDATIONS

Although the USDA does not currently require facilities to have clear protocols for removing infants from their mothers and rearing them in nurseries, facilities that do not have such protocols would be well advised to prepare them. Research institutions need to ensure that projects involving nursery rearing have protocols approved by the IACUC. Institutional environmental enhancement plans should include a separate section or document relevant to infants. For example, very young infants in the nursery setting might be exempted from toys and food treats. Those moved from incubator to cage might be provided with cloth-covered mobile hanging surrogates (Clarke *et al.*, 1989; Duijghuisen *et al.*, 1992) rather than perches. Toys and treats should be of appropriate size and texture for infants. To promote normal development, tactile social contact involving opportunities for play with conspecific peers should be provided, but (at least for macaques) not on a continuous basis (Ruppenthal *et al.*, 1991). Reference materials and publications supporting the plan and other procedures should be available for inspection. Documentation of nursery practices is important both for training personnel and for achieving uniform care within the facility. For an example, see Ruppenthal and Sackett's (1992) *Research Protocol and Technician's Manual: A Guide to the Care, Feeding, and Evaluation of Infant Monkeys*, available on-line at http://www.rprc.washington.edu/iprl/index.htm.

Facilities should also have a plan for evaluating the psychological well-being of their primates. Inspectors can request behavioral as well as medical and experimental records of animals in regulated facilities. Criteria for adequate psychological well-being include physical health, a range of species-typical behaviors with few or no abnormal behaviors, appropriate responses to environmental challenges such as cage change, and absence of chronic distress (Bayne and Novak, 1998). At the University of Washington National Primate Research Center, the Psychological Well-being (PWB) Program systematically evaluates animals referred for assessment of possible behavior problems. Normal and abnormal behaviors during a 10-min observation period are scored at 30-sec intervals (instantaneous scan samples) (Crockett, 1996) to estimate the percentage of time spent on each behavior. In addition, any abnormal behavior is also scored on a 1–0 basis (i.e., scored as "1" during any interval when it occurred) and informal notes taken on possible eliciting factors. Behavior profiles based on four 10-min samples were adequate to distinguish between 29 adult male pigtailed macaques

(*Macaca nemestrina*) previously referred for assessment of abnormal behavior and 58 that were a control group, never referred for abnormal behavior (Bellanca and Crockett, 2002). Referred animals are scored for the presence or absence of abnormal behaviors during weekly visits. Animals at the UW Infant Primate Research Laboratory (IPRL) are observed daily by IPRL staff, and most infants are scored several days a week using a home-cage scoring system (Ruppenthal and Sackett, 1992). Animals with abnormal behaviors are reported to PWB Program staff to ensure that they have a complete listing of behavior problem cases to check on weekly. The PWB Program, in turn, provides the listing of the cases to IPRL staff whenever new animals are added so that everyone is aware of individuals that might need special consideration. Veterinary staff provide information on cases in which unusual behaviors might have a medical basis, such as disease or birth defect. Behaviors of particular concern in infants include self-biting and self-hitting (which may escalate to actual self-injurious behavior later in juvenility), excessive self-clasping or self-sucking, high levels of locomotor stereotypy (e.g., constant pacing), and cowering or other evidence of chronic fear or distress.

10. CONCLUSIONS

Although very few current USDA animal welfare regulations apply specifically to infants or nursery rearing, potential changes may increase the number of regulations or policies. By identifying infants in the class of primates requiring "special considerations," the USDA inspectors are already on the lookout for conditions that fail to measure up to their scrutiny. This, coupled with research (primarily in rhesus macaques) associating social privation and partial isolation rearing with the development of behavioral disorders (Bayne and Novak, 1998), makes nursery rearing especially vulnerable to criticism. It behooves administrators and managers to be aware of the regulations, policies, and currently accepted husbandry practices for nursery rearing. Most veterinarians and behaviorists would agree that infant primates require additional attention and specialized practices in the interests of promoting their psychological and physical well-being.

ACKNOWLEDGMENTS

Thanks go to Jim Sackett for inviting me to contribute to this volume based on a very informative ASP workshop. The constructive comments of Jim Sackett, Steve Kelley, Rita Bellanca, and Kate Elias improved this chapter. I appreciate the efforts of Michael Hutchins and Andy Baker in providing information on AZA animal husbandry draft documents. Supported by NIH RR00166 (Washington National Primate Research Center).

REFERENCES

American Society of Primatologists. (2002). ASP's role in AAALAC. *ASP Bull.* 26(4):5.

American Society of Primatologists ad hoc committee, Capitanio, J.P., Bloomsmith, M., Crockett, C.M., Novak, M.A., and Rasmussen, K.L. (1999). ASP—USDA Response. American Society of Primatologists Web Page, http://www.asp.org/resolutions/usda_response.html.

Anonymous. (2004). Court dismisses primate case. *Lab. Primate Newsl.* 43(2):20.

Bayne, K., and deGreeve, P. (2003). An overview of global legislation, regulation, and policies on the use of animals for scientific research, testing, or education. In: Hau, J. and van Hoosier, G.L., Jr. (eds.), *Handbook of Laboratory Animal Science 2nd ed. Volume I: Essential Principles and Practices.* CRC Press, Boca Raton, FL, pp. 31–50.

Bayne, K., and Novak, M.A. (1998). Behavior disorders. In: Bennett, B.T., Abee, C.R., and Henrickson, R. (eds.), *Nonhuman Primates in Biomedical Research: Diseases.* Academic Press, San Diego, pp. 485–500.

Bellanca, R.U., and Crockett, C.M. (2002). Factors predicting increased incidence of abnormal behavior in male pigtailed macaques. *Am. J. Primatol.* 58:57–69.

Champoux, M., Shannon, C., Airoso, W.D., and Suomi, S.J. (1999). Play and attachment behavior of peer-only reared and surrogate/peer-reared rhesus monkey infants in social groups. In: Reifel, S. (ed.), *Play and Culture Studies,* Vol. II. Ablex Publishing Corp., Norwood, NJ, pp. 209–217.

Clarke, M.R., Koritnik, D.R., Martin, L.N., and Baskin, G.B. (1989). Cage enrichment, physiology, and behavior in nursery-reared rhesus monkeys. *Am. J. Primatol. Suppl.* 1:53–57.

Crockett, C.M. (1996). Data collection in the zoo setting, emphasizing behavior. In: Kleiman, D.G., Allen, M.E., Thompson, K.V., Lumpkin, S., and Harris, H. (eds.), *Wild Mammals in Captivity: Principles and Techniques.* University of Chicago Press, Chicago, pp. 545–565.

Duijghuisen, J.A.H., Timmermans, P.J.A., Vochteloo, J.D., and Vossen, J.M.H. (1992). Mobile surrogate mothers and the development of exploratory behavior and radius of action in infant long-tailed macaques (*Macaca fascicularis*). *Dev. Psychobiol.* 25(6):441–459.

National Research Council (Institute for Laboratory Animal Research). (1996). *Guide for the Care and Use of Laboratory Animals.* National Academy Press, Washington, D.C.

National Research Council (Institute for Laboratory Animal Research). (1998). *The Psychological Well-Being of Nonhuman Primates.* National Academy Press, Washington, D.C.

Ruppenthal, G.C., and Sackett, G.P. (1992). *Research Protocol and Technician's Manual: A Guide to the Care, Feeding, and Evaluation of Infant Monkeys*, 2nd ed. University of Washington, Seattle.

Ruppenthal, G.C., Walker, C.G., and Sackett, G.P. (1991). Rearing infant monkeys (*Macaca nemestrina*) in pairs produces deficient social development compared with rearing in single cages. *Am. J. Primatol.* 25:103–113.

U.S. Department of Agriculture. (1991). Animal Welfare, Standards, Final Rule (Part 3, Subpart D: Specifications for the humane handling, care, treatment, and transportation of nonhuman primates). *Fed. Reg.* 56(32):6495–6505.

U.S. Department of Agriculture. (1999a). Animal Welfare: Draft Policy on Environment Enhancement for Nonhuman Primates. *Fed. Reg.* 64(135): 38145–38150.

U.S. Department of Agriculture. (1999b). *Final Report on Environment Enhancement to Promote the Psychological Well-being of Nonhuman Primates.* USDA, APHIS, Riverdale, MD.

CHAPTER FOUR

Data Management for the Nonhuman Primate Nursery

James C. Ha and Arthur E. Davis

1. INTRODUCTION

In any facility that manages the health and welfare of nonhuman primates, it quickly becomes obvious that record keeping, and therefore data management, is required. In fact, increasing levels of record keeping and data management are now required by various facility-monitoring organizations. Much less concern has been placed on the uses for such data beyond the legal and immediate, day-to-day needs. Furthermore, many data-management systems have been built gradually, and no standards currently exist for what information such systems should contain. Our goal in this chapter is to describe the dimensions of data-management systems for nonhuman primate colonies, to provide a brief history of previous efforts to develop some structure and standards for these systems, and then to bring these efforts up to date with suggestions that have worked to satisfy the more modern, multiple needs of clinicians, researchers, and administrators.

Early schemes for managing data on nonhuman primates were designed to satisfy the immediate needs of clinical care and to provide gross colony-management data such as census reporting. An additional requirement in modern biomedical research is the availability of short-term records for financial management. More recently, the considerable value of long-term, archival records also has begun to be realized (e.g., Ha *et al.*, 1999, 2000a,b, 2002). In reality there are many short- and long-term uses for these systems. While the short-term uses tend to be more obvious at the outset, long-term uses and the attendant value of information that can be harvested from these systems frequently do not become apparent until a significant amount of data has been accumulated over time.

The multiple dimensions or demands on such data systems are illustrated in Fig. 4–1. On one axis, there is a time dimension: the distinction between immediate, day-to-day demands for clinical data versus the long-term archival value of data that have been collected over many months or years. On another dimension, there is a distinction between the need to obtain data on single animals (again, primarily a clinical need or research need) and the value of being able to answer questions at a population level with data on, for example, inbreeding, population fertility, and mortality rates. Finally, a more recent desire—even requirement in some cases—is the ability to exchange and compare data across populations, whether within the same facility (e.g., research versus breeding populations) or across facilities. This demand to compare data across

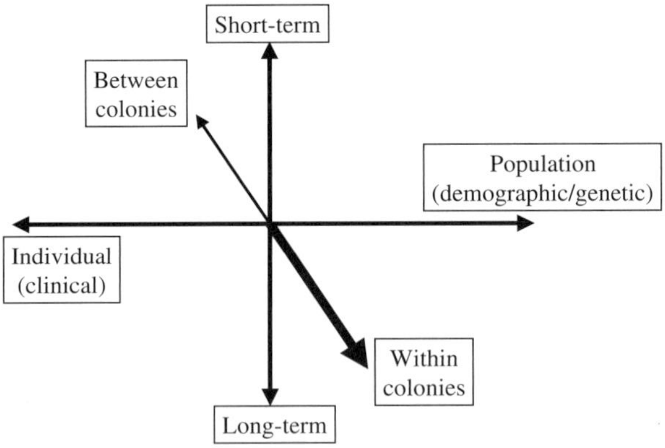

Figure 4–1. Multidimensional depiction of the applications for stored data in nonhuman primate colonies.

federally funded facilities is currently expanding, and presents some very difficult issues to policymakers and colony managers as well as information technology managers.

It is important to note that the perceived value of different uses of these data systems depends on the user's viewpoint. For example, veterinary technicians are interested in keeping the current set of animals healthy and well cared for and need to consult only the immediate records for individual animals. After an animal has left the colony the veterinary technician has no further use for its records. On the other hand, a researcher who is studying the genetic makeup of a breeding colony needs to see the known genetic relationships (genealogy) of the colony for as many generations as possible. An important task for information-technology managers is to design and implement systems that can meet all of these needs.

In this chapter we will focus on the issues raised above, in the hope that the information we present here will continue a dialogue among the policymakers, information-technology specialists, and other concerned personnel of nonhuman primate facilities. Our goal is to improve the quality of information technology for the care and rearing of nonhuman primates, whether in biomedical research or in conservation biology contexts.

1.1. Historical Standards

There have been two major efforts to standardize the minimum data kept in nonhuman primate record systems. In 1979 the Institute for Laboratory Animal Research (ILAR) published standards for biomedical primate colonies. In this effort, ILAR established four basic uses for primate colony data: (1) to be stored for immediate and/or later use, (2) to serve as a basis for the evaluation of breeding programs, (3) to be sent along with animals that are being moved between facilities, and (4) to provide reporting standards for national management of primate populations. ILAR suggested that primate colony data systems should be designed to meet these four requirements.

In 1993, an ad hoc committee of the American Society of Primatologists (ASP) reviewed the ILAR effort to establish standards for primate

Table 4–1. American Society of Primatologists Standard Registry Record for Primate Facility Data Systems

ASP Standard Registry Record
Required
ID
Sire
Dam
Sex
Birth date
Entry date
Acquisition code
Exit date
Exit code
Optional
Taxonomic code
Institution code
Local subgroup code
Current location code

data systems, and proposed a system of more specific information to be maintained by primate facilities (Table 4–1; Dyke, 1993). The Standard Registry Record, which was adopted after extensive discussion, included very basic information that could be used to characterize the animal (facility identification code; species; sex; dates of birth, entry, death, and exit) and information that could be used to calculate pedigree-based measures of inbreeding (sire and dam identification) and summaries of population demographics (fertility and mortality). It was assumed that additional information would be kept on each animal (e.g., clinical or research records), but the hope was that the minimum information in the Standard Registry Record would be available for animals moving between facilities and for purposes of centralized reporting on stock levels and basic demographics.

The ASP system was loosely based on the record system established many years earlier for zoo animals [International Species Information System's Animal Record-Keeping System (www.isis.org)]. The ISIS system is in widespread (but not 100%) use throughout the zoo community. Standardization in the zoos was established early from a central standardized effort, before multiple systems arose independently, as has happened in the captive primate community.

2. MODERN ISSUES AND APPROACHES

Computer technology has advanced tremendously in the past two decades. Information specialists in modern primate facilities are aided by database advancements that were unimaginable in previous decades. Past efforts at standardization of data management had to be concerned with exact details of data storage, including variable names and the order and specifics of data format and storage protocols. Evolving database software has helped to minimize concerns about the internal details of database architecture and interoperability between different types of database systems. As described below, software solutions now obviate the need for data storage in specific physical formats. The issue of *how* data are stored is no longer so important. Of greater concern is what *kind* of data are stored, and how *well*: how efficiently are records stored, both for immediate use and for the long term, and how can authorized users obtain the data via internal or external (Internet) routes?

Two efforts are beginning to deal with the latter questions of electronic storage and Internet availability of data. *Ethosource*, funded by the National Science Foundation (NSF), is an effort to make behavioral data (for primates and any other species) available at some level of access to the wider ethological community (php.indiana.edu/~emartins/Ethosource/). Begun by a small group of animal behaviorists and database software researchers at a workshop at Indiana University in April 2002, the effort has acquired the support of numerous professional organizations. A similar effort sponsored by the National Institutes of Health (NIH) specifically targets federally funded primate facilities. Led by a group of researchers from the National Primate Research Centers (NPRCs), three Colony Records Analysis Workshops have led to a more focused effort by the NIH National Center for Research Resources to develop standards for data sharing among these primate facilities (www.sfbr.org/external/conferences.html).

Regardless of efforts to standardize data availability and sharing among facilities, the questions of what data should be stored and standards for data quality control remain important. In the rest of this chapter, we describe some new standards for minimum data storage, based on our experience of dealing with these issues over the past decade, and outline some techniques to maintain and improve data quality. Finally, we outline some state-of-the-art approaches to data collection and storage/retrieval that are becoming available and affordable.

3. ANIMAL-RECORD CONTENTS

The decision about the type and amount of data to be stored is critical to the success of any data-management system. Most data systems are developed through what we term "evolved" decisions: some information to be stored is obvious from the start, but other items are added as the need is discovered, and the system therefore evolves over time. Depending on the skill with which this process is accomplished, the result can be a hodge-podge of data that is difficult to access or document. Entire system redesigns may be required. For instance, the Washington NPRC's Animal Record System (ARS) has gone through at least two major restructurings. This process of evolving a major data system is unavoidable because foresight is not perfect and because new requirements develop over time. We hope that some of the ideas presented in this chapter might help to minimize the pain involved in such a process.

There is very little question about the need to maintain minimum records at the level of the ASP minimum data standard (animal identification code, sex, species, sire, dam, birth/entry date, death/departure date). We will refer to these fields as the Minimum Data Set (Table 4–2). These fields are examples of single-entry data fields: fields that require only a single entry per animal and will not change during the animal's lifetime. The alternative will be termed multiple-entry data fields, in which there is a chronological component and possibly multiple entries for each animal on different dates. Examples of this type of data include weights and breeding records.

When we move past the Minimum Data Set, the needs for data storage become less well defined. Several important forms of animal history, however, are becoming increasingly critical to good animal husbandry. These are summarized in Table 4–2 as History Data. These are all multiple-entry forms of data because the record for an animal may change over the lifetime of the subject.

Table 4–2. Potential Data Fields for a Nonhuman-Primate-Data Management System

Field	Description	Type
Minimum data		
Identification	Individual identification code	Single entry
Sex	Self-explanatory	Single entry
Species	Self-explanatory	Single entry
Sire	Identification code of sire	Single entry
Dam	Identification code of dam	Single entry
Birth date	Self-explanatory	Single entry
Entry date	Date of entry into colony/data system	Single entry
Exit date	Date of departure from records for any reason	Single entry
Exit code	Reason for departure from records	Single entry
History data		
Clinical records	Procedure records: blood draws, surgery	Multiple entry
	Diagnostic testing records: blood hematology, microbiology, urinalysis, virology assays	Multiple entry
	Diagnosis and treatment records	Multiple entry
Weights	Self explanatory	Multiple entry
Movement records	Movements between sites, locations, or specific cages	Multiple entry
Psychological well-being	Monitoring of PWB protocols and treatment of PWB cases	Multiple entry
Breeding records	Records of breeding opportunity, pregnancy, and outcome	Multiple entry
Administrative data		
Project assignment	Responsible individuals, emergency protocols, cost-recovery information	Multiple entry
Institutional or animal care protocols and approvals	Monitoring of up-to-date approvals and animal assignments and procedures	Multiple entry
Financial record-keeping	Per diem calculation, procedure costs, billing, accounts payable and receivable	Multiple entry

Of course, clinical records are critical to veterinarians for maintaining proper veterinary care, but there is increasing use of archival clinical records for monitoring potential risk factors for clinical syndromes as well as the efficacy of treatments. Weight records are a common measure of animal health and a longitudinal history of weights can provide a better measure of an animal's status, especially for infants, whose growth is rapid and critical to health.

Movement records provide several forms of information. Housing locations, of course, allow animal husbandry and clinical staff to locate animals efficiently. However, archival records of housing and animal movements can also provide information about levels of social housing, which animals have been in contact, and specific aspects of the rearing environment. This information can be important for later analyses of the influence of rearing environment, for reporting and evaluating compliance with animal-housing protocols, and, more recently, for documenting disease exposure status in colonies of specific pathogen-free animals.

All facilities that house nonhuman primates have become increasingly concerned about the psychological well-being (PWB) of their animals as they have learned more about the factors that influence development and behavior. Archival records of PWB-monitoring programs provide a record of current PWB efforts for specific animals, a readily available record of PWB protocols either for specific animal(s) or across species, and methods for reporting on PWB efforts and efficacy to monitoring organizations. Very importantly, a record of problems and attempted solutions and their outcomes will gradually provide a significant amount of data for analysis of risk factors and remedies for PWB problems, which, being relatively rare, have proven difficult to study experimentally.

The final form of History Data is breeding information for both males and females. This, of course, is not an issue in a facility that is only a primate nursery, but most such facilities are generally part of a larger facility that is involved in breeding (hence the presence of infants!). Breeding records provide information about individual and colony-wide fertility rates, risk factors leading to reproductive failure, and surgical interventions (e.g., cesarian sections) that may be limited by protocols. In combination with housing and movement data, breeding data can provide information about the relation between social factors and reproduction (e.g., Ha *et al.*, 1999).

Finally, we define a third class of data as Administrative, data used primarily for colony management. These data types include project assignments, data on animal care and use protocols, any per-animal exceptions to protocols, and financial management data for billing and budgeting.

Each facility has to decide on the depth and breadth of the data that it deems necessary to maintain in its records system. But most of the forms of data that we describe have demonstrated importance to animal health and well-being and to both day-to-day and long-term colony management. New standards for data reporting and exchange for federally funded facilities are being developed and they will probably include our Minimum Data Set as well as most, if not all, of our History Data Set. It has been suggested in discussions with the NIH, for instance, that records of basic health, experimental exposure, and social exposure should travel with each animal across facilities and that these records, along with breeding information, should be made available for federal reporting.

4. QUALITY CONTROL

Regardless of the type and amount of data stored in a data-management system, data quality remains an important issue. In many primate record systems, the data are intended primarily for short-term use and the emphasis on quality control is therefore limited. While the quality of data with a short lifespan is important, when the additional dimension of archival data is added to a system design, additional requirements for data quality are added. In this section, we describe a set of basic guidelines for maintaining and improving data quality in any kind of data-management system, regardless of the type or quantity of the data.

A set of basic data validity techniques is summarized in Table 4–3. These techniques are now readily available in modern database applications. They fall into several categories: techniques for improving initial data entry (Code List Tables and Generated Fields), techniques for immediately checking the validity of data and providing immediate user feedback (Primary and Foreign Key Constraints, Simple Edit Checks, and Logical Edit Checks), and techniques for reviewing the efficacy of validity techniques (Periodic Consistency Checks).

One simple and very effective rule for improving data entry is to maintain and monitor data quality as early as possible (as close to the "producer" of the data as possible). This reduces the time lag for producing high-quality data and allows corrections to be made by the individuals who are first generating the data. The use of Code List Tables whenever possible is one method for simplifying data entry while improving data

Table 4–3. Techniques for Checking the Validity of Data

Technique	Description	Examples
Code list tables	Restrict free-form data entry; require choices from preplanned lists	Choice from Species or Location Lists
Generated fields	Allow database system to generate record entries	Date, time, sequential animal identification numbers
Primary key constraints	Require checks for unique record identification within the current table	"Animal ID must exist" "Animal ID must not exist"
Foreign key constraints	Require checks for unique record identification using information from other tables of data	Animal number in weights record must also exist in a master animal identification record
Simple Edit checks	Check for proper data format and range	No negative weights Year must be four digits
Logical Edit checks	Check validity of data entered against parts of record	Date of weight measurement must be after birth or entry date and before death date
Periodic consistency checks	Review of records to check for proper use of codes, contents of Other categories	Establish additional code list contents for frequently used entries in Other category

quality. These database tables contain a set of predefined entries for a specific field. The data-entry user is required to choose from a menu of these choices when entering data into the field. Examples include species, sex, perhaps housing locations or sites, and source location. This option reduces error and makes data retrieval at a later time much simpler.

A second option for reducing errors in initial data entry is to use Generated Fields. Again, requiring the computer system to supply information automatically for certain fields will reduce errors from the beginning of the data-entry process. Examples include date and time (perhaps with the possibility of override and manual entry, if necessary) and the automatic assignment of sequential animal identification codes. This latter technique is currently used in the ARS at the Washington NPRC.

The previous two techniques control data-entry errors as data are entered into the system. The next two methods enforce relational consistency of the database by checking key fields before new records are created or existing records are deleted. These are the Primary Key Constraint and the Foreign Key Constraint. The Primary Key Constraint specifies a combination of one or more ordered fields (columns in relational database parlance) on which each record (row) in a table must be unique. The database "machine" will not allow duplicate records having the same Primary Key into the system. Foreign Key Constraints are similar to Primary Key Constraints. They specify a combination of one or more ordered fields whose combined values must match a corresponding set of values in a row in a different (foreign) table. An example system would have a table that holds the single-entry data fields for each animal and a multiple-entry data table holding all of the weights for all animals. The logical choice for the Primary Key for the single-entry table would be the animal ID. The system would then allow only one such record for each distinct animal. For the multiple-entry weights table the Primary Key might be the combination of the animal ID and the date of the weight so that each animal could have weights on many different dates but only a single weight record for any specific date. An additional Foreign Key Constraint could be placed on the weights table to specify that the animal ID in each weight record must exist as an animal ID in the single-entry data table. No entries for a given animal would be allowed in the weights table unless there was a corresponding entry for the same animal in the single-entry data table. In addition, the data-entry user would be prevented from deleting an existing single-entry data record for any animal where there are existing related (detail) weights records for that animal. Key constraints are an important tool for building and preserving the integrity of any database. These constraints can be enforced automatically in the database system and/or programmatically at the user interface. Many well-designed systems do both so that the data-entry user gets immediate feedback and so that programming errors and automated processes cannot inadvertently violate database integrity.

An additional immediate-check technique is a Simple Edit Check. In this technique, the data-entry application is designed to check the datum for simple formatting rules as soon as it is entered. For instance, the system can be "told" that the number in a weight field cannot be negative or that a year entry must contain four digits.

A related technique for immediate feedback is a Logical Edit Check, in which the datum entered into a field is checked immediately for logical validity with other parts of the record system. For example, the date of a weight entry must be later than a birth or entry date and earlier than a death or departure date. Likewise, a positive pregnancy check must be entered into the record of only a female animal. This technique differs from the previous method because it uses logical relationships (AND, OR, NOT) to check the validity of the entry in question against other information in the system. The Simple Edit Check performs within-field checks against formatting rules.

Finally, a less common method of maintaining data integrity is the use of Periodic Consistency Checks. These are basically reports that are routinely run by the system administrator, or by the system itself, checking for the internal consistency of the records to make sure that no errors are slipping through the initial data-entry checks described above. One use for these types of checks in the Washington NPRC ARS is to examine entries in the "Other" category of records that allow it. Frequently "Other" is included in a code list to catch any entries that are not adequately coded by one of the other existing codes. A text field may be provided in which the data-entry user can enter a description that defines what Other means. We have found that over time, basically the same entry (with varied spelling, capitalization, and word order) will appear in the Other category repeatedly (e.g., a new blood assay is available and entered into a Type-of-Assay field as "Other"). When this happens, an addition to the established Code List (of assay types, in our example) is made, to reduce the use of the Other category. Other examples include checking that a project is shown as Closed after the last animal is unassigned or Institutional Animal Care and Use Committee approval ends, or checking for gaps in data that should be continuous (e.g., housing assignment). When inconsistencies or errors are found, attempts can be made to correct such errors. Constantly monitoring for and reporting on various types of consistency errors allows problems to be detected and corrected while the data trail is still hot. A general rule is that the longer it takes to detect a consistency violation, the harder it is to correct.

One final point in any discussion of data quality control is that user feedback is critical. The responses of system users, either at the data-entry or the data-retrieval end, are critical to the shaping of a high-quality system. At the Washington NPRC, monthly meetings are held with the ARS data-entry and programming staff and clinical, husbandry, and research users of the system. These meetings provide a tremendous amount of valuable information to the ARS management staff, and have greatly improved the system's quality and both the short- and long-term utility of the data.

5. NEW TECHNIQUES FOR DATA COLLECTION, STORAGE, AND RETRIEVAL

In this section, we will describe the application of leading-edge research on information technology to data collection and the storage and retrieval of animal records. Our goal is to introduce techniques that are being used in some facilities and, in some cases, still in experimental form. Our hope is that by introducing some developing techniques, we might hasten their application in the nursery, to the benefit of both the animals and researchers.

The inexpensive availability of powerful laptop computers and commercially available behavior-coding software and database applications has revolutionized the collection of data in many primate nurseries. The most widely available software for collecting behavioral data is a sophisticated suite of programs called The Observer (Noldus Technologies). EVENT-PC, written by J.C. Ha, is also available and in use in a number of research, zoo, and classroom settings (jcha@u.washington.edu). Other systems specifically designed for data on infant primates are available from individual laboratories, such as the OS-4 data-collection system used in the Infant Primate Research Laboratory (IPRL) at the University of Washington (iprl@bart.washington.edu or J.C. Ha). For other types of data, the IPRL now uses laptop computers with direct network connections to a custom-designed Microsoft Access database system.

We can now go one step further. The availability of inexpensive Palm-based and Microsoft CE-based Personal Digital Assistants (PDAs) has made portable data collection even more feasible. EVENT-PC is being rewritten for the Palm operating system, and the IPRL now has several data-collection programs already in use on Palms, written in a simple-to-learn programming language called NS-BASIC (nsbasic.com).

Voice-recognition software provides an even more portable option. In Meredith West's laboratory at Indiana University, the coding of complex behavioral interactions is performed with the use of a commercial voice-recognition system (David White and Shan Duncan). This hands-free system allows the recorder to maintain eye contact with the subjects at all time. Data are linked to a nearby desktop computer via wireless transmission.

Touchscreens and joysticks have been used experimentally for a number of years in operant conditioning tasks of learning and memory, and are now becoming much easier to design and operate. These systems, unlike the more traditional systems based on lit key or slide projector, provide significantly greater flexibility in stimulus presentation and improved data quality (and quantity!). Data are recorded directly to an electronic form and summaries of results and status are available almost continuously. In some facilities, such systems are under consideration as a form of environmental enrichment in established psychological well-being programs.

Data retrieval also has advanced in recent years, though perhaps not to the degree of data-collection options. Wireless networks and PDAs provide readily accessible data. For example, clinicians at the Southwest Foundation for Biomedical Research and Southwest NPRC have developed a Palm-based system for maintaining much of the Minimum and History data described in this chapter at their fingertips (or in their palms!) for immediate referral. The provision of connections between datasets, even across facilities and across data stored in different formats, is now an area of very active research, and the results are beginning to trickle down to practical applications. XML, a new standard format for data exchange across the Internet, is becoming widely supported by many applications (www.xmldb.org). New software such as Storage Resource Broker (www.npaci.edu/DICE/SRB) and Ontobuilder (www.cs.msstate.edu/~gmodica/Education/OntoBuilder/) allow users of data stored on different platforms, in different formats, and, in some cases, even with different definitions, to exchange data across the Internet with full control of data access. Some form of these new data-exchange developments will doubtless be used in the implementation of the NSF initiative for the exchange of behavioral data and the NIH program for the exchange of data between primate colonies.

Finally, an intriguing new research area for expanding the value of large data sets is a field called "data mining." This research is focused on new techniques for making complex connections within data, often using technologies such as fuzzy logic, artificial intelligence, and neural nets. A practical application, and one of the first commercial applications of this research, is Theme 4.0 from Noldus Technology. To quote the manufacturer's materials, "During pattern detection, Theme considers both the order and the time between behavioral events as well as their hierarchical organization. Once patterns have been detected, Theme offers various tools for the filtering, analysis and presentation of the patterns, on the basis of their behavioral content or quantitative criteria."

6. CONCLUSIONS

Our goal in this chapter was to describe some of the properties and values of a data-management system for a nonhuman primate nursery or colony, to outline some of the types of data that should be stored in a modern system, to describe methods to improve and maintain the quality of the data, and to introduce some new technologies for improved data-collection and storage systems. We have emphasized several points that may not be reflected in design features built into original data systems: the multiple uses of archival data in addition to the obvious day-to-day value of such data, an emphasis on data quality as much as data quantity, and technologies that have already revolutionized data collection and analysis in some facilities. We hope that the information in this chapter leads to a broader discussion of the value, collection, and use of long-term data sets in the nonhuman primate nursery environment. Clearly, many valuable software and hardware tools and database designs are in use, and greater exchange of ideas and methods can only benefit the nonhuman primate community.

ACKNOWLEDGMENT

Preparation of this chapter was supported by NIH grant RR-00166 to the Washington National Primate Research Center.

REFERENCES

Dyke, B. (1993). Basic data standards for primate colonies. *Am. J. Primatol.* 29:125–143.

Ha, J.C., Robinette, R.L., and Sackett, G.P. (1999). Social housing and pregnancy outcome in captive pigtailed macaques. *Am. J. Primatol.* 47:153–163.

Ha, J.C., Robinette, R.L., and Davis, A. (2000a). Survival and reproduction in the first two years following a large-scale primate colony move and social reorganization. *Am. J. Primatol.* 50:131–138.

Ha, J.C., Robinette, R.L., and Sackett, G.P. (2000b). A demographic analysis of the Washington Regional Primate Research Center pigtailed macaque colony, 1962–1997. *Am. J. Primatol.* 52:187–198.

Ha, J.C., Robinette, R.L., Almasy, L., and Dyke, B. (2002). Heritability of birthweight in pigtailed macaque monkeys. *Am. J. Primatol.* 56:207–213.

ILAR/National Research Council. (1979). *Laboratory Animal Records.* Washington, D.C.

CHAPTER FIVE

Very Early Rearing Experience: Rationale and Methodologies for Studying Prenatal Development in Nonhuman Primates

Matthew Francis Stuart Xavier Novak

1. INTRODUCTION

At first glance, the study of fetal development may seem tangential in a book about the nursery rearing of infant monkeys. The nursery environment is essential, however, to a discipline that seeks to understand the complex mechanisms of prenatal development. As in studies of postnatal development, the subjects of prenatal research need an environment in which they can thrive and in which they can be accessed at specific time points so that their development can be studied and manipulated systematically. Such studies are necessary if we are to understand the genetic and environmental influences underlying normal and abnormal development of the human fetus. Whereas ethical and practical limitations preclude the use of humans in these studies, nonhuman primates offer an excellent opportunity to model many human developmental processes (Newell-Morris and Fahrenbruch, 1985). And, these nonhuman primates frequently need nursery care, especially if the effect of prenatal manipulations are unknown, putting them at risk for poor postnatal development.

Experimentation to understand the induction, facilitation, and maintenance of developmental processes by environmental factors is the focus of this chapter. The approach is based on the premise that species and individual differences in development originate in the period of life before birth. Contrary to ideas about causation, with respect to behavior and development created by the human genome project, the genes that account for differences among individuals are only one piece of the puzzle. Another important piece involves the way in which the environment of the embryo and fetus interacts with genetic instructions to produce typical and atypical development.

Models and technologies have been developed to study the role of environmental influences during prenatal development. For humans there is a considerable literature about fetal development resulting from studies using ultrasound technologies (Lecanuet *et al.*, 1995). However, many questions about developmental processes require experimental manipulations not possible during human pregnancy. An exquisite rodent model has been developed in which the rat fetus can be exteriorized and yet remain physiologically attached to the mother (Smotherman and Robinson, 1986; Smotherman *et al.*, 1986). With the use of this preparation, fetal behavior has been characterized in response to both tactile and chemical stimuli (Smotherman and Robinson, 1988; Smotherman, 2002) and to classical conditioning (Robinson and Smotherman, 1995). However, the ability to generalize from rodent species to human species is limited, and nonhuman primate models of prenatal development are essential to answer many questions of interest to science and human health.

The conclusions of many studies about postnatal environmental influences on behavior and development emphasize a need for knowledge about early physiological and behavioral developmental processes in the prenatal period (Sackett, 1991; Sackett *et al.*, 1992). Until recently, there have been few technologies and procedures to answer many of these questions. In this chapter, I will review the status and potential of technologies and methods for studying the prenatal development of nonhuman primates. In Chapter 24, I will explore one of the technologies, tethering with maternal and fetal catheterization, in some depth. This will include discussion of some theoretical applications of methodologies designed to assess the acute maternal and fetal effects of environmental manipulations in pigtailed macaques (Novak, 2002).

Although I have attempted to present a comprehensive overview of the species and technologies used in studies of prenatal development, this chapter is not exhaustive. The literature reviewed will primarily emphasize (1) methods and technologies that may be used to examine the psychosocial aspects of development and (2) studies involving both fetal and postnatal development. My aim is to provide a sufficiently broad reference to serve as a start for pursuing studies of the prenatal development of physiology and behavior in nonhuman primates.

2. METHODOLOGIES AND DATASETS

2.1. Happenstance and Evolutionary Byproduct

Taking advantage of evolution or known quirks in nature can be a powerful tool for trying to understand prenatal development. Nonhuman primate species that twin, such as marmosets, offer an opportunity to study the postnatal development of two different animals produced by the same pregnancy, a situation that may reduce error variance. For prenatal studies, however, this is more complicated. Without directly manipulating one fetus, it is difficult to use one twin as an experimental subject and the other as a control. Even when direct manipulation is possible, many manipulations affecting one fetus may indirectly affect the other during the remaining gestation. In addition, the species of nonhuman primate that commonly twin may not be the best models of human developmental phenomena, and the multifetal environment suffers from a weakness similar to that in rodent models in that it is considerably different from that of typical human pregnancy. Furthermore, species such as the marmoset most commonly produce fraternal twins, whose genetic relatedness is only that of siblings, although two fetuses sharing a uterine environment may be useful for some experimental purposes. One future use of twinning in marmosets may involve combining reproductive technologies, such as embryo transfer, to gestate two different subjects in the same uterus. For example, two embryos that differ by a known specific genotype or manipulation could be gestated in the same uterus to reduce the differential effects of uterine environment on the phenotypic expression of the genotype of interest.

A good example of happenstance methodology is the study of an infant that was considerably slower to mature than its peers. Testing revealed that the animal has a trisomy 16 karyotype (Ruppenthal *et al.*, 2003). From a clinical point of view, such an animal may present as little more than a nuisance because of its inability or unwillingness to progress through established housing and feeding protocols. However, from a research point of view this infant offers a rare and invaluable opportunity to study trisomy, which has not yet been produced experimentally. During assessment of this trisomic individual, another naturally occurring primate model also was discovered. The sire, which has a history of producing infants with atypical developmental patterns, was found to have an abnormally small anterior commissure. The behavior and brain anatomy of this animal's living offspring are currently under study. A fetal brain imaging or a sequential end-product study of future offspring would be of particular interest in trying to understand how the brains of this male's offspring follow different developmental pathways than those of normal control offspring.

2.2. Terminal Methodologies

2.2.1. Accidents

It is seldom efficient to design studies based on rarely occurring phenomena, but when these phenomena do happen they can yield extremely valuable and heuristic data. In studies of prenatal developmental, this takes three forms. Upon recognizing a pregnancy-related anomaly, it is possible to perform periodic ultrasounds to monitor fetal development. If the anomaly results in termination of the pregnancy, the fetus can be assessed. And finally, even if ultrasound or other imaging techniques are not feasible, once an anomaly has been detected the subject can be maintained in optimum conditions to ensure delivery of a live-born infant so that the effects on prenatal development can be studied indirectly by inferences made from postnatal data.

An excellent example of a naturally occurring phenomenon was a pair of conjoined rhesus macaques that were stillborn at full term (Canfield *et al.*, 2000). The cause of death was attributed to asphyxia and trauma experienced during parturition. Both infants' birth weights were normal for gestational age. Recovery of these infants allowed researchers a rare glimpse into the conjoining phenomenon. This example illustrates once again the usefulness of the primate nursery: when a pregnant dam is housed in the nursery so that its delivery can be monitored and protected, personnel can intervene and recover the infant immediately if there are problems with parturition (Mahoney *et al.*, 1977).

2.2.2. End product

As opposed to "accidents," which are not controlled, studies can be designed to intervene at set points during pregnancy to study developmental status at that time point. With the exception of Smotherman and Robinson's type of experimentation in the rodent model, not currently being used in nonhuman primates, this type of study cannot produce behavioral data because it involves the death of the fetus. Age differences, but not developmental change, can be studied by terminating pregnancies at sequential or progressive time points, but in essence, all that is achieved is a snapshot of a moment. Examples of this approach are studies of the differentiation of fetal gonads in prosimian bush babies (*Galago crassicaudatus crassicaudatus*), in which serial sacrifices revealed age-related differences, but not developmental changes (Butler, 1983; Yoshinaga *et al.*, 1988, 1990).

Despite the commonness of this approach, interpreting developmental trajectories from cross-sectional data is fraught with cohort and time-of-testing confounds (Schaie, 1965). An end-product design was used to describe the role of peptides in gonadal regulation in rhesus macaques (Jaffe *et al.*, 1993), but the authors presented the results as developmental changes while using cross-sectional data. This developmental argument is necessarily confounded by individual differences between the animals studied at the two time points, as well as any differences in the time of measurement or birth cohort. This is an unavoidable constraint of end-product studies.

One solution is to combine end-product research with other designs such as direct fetal manipulation. A combination of serial sacrifices with postnatal follow-up of selectively timed doses of irradiation provided a view of the origin and development of monoamine neurons in the brain stem of rhesus macaques (Levitt and Rakic, 1982). Similarly, behavioral effects of prenatal androgen administration have been observed in rhesus macaques, including developmental changes in hormone uptake by the brain (Michael and Zumpe, 1998). Although most studies of hormone development have required sacrifice of the fetus to gain access to the dependant variable, some have followed development postnatally (e.g., Eaton *et al.*, 1990).

A second solution is not to make developmental claims from this type of data. Age differences in toxicity were studied in patas monkeys (*Erythrocebus patas*) to assess the potential for long-term developmental effects (Gerschenson and Poirier, 2000). Pregnant dams were given human-equivalent doses of 3′-azido-3′deoxythymidine (AZT) during the second half of gestation. Then their fetuses were sacrificed and mitochondria were isolated immediately from fetal tissues such as heart and skeletal muscle and from the placenta. Similarly, patas monkeys were used to study the effects of AZT together with other combinations of AIDS-fighting drugs to investigate age-related differences at the genetic level (Olivero *et al.*, 2002). In both studies, the investigators correctly refrained from the temptation to discuss the differences as changes in vulnerability within fetuses.

Rhesus macaques have been used to study a variety of teratogens, including the effect of prenatal nicotine exposure on lung functioning at 134 and 160 gestational days (GD) (Sekhon *et al.*, 2002) and the effects of dexamethasone exposure on the development of lungs (Engle *et al.*, 1996) and hippocampus (Uno *et al.*, 1990).

An often overlooked phenomenon in prenatal developmental studies, and in nonhuman primate studies as a whole, is the presence of species differences. For example, end-product studies demonstrated that Depo Provera in high doses is teratogenic in both male and female longtailed macaque fetuses (Prahalada *et al.*, 1985), but not in baboons (Tarara, 1984). The most dramatic accidental prenatal demonstration of this phenomenon is the thalidomide disaster. Although studies with nonprimate species had failed to detect the teratogenic effects later observed in humans, thalidomide was found to be similarly teratogenic in nonhuman primate models (Hendrickx, 1973; Hendrickx and Newman, 1973; Hendrickx and Sawyer, 1978; Hendrickx and Helm, 1980). In this example, hindsight offers a powerful admonishment that care must be taken to identify the most appropriate models, especially when species differences are likely. Tragic experience has shown that incorrect inferences and conclusions abound when species differences are ignored. On the other hand, species differences offer an excellent opportunity to model the origins of individual differences.

2.3. Indirect Methodologies

With indirect methodologies it is not necessary to terminate the pregnancy to access the fetus. Instead, it is possible to infer the status of fetal development on the basis of (1) known information about the relation between maternal variables and fetal state or (2) information gained by performing a prenatal manipulation of the mother followed by postnatal measurement of the infant. The advantage of these methods compared with those already discussed is that longitudinal assessment is possible. However, for the first option, the amount of information that can be gleaned is limited by current knowledge. Data will also be skewed by the strength or weakness of the correlation between maternal and fetal processes. For example, low estrogen level during human pregnancy is often regarded as an indicator of fetal distress (Diczfalusy, 1974), but in rhesus macaques several factors in addition to fetal distress may account for fluctuations in estrogen level (Novy and Walsh, 1981).

The second option, indirect measurement using a prenatal manipulation and postnatal follow-up, is used more commonly than the first. It has the added advantage that inferences are made directly from the infant to the fetus, so there is no need to rely on maternal–fetal correlations. There is a substantial difference between identifying factors that can affect the fetus, and identifying factors that do affect fetal developmental pathways. Questions about the acute effects of prenatal manipulations cannot be asked because neither of these methods assesses the fetus directly. In the best studies, infants are assessed at birth, thereby reducing the likelihood of postnatal interference. However, this is not always possible. As the amount of developmental time passes between the manipulation and measurement, the possibility of intervening variables that may either exacerbate or mask the effects of the experimental manipulation becomes more likely. Furthermore, as the length of exposure increases to include multiple developmental stages, it becomes more difficult to make any inferences about the specific mechanism of any effect.

One way to strengthen inferences in these types of studies is to combine methodologies. For example, by combining directed fetal manipulations at specific developmental time points with serial sacrifices (Levitt and Rakic, 1982), it is possible to create a powerful design for studying development (Schaie, 1965). The drawback lies in the sample sizes required to perform serial sacrifices in addition to maintaining longitudinal samples.

Most relevant to a book on nursery rearing is the confounding effect of the infant's postnatal experience with its mother. This issue is particularly germane in studies that involve prenatal manipulation followed by postnatal examination. To isolate prenatal experience as a causal factor in these studies, it may be necessary to separate infants from their birth mothers and rear them with a "foster mother" or in a primate nursery.

2.3.1. Chemical and Environmental Teratogens

Indirect methodology has been used in a number of studies to examine the effects of chemical and environmental teratogens on a variety of dependent variables such as learning, social behavior, cognitive development, and the central nervous system (for a review see Burbacher and Grant, 2000). The nature of the effects on fetal and infant development depends on the type and amount of the teratogen, the timing and duration of the exposure, and the health parameters of the mother (Vahter *et al.*, 1994).

In utero exposure to alcohol during early gestation, an issue of particular importance for human health and development, has been studied in the rhesus macaque model. Although moderate doses of alcohol were not associated with birth weight, length of gestation, or postnatal anthropometric measures of fetal development, consumption of alcohol during early gestation was related to poor visual orientation and delayed motor maturity of the infant (Schneider *et al.*, 2001a). Infants in the moderate-consumption group needed more trials of a nonmatching-to-sample test for the initial learning, but then had no difficulty adjusting to delays before performance. Prenatal alcohol was also associated with increased behavioral stereotypies during testing but was not related to performance on the cognitive task. However, this study did not isolate prenatal experience as the causal mechanism because the infants were reared with their mothers for the first 6 months of life. It is entirely possible that alcohol consumption during pregnancy alters postnatal maternal behavior, and it is the differences in maternal behavior that account for the postnatal effects observed. Prenatal exposure to alcohol also has been studied in pigtailed macaques, both by direct fetal measurements via ultrasound and by postnatal follow-up (Bowden *et al.*, 1983; Clarren and Bowden, 1984). In addition to macaque species, squirrel monkeys have been investigated as a potential model for studying fetal alcohol syndrome (Kaplan *et al.*, 1982). Despite significant limitations on how much alcohol the pregnant females would drink, the authors concluded that the model has potential. To date, however, a squirrel monkey model of fetal alcohol syndrome has not been published.

Prenatal exposure to marijuana has been associated with deficits in visual attention at 2 years of age (Golub *et al.*, 1982) and an increase in mother–infant interactions (Golub *et al.*, 1981). Infants were reared with their mothers, and exposure to marijuana also continued after birth, making a postnatal mechanism likely. Were the mechanisms for changes in maternal–infant interaction and visual attention pre- or postnatally driven?

Prenatal exposure to cocaine is another serious issue for human health. An early study with a small sample found no negative effects on cocaine-exposed infants (Morris *et al.*, 1996), but a second study showed that cocaine-treated infants weighed less at birth, were shorter, and had a smaller head circumference than untreated controls (Morris *et al.*, 1997). In addition to confounding the pre- and postnatal environments by rearing infants with their mothers for the first 4 months postnatally, the second study also confounded prenatal environment with the preconception environment by treating the females with cocaine even before they became pregnant. The preconception doses were smaller, to decrease the likelihood of this confound, but the dosing continued for 27 weeks so that the cumulative preconception dose was actually higher than the dose during pregnancy. In a study in which the confounding element of postnatal maternal infant interaction was eliminated, prenatal cocaine exposure caused no significant differences between experimental and control infants on any measures (Howell *et al.*, 2001). The cocaine-exposed group did experience decreased survivorship, but this result was no longer significant when animals with multiple pregnancies were eliminated from the analyses. In another study, the prenatal environment was isolated from preconception and postnatal confounds to test the hypothesis that cocaine exposure from gestational days 40 to 120 would specifically affect neurons migrating to cortical layers II–IV (Rakic, 1982, 1994). Changes in neurogenesis did occur, but the overall numbers of neurons generated did not differ from the norm (Lidow and Song, 2001). Instead, cocaine was associated with increased cell death (He *et al.*, 1999).

The effects of prenatal exposure to methylmercury, an important environmental toxin, have been studied in longtailed macaques that were reared in a nursery to eliminate the postnatal maternal confound. Testing revealed delays in the performance of object permanence (Burbacher *et al.*, 1990a), disturbed visual recognition memory (Gunderson *et al.*, 1986, 1988), and disrupted social behavior, with reduced social play and more nonsocial passive behavior (Burbacher *et al.*, 1990b).

2.3.2. Steroids

The effects of prenatal exposure to steroids have been studied in rhesus macaques. Genetically female fetuses were androgenized by exposing mothers to testosterone propionate (Goy *et al.*, 1988) and the synthetic estrogen diethylstilbestrol (Goy and Deputte, 1996). Although the authors argued against postnatal confounds in their data using postnatal behavioral observation of the mother, nursery rearing would have assisted them in this endeavor. Similarly, infants were reared with their mothers in studies to investigate prenatal androgenization in both male and female offspring. These infants displayed alterations in both anatomy (Herman *et al.*, 2000) and vocalizations (Tomaszycki *et al.*, 2001). While it seems likely that vocalizations are more malleable to postnatal environment than anatomy would be, any indirect study that does not eliminate postnatal maternal influence risks confounds.

2.3.3. Prenatal Psychosocial Stress

The effects of prenatal exposure to psychosocial stress have been studied primarily in macaque species. In rhesus macaques, the reported effects include low birth weight (Schneider, 1992a), impaired neuromotor development and reduced postrotatory nystagmus (Schneider and Coe, 1993), lower levels of gross motor and exploratory behavior (Schneider, 1992b), delays in the performance of object permanence tasks (Schneider, 1992c), and abnormal social behavior (Clarke and Schneider, 1993). Sex-of-fetus by stress interactions were also found for immune status (Coe and Crispen, 2000). In pigtailed macaques, some studies have also found abnormal social behavior, but have failed to find associations between stress and birth weight (Worlein and Sackett, 1995). This species does show sex-of-fetus by stress interactions for neonatal gross motor behavior and neuromotor development (Novak and Sackett, 1996; M.F.S.X. Novak, unpublished observations).

Because the literature is not large and needs replication, the discrepancies may originate from several sources, including differences in species, age, the assessments used to measure postnatal stress effects, the type of stressor used, the length and timing of stress, the presentation of the stressor, the premating environments of the mothers, and postnatal rearing. For example, one researcher used a noise stressor between days 90 and 145 of gestation (Schneider, 1992b), while another used unpredictable capture between days 30 and 130 of gestation (Sackett, 1991). Although stress responses may tend to follow a generalized pattern regardless of the stimulus (Selye, 1980), this remains to be demonstrated in the nonhuman primate model where sensitive, controlled measurement of multiple dependent variables is possible over time.

Further clouding the picture is a study in which squirrel monkeys were stressed by having their social relationships disturbed at a single time point versus several time points both before and during pregnancy (Schneider and Coe, 1993). As in some of the cocaine exposure studies cited above (Section 2.3.1), the effects detected in infants whose mothers were stressed at multiple time periods may be entirely due to the first stressor period, presented before conception.

A further difficulty of the study with a noise stressor is that the effects of maternal psychological stress cannot be separated from direct fetal stress: it is entirely possible that the fetus heard the noise directly and that the effects in the study are unrelated to any changes due to prenatal psychological stress to the mother. The significance of this potential confound depends upon the goal of the research, as has been discussed at length in the rodent literature (Archer and Blackman, 1971).

2.4. Selective Breeding Studies

Selective breeding studies are most typically a subset of indirect studies, but also involve direct manipulation of prenatal developmental processes. They belong to the "reproductive technologies" category as a prenatal research strategy for examining genetic or nongenetic intergenerational effects. Experiments designed to study intergenerational differences or similarities in animal groups have been used extensively in rodents, but less so in nonhuman primates. This is most likely due to the considerable resources, especially the amount of time, necessary to produce a study with even moderate power in most species. Most of the studies that do exist are not focused on prenatal developmental processes. An exception was a large-scale project ($n > 200$) designed to isolate the source of pregnancy-related difficulties (Fahrenbruch *et al.*, 1979; Holm, 1979; Sackett, 1980, 1984, 1991). This project, which involved the interbreeding of both males and females stratified according to their reproductive histories, demonstrated that males contribute at least as much as females to the success of pregnancy. Given the strength of the sire effect, it is surprising that teratology research into prenatal contaminants such as alcohol and cocaine, discussed earlier, has neglected paternal influences. The exact mechanism of these paternal effects was sought, but never isolated (Klein *et al.*, 1982). Genetic explanations are tempting, but the degree of genetic influence versus nongenetic intergenerational influence is not known.

Similarly, heritability studies in general often confound genetic and prenatal environmental influences when determining the strength of intergenerational transfer between parents. This problem emphasizes the necessity for selective breeding studies focusing on prenatal development, especially in the domain of reproductive technologies, where the potentially deleterious long-term effects of reproductive procedures are not understood in either nonhuman primates or humans. Despite the considerable cost of conducting these long-term intergenerational studies, more effort needs to be devoted to them. Crucial to the success of such studies is the availability of complete animal records and the kind of neonatal care provided by the primate nursery facility.

2.5. Direct Methodologies

2.5.1 Ultrasound

Noninvasive imaging technologies, particularly ultrasound, are used extensively in studies of prenatal development because they do not require termination of the pregnancy to yield data and they permit postnatal follow-up. However, comparisons of prenatal growth and behavior with postnatal variables are uncommon. Concerns have been raised about whether multiple ultrasound procedures have an adverse effect on the developing fetus. In one study, longtailed macaque fetuses were exposed to more than 20 relatively short ultrasound examinations, decreasing in frequency from conception through cesarean delivery at 153 GD. Fetal blood parameters were collected using cardiac puncture, and infants were followed for the first 30 days of life. The data revealed short-lived developmental effects and, the authors argued, biologically significant but not statistically significant reduced birth weights (Tarantal *et al.*, 1995). However, in light of the stressfulness of restraint and the use of cardiocentesis, it is difficult to argue that any effects observed were due solely to multiple ultrasounds despite the high number of procedures.

Basic research using ultrasound has yielded a wealth of information about pregnancy and fetal morphological development. Normative development has been described in 454 longtailed and rhesus macaques (1267 total scans; Tarantal and Hendrickx, 1988), 37 pigtailed macaques (710 scans; Conrad *et al.*, 1995), and 12 rhesus, 15 longtailed, and 21 Japanese macaques (Shimizu, 1988). In a study of pregnancy in marmosets, a species with multifetal pregnancies, 15 females of varying parity were assessed from ovulation to parturition (Nubbemeyer *et al.*, 1997). In vervet monkeys, ultrasound was used to describe early pregnancy in 20 animals (Seier *et al.*, 2000). It is possible that comparisons of the differences and similarities among species of varying genetic relatedness could be used as a model to understand individual differences in human prenatal development.

Ultrasound is also useful for applied research on the developing fetus. Examples include AIDS research to determine the teratogenic effects of potentially toxic drugs administered to pregnant females (Tarantal *et al.*, 2002) and development of methods for improving gestational estimates and fetal surgical techniques (Farine *et al.*, 1988).

Although it is technically possible to observe fetal behavior via ultrasound, in practice this possibility is limited by difficulty in immobilizing the abdomen. Unlike humans, nonhuman primates often are anesthetized for ultrasound procedures, and this of course limits the usefulness of behavioral parameters. Unanesthetized animals can be trained to accept restraint chairs for the procedure, or can be restrained by technicians, but the stress of both kinds of restraint precludes both typical and subtle measurement. Ultrasound data have been reported for restrained, unanesthetized rhesus and longtailed macaques (Tarantal and Hendrickx, 1988) and for marmosets (Nubbemeyer *et al.*, 1997), but these reports have not included behavioral data.

Doppler ultrasound techniques for measuring blood flow provide another means of studying the fetal environment. Although Doppler technology has been used primarily in studies of uteroplacental blood flow (Nimrod *et al.*, 1996; Simpson *et al.*, 1998), it can also be used to assess peripheral and central fetal structures. The fundamental limitation of these assessments involves the lack of standardized and objective measurements.

2.5.2. *Imaging*

Future directions for the use of ultrasound, in terms of imaging the fetus, may include intravascular ultrasound, developed for cardiovascular medicine (Mintz *et al.*, 2001; Sheifer *et al.*, 2000). In current use, a miniature transducer (<1 mm in diameter) at the tip of a catheter is inserted into coronary arteries to visualize blockages and the integrity of blood vessels. Similarly, imaging catheters with optical fibers have been developed to produce video images of artery walls and fluids (Parker, 2000). In combination with fetal surgery, telemetry, and/or fetal catheterization, these tiny catheters can be developed to help visualize the fetus. The usefulness of these imaging catheters is compromised by their limited visual field and the difficulty in maintaining navigational control to visualize the target area of the fetus. These limitations notwithstanding, visualization of the fetus in an unrestrained, unanesthetized mother would be invaluable for many developmental questions. It would also allow replication of human fetal ultrasound studies in nonhuman primates without the need for stressful restraint procedures.

Magnetoencephalography uses magnetic fields to measure changes in brain function. This technology has already been adapted for studies of human fetal development and improved methods are still being explored (Eswaran *et al.*, 2000). Although it has yet to be used with nonhuman primates, it has been used to study brain function (Blum *et al.*, 1985) and auditory evoked cortical fields (Lengle *et al.*, 2001; Schneider *et al.*, 2001b; Wakai *et al.*, 1996) in the human fetus, as well as brain functioning of the human neonate during sleep (Wakai *et al.*, 1999).

Position emission tomography (PET), computed tomography (CT) and magnetic resonance (MR) imaging are potentially useful tools for observing fetal developmental processes and are the only direct ways of viewing the fetal brain and other organ systems *in vivo*. Similar to ultrasound in their strengths and weaknesses, they are used much less often but may become more commonplace as the resolution increases and exposure times decrease. In recent studies, MR imaging was used to detect the results after fetuses were irradiated during and after the neurogenic period for thalamic cells (Gelowitz *et al.*, 2002; Schindler *et al.*, 2002). Volume loss and changes in shape of the thalamus were observed, and these thalamic changes were associated with craniofacial differences. Together, these two studies suggest that rhesus macaque fetuses irradiated during the first trimester may be a good nonhuman primate model of human schizophrenia. An invaluable application of MR imaging, more relevant to the purposes of this chapter, would be to replace the serial sacrifices used in earlier studies (e.g., Levitt and Rakic, 1982) with serial fetal MR or magnetoencephalography scans.

The potential exists, in the next few years, for a system in which a pregnant dam can be adapted to a nonmetallic cage enveloped in an MR machine, thereby allowing real-time imaging of both the mother and the fetus during periods of stillness or even during slow movement on the part of the mother and/or the fetus. Whether this technology is actually developed depends on whether investigators wish to pursue questions about fetal development that might best be studied in this manner. The advantage of such technology is that it would allow nonsurgical access to the fetus and, except for blood draws, would accomplish everything that tethering with maternal and fetal catheterization provides. It would even permit a view of fetal behavior, something that is still being developed for the catheterization model.

2.5.3. Reproductive Technologies

Many of the reproductive technologies for producing pregnancy in endangered species or in livestock have not been used extensively with nonhuman primates. In humans, the use of reproductive technologies is driven not by science but by the desire to produce biological offspring. Nonhuman primates offer an excellent opportunity to conduct long-term intergenerational studies that are necessary to anticipate and understand how these procedures affect human well-being. In addition, newer reproductive technologies, such as embryo splitting and cloning, provide an opportunity to isolate prenatal developmental experience from genetic factors and postnatal experience. To date, one of the most pervasive theoretical questions in studies of organismic development concerns the nature and nurture of behavior. With notable exceptions, most research confounds the genetic and prenatal environmental determinants of behavior. Considering the fundamental impact of the early postnatal environment, the impact of the prenatal environment also needs to be assessed.

For many studies, reproductive technologies can be combined with end-product methodologies or with indirect or direct fetal manipulation and/or measurement. In addition, reproductive technologies can now produce twins in species that do not typically twin. Combined with embryo transfer, this technology may prove extremely powerful.

Procedures for embryo transfer in nonhuman primates are always being improved (Meng *et al.*, 1997; Wolf *et al.*, 1999; Sankai, 2000; Chan *et al.*, 2001). Recent experiments involving the transfer of 86 rhesus macaque embryos to 46 unrelated recipient females resulted in 29 full-term infants, 2 biochemical pregnancies, 10 blighted ova, and 7 spontaneously aborted fetuses. In addition, however, these experiments resulted in the births of the world's first transgenic nonhuman primates (Chan *et al.*, 2000). These technologies offer an excellent opportunity for experimenters to separate genetic and prenatal experiential processes by gestating fetuses with one genetic blueprint in a uterine environment with a different genetic blueprint.

Although many of these technologies are just being developed for practical use, within the next 50 years it is likely that laboratory research on nonhuman primates will be done entirely with twins and genetically engineered animals. This situation would be analogous to that seen with inbred strains of rats and mice. The question that remains is how we get from here to 50 years from now? Again, we need intergenerational selective breeding studies to understand the long-term implications of using these technologies for nonhuman primate behavior and for human health.

2.5.4. Direct Fetal Access

Experiments involving instrumentation or manipulation of the fetus require direct access, which can be achieved with or without invading the uterine compartment. The efficacy of direct fetal manipulation rests on the success of surgical procedures. Fetal endoscopy techniques, perfected in midgestation rhesus macaques for use in monitoring human pregnancy (Feitz *et al.*, 1996), not only contribute to human health, they also improve nonhuman primate research. Consistent with improving techniques, care must be taken to assess their invasiveness. Depending on the type of manipulation, direct fetal access can facilitate the collection of longitudinal fetal data and postnatal follow-up, as long as the proper control groups are used. Surgeries that invade the uterine compartment significantly increase the potential for damaging fetal developmental processes (Nathanielsz *et al.*, 1984). However, implanting telemetry devices directly on the fetus increases the types of data and degree of subtlety that can be achieved.

Radio telemetry can be used to measure heart rate, blood pressure, and muscle contractions. In the most exquisite example of radio telemetry, adult baboons and rhesus macaques were instrumented to allow long-term collection of physiological data without the need for anesthesia, tethers, or restraint (Astley *et al.*, 1979). Such technologies could be adapted to the assessment of pregnancy in free-ranging animals, although transmitters directly attached to the fetus need to be miniature enough to avoid complicating pregnancy and must somehow "grow" with the fetus so as to not obstruct development. Despite these technical difficulties, research on fetal development would benefit greatly from radio telemetry.

Catheterization is another technique that allows direct access to the fetus, permitting simultaneous collection of maternal and fetal blood and amniotic fluid. Pressure transducers and flow meters attached to catheters can monitor a range of cardiovascular variables. Wired electrodes can also be implanted in various tissues and then exited without the need for telemetry. Catheterization models also make it possible to manipulate the fetus directly by delivering experimental vehicles directly to the fetal bloodstream.

Direct fetal manipulation can also be achieved by surgery. In one example, normal and castrated rhesus macaque fetuses were treated with 5 mg of testosterone propionate to induce differentiation of the epididymidis (Alexander, 1972). Differentiation processes were blocked with castration, but then restored with prenatal administration of androgen. In addition, fetal gonadectomy abolished differences between males and females, supporting the hypothesis that occupation of estrogen receptors in the hypothalamus by aromatized-endogenous testosterone is greater in intact males than in intact females around 122 GD (Michael *et al.*, 1992). In another study involving fetal surgery, rhesus macaque fetuses underwent hypophysectomy at midterm so that the influence of the endocrine system on fetal development could be examined (Kittinger, 1977).

2.5.5. *Viral Vectors and Nanotechnology*

Longitudinal studies with direct sampling across both fetal and infant periods are possible without invasive surgical procedures involved in direct fetal access. Currently the most commonly used of these technologies is gene therapy. Nanomolecular machines including both viral and nonviral vectors are already in use to deliver genetic material to target cells. Methodological questions surrounding the use of these technologies include the type of virus, retrovirus or nonvirus that is most appropriate for delivering a new gene to specific target organs and cells, the method of administering it to the fetus, and the efficacy of infection at different gestational ages.

In many cases, fetal gene therapy may be preferred over postnatal models. First, therapy can be applied before the disease develops, so treatment of the disorder could be decreased or avoided. Second, depending on the stage of prenatal development, cells are still dividing, thereby facilitating the spread of treatment to new cells. Third, at least early in gestation the undeveloped immune system is not likely to react against antigen viral vectors and/or genetic material (Bennett *et al.*, 2001).

In one example of gene therapy, retroviral vector carrying a producer cell line was injected into the amniotic cavity of pregnant bonnet macaques (Bennett *et al.*, 2001). Targeting the fetal throat and stomach, the investigators hypothesized that producer cells would create an amplification effect and the particles produced would be transmitted more effectively via swallowing and breathing motions of the fetus. The amplification effect of using a producer cell line was necessary to counter the finding that amniotic fluid had an increasingly deleterious effect on transduction rates as development progressed. In addition, retroviral vectors had produced low concentrations of the viral vector particles. Two of four fetuses were successfully transduced via this method, but it is unclear what percentage of target cells must express the new gene for clinical improvement to occur.

In another study, rhesus macaque fetuses were infected with adenoviruses directed at the lung and intestine via the amniotic fluid (Larson *et al.*, 2000). Reporter genes, which produce unique assayable proteins upon insertion, were used to evaluate success rates and timing. Unexpectedly large amounts of the reporter protein were produced, and injection at 110 GD resulted in unintended gene activity in the kidneys. In addition, the gene product entered the bloodstream and traveled to untargeted organs. The reporter gene protein was still detectable up to 30 days postinfection and no adverse maternal effects were reported. Serial sacrifices precluded long-term follow-up in the infants. These results differed from those of identical studies in rodents, thereby demonstrating the importance of nonhuman primate models for gene therapy.

A longitudinal study with serial sacrifices addressed several issues relating to the efficiency and safety of *in utero* gene transfer (Tarantal *et al.*, 2001). Rhesus macaque fetuses were infected with various retroviral-based vectors via intrapulmonary or intrahepatic injection, and then followed with ultrasound examinations and ultrasound-guided blood and liver sampling. Postnatally, some of the infants were examined for Apgar scores, morphometrics, health, food intake, and body weight until 6 months of age. Unfortunately, data on the infants' development are not reported, but inserted sequences were detected even 10 months after transduction. Furthermore, transduction was accomplished to multiple organ systems without placental transfer to the gene sequence of the mother.

At least in theory, future studies will be able to use other kinds of nanotechnology to deliver genetic and other material to target cells. Possibilities for this technology include nanomolecules that can be combined with inserted gene products to label successfully infected cells so that subsequent assessment of transduction proficiency can be accomplished *in vivo* with imaging technology. Additionally, molecular vehicles may be designed for sampling fetal blood and tissues and for monitoring fetal behavior through the metabolism of muscle tissues. Injected nanomolecules can be excreted or removed from an organism and separated from body fluids via centrifugation so that they can be measured or assayed.

Various nanotechnologies are currently being developed. At the Center for Biologic Nanotechnology at the University of Michigan, devices are in development, potentially for the space program, to monitor radiation and radiation-induced illnesses. The devices, <5 nm in diameter, break upon detection of radiation and emit a fluorescent dye that can be seen and measured (J.R. Baker, Jr., as cited by Menzie, 2002). These devices could be adapted to prenatal developmental questions as well. Ultrasound or other imaging technology could be adapted to detect fluorescence.

Nanomolecular devices could be used to test Mendelian genetic theory. If genes in sperm and eggs could be labeled differentially, the paternal, maternal, or mitochondrial origin of specific alleles could be identified. Is it possible to turn a gene on or off during a given period of development, and then reverse the effect? Which genes are necessary for normal development? Which genes are sufficient to produce normal development? To what degree are there redundancies maintained in the genome? The answers to many of these questions would benefit greatly from fetal and/or embryonic applications. Development of these technologies may prove to be the most powerful tool yet for understanding fetal development. Every type of data discussed in this chapter could be collected with this technology, without any of the limitations or confounds of sacrifice, surgery, restraint, anesthesia, or magnetic fields. Use of these technologies is limited only by a researcher's patience, determination, and imagination.

3. BASIC REQUIREMENTS

The methodologies for studying prenatal development present several implications for research and husbandry. First, information about the subject's parity and breeding history is essential. In addition, breeding practices must permit the time of conception to be isolated independent of the data being collected. Colonies with nursery facilities designed to handle both the pregnancy and the early postnatal development of the infant can facilitate the development and maintenance of this necessary information.

Reproductive technologies and many of the monitoring technologies, particularly magnetoencephalography and tethering with maternal and fetal catheterization, are extremely expensive and time-consuming. The infants that result from these programs are therefore particularly valuable and often require intensive care in a nursery environment. The need to observe and support these animals 24 hr a day produces a large body of clinical and research data, thus maximizing the use of these research subjects and minimizing the number of animals needed for research on prenatal development. In an age when there is growing resistance to intensive nursery rearing, the need for laboratories of this type is in fact increasing. Many types of research with nonhuman primates can indeed be conducted in the wild and in social situations. However, the types of prenatal developmental questions discussed in this chapter often present significant risk to these valuable subjects, and the clinical support provided by a nursery environment is vital to the welfare of these animals.

REFERENCES

Alexander, N.J. (1972). Prenatal development of the ductus epididymidis in the rhesus monkey: The effects of fetal castration. *Am. J. Anat.* 135:119–134.

Archer, J., and Blackman, D. (1971). Prenatal psychological stress and offspring behavior in rats and mice. *Dev. Psychobiol.* 4:193–248.

Astley, C.A., Hohimer, A.R., Stephenson, R.B., Smith, O.A., and Spelman, F.A. (1979). Effect of implant duration on in vivo sensitivity of electromagnetic flow transducers. *Am. J. Physiol.* 236:H508–H512.

Bennett, M., Galan, H., Owens, G., Dewey, R., Banks, R., Hobbins, J., Accurso, F., and Schaack, J. (2001). In utero gene delivery by intraamniotic injection of a retroviral vector producer cell line in a nonhuman primate model. *Hum. Gene Ther.* 12:1857–1865.

Blum, T., Saling, E., and Bauer, R. (1985). First magnetoencephalography recordings of the brain activity of a human fetus. *Br. J. Obstet. Gynaecol.* 92:1224–1229.

Bowden, D.M., Weathersbee, P.S., Clarren, S.K., Fahrenbruch, C.E., Goodlin, B.L., and Caffery, S.A. (1983). A periodic dosing model of fetal alcohol syndrome in the pig-tailed macaque (*Macaca nemestrina*). *Am. J. Primatol.* 4:143–157.

Burbacher, T.M., and Grant, K.S. (2000). Methods for studying nonhuman primates in neurobehavioral toxicology and teratology. *Neurotoxicol. Teratol.* 22:475–486.

Burbacher, T.M., Grant, K.S., and Mottet, N.K. (1990a). Retarded object permanence development in methylmercury exposed *Macaca fascicularis* infants. *Dev. Psychol.* 22:771–776.

Burbacher, T.M., Sackett, G.P., and Mottet, N.K. (1990b). Methylmercury effects on the social behavior of Macaca fascicularis infants. *Neurotoxicol. Teratol.* 12:65–71.

Butler, H. (1983). *The Embryology of the Lesser Galago (Galago senegalensis).* Karger, Basel.

Canfield, D., Brignolo, L., Peterson, P.E., and Hendrickx, A.G. (2000). Conjoined twins in a rhesus monkey. *J. Med. Primatol.* 29:427–430.

Chan, A.W.S., Luetjens, C.M., Dominko, T., Ramalho-Santos, J., Simerly, C.R., Hewitson, L., and Schatten, G. (2000). Foreign DNA transmission by ICSI: Injection of spermatozoa bound with exogenous DNA results in embryonic GFP expression and live rhesus monkey births. *Mol. Hum. Reprod.* 6:26–33.

Chan, A.W.S., Chong, K.Y., Martinovich, C., Simerly, C., and Schatten, G. (2001). Transgenic monkeys produced by retroviral gene transfer in mature oocytes. *Science* 290:309–311.

Clarke, A.S., and Schneider, M.L. (1993). Prenatal stress has long term effects on behavioral responses to stress in juvenile rhesus monkeys. *Dev. Psychobiol.* 26:293–304.

Clarren, S.K., and Bowden, D.B. (1984). Measures of alcohol damage in utero in the pigtailed macaque (*Macaca nemestrina*). *Ciba Found. Symp.* 105: 157–172.

Coe, C.L., and Crispen, H.R. (2000). Social stress in pregnant squirrel monkeys (Saimiri boliviensis peruviensis) differentially affects placental transfer of maternal antibody to male and female infants. *Health Psychol.* 19:554–559.

Conrad, S., Ha, J., Lohr, C., and Sackett, G. (1995). Ultrasound measurement of fetal growth in *Macaca nemestrina. Am. J. Primatol.* 36:15–35.

Diczfalusy, E. (1974). Endocrine function of the human fetus. *Am. J. Obstet. Gynecol.* 119:419–433.

Eaton, G.G., Worlein, J.M., and Glick, B.B. (1990). Sex differences in Japanese macaques (*Macaca fuscata*): Effects of prenatal testosterone on juvenile social behavior. *Horm. Behav.* 24:270–283.

Engle, M.J., Kemnitz, J.W., Rao, T.J., Perelman, R.H., and Farrel, P.M. (1996). Effects of maternal dexamethasone therapy on fetal lung development in the rhesus macaque. *Am. J. Perinatol.* 13:399–407.

Eswaran, H., Lowery, C.L., Robinson, S.E., Wilson, J.D., Cheyne, D., and McKenzie, D. (2000). Challenges of recording human fetal auditory-evoked response using magnetoencephalography. *J. Matern. Fetal Med.* 9:303–307.

Fahrenbruch, C.E., Burbacher, T.M., and Sackett, G.P. (1979). Gross morphology and pregnancy outcome in *Macaca nemestrina*. In: Ruppenthal, G.C., and Reese, D.J. (eds.), *Nursery Care of Nonhuman Primates.* Plenum Press, New York, pp. 27–34.

Farine, D., MacCarter, G.D., Timor-Tritch, I.E., Yeh, M.N., and Stark, R.I. (1988). Real-time ultrasonic evaluation of the baboon pregnancy: Biometric measurements. *J. Med. Primatol.* 17:215–221.

Feitz, W.F., Steegers, E.A., Aarnink, R.G., Arts, T., De Vries, J.D., and Van der Wildt, B. (1996). Endoscopic intrauterine fetal therapy: A monkey model. *Urology* 47:118–119.

Gelowitz, D.L., Rakic, P., Goldman-Rakic, P.S., and Selemon, L.D. (2002). Craniofacial dysmorphogenesis in fetally irradiated nonhuman primates: Implications for the neurodevelopmental hypothesis of schizophrenia. *Biol. Psychiatry* 52:716–720.

Gerschenson, M., and Poirier, M.C. (2000). Fetal patas monkeys sustain mitochondrial toxicity as a result of in utero zidovudine exposure. *Ann. N.Y. Acad. Sci.* 918:269–281.

Golub, M.S., Sassenrath, E.N., and Chapman, L.F. (1981). Mother-infant interaction in rhesus monkeys treated chronically with delta-9-tetrahydrocannabinol. *Child Dev.* 52:389–392.

Golub, M.S., Sassenrath, E.N., and Chapman, L.F. (1982). An analysis of altered attention in monkeys exposed to delta-9-tetrahydrocannabinol during development. *Neurobehav. Toxicol. Teratol.* 4:469–472.

Goy, R.W., and Deputte, B.L. (1996). The effects of diethylstilbesterol (DES) before birth on the development of masculine behavior in juvenile rhesus monkeys. *Horm. Behav.* 30:379–386.

Goy, R.W., Bercovitch, F.B., and McBrair, M.C. (1988). Behavioral masculinization is independent of genital masculinization in prenatally androgenized female rhesus macaques. *Horm. Behav.* 22:552–571.

Gunderson, V.M., Grant, K.S., Burbacher, T.M., Fagan, J.F., 3rd., and Mottet, N.K. (1986). The effect of low-level prenatal methylmercury exposure on visual recognition memory in infant crab-eating macaques. *Child Dev.* 57:1076–1083.

Gunderson, V.M., Grant-Webster, K.S., Burbacher, T.M., and Mottet, N.K. (1988). Visual recognition memory deficits in methylmercury-exposed *Macaca fascicularis* infants. *Neurotoxicol. Teratol.* 10:373–379.

He, N., Song, Z.M., and Lidow, M.S. (1999). Cocaine induces cell death within the primate fetal cerebral wall. *Neuropathol. Appl. Neurobiol.* 25:504–512.

Hendrickx, A.G. (1973). The sensitive period and malformation syndrome produced by thalidomide in crab-eating monkey (*Macaca fascicularis*). *J. Med. Primatol.* 2:267–276.

Hendrickx, A.G., and Helm, F.C. (1980). Nonteratogenicity of a structural analog of thalidomide in pregnant baboons (*Papio cynocephalus*). *Teratology* 22:179–182.

Hendrickx, A.G., and Newman, L. (1973). Appendicular skeletal and visceral malformations induced by thalidomide in bonnet monkeys. *Teratology* 7:151–159.

Hendrickx, A.G., and Sawyer, R.H. (1978). Developmental staging and thalidomide teratogenicity in the green monkey (*Cercopithecus aethiops*). *Teratology* 18:393–404.

Herman, R.A., Jones, B., Mann, D.R., and Wallen, K. (2000). Timing of prenatal androgen exposure: Anatomical and endocrine effects on juvenile male and female rhesus monkeys. *Horm. Behav.* 38:52–66.

Holm, R.A. (1979). Effects of parental risk and prenatal stress on pregnancy outcome. In: Ruppenthal, G.C., and Reese, D.J. (eds.), *Nursery Care of Nonhuman Primates.* Plenum Press, New York, pp. 21–26.

Howell, L.L., Schama, K.F., Ellis, J.E., Grimley, P.J., Kitchens, A.J., and Byrd, L.D. (2001). Fetal development in rhesus monkeys exposed prenatally to cocaine. *Neurotoxicol. Teratol.* 23:133–140.

Jaffe, R.B., Spencer, S.J., and Rabinovici, J. (1993). Activins and inhibins: Gonadal peptides during prenatal development and adult life. *Ann. N.Y. Acad. Sci.* 687:1–9.

Kaplan, J.N., Henessey, M.B., and Howd, R.A. (1982). Oral ethanol intake and levels of blood alcohol intake in the squirrel monkey. *Pharmacol. Biochem. Behav.* 17:111–117.

Kittinger, G.W. (1977). Endocrine regulation of fetal development and its relation to parturition in the rhesus monkey. *Ciba Found. Symp.* 47:235–257.

Klein, N.W., Plenefisch, J.D., Carey, S.W., Fredrickson, W.T., Sackett, G.P., Burbacher, T.M., and Parker, R.M. (1982). Serum from monkeys with histories of fetal wastage causes abnormalities in cultured rat embryos. *Science* 215(4528):66–69.

Larson, J.E., Morrow, S.L., Delcarpio, J.B., Bohm, R.P., Ratterree, M.S., Blanchard, J.L., and Cohen, J.C. (2000). Gene transfer into the fetal primate: Evidence for the secretion of transgene product. *Mol. Ther.* 2:631–639.

Lecanuet, J.P., Fifer, W.P., Krasnegor, N.A., and Smotherman, W.P. (eds.). (1995). *Fetal Development: A Psychobiological Perspective.* Lawrence Erlbaum Associates, Hillsdale, NJ.

Lengle, J.M., Chen, M., and Wakai, R.T. (2001). Improved neuromagnetic detection of fetal and neonatal auditory evoked responses. *Clin. Neurophysiol.* 112:785–792.

Levitt, P., and Rakic, P. (1982). The time of genesis, embryonic origin and differentiation of the brain stem monoamine neurons in the rhesus monkey. *Dev. Brain Res.* 4:35–57.

Lidow, M.S., and Song, Z.M. (2001). Primates exposed to cocaine in utero display reduced density and number of cerebral cortical neurons. *J. Comp. Neurol.* 435:259–262.

Mahoney, C.J., Eisele, S., and Capriolo, M. (1977). Clinical indications for caesarian section in the rhesus monkey. In: Ruppenthal, G.C., and Reese, D.J. (eds.), *Nursery Care of Nonhuman Primates.* Plenum Press, New York, pp. 3–20.

Meng, L., Ely, J.J., Stoufer, R.L., and Wolf, D.P. (1997). Rhesus monkeys produced by nuclear transfer. *Biol. Reprod.* 57:454–459.

Menzie, A.J. (2002). "Star Trek" sensors monitor health. Associated Press [on-line]. August 23, 2002, http://www.msnbc.com/news/796105.asp.

Michael, R.P., and Zumpe, D. (1998). Developmental changes in behavior and in steroid uptake by the male and female macaque brain. *Dev. Neuropsychol.* 14:233–260.

Michael, R.P., Zumpe D., and Bonsall, R.W. (1992). The interaction of testosterone with the brain of the orchidectomized primate fetus. *Brain Res.* 570:68–74.

Mintz, G.S., Weissman, M.D., and Fitzgerald, P.J. (2001). Intravascular ultrasound assessment of the mechanisms and results of brachytherapy. *Circulation* 104:1320–1325.

Morris, P., Binienda, Z., Gillam, M.P., Harkey, M.R., Zhou, C., Henderson, G.L., and Paule, M.G. (1996). The effect of chronic cocaine exposure during pregnancy on maternal and infant outcomes in the rhesus monkey. *Neurotoxicol. Teratol.* 18:147–154.

Morris, P., Binienda, Z., Gillam, M.P., Klein, J., McMartin, K., Koren, G., Duhart, H.M., Slikker, W., and Paule, M.G. (1997). The effect of chronic cocaine exposure throughout pregnancy on maternal and infant outcomes in the rhesus monkey. *Neurotoxicol. Teratol.* 19:47–57.

Nathanielsz, P.W., Frank, D., Gleed, R., Dillingham, L., Figueroa, J.P., and Poore, E.R. (1984). Methods for investigation of the chronically instrumented pregnant rhesus monkey preparation maintained on a tether and swivel system. In: Nathanielsz, P.W. (ed.), *Animal Models in Fetal Medicine.* Perinatology Press, New York, pp. 109–159.

Newell-Morris, L., and Fahrenbruch, C. (1985). Practical and evolutionary considerations for use of the nonhuman primate model in prenatal research. In: Watts, E.S., (ed.), *Nonhuman Primate Models for Growth and Development.* Liss, New York, pp. 9–40.

Nimrod, C., Simpson, N., Hafner, T., de Vermette, R., Fournier, J., Coady, L., and Baccanale, C. (1996). Assessment of early placental development in the cynomolgus monkey (*Macaca fascicularis*) using colour and pulsed wave Doppler sonography. *J. Med. Primatol.* 25:106–111.

Novak, M.S. (2002). A model experimental system for studying prenatal stress in pigtailed macaque monkeys. Doctoral Dissertation. University of Washington, Seattle, WA.

Novak, M.F.S.X., and Sackett, G.P. (1996). Reflexive and early neonatal development in offspring of pigtailed macaques exposed to prenatal psychosocial stress. *Dev. Psychobiol.* 29:294.

Novy, M.J., and Walsh, S.W. (1981). Regulation of fetoplacental steroidogenesis in rhesus macaques. In: Novy, M.J., and Resko, J.A. (eds.), *Fetal Endocrinology.* Academic Press, New York, pp. 65–94.

Nubbemeyer, R., Heistermann, M., Oerke, A-K., and Hodges, J.K. (1997). Reproductive efficiency in the common marmoset (*Callithrix jacchus*): A longitudinal study from ovulation to birth monitored by ultrsonography. *J. Med. Primatol.* 26:139–146.

Olivero, O.A., Fernandez, J.J., Antiochos, B.B., Wagner, J.L., St. Claire, M.E., and Poirier, M.C. (2002). Transplacental genotoxicity of combined antiretroviral nucleoside analogue therapy in *Erythrocebus patas* monkeys. *J. Acquir. Immune Defic. Syndr.* 29:323–329.

Parker, A. (2000). Imaging catheter gives surgeons the inside picture. *Science and Technology Review.* [On-line]. Lawrence Livermore National Laboratories. www.llnl.gov/str/daSilva.html.

Prahalada, S., Carroad, E., Cukierski, M., and Hendrickx, A.G. (1985). Embryotoxicity of a single dose of medroxyprogesterone acetate (MPA) and maternal serum MPA concentrations in cynomolgus monkey (*Macaca fascicularis*). *Teratology* 32:421–432.

Rakic, P. (1982). Early developmental events: Cell lineages, acquisition of neuronal positions, and aerial and laminar development. *Neurosci. Res. Prog. Bull.* 20:439–451.

Rakic, P. (1994). Corticogenesis in human and non-human primates. In: Gazzaniga, M.S. (ed.), *Cognitive Neuroscience: A Handbook for the Field.* MIT Press, Boston, pp. 127–145.

Robinson, S.R., and Smotherman, W.P. (1995). Habituation and classical conditioning in the rat fetus: Opioid involvements. In: Lecanuet, J.P., Fifer, W.P., Krasnegor, N.A., and Smotherman, W.P. (eds.), *Fetal Development: A Psychobiological Perspective.* Lawrence Erlbaum Associates, Hillsdale, NJ, pp. 295–314.

Ruppenthal, G.C., Moore, C.M., Best, R.G., Walker-Gelatt, C.G., Delio, P.J., and Sackett, G.P. (2004). Trisomy 16 in a pigtailed macaque (*M. nemestrina*) with multiple anomalies. *Am. J. Ment. Retard.* 109:9–20.

Sackett, G.P. (1980). A nonhuman primate model for studying the causes and effects of poor pregnancy outcome. In: Friedman, S., and Sigman, M. (eds.), *Pre-term and Post-term Birth: Relevance to Optimal Psychological Development.* Academic Press, San Diego, pp. 41–63.

Sackett, G.P. (1984). A nonhuman primate model of risk for deviant development. *Am. J. Ment. Defic.* 88:469–476.

Sackett, G.P. (1991). Toward a more temporal view of organism-environment interaction. In: Wachs, T.D., and Plomin, R. (eds.), *Conceptualization and Measurement of Organism Enviroment Interaction.* American Psychological Association, Washington D.C., pp. 11–28.

Sackett, G.P., Sameroff, A.J., Cairns, R.B., and Suomi, S.J. (1992). Continuity in behavioral development: Theoretical and empirical issues. In: Immelmann, K., Barlow, G., Petrinovich, L., and Main, M. (eds.), *Behavioral Development: The Bielefeld Interdisciplinary Project.* Cambridge University Press, London, pp. 23–57.

Sankai, T. (2000). Embryo production and embryo transfer. *Exp. Anim.* 49:69–81.

Schaie, K.W. (1965). A general model for the study of developmental problems. *Psychol. Bull.* 64(2):92–107.

Schindler, M.K., Wang, L., Selemon, L.D., Goldman-Rakic, P.S., Rakic, P., and Csernansky, J.G. (2002). Abnormalities of thalamic volume and shape detected in fetally irradiated rhesus monkeys with high dimensional brain mapping. *Biol. Psychiatry* 51:827–837.

Schneider, M.L. (1992a). The effect of mild stress during pregnancy on birthweight and neuromotor maturation in rhesus monkey infants (*Macaca mulatta*). *Infant Behav. Dev.* 15:389–403.

Schneider, M.L. (1992b). Prenatal stress exposure alters postnatal behavioral expression under conditions of novelty challenge in rhesus monkey infants. *Dev. Psychobiol.* 25:529–540.

Schneider, M.L. (1992c). Delayed object permanence development in prenatally stressed rhesus monkey infants *(Macaca mulatta)*. *Occup. Ther. J. Res.* 12(2): 96–112.

Schneider, M.L., and Coe, C.L. (1993). Repeated social stress during pregnancy impairs neuromotor development of the primate infant. *J. Dev. Behav. Pediatr.* 14(2):81–87.

Schneider, M.L., Moore, C.F., and Becker, E.F. (2001a). Timing of moderate alcohol exposure during pregnancy and neonatal outcome in rhesus monkeys (*Macaca mulatta*). *Alcoholism Clin. Exp. Res.* 25:1238–1245.

Schneider, U., Schleussner, E., Haueisen, J., Nowak, H., and Seewald, H. (2001b). Signal analysis of auditory evoked cortical fields in fetal magnetoencephalography. *Brain Topogr.* 14:69–80.

Seier, J.V., van der Horst, G., de Kock, M., and Chwalisz, K. (2000). The detection and monitoring of early pregnancy in the vervet monkey (*Cercopithecus aethiops*) with the use of ultrasound and correlation with reproductive steroid hormones. *J. Med. Primatol.* 29:70–75.

Sekhon, H.S., Keller, J.A., Proskocil, B.J., Martin, E.L., and Spindel, E.R. (2002). Maternal nicotine exposure upregulates collagen gene expression in fetal monkey lung. *Am. J. Respir. Cell. Mol. Biol.* 26:31–41.

Selye, H. (ed.). (1980). *Selye's Guide to Stress Research*. Van Nostrand Reinhold, San Francisco.

Sheifer, S.E., Canos, M.R., Weinfurt, K.P., Arora, U.K., Mendelsohn, F.O., Gersh, B.J., and Weissman, N.J. (2000). Sex differences in coronary artery size assessed by intravascular ultrasound. *Am. Heart J.* 139:649–653.

Shimizu, K. (1988). Ultrasonic assessment of pregnancy and fetal development in three species of macaque monkeys. *J. Med. Primatol.* 17:247–256.

Simpson, N.A.B., Nimrod, C., De Vermette, R., Leblanc, C., and Fournier, J. (1998). Sonographic evaluation of intervillous flow in early pregnancy: Use of echo-enhancement agents. *Ultrasound Obstet. Gynecol.* 11:204–208.

Smotherman, W.P. (2002). Classical conditioning in the rat fetus: Involvement of mu and kappa opioid systems in the conditioned response. *Dev. Psychobiol.* 40:104–115.

Smotherman, W.P., and Robinson, S.R. (1986). A method for endoscopic visualization of rat fetuses in situ. *Physiol. Behav.* 37:663–665.

Smotherman, W.P., and Robinson, S.R. (1988). Behavior of the rat fetus following chemical and tactile stimulation. *Behav. Neurosci.* 102:24–34.

Smotherman, W.P., Robinson, S.R., and Miller, B.J. (1986). A reversible preparation for observing the behavior of fetal rats in utero: Spinal anesthesia with lidocaine. *Physiol. Behav.* 37:57–60.

Tarantal, A.F., and Hendrickx, A.G. (1988). The use of ultrasonography for evaluating pregnancy in macaques. In: Neubert, D., Merker, H., and Hendrickx, A.G. (eds.), *Non-Human Primates—Developmental Biology and Toxicology.* Ueberreuter Wissenschaft, Berlin, pp. 91–111.

Tarantal, A.F., Gargosky, S.E., Ellis, D.S., O'Brien, W.D., Jr., and Hendrickx, A.G. (1995). Hematologic and growth-related effects of frequent prenatal ultrasound exposure in the long-tailed macaque (*Macaca fascicularis*). *Ultrasound Med. Biol.* 21(8):1073–1081.

Tarantal, A.F., O'Rourke, J.P., Case, S.S., Newbound, G.C. Li, J., Lee, C.I., Baskin, C.R., Kohn, D.B., and Bunnell, B.A. (2001). Rhesus monkey model for fetal gene transfer: Studies with retroviral-based vector systems. *Mol. Ther.* 3:128–138.

Tarantal, A.F., Castillo, A., Ekert, J.E., Bischofberger, N., and Martin, R.B. (2002). Fetal and maternal outcome after administration of tenofovir to gravid rhesus monkeys (*Macaca mulatta*). *J. Acquir. Immune Defic. Syndr.* 29:207–220.

Tarara, R. (1984). The effect of medroxyprogesterone acetate (Depo-Provera) on prenatal development in the baboon (*Papio anubis*): A preliminary study. *Teratology* 30:181–185.

Tomaszycki, M.L., Davis, J.E., Gouzoules, H., and Wallen, K. (2001). Sex differences in infant rhesus macaque separation-rejection vocalizations and effects of prenatal androgens. *Horm. Behav.* 39:267–276.

Uno, H., Lohmiller, L., Thieme, C., Kemnitz, J.W., Engle, M.J., Roecker, E.B., and Farrell, P.M. (1990). Brain damage induced by prenatal exposure to dexamethasone in fetal rhesus macaques, I: Hippocampus. *Dev. Brain Res.* 53:157–167.

Vahter, M., Mottet, N.K., Friberg, L., Lind, B., Shen, D.D., and Burbacher, T. (1994). Speciation of mercury in the primate blood and brain following long-term exposure to methyl mercury. *Toxicol. Appl. Pharmacol.* 124:221–229.

Wakai, R.T., Leuthold, A.C., and Martin, C.B. (1996). Fetal auditory evoked responses detected by magnetoencephalography. *Am. J. Obstet. Gynecol.* 74:1484–1486.

Wakai, R.T., Leuthold, A.C., Weber, S., and Martin, C.B. (1999). MEG of the neonate during sleep. In: Yoshimoto, T., Kotani, M., Kuriki, S., Karibe, H., and Nakasato, N. (eds.), *Recent Advances in Biomagnetism.* Tohoku University Press, Sendai, pp. 978–981.

Wolf, D.P., Meng, L., Ouhibi, N., and Zelinski-Wooten, M. (1999). Nuclear transfer in the rhesus monkey: Practical and basic implications. *Biol. Reprod.* 60:199–204.

Worlein, J.M., and Sackett, G.P. (1995). Maternal exposure to stress during pregnancy: Its significance for infant behavior in pigtail macaques. In: Pryce, C.R., and Martin, R.D. (eds.), *Motherhood in Human and Nonhuman Primates: Biosocial Determinants.* Karger, Basel, pp. 142–151.

Yoshinaga K., Hess, D.L., Hendricks, A.G., and Zamboni L. (1988). The development of the sexually indifferent gonad in the prosimian, *Galago crassicaudatus crassicaudatus. Am. J. Anat.* 181:89–105.

Yoshinaga, K., Hess, D.L., Hendrickx, A.G., and Zamboni, L. (1990). Germinal cell ectopism in the strepsirhine prosimian *Galago crassicaudatus crassicaudatus. Am. J. Anat.* 187:213–231.

SECTION TWO

Introduction to Section 2: Methods and Outcomes for Infrequently Hand-Reared Species

Most primate species have not been hand reared in large numbers owing in part to the use of social group rearing for most prosimians and for smaller New World species. We were hard pressed to find even a few facilities that had reared captive prosimians and marmosets in a nursery environment. The three chapters, 6–8, in this section are the exception, offering information on hand-rearing methods and outcomes for several prosimian species and for marmosets. The data appear to show that many of these species are difficult to rear by hand, with outcomes ranging across species from good to very poor. As many of these species are endangered in their natural habitats, it is likely that care and research in a captive environment will be required to preserve these primates.

CHAPTER SIX

The Effect of Hand Rearing on the Sexual and Maternal Competence of Three Species of Lemurs, *Varecia variegata*, *Varecia rubra*, and *Eulemur macaco*

Kelli Niebruegge and Ingrid Porton

1. INTRODUCTION

Breeding populations of black-and-white ruffed lemurs (*Varecia variegata*), red ruffed lemurs (*Varecia rubra*), and black lemurs (*Eulemur macaco macaco*) have been housed in zoos since the early to mid 1970s. The status and reproductive history of the three taxa in zoos and primate centers are documented in The International Studbook for the Ruffed Lemur (Porton, 2000) and the International Studbook for the Black Lemur (Porton, 1997).

As in the wild, captive black lemurs and ruffed lemurs are seasonally reproductive. The onset of the breeding season corresponds to a decreasing photoperiod in the black lemur and an increasing photoperiod in ruffed lemurs (Rasmussen, 1985). In both species the testicles of males increase in size approximately 2 months before females initiate estrous cycling. Both sexes increase scent-marking behavior and males inspect and groom female's anogentital region more often during the breeding season (Colquhoun, 1997; Moreland, 1991). Mounting attempts by males can be met with cackling, cuffing, chasing, or sitting with tucked tail if females are not receptive. Estrous females may solicit male attention and stand or partially crouch and deflect their tail to facilitate copulation.

Ruffed lemurs are the only diurnal primate species that gives birth and raises offspring in nests. In the wild, females construct nests by adding small-diameter leafy branches to liana tangles (Moreland, 1991). Females and offspring are well hidden in the nest (Moreland, 1991; Britt, personal communication; I. Porton, personal observations). In captivity, pregnant females may exhibit nest-building behaviors such as chewing on branches, rope, or other nesting materials, orally and or manually manipulating the items, transporting them by mouth, and placing them in a provided nestbox, on the ground in a pile, or haphazardly about the enclosure (Pereira *et al.*, 1987; I. Porton, personal observations). Ruffed lemur females have litters of one to three in the wild and up to six in captivity. Ruffed lemur mothers thoroughly clean their newborns, chew the umbilical cord, tuck the infants close to their bodies for thermoregulation, groom the infants' anogenital region, and transport their infants by mouth.

Black lemur females typically give birth to a singleton, although some captive births are twins. The female immediately grooms the neonate dry and consumes the placenta. Newborns randomly crawl around the mother's body searching for her nipples. Inexperienced mothers pull nervously at the infant whereas more experienced females adjust their arms to facilitate the infant's placement for nursing (I. Porton, personal observations). Infants are carried transversally on the abdomen and, in captivity, begin to leave the mother's body at about 6 weeks (Frueh, 1979).

Zoo personnel regularly face the need to hand raise infants that have been rejected by their mothers or are at risk owing to low birth weight or other health factors. Such infants have been hand reared and resocialized according to a variety of protocols. Some have been raised in complete isolation from conspecifics, some have been raised with peers or in proximity to adults, and others have experienced a combination of approaches. Physical access and resocialization with one or more adults has occurred as early as 4 weeks and as late as 15 months. Although hand rearing nonhuman primates can affect their ability to socialize appropriately with conspecifics (Harlow *et al.*, 1971; Meder, 1989; King and Mellen, 1994), the extent to which lemur species are affected has not been quantified. In this retrospective study, we investigated whether hand-reared individuals of these three lemur species had lower reproductive success than parent-reared individuals.

2. METHODS

The Black Lemur and the Ruffed Lemur International Studbooks contain information on the rearing status (hand reared, parent reared, or unknown) and reproductive history of individuals housed in zoos worldwide. This study was limited to lemurs from North American zoos that survived to breeding age and were not transferred to animal dealers and thereby lost to follow-up. Our initial step was to contact owning institutions to clarify the "unknown" rearing status of individuals, as the information is often known but not reported to the studbook keeper. A total of 33 red ruffed lemurs from 8 zoos, 31 black-and-white ruffed lemurs from 10 zoos, and 9 black lemurs from 4 zoos were identified as hand reared.

E-mail and telephone surveys were conducted to obtain information on the methods used to hand rear the above 73 lemurs. The questions included (1) age at which the infant was removed for hand rearing, (2) whether the infant was raised alone or with peers, (3) whether the infant was raised within visual, olfactory, and auditory contact of adult conspecifics, and (4) the age at which the individual was socialized with one or more adults.

Data on the reproductive and maternal competence of parent-reared lemurs were collected for 91 red ruffed and 91 black-and-white ruffed lemurs, all housed at the Duke University Primate Center, and 76 black lemurs housed at the Saint Louis Zoo. These two locations were chosen for their large historical populations of lemurs and the concomitant availability of accurate records on social housing and maternal care.

The breeding histories of reproductively successful adults were obtained from studbook data. For all lemurs that had not produced offspring, surveys were conducted to determine (1) if the individual was given an opportunity to breed, e.g., housed with an opposite-sex conspecific of breeding age in the absence of any contraceptive measures, and (2) if the individual was observed to copulate but a known pregnancy did not result.

Maternal competence was defined as a female that raised at least one offspring. Thus a ruffed lemur female that raised one of three infants in a litter was considered maternally competent, as was a female that lost her first but raised subsequent offspring. Females that produced only one litter and the infant(s) died at 0–1 days of unknown cause were excluded from the maternal care evaluation.

3. RESULTS

To increase sample size, red ruffed and black-and-white ruffed lemurs were combined into one group. Comparisons were made between the reproductive success of hand-reared versus mother-reared lemurs and between peer-reared versus solitarily reared ruffed lemurs. None of the black lemurs was peer reared. The age at which hand-raised ruffed lemur and black lemurs are resocialized with adult conspecifics may have an impact on their reproductive behavior as adults. Unfortunately, many of the records detailing the age and process used to resocialize lemurs were missing, thereby preventing the evaluation of this factor. Table 6–1 presents a summary of the findings.

3.1. Reproductive Success: Parent-Reared and Hand-Reared Ruffed Lemurs

Of the 45 hand-reared males, 23 had an opportunity to breed and 14 (61%) were successful. Of the 95 parent-reared males, 72 had an opportunity to breed and 57 (79%) were successful. The difference neared sig-

Table 6–1. A Comparison of Reproductive and Maternal Success between Parent-Reared and Hand-Reared Ruffed Lemurs and Black Lemurs and Peer-Reared versus Singly Reared Ruffed Lemurs

Behavior	% Parent reared (*n*)	% Hand reared (*n*)	% Peer reared (*n*)	% Singly reared (*n*)
Ruffed lemur male reproductive success	79% (72)	61% (23)	63% (16)	60% (5)
Ruffed lemur female reproductive success	89% (64)	100% (14)	100% (12)	100% (2)
Ruffed lemur maternal success	94% (51)	75% (12)	80% (10)	50% (2)
Black lemur male reproductive success	86% (28)	0% (4)	NA	NA
Black lemur female reproductive success	86% (29)	75% (4)	NA	NA
Black lemur maternal success	95% (22)	67% (3)	NA	NA

nificance ($\chi^2 = 3.08$; $p < 0.1$). Of the 19 hand-reared females, 14 had an opportunity to breed and all were successful (100%). Of 87 parent-reared females, 64 had a chance to breed and 57 (89%) were successful. The difference was not significant ($\chi^2 = 1.676$; $p < 0.20$).

3.2. Maternal Competence: Parent-Reared and Hand-Reared Ruffed Lemurs

Maternal care data were available for 12 of the 14 hand-reared females that became pregnant. Nine (75%) raised offspring and three (25%) did not. Maternal care data for 51 parent-reared females showed that 48 (94%) reared their young whereas three (6%) did not. The difference was significant ($\chi^2 = 4.119$; $p < 0.05$).

3.3. Reproductive Success: Peer versus Solitarily Hand-Reared Ruffed Lemurs

Thirty of the 39 hand-reared males for which social rearing status is known were raised with one or more peers. Of these, 16 had an opportunity to breed and 10 (63%) were successful. Nine males were hand reared away from peers; of these, five had a chance to breed and three (60%) were successful. The difference was not significant ($\chi^2 = 0.032$; $p < 0.95$). Of 19 hand-reared females, 17 were reared with peers. Twelve of these had a chance to breed and all were successful (100%). Two females were raised alone and both reproduced (100%).

3.4. Maternal Competence: Peer versus Solitarily Hand-Reared Ruffed Lemurs

Ten of the ruffed lemur females that produced offspring were hand reared with peers. Of these, eight (80%) raised offspring and two (20%) did not. Of the two females hand reared alone, one raised her litter and the other did not.

3.5. Reproductive Success: Hand-Reared and Parent-Reared Black Lemurs

Four of the five hand-reared males were raised in a nursery away from conspecifics; data for the fifth male are lacking. The four nursery-reared males had an opportunity to breed, but all were unsuccessful. The only male not nursery reared and integrated into a group at age 42 days has not had a chance to breed. Of the 38 parent-reared males, 28 had an opportunity to breed and 24 (86%) were successful. The difference was significant but the sample size is small ($\chi^2 = 13.713$; $p < 0.001$).

All four hand-reared females had an opportunity to breed and three (75%) were successful. The three successful females were reintegrated into a social group with adult conspecifics between 42 and 90 days while the fourth female was raised in a nursery setting for 7.5 months. Of the 38 parent-reared females, 29 had an opportunity to breed and 25 (86%) were successful. The difference was not significant ($\chi^2 = 624$; $p < 0.50$).

3.6. Maternal Competence: Hand-Reared and Parent-Reared Black Lemurs

Two of the three hand-reared females (67%) raised their offspring. Maternal history was available for 22 of the 25 parent-reared females that conceived; of these, 21 (95%) raised their offspring and one (5%) did not.

4. DISCUSSION

More ruffed lemurs and black lemurs have been hand reared in North American zoos than are represented in these data. However, these animals could not be included as they were sent to dealers and their reproductive histories were lost to follow-up. King and Mellen (1994) found the same to be true for chimpanzees and postulated that at least some of the individuals were purposefully removed from the zoo population due to a history of reproductive failure. This is likely to be true for lemurs as well. Some individuals with incomplete records could not be included in the analysis because lack of reproductive opportunity could not be reliably eliminated as a reason for lack of offspring. Because records describing the reintroduction process were incomplete, we were unable to determine whether age at resocialization had an impact on adult breeding and parental behavior. Nevertheless, the results of the analysis do reveal trends that may prove useful for the reproductive management of these and related lemur species.

Our data show that hand rearing ruffed lemurs and black lemurs in a zoo setting does not exclude the possibility that the individuals can become reproductively competent adults. With the exception of male black lemurs, over 60% of the hand-reared lemurs succeeded in breeding and, in the case of females, raising their offspring. The data do suggest that the negative impact of hand rearing on future mating ability may be greater in males than females. Within ruffed lemurs the difference in the reproductive competence between parent-reared versus hand-reared male ruffed lemurs neared statistical significance. Equivalent data for black lemurs of both sexes are extremely slim, but they suggest that hand rearing may be more detrimental to adult copulatory behavior in males than in females. However, in this instance, age at reintroduction cannot be ruled out as a factor. The female black lemurs were resocialized at an earlier age than the males and the one male socialized at 6 weeks has not yet been placed in a reproductive situation.

Parent-reared ruffed lemur and black lemur females show equivalent success rates in raising offspring (94% and 95% respectively). In both species, hand rearing decreased maternal success, significantly so for ruffed lemurs. Sample size for black lemurs was too small to test statistical significance. Nevertheless, 2 of the 3 and 9 of the 12 hand-reared black and ruffed lemur females, respectively, were able to raise offspring. Incomplete hand-rearing records preclude our identifying resocialization methods that improved maternal care, but we do know that in the case of back lemurs the two competent females were introduced to a foster mother at 45 days and shortly thereafter integrated into a larger group. Both females, one of which had twins, raised their first-born offspring. The influence of peer-rearing on ruffed lemur maternal competence can not be satisfactorily teased apart as only two females were not peer reared. It would appear, however, on the basis of these results, that peer rearing is beneficial. Peer companionship is not unusual as ruffed lemurs typically give birth to multiple infants. Limited records from several zoos show that many ruffed lemurs were resocialized by approximately 6 months but whether this was more true for successful mothers cannot be determined.

That hand-reared ruffed and black lemurs can become reproductively competent adults should not be surprising as this has been shown to be true for other primates such as gorillas (Beck and Power, 1988; Ryan *et al.*, 2002), chimpanzees (King and Mellen, 1994), golden-lion tamarins (Rettberg-Beck and Ballou, 1988), and macaques (Harlow *et al.*, 1971). Brockman *et al.* (1987) reported that ruffed lemurs hand reared at the San Diego Zoo had reproduced as adults but no data on percentage of success were presented. Our data indicate a trend for the reproductive and maternal success of hand-reared lemurs to be lower than that of their parent-reared counterparts. This can be a problem as limited housing within zoos restricts the total number of breeding programs that can be undertaken. Therefore, breeding programs, especially those for endangered species, benefit when all individuals within the captive population have the potential to contribute to the program's genetic and demographic goals. Even the inclusion of a few behaviorally dysfunctional individuals has negative consequences for the program by changing the ratio of effective population size to population size (Ne/N). Although we were unable to quantitatively distinguish the effect of resocialization methods on adult behavior, qualitative information indicates that reintroduction to a social group at a young age benefits lemurs. Absent quantitative data, it is nevertheless logical to posit that early resocialization will improve the likelihood of producing reproductively competent adults. Experience at the Saint Louis Zoo has shown that hand-reared ruffed and black lemurs can be integrated into adult groups at 6 weeks of age (Knobbe, 1991; I. Porton and K. Neibruegge, Chapter 2, this volume). Therefore, if lemurs have to be hand reared, animal managers should structure their program to resocialize youngsters as early as possible.

ACKNOWLEDGMENTS

We gratefully acknowledge all the zoos that responded to our surveys and emails: Central Florida Zoological Park, Cheyenne Mountain Zoo, Cincinnati Zoo and Botanical Gardens, Duke University Primate Center, Gladys Porter Zoo, Houston Zoological Gardens, Jacksonville Zoological Gardens, Lincoln Park Zoo, Los Angeles Zoo, Micke Grove Zoo, Milwaukee County Zoological Gardens, Omaha's Henry Doorly Zoo, San Diego Zoo, Toledo Zoological Gardens, St. Catherine's Wildlife Survival Center, Utah's Hogle Zoo, and Wildlife Safari, Zoo World. We thank Alice Seyfried for her thoughtful review of the manuscript.

REFERENCES

Beck, B.B., and Power, M.L. (1988). Correlates of sexual and maternal competence in captive gorillas. *Zoo Biol.* 7:339–350.

Brockman, D.K., Willis, M.S., and Karesh, W.B. (1987). Management and husbandry of ruffed lemurs, *Varecia variegata*, at the San Diego Zoo. II. Reproduction, pregnancy, parturition, litter size, infant care, and reintroduction of hand-raised infants. *Zoo Biol.* 6:349–363.

Colquhoun, I.C. (1997). A predictive socioecological study of the black lemur (*Eulemur macaco macaco*) in northwestern Madagascar. Ph.D. Dissertation. Washington University, St. Louis, MO.

Frueh, R.J. (1979). The breeding and management of black lemurs *Lemur macaco macaco* at St. Louis Zoo. *Int. Zoo Yearb.* 19:214–217.

Harlow, H.K., Harlow, M.K., and Suomi, S.T. (1971). From thought to therapy: Lessons from a primate laboratory. *Am. Sci.* 59:538–549.

King, N.C., and Mellen, J.D. (1994). The effects of early experience on adult copulatory behavior in zoo-born chimpanzees (*Pan troglodytes*). *Zoo Biol.* 13:51–59.

Knobbe, J. (1991). Early resocialization of hand-reared primates. In: *AAZPA Regional Conference Proceedings*. American Association of Zoological Parks and Aquariums, Wheeling, WV, pp. 763–770.

Meder, A. (1989). Effects of hand-rearing on the behavioral development of infant and juvenile gorillas (*Gorilla g. gorilla*). *Dev. Psychobiol.* 22(4):357–376.

Moreland, H.S. (1991). Social organization and ecology of black and white ruffed lemurs (*Varecia variegata variegata*) in lowland rain forest, Nosy Mangabe, Madagascar. Ph.D. Dissertation. Yale University, New Haven, CT.

Pereira, M.E., Klepper, A., and Simmons, E.L. (1987). Tactics of care for young infants by forest-living ruffed lemurs (*Varecia variegata variegata*): Ground nests, parking, biparental guarding. *Am. J. Primatol.* 13:129–144.

Porton, I. (1997). International Studbook for the Black Lemur, *Eulemur macaco macaco* and *Eulemur macaco flavifrons*. Saint Louis Zoo, St. Louis, MO.

Porton, I. (2000). International Studbook for the Ruffed Lemur, *Varecia variegata variegata* and *Varecia variegata rubra*. Saint Louis Zoo, St. Louis, MO.

Rasmussen, D.T. (1985). A comparative study of breeding seasonality and litter size in eleven taxa of captive lemurs (*Lemur* and *Varecia*). *Int. J. Primatol.* 6(5):501–517.

Rettberg-Beck, B., and Ballou, J.D. (1988). Survival and reproduction of hand-reared golden lion tamarins. In: Ballou, J.D. (ed.), *1987 Golden Lion Tamarin Studbook*. National Zoological Park, Washington, D.C.

Ryan, S., Thompson, S.D., Roth, A.M., and Gold, K.C. (2002). Effects of hand-rearing on the reproductive success of western lowland gorillas in North America. *Zoo Biol.* 21:389–401.

CHAPTER SEVEN

Nursery-Reared Prosimian Primates

M. Kay Izard

1. INTRODUCTION

The prosimian group encompasses a wide range of primitive primate species that are distinct from the simian group of monkeys, apes, and humans. There are more than 40 species of prosimians in seven families (Kavanagh, 1983; Napier and Napier, 1985). Activity cycles of these species can be nocturnal, diurnal, or cathemeral (Tattersall, 1982, 1988; Overdorff and Rasmussen, 1995). Prosimians also show variety in their patterns of maternal care. Depending on the species, dams may leave infants in a nest, "park" their offspring in foliage and return to nurse or groom them, or carry their infants on their bodies (Ross, 2001). The diurnal lemurs (Lemuriformes) tend to live in larger social groups and are generally more gregarious than the nocturnal lorisids and cheirogaleids (Lorisiformes; Alterman *et al.*, 1995; but see Sterling and Richard, 1995).

Prosimians are not often nursery reared in great numbers, probably because there are fewer prosimians in captivity compared with many other primate species, and not many are used in laboratory research. In addition, many species of prosimians are endangered. Most prosimians, especially those from Madagascar, are seasonal breeders (Tattersall, 1982), and with enough breeding females of a particular species, peer groups of infants of close age could be established in a nursery setting. However, few institutions have enough breeding females of a single species to warrant a nursery per se. Usually, only one or two infants at a time need human care to replace maternal care. For this reason, the term "hand reared" is more appropriate than "nursery reared," and the former term is used throughout this chapter.

The Duke University Primate Center (DUPC), the largest institution in the world devoted exclusively to prosimians, houses representatives of six of the seven prosimian families (Izard, 1989). Over the 35 years that the DUPC has operated, there have been instances when an infant must be hand reared. Reasons for hand rearing are many, but the main problems that result in the decision to remove an infant from its dam are common to all primates: maternal abuse or neglect, maternal failure to lactate, and infant hypothermia.

Much of the information available on hand rearing of prosimians comes from zoos (see I. Porton and K. Niebruegge, Chapter 2, this volume) and the DUPC (Haring and Wright, 1989). From anecdotal accounts, personal experience at the DUPC, and published reports (Valerio *et al.*, 1972; Eaton *et al.*, 1973), it appears that, as adults, many hand-reared lemurs (Lemuriformes, family Lemuridae) have poor social skills, are difficult to introduce into a social group, and are aggressive to caretakers. On the other hand, hand-reared prosimians in the families Lorisidae and Cheirogaleidae (Lorisiformes) appear to socialize well and are relatively easy to introduce to conspecifics.

What could account for this variation in response to hand rearing? There are some major differences between the Lemuridae on the one hand and the Cheirogaleidae and Lorisidae on the other. For example, most prosimians in the Lemuridae family are diurnal (Napier and Napier, 1985) or cathemeral (Tattersall, 1988). With the exception of the ruffed lemur (*Varecia variegata*) and the bamboo lemur (*Hapalemur griseus*), female lemurs carry their infants on their bodies (Ross, 2001), where they can nurse on demand, ingesting small amounts at a time. For this reason, lemurs usually have dilute milk, which is low in energy, fat, and protein (Tilden and Oftedal, 1997). In contrast, members of the families Cheirogaleidae and Lorisidae are nocturnal (Alterman *et al.*, 1995). Infants are left in a nest alone or "parked" while the dam is active and foraging (Kappeler, 1998), and are able to nurse only when she returns.

Consequently, these species produce milk that is high in gross energy, fat, and protein (Tilden and Oftedal, 1997).

These observations led to the hypothesis that infants of diurnal prosimians, most of which are social and carry their infants, need socialization at an early age to interact appropriately with conspecifics. In contrast, offspring of nocturnal prosimians, most of which are relatively solitary and leave their infants either "parked" or in a nest, are buffered against the effects of the absence of conspecifics during infancy, and can interact appropriately even if hand reared without social group members.

As a measure of appropriate social behavior, I used the criterion of breeding competence, because it is a rigorous, objective measure that can be confirmed. I performed a retrospective analysis of DUPC records of hand-raised prosimians to determine whether there was a difference in reproductive competency between diurnal and nocturnal species and between species that carry their infants clinging to their body (carriers) and those that leave their infants parked or in a nest when active (nesters/parkers).

2. MATERIALS AND METHODS

I examined records of all infant prosimians that had been hand reared at the DUPC between 1976 and 2002. Infants were categorized by dam's activity pattern (diurnality versus nocturnality) according to Martin (1972) and Tattersall (1982) and by type of maternal care (carrier versus nest/parker), according to Ross (2001). Species with a cathemeral activity pattern were included in the diurnal category. The nocturnal slow loris (*Nycticebus coucang*) exhibited both patterns of maternal care, and therefore was excluded from the analyses of differences in maternal care as they relate to mortality and reproductive competence.

Reproductive competency was defined for males as siring at least one offspring, and for females as giving birth to at least one infant. If sample sizes were larger than 20, a one-tailed Chi-square test corrected for continuity was used for comparisons (Snedecor and Cochran, 1967). If sample sizes were less than 20, the observed proportions were compared using 2×2 contingency tables for small, unequal samples (Natrella, 1963). Level of significance was $p < 0.05$.

Table 7–1. Prosimian Species Hand Reared at the Duke University Primate Center, 1976–2002

Genus and species	Common name	Male	Female	Unknown	Total
Daubentonia madagascariensis	Aye-aye	1			1
Tarsius syrichta	Philippine tarsier		1		1
Mirza coquereli	Coquerel's greater mouse lemur	1	1		2
Cheirogaleus medius	Dwarf lemur	1			1
Microcebus murinus	Mouse lemur	1			1
Galago moholi	Lesser bushbaby	1			1
Otolemur garnettii	Garnett's bushbaby	1	1		2
Otolemur crassicaudatus	Thick-tailed bushbaby	2	5	2	9
Nycticebus coucang	Slow loris	1			1
Hapalemur griseus	Bamboo lemur	1	1		2
Varecia variegata	Ruffed lemur	7	3	1	11
Eulemur fulvus	True lemurs	2	5	3	10
Eulemur coronatus	Crowned lemur		1		1
Lemur catta	Ring-tailed lemur	2	1		3
Propithecus tattersalli	Golden-crowned sifaka		1		1
Totals		21	20	6	47

3. RESULTS

Forty-seven prosimians, representing almost equal numbers of males and females, were hand reared at the DUPC between 1976 and 2002 (Table 7–1). Of these, 77% were hand reared in the 1980s (Fig. 7–1), when the population at the DUPC was increasing. One infant that was hand reared

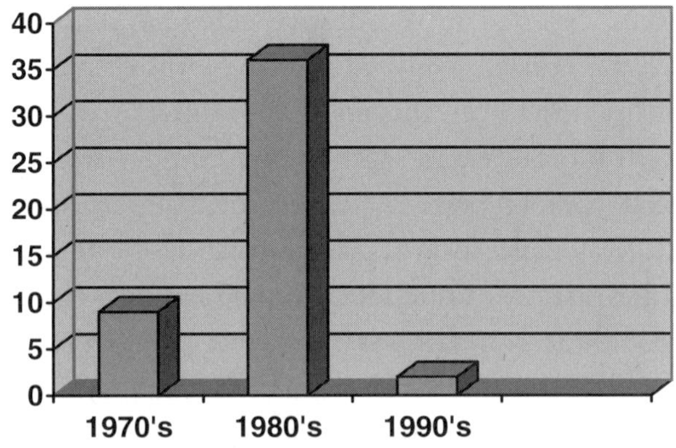

Figure 7–1. Number of prosimians that were hand reared at the DUPC.

after 1989 was a member of a rare species (*Daubentonia madagascariensis*), and the other was of important genetic stock (*Varecia variegata rubra*). Six families and 15 species of prosimians were represented (Table 7–1), of which nine were nocturnal and six were either diurnal or cathemeral (Table 7–2). Ten species were nesters/parkers and four were carriers (Table 7–2).

Of the 47 infants that were hand reared, 31 (66%) lived to adulthood (Table 7–3). Ten of these were male and 21 were female. There was no statistical difference in survivability between nocturnal and diurnal prosimians or between nesters/parkers and carriers.

Of the 31 prosimians that lived to adulthood, 26 had an opportunity to breed (Table 7–3). Of these, only 12 (39%) had records of siring or conceiving offspring (Table 7–3). I was unable to determine whether

Table 7–2. Classification of Prosimian Species by Active Period and Maternal Pattern[a]

Genus/species	Active period	Maternal pattern	References[b]
Daubentonia madagascariensis	Nocturnal	Nest	1, 2, 3
Tarsius syrichta	Nocturnal	Park	4, 5
Mirza coquereli	Nocturnal	Nest	3, 6, 7
Cheirogaleus medius	Nocturnal	Nest	3, 8
Microcebus murinus	Nocturnal	Nest	3, 9
Galago moholi	Nocturnal	Nest	10, 11
Otolemur garnettii	Nocturnal	Nest/park	12, 13
Otolemur crassicaudatus	Nocturnal	Nest	12, 14
Nycticebus coucang	Nocturnal	Carry/park	15, 16
Hapalemur griseus	Diurnal	Nest/park	17
Varecia variegata	Diurnal	Nest	3, 18, 19
Eulemur fulvus	Diurnal	Carry	3, 20
Eulemur coronatus	Diurnal	Carry	3, 20
Lemur catta	Diurnal	Carry	3, 20
Propithecus tattersalli	Diurnal	Carry	21

[a] *Sources:* Kappeler (1998) and Ross (2001).
[b] (1) Feistner and Ashborne (1994); (2) Winn (1994); (3) Tattersall (1982); (4) Niemitz (1979); (5) Haring and Wright (1989); (6) Petter-Rousseaux (1980); (7) Pages (1980); (8) Foerg (1982); (9) Martin (1972); (10) Doyle *et al.* (1969); (11) Doyle (1979); (12) Welker and Schaefer-Witt (1988); (13) Ehrlich and MacBride (1990); (14) Rosenblum (1972); (15) Bearder (1987); (16) Ehrlich and MacBride (1989); (17) Taylor and Feistner (1996); (18) Klopfer and Dugard (1976); (19) Pereira *et al.* (1988); (20) Klopfer and Boskoff (1979); (21) Jolly (1966).

Table 7–3. Viability and Subsequent Reproduction among Hand-Reared Prosimians

Category	Total number of hand-reared infants	Number of infants that died[a]	Number (%) of infants that lived to adulthood	Number with opportunity to breed	Number (%) that successfully reproduced[b]
Nocturnal	19	4	15 (79%)	12	7 (58%)
Diurnal	28	12	16 (57%)	14	5 (36%)
Total	47	16	31 (66%)	26	12 (46%)
Nest/park	31	9	22 (71%)	17	11 (65%)
Carry	15	7	8 (53%)	8	1 (12%)
Total[c]	46	16	30 (65%)	25	12 (48%)

[a] Died before reaching the age at which each species first conceives offspring in captivity.
[b] For males, sired at least one offspring. For females, gave birth to at least one infant.
[c] Does not include *N. coucang* infants, which cling to dam's fur and are also parked.

three of the nocturnal individuals (all nester/parker species) and two of the diurnal animals (both nester/parker species) had ever bred. When these five animals were excluded from the analysis of the difference in reproductive competence between nocturnal and diurnal hand-reared animals, no statistical difference was observed. However, when these five animals along with the *N. coucang* (see Materials and Methods) were excluded from the analysis of differences in reproductive competence between hand-reared nester/parker and carrier offspring, significantly more of the nester/parker offspring reproduced successfully than the carrier offspring (65% versus 12%; $p < 0.05$).

4. DISCUSSION

At the DUPC, infants of a number of prosimian species were hand reared prior to 1989. Because the reason for hand rearing was neglect or abnormal behavior on the part of the dam, these infants were already at risk and had a low survival rate. Thus, a decision was made in 1989 to abandon hand rearing, except for infants that were of rare species or important genetic stock. With the low survival rate (65%) and the even lower rate of reproduction (25% of the total number of prosimians that were hand reared), that decision appears to have been warranted.

There was no difference in reproductive competency between nocturnal and diurnal hand-reared prosimians. However, infants of nester/parker species had a significantly higher incidence of reproductive competency than did infants of carrier species. These findings are congruent with those of Porton and Niebruegge (Chapter 2, this volume). On the basis of these results, I suggest that the infants of maternal carriers need social interaction with conspecific adults and peers to develop proper social behavior. Hand rearing of these prosimians results in animals with deficits in social behavior, which accounts for the poor reproductive competency in that group. On the other hand, infants of maternal nesters/parkers can be hand reared with less of an influence on adult social behavior and subsequent reproduction. On a cautionary note, the numbers in this retrospective study were small, and confirmation of the above notion is warranted.

ACKNOWLEDGMENTS

Many thanks to the staff of the DUPC, especially Andrea Katz and David Haring. My appreciation to Ingrid Porton for providing information from the SSP for *V. variegata*. DUPC Publication No. 776.

REFERENCES

Alterman, L., Doyle, G., and Izard, M.K. (eds.). (1995). *Creatures of the Dark: The Nocturnal Prosimians.* Plenum Press, New York.

Bearder, S.K. (1987). Lorises, bushbabies, and tarsiers: Diverse societies in solitary foragers. In: Smuts, B.B., Cheney, D.L., Seyfarth, R.M., Wranghan, R.W., and Struhsaker, T.T. (eds.), *Primate Societies.* University of Chicago Press, Chicago, IL, pp. 11–24.

Doyle, G.A. (1979). Development of behaviour in prosimians with special reference to the lesser bushbaby, *Galago senegalensis moholi.* In: Doyle, G.A. and Martin, R.D. (eds.), *The Study of Prosimian Behavior.* Academic Press, New York, pp. 158–206.

Doyle, G.A., Anderson, A., and Bearder, S.K. (1969). Maternal behavior in the lesser bushbaby (*Galago senegalensis moholi*) under semi-natural conditions. *Folia Primatol.* 11:215–238.

Eaton, G.G., Slob, A.K., and Resko, J.A. (1973). Cycles of mating behavior, oestrogen and progesterone in the thick-tailed bushbaby Galago crassicaudatus under laboratory conditions. *Anim. Behav.* 21:309–315.

Ehrlich, A., and MacBride, L. (1989). Mother-infant interactions in captive slow lorises *(Nycticebus coucang)*. *Am. J. Primatol.* 19:217–228.

Ehrlich, A., and MacBride, L. (1990). Mother-infant interactions in captive thick-tailed galagos (*Galago garnettii*). *J. Mammal.* 71:198–204.

Feistner, A.T.C., and Ashborne, C.J. (1994). Infant development in a captive-bred aye-aye (*Daubentonia madagascariensis*) over the first year of life. *Folia Primatol.* 62:74–92.

Foerg, R. (1982). Reproduction in *Cheirogaleus medius. Folia Primatol.* 38:49–62.

Haring, D.M., and Wright, P.C. (1989). Hand-raising a Phillipine tarsier, *Tarsius syrichta. Zoo Biol.* 8:265–274.

Izard, M.K. (1989). Duke University Primate Center. *J. Med. Primatol.* 18:137–147.

Jolly, A. (1966). *Lemur Behavior.* Chicago University Press, Chicago, IL.

Kappeler, P.M. (1998). Nests, tree holes, and the evolution of primate life histories. *Am. J. Primatol.* 46:7–33.

Kavanagh, M. (1983). *A Complete Guide to Monkeys, Apes and Other Primates.* Viking Press, New York.

Klopfer, P.H., and Boskoff, K.L. (1979). Maternal behavior in prosimians. In: Doyle, G.A., and Martin, R.D. (eds.), *The Study of Prosimian Behavior.* Academic Press, New York, pp. 123–156.

Klopfer, P.H., and Dugard, J. (1976). Patterns of maternal care in lemurs. III. *Lemur variegates. Z. Tierpsychol.* 27:984–996.

Martin, R.D. (1972). A preliminary study of the lesser mouse lemur (*Microcebus murinus* J.F. Miller, 1777). *Z. Tierpsychol.* Suppl. 9:43–89.

Napier, J.R., and Napier, P.H. (1985). *The Natural History of the Primates.* MIT Press, Cambridge, MA.

Natrella, M.G. (1963). *Experimental Statistics.* John Wiley & Sons, New York.

Niemitz, C. (1979). Outline of the behavior of *Tarsius bancanus.* In: Doyle, G.A., and Martin, R.D. (eds.), *The Study of Prosimian Behavior.* Academic Press, New York, pp. 631–660.

Overdorff, D.J., and Rasmussen, M.A. (1995). Determinants of nighttime activity of "diurnal" lemurid primates. In: Alterman, L., Doyle, G., and Izard, M.K. (eds.), *Creatures of the Dark: The Nocturnal Prosimians.* Plenum Press, New York, pp. 61–74.

Pages, E. (1980). Ethoecology of *Microcebus coquereli* during the dry season. In: Charles-Dominique, P., Cooper, H.M., Hladik, A., Hladik, C.M., Pages, E., Pariente, G.F., Petter-Rousseaux, A., and Schilling, A. (eds.), *Nocturnal Malagasy Primates: Ecology, Physiology, and Behavior.* Academic Press, New York, pp. 97–116.

Pereira, M.E., Klepper, A., and Simons, E.L. (1988). Tactics of care for young infants by forest-living ruffed lemurs (*Varecia variegata variegata*): Ground nests, parking, and biparental guarding. *Am. J. Primatol.* 13:129–144.

Petter-Rousseaux, A. (1980). Seasonal activity rhythms, reproduction, and body weight vvariations in five sympatric prosimians, in simulated light conditions. In: Charles-Dominique, P., Cooper, H.M., Hladik, A., Hladik, C.M. Pages, E., Pariente, G.F. Petter-Rousseaux, A., and Schilling, A. (eds.), *Nocturnal Malagasy Primates: Ecology, Physiology, and Behavior.* Academic Press, New York, pp. 137–152.

Rosenblum, L. (1972). Observation of maternal behaviour of two captive bushbabies (*Galago crassicaudatus argentatus*). *Anim. Behav.* 20:677–688.

Ross, C. (2001). Park or ride? Evolution of infant carrying in primates. *Int. J. Primatol.* 22:749–771.

Snedecor, G.W., and Cochran, W.G. (1967). *Statistical Methods.* Iowa State University Press, Ames, IA.

Sterling, E.J., and Richard, A. (1995). Social organization in the aye-aye (*Daubentonia madagascariensis*) and the perceived distinctiveness of nocturnal primates. In: Alterman, L., Doyle, G., and Izard, M.K. (eds.), *Creatures of the Dark: The Nocturnal Prosimians.* Plenum Press, New York, pp. 439–451.

Tattersall, I. (1982). *The Primates of Madagascar.* Columbia University Press, New York.

Tattersall, I. (1988). Cathemeral activity in primates: A definition. *Folia Primatol.* 49:200–202.

Taylor, T.D., and Feistner, A.T.C. (1996). Infant rearing in captive *Hapalemur griseus alaotrensis*: Singleton versus twins. *Folia Primatol.* 67:44–51.

Tilden, C.D., and Oftedal, O.T. (1997). Milk composition reflects pattern of maternal care in prosimian primates. *Am. J. Primatol.* 41:195–211.

Valerio, D.A., Johnson, P.T., and Thompson, G.E. (1972). Breeding the greater bushbaby in a laboratory environment. *Lab. Anim. Sci.* 22:203–206.

Welker, C., and Schaeffer-Witt, C. 1988. Preliminary observations on behavioural differences among thick-tailed bushbabies. *Int. J. Primatol.* 9:507–518.

Winn, R.M. (1994). Development of behaviour in a young aye-aye in captivity. *Folia Primatol.* 62:93–107.

CHAPTER EIGHT

Hand Rearing of Infant Common Marmosets (*Callithrix jacchus*)

Bernhard Voelkl and Ludwig Huber

1. INTRODUCTION

Marmoset litters usually contain twins, but can contain triplets. In the latter case, usually only two infants survive, while the third often starves to death within the first week. To maximize breeding success by saving the life of the third infant, a common practice is to remove one infant from the mother, rear it by hand, and reintroduce it into the group as soon as it can survive on its own. Hand rearing is also required if a mother is unable to nurse its infants or if a research protocol requires subjects that are free of specific pathogens. Early descriptions of hand rearing have been published for *Saguinus* (Hampton and Hampton, 1967; Deinhardt, 1970; Wolfe *et al.*, 1972; Pook, 1974; Kingston, 1975; Dronzek *et al.*, 1986), *Leontopithecus* (Rohrer, 1979), *Callithrix* (Kingston, 1975; Ingram, 1975; Stevenson, 1976; Pook, 1974; Winter and Rothe, 1978; Hearn and Burden, 1979), and *Callimico* (Beck *et al.*, 1982; Sodaro *et al.*, 1994). For a recent review of the existing literature, see Sodaro and Crissey (1999). In this chapter, we describe the successful hand rearing of three common marmosets (*Callithrix jacchus*).

2. THERMOREGULATION

Newborn marmosets cling to the body of their mothers or other group members and are warmed by the body heat of their carriers. Left alone, the neonates soon suffer from hypothermia. Thus when a neonatal marmoset is being hand reared, its body temperature must be maintained by artificial means such as an incubator (Hampton and Hampton, 1967), heating lamp (Stevenson, 1976), heating pad, or heated surrogate (Winter and Rothe, 1978). While standard incubators used for human infants are ideal to keep the neonates under constant optimal conditions, such an apparatus may not always be available, especially in smaller zoos or institutions. An acceptable low-budget alternative is a heating pad. The three marmosets in this report were reared with a heating pad that was wrapped in a towel and placed in a small, fur-lined box. A digital thermometer with a minimum–maximum temperature alarm was used to ensure that the surface temperature of the pad was always between 35°C and 42°C.

3. SURROGATES AND HOUSING

From birth, infant marmosets cling to the bodies of other group members and are carried constantly for the first weeks. If an infant is separated from its carrier it immediately starts uttering distress calls while searching for a monkey to cling to. Thus when an infant is being hand reared it should be offered a carrier replacement. Various surrogates have been employed, such as a simple piece of towel (Stevenson, 1976), a plush toy animal, or heating pads covered with real or artificial fur (Winter and Rothe, 1978). We offered our three infants a plush monkey and strips of marten fur. Both surrogates were equally well accepted. The surrogate and the infant were placed in a 25 × 18 × 15-cm box that was lined with marten fur and had a heating pad at the bottom.

4. FEEDING

Infant marmosets can be fed with milk substitutes formulated for human infants. Several different products are available and brand names vary between countries, but the main ingredients are more or less the same. Table 8–1 shows the composition of the milk of common marmosets (*Callithrix jacchus*), golden lion tamarins (*Leontopithecus rosalia*), cotton-top tamarins (*Saguinus oedipus*), humans, and the formulase suc-

Table 8–1. Major Components of Primate Milk and Milk Substitutes Used for Hand Rearing

Species/formula	Crude protein (g/100 ml)	Lactose (g/100 ml)	Total lipids (g/100 ml)	Calcium (mg/100 ml)	Reference
Callithrix jacchus	3.1–4.0	6.9–8.1	5.0–9.5	88.8–95.5	Turton *et al.*, 1978
Leontopithecus rosalia	5.7	6.9	5.8		Buss, 1975
Saguinus oedipus	3.8	5.8	3.1		Buss, 1971
Human	1.5	6.8	4.0		Davidson *et al.*, 1975
Aptamil Pre	1.6	7.1	3.6	53.0	Milupa[a]
LAC formula	3.3	5.1	7.6	85.5	Hobbs *et al.*, 1977
Aberystwyth formula	6.8	5.7	3.7	78.5	Stevenson, 1976
SMA/Sustagen formula	2.7	11.2	4.8	76.0	Sodaro and Crissey, 1999
Primilac formula	3.6	8.1	3.8	75.0	Sodaro and Crissey, 1999

[a] Data taken from product information, Milupa AG, Puch bei Salzburg, Austria.

cessfully used for hand rearing marmosets. While carbohydrates and the total amount of lipids are within the same range for marmoset and human milk substitutes, marmoset milk contains more crude protein and polyunsaturated fatty acids (Turton *et al.*, 1978) and has a higher concentration of calcium. Callitrichids are not able to utilize vitamin D_2, necessitating a higher amount of vitamin D_3 (Hampton *et al.*, 1966).

Because human formula is not a perfect match for marmoset milk, it is advisable to add some ingredients to the milk substitutes to obtain a mixture that more closely resembles the natural milk of marmosets. The amount of crude protein can be increased by the addition of Sustagen™, a protein powder, or Primilac™, a commercially distributed formula for rearing primates (Crissey, 1993). Vitamin D_3 is provided by adding D_3 supplements (e.g., Oleovit™ or Vignatol™) in amounts of 600–2000 IU/kg body weight/day, with 1000 IU/kg body weight/day being the most common recommendation (e.g., Nicely and Banks, 1978; Winter and Rothe, 1978; Moore, 1989). While some recommend vitamin D_3 supplements from the first week onward, others begin only after 3 weeks (Hampton and Hampton, 1967), 4 weeks (Kingston, 1975), or 9 weeks (Nicely and Banks, 1978). If the calcium concentration of the milk substitute is much lower than the marmoset average of 88–95 mg/100 ml (Turton *et al.*, 1978), calcium caseinate can be added (Stevenson, 1976; Hobbs *et al.*, 1977). We added Calcipot™ and vitamin D_3 supplements to Aptamil Pre (Milupa™) baby formula beginning at 6 weeks. The composition of fatty acids can be modified by the addition of 450–500 mg cod liver oil to 100 ml milk (Turton *et al.*, 1978). Although the three infants in this study refused to take the milk after the oil was added, they did accept it when the mixture was sweetened with small amounts of Hipp™ fruit juice for babies. Various multivitamin supplements have been added to the milk formula: Abidec™ (Kingston, 1975; Stevenson, 1976; Moore, 1989), Multibionta™ (Winter and Rothe 1978), Sanostol™, and Hipp-Multivitamin™ (the authors).

Our infants first tasted solid food at 3 weeks of age and began taking it regularly at 5 weeks. The first solid food offered was baby cereal, baby formula, or small amounts of soaked white bread, followed by various fruits, such as apple, melon, and banana. At the beginning, the baby formula was fed with a spoon and fruit was offered by hand, while later on the food was given in a small bowl. One criterion for reintroduction to the natal group was that the infant be ready to feed on its own from the bowl. We fed the same fruit to the infant as to the group, so that at the time of reintroduction the infants were familiar with the provided food. For the first 2 weeks after reintroduction the infants continued to be hand fed twice a day with baby formula enriched with 400 IU vitamin D_3, calcium caseinate, and 1 drop of Sanostol™.

5. FEEDING SCHEDULE

On the infant's first day we provided a diluted milk substitute that had half the concentration of the Aptamil formula given in Table 8–1. We then increased the concentration of the milk stepwise and on the fourth day of hand rearing used the full formula. During the first week we offered a glucose solution between milk feeding bouts, to ensure that the infants were supplied with an adequate amount of liquid. Such a procedure was widespread within the institutions surveyed by Crissey (1993), with some using pediatric electrolyte solution or a dextrose solution instead of glucose solution. During the first 2 weeks the infants were fed in 2- to 3-hr intervals. From the third week onward the intervals were increased continually: at 4 weeks there were about six feeding bouts during the day and none during the night. This schedule is similar to others reported for marmosets (Winter and Rothe, 1978; Moore, 1989; Sodaro and Crissey, 1999). The number of feeding sessions during the night and the speed at which the feeding intervals are increased vary slightly; for example, Winter and Rothe (1978) provided no food between 11 pm and 7 am. At each feeding session we provided milk *ad libitum*. During the first 2 weeks the amount of milk consumed during a feeding bout varied from 0.5 to 2.5 ml, for a daily consumption of 5.0–15.5 ml.

6. FEEDING TECHNIQUE

For feeding, the milk substitute was put in a pipette with a rubber teat and warmed to 40°C. The infant was held upright with one hand and fixed between index finger and thumb. The milk was then slowly dropped into the mouth of the infant. If the milk was given too fast, it flowed out of the infant's mouth and the infant started to sneeze, shook its head, and refused to take the pipette in its mouth for a short time afterward. Initially we tried to use kitten-rearing bottles and small glass tubes with hand-made rubber nipples, but without success. The infants did not suckle on these devices on their own, a problem also reported by others (Ingram, 1975; Kingston, 1975; Winter and Rothe, 1978; Moore, 1989). After feeding, we wiped the anogenital region with a moist piece of cloth to stimulate the infant to urinate and defecate.

7. HEALTH

The only health problems we faced were some light cases of diarrhea and one case of constipation. Diarrhea was treated by (1) diluting the milk formula, (2) feeding no milk but only glucose solution, or (3) reducing the amount of cod liver oil. In all cases diarrhea stopped within the same day or the next morning. More serious cases of diarrhea can be treated with Arobon (1 g in 30 ml milk substitute; Winter and Rothe, 1978) or tetracycline (oral, 5–10 mg/kg body weight/day; Hampton and Hampton, 1967). At the age of 5 weeks one infant suffered from constipation, which was treated by canceling the feeding of baby formula and solid food and by massaging the infant's belly. The symptoms vanished on the following day. Constipation can also be treated with milk of magnesia (Kingston, 1975).

8. WEIGHT DEVELOPMENT

When we removed the infants from their natal groups, they weighed 28.5 g, 29.3 g, and 29.1 g. During hand rearing their weight increased steadily at a rate of 0.9 g per day. Even though it is not possible to draw statistical conclusions with only three infants, the weight of the hand-reared infants was always very close to the weight of their siblings that were left with their mother. The development of body weight was also consistent with data given by Winter and Rothe (1978), who studied the development of 11 hand-raised common marmosets.

9. REINTRODUCTION

The final goal of any hand-rearing program should be the integration of the infant into either its original group or a foster group. While the rearing of infants can be considered relatively easy, apart from its time-consuming nature, reintroduction has proven to be more difficult. Infant marmosets start to interact with other group members at an early age. In the second week the infant answers twitter and phee calls by replicating them and in the fourth week it begins to engage in social play bouts with elaborated social signals (e.g., "open-mouth display"). From the fifth week onward interactions with older group members increase rapidly, and in the eighth week the first social grooming attempts can be observed (Stevenson, 1978; Missler *et al.*, 1992). Thus, to ensure that the infant gathers all the relevant social experiences, which will be important for its further socialization, reintroduction should be tried as early as possible.

Once reintroduced into its natal group the infant may not feed on milk from its mother. In fact, this was the case with all three of our infants. Therefore, the infant must be developed enough to survive on its own at the time of reintroduction. Missler *et al.* (1992) reported that the youngest survival age without human intervention was about 8.5 weeks for *Callithrix jacchus*. We reintroduced the infants at the ages of 35, 38, and 45 days, but continued feeding the infants in their home cage for 2 more weeks.

Before reintroduction, food was offered from the same kind of bowls as used for the colony. Infants were trained to drink from standard water dispensers, and they were given climbing exercises on wooden branches and ropes to prepare them for their life in the colony cage. During the hand-rearing period the infants were taken to the animal laboratory at regular intervals to allow visual, auditory, and olfactory contact. The infants were also returned to the colony for a number of hours. The reaction of the colony members was always positive, but as it got more and more difficult to retrieve the infants again, these contact sessions were conducted only two to six times per infant. Some authors put the infants into a "howdy" cage, which is placed near or within the cage of the colony for several days before reintroduction (e.g., Kuster, 1978; Watson and Petto, 1988; Sodaro *et al.*, 1994).

On the day of reintroduction we put the infant into the colony cage, together with its sleeping box and its surrogate. Sodaro and Crissey (1999) emphasize that the surrogate shall not be removed until the infant starts to huddle and sleep with the other group members. Two of our infants exhibited fear of conspecifics, a problem also reported by Kuster (1978) and Sodaro and Crissey (1999). This fear vanished after some days, but it can also happen that the fear persists and even becomes more intense. In such cases the reintroduction has to be considered unsuccessful. Pook (1974) reported serious aggression of a pair of *Saguinus oedipus* toward their reintroduced infant; this reintroduction also was unsuccessful and had to be terminated.

REFERENCES

Beck, B.B., Anderson, D., Ogden, J., Rettberg, B., Brejla, C., Scola, R., and Warneka, M. (1982). Breeding the Goeldi's monkey Callimico goeldii at Brookfield Zoo, Chicago. *Int. Zoo Yearb.* 22:106–114.

Buss, D.H. (1971). Mammary glands and lactation. In: Hafez, E.S.E. (ed.), *Comparative Reproduction of Nonhuman Primates.* C.C. Thomas, Springfield, pp. 315–333.

Buss, D.H. (1975). Composition of milk from a golden lion marmoset. *Lab. Primate Newsl.* 14:17–18.

Crissey, S. (1993). *Infant. Diet. Care Notebook.* American Zoo and Aquarium Association, Wheeling, WV.

Davidson, S., Passmore, R., Brock, J.F., and Turvey, A. (1975). *Human Nutrition and Dietetics.* Churchill-Livingstone, Edinburgh.

Deinhardt, F. (1970). Nutritional requirements of marmosets. In: Harris, R.S. (ed.), *Feeding and Nutrition in Nonhuman Primates.* Academic Press, New York, pp. 175–182.

Dronzek, L.A., Savage, A., Snowdon, C.T., Whaling, C.S., and Ziegler, T.E. (1986). Technique for hand-rearing and reintroduction of rejected cotton-top tamarin infants. *Lab. Anim. Sci.* 36:243–247.

Hampton, S.H., and Hampton, J.K., Jr. (1967). Rearing marmosets from birth by artificial laboratory techniques. *Lab. Anim. Care* 17:1–10.

Hampton, J.K., Jr., Hampton, S.H., and Landwehr, B.T. (1966). Observations on a successful breeding colony of the marmoset. *Oedipomidas oedipus. Folia Primatol.* 4:265–287.

Hearn, J.P., and Burden, F.J. (1979). "Collaborative" rearing of marmoset triplets. *Lab. Anim.* 13:131–133.

Hobbs, K.R., Clough, C., and Bleby, J. (1977). The establishment of specified-pathogen-free marmosets (*Callithrix jacchus*). *Lab. Anim.* 11:29–34.

Ingram, J. (1975). Husbandry and observation methods of a breeding colony of marmosets (*Callithrix jacchus*) for behavioural research. *Lab. Anim.* 9:249–259.

Kingston, W.R. (1975). The breeding of endangered species of marmosets and tamarins. In: Martin, R.D. (ed.), *Breeding Endangered Species in Captivity.* Academic Press, London, pp. 213–222.

Kuster, J. (1978). Preliminary investigations on the integration/reintroduction of hand reared common marmosets (*Callithrix jacchus*) into family groups. In: Rothe, H., Wolters, H.-J., and Hearn, J. (eds.), *Biology and Behavior of Marmosets.* Eigenverlag, Göttingen, pp. 149–152.

Missler, M., Wolff, J.R., Rothe, H., Heger, W., Merker, H.-J., Treiber, A., Scheid, R., and Crook, G.A. (1992). Developmental biology of the common marmoset: Proposal for a postnatal staging. *J. Med. Primatol.* 21:285–298.

Moore, M. (1989). *Marmosets in Captivity.* Basset Publications, Plymouth.

Nicely, C., and Banks, S. (1978). Hand rearing of a golden lion marmoset. Unpublished manuscript, cited in: Winter, M., and Rothe, H. (1978). Einige Aspekte zur künstlichen Aufzucht und körperlichen Entwicklung von Weissbüscheläffchen (*Callithrix jacchus*). *Z. Kölner Zoo* 20:129–141.

Pook, A.G. (1974). Hand rearing and reintroduction to its parents of a saddleback tamarin. *Jersey Wildlife Preservation Trust Annu. Rep.* 11:35–59.

Rohrer, M.A. (1979). Handrearing golden lion marmosets, *Leontopithecus rosalia* at the Oklahoma City Zoo. *Anim. Keepers' Forum* 6:33–39.

Sodaro, V., and Crissey, S. (1999). Handrearing callithrichids and callimico in zoological parks. In: Sodaro, V., and Saunders, N. (eds.), *Callitrichid Manual.* Chicago Zoological Park, Chicago. December 20, 2002: http://www.csem.com/newworldprimate/callithrichid%20Manual.htm

Sodaro, V., Pingry, K., and Snyder, K. (1994). Changes in hand rearing procedures for *Callimico goeldii* at Brookfield Zoo. In: *AZA Regional Conference Proceedings.* American Zoo and Aquarium Association, Wheeling, WV, pp. 404–407.

Stevenson, M.F. (1976). Maintenance of breeding groups of the common marmoset (*Callithrix jacchus jacchus*) with details of hand rearing infants. *Int. Zoo Yegrb.* 16:110–115.

Stevenson, M.F. (1978). Ontogeny of playful behaviour in family groups of the common marmoset. In: Chivers, D.J., and Herbert, J. (eds.), *Recent Advances in Primatology,* Vol. 1: *Behaviour.* Academic Press, London, pp. 139–143.

Turton, J.A., Ford, D.J., Bleby, J., Hall, B.M., and Whiting, R. (1978). Composition of the milk of the common marmoset (*Callithrix jacchus*) and milk substitutes used in hand-rearing programmes, with special references to fatty acids. *Folia Primatol.* 29:64–79.

Watson, L.M., and Petto, A.J. (1988). Infant adoption and reintroduction in common marmosets (*Callithrix jacchus*). *Lab. Primate Newsl.* 27(3):1–3.

Winter, M., and Rothe, H. (1978). Einige Aspekte zur künstlichen Aufzucht und körperlichen Entwicklung von Weissbüscheläffchen (*Callithrix jacchus*). *Z. Kölner Zoo* 20:129–141.

Wolfe, L.G., Ogden, J.D., Deinhardt, J.B., Fisher, L., and Deinhardt, F. (1972). Breeding and handrearing marmosets for viral oncogenesis studies. In: Beveridge, W.I.B. (ed.), *Primate Breeding.* Karger, Basel, pp. 145–157.

SECTION THREE

Introduction to Section 3: Methods and Outcomes for Frequently Hand-Reared Species

Most of the available information on nursery-rearing methods and outcomes is based on just a few species of great apes, Old World monkeys, and baboons. The eight chapters, 9–16, in this section present methods for rearing and developmental testing, as well as outcome data, for the species that are reared most frequently under nursery conditions. Each chapter presents comparisons between nursery- and mother-reared animals involving one or more aspects of growth, behavioral development, or physiology. Specific outcome measures include immunological development and competence, typical social and nonsocial behavior and temperament, neurobehavioral development of neonates and young infants, ponderal growth and skeletal development, mortality and morbidity, and hematology and serum chemistry. Taken together, these chapters detail a variety of rearing methods and identify a number of dimensions on which nursery-reared primates do and do not differ from mother-raised individuals. The chapters also present up-to-date methods for performing many developmental assessments, as well as some important modern statistical approaches for assessing development and growth. Many of these chapters also document changes that have been made in nursery-rearing practices over the past 35 years, as well as a number of positive changes in the developmental outcomes of nursery-reared primates over this time period.

CHAPTER NINE

Immunological Consequences of Nursery Rearing

Gabriele R. Lubach and Christopher L. Coe

1. INTRODUCTION

The concept of rearing primate infants in a nursery fell out of favor over the past two decades, after research showed the detrimental behavioral effects of restricted rearing and raised concerns about how best to care for animals with these abnormalities (Harlow and Harlow, 1969; Ruppenthal and Sackett, 1979; Capitanio, 1985). In addition to behavioral effects, several investigators have described physiological differences between mother-reared (MR) and nursery-reared (NR) infants (Sackett *et al.*, 1973; Champoux *et al.*, 1989; Kraemer *et al.*, 1989). Some of these differences were resolved as the infants grew older, while others persisted (Novak and Harlow, 1975; Mason *et al.*, 1991). Today, however, many institutions have changed or expanded their research focus and once again require some degree of nursery rearing. In some cases, nursery rearing is needed either to keep maternal variables from interfering with certain aspects of the infant's development, such as in AIDS research, or to keep the animals free from specific pathogens (Marthas *et al.*, 1995; Lindberg *et al.*, 1997).

It is imperative, therefore, to continue to educate researchers and care staff about the important differences between MR and NR monkeys in order to compare data generated from monkeys derived from different rearing conditions. Over the years, we have learned how to normalize the behavior of NR monkeys (Suomi and Harlow, 1972; Sackett *et al.*, 2002), but some of the physiological abnormalities cannot be so easily corrected. The immune system appears to be particularly vulnerable to perturbations early in development, both from the point of view of exposure to infectious pathogens and illness during infancy, as well as following disruptions of maternal care. Like the brain, it depends on feedback from the environment to mature normally. Both systems appear to have evolved with biological expectations of specific and sequential input for appropriate maturation (Rook and Stanford, 1998). In fact, we have known for about 50 years that animals raised in germ-free environments ultimately develop abnormal immune systems because of the absence of antigenic stimulation to prime immune responses and ontogeny of lymphoid tissue and cells.

The immunological environment of NR infants differs from that of MR infants in two important ways: the lack of exposure to breast milk and the absence of physiological regulation through maternal contact. Breast milk is of major significance to the infant because of the growth factors, immunoglobulins, cytokines, antimicrobials, and nourishment that it contains. Maternal contact provides warmth and tactile stimulation, which enables the infant to devote more energy to growth and maintenance rather than thermoregulation. The mother's activity and sleep patterns are also important sources of information about day/night cycles, essential for entrainment of the infant's developing biorhythms (Lubach *et al.*, 1992).

We have found that NR and MR rhesus macaque (*Macaca mulatta*) infants differ in several aspects of their immunity: number and proportions of T cell subsets, lymphocyte proliferation responses, and natural killer cell cytotoxicity (Lubach *et al.*, 1995). In this chapter we will review the immune-related data and possible explanations for these differences. In the decade since these studies were completed, many immunological methods and reagents have improved. New primate-specific reagents, and new techniques, could provide additional information on the mechanisms underlying the differences we found.

2. DEVELOPMENT OF THE INFANT IMMUNE SYSTEM

The ontogeny of the immune system begins early in fetal development (Cauchi, 1981; Osburn *et al.*, 1982). In humans, lymphocytes appear in the peripheral blood at 7 weeks after conception, and by 11 weeks, the fetus has the potential to respond to foreign antigens. Despite the immunological developments that occur during gestation, and the capacity for rudimentary responses to pathogens, both human and monkey neonates are born in an immunologically naive state (Martin *et al.*, 1973; Miller, 1980). They are prepared to fight off some infectious organisms, but the more sophisticated aspects of adaptive immunity are not fully developed (Klein *et al.*, 1977; Hayward and Kurnick, 1981).

Developmental delays in immune competence during the infancy period are common in most mammals. These delays, or deficits, may be accommodated by supplementary maternal factors transferred via the placenta *in utero*, and by breast milk postpartum (Kelleher and Lonnerdal, 2001; Goldman, 2002). For example, the primate fetus receives important immunological protection from the mother through the placental transfer of immunoglobulin G (IgG), a process that becomes increasingly elaborated from prosimians to Old World monkeys and the great apes (Coe *et al.*, 1994; Packard, 1982). In the case of rhesus macaques, placental transfer enables IgG levels in the neonate to be almost identical to those found in adult monkeys. This maternally derived IgG has a half-life of about 3 weeks, and thus declines progressively over time. In both human and monkey infants, maternally derived IgG reaches a nadir at about 3 months of age, at which point the infant begins to synthesize its own antibody. Human infants also show developmental delays in the production of other classes of immunoglobulin (A, E, M), as well as cytokines, growth factors, and immune-related substances (Goldman *et al.*, 1998). The same is probably true for nonhuman primates. In addition to the prenatal contribution from the mother, breast milk protects infants from other types of immune challenges by providing some of the missing agents to the infant until it is able to mount its own appropriate immune responses. Specifically, the substances in breast milk, such as antibody of the A class (IgA), are critical for protecting the epithelial surfaces of the gastrointestinal (GI) tract from bacterial pathogens.

Because nonhuman primate infants develop in much the same way as human infants, they are excellent models for developmental studies (Bailey and Coe, 1999; Burbacher and Grant, 2000; Buse *et al.*, 2003). Rhesus neonates, like human neonates, have an intact immune system at birth (Makori *et al.*, 2003). The system, albeit immature, is comprised of all the major immune sites such as the thymus and spleen, and all the precursors to the major leukocyte subsets. However, neither human nor monkey neonates can mount fully mature immune responses. Therefore, the protective aspects of breast milk are likely to be very similar in both monkey infants and children. Indeed, the composition of monkey milk is very similar to that of human milk, even though milk composition in general can be very species specific (Lonnerdal *et al.*, 1984). As Dahlgren *et al.* (2001) wrote, breast milk has "been carefully designed by evolution to promote proper maturation and tuning of the immune system" (p. 311).

3. IMMUNE RESPONSES OF NURSERY-REARED INFANTS

The research reviewed here was prompted by earlier cross-sectional surveys in which differences in lymphocyte proliferation responses were found to vary in juveniles that had been raised under different maternal conditions (Coe *et al.*, 1989, 1992). As juveniles, rhesus infants reared in a nursery or with limited maternal contact were found to have significantly higher lymphocyte proliferation than monkeys that had at least 6 months of mother rearing. In fact, the cellular activity was paradoxically higher than any responses normally seen in other rhesus macaques, even in adulthood. For some infants, these elevated cellular responses to stimulation with mitogen *in vitro* persisted until they were 3 years old or older.

Those intriguing observations led us to assess the ontogeny of the divergence by comparing immunity in 8 MR and 8 NR rhesus macaque infants from birth through 2 years of age (Lubach *et al.*, 1995). All MR infants lived as mother–infant dyads in the same colony room and remained with their mothers until ~1 year of age, when the yearlings were transferred into a peer group comprised of the eight yearlings and one adult female. Both mothers and infants were provided with Purina

Monkey Chow (Ralston Purina, St. Louis, MO), allowing the infants access to solid food from birth. The NR infants were separated from their mothers within 2 days after birth and raised by human caregivers in a nursery. Nursery rearing followed established guidelines, which included feeding with Similac formula (Blomquist and Harlow, 1961; Anderson, 1986; Ruppenthal and Reese, 1979). Infants were introduced to Purina chow at about 3 weeks of age. They were gradually weaned off formula and were eating only chow by 6 months of age.

To provide a comprehensive understanding of the physiological systems that might be most responsive to rearing conditions, we investigated several aspects of immune development. Every 2 months, between 8 and 10 am, we collected small blood samples from each infant and conducted three different assays on each sample. In the first assay, to examine the development of lymphocyte populations, separated mononuclear cells were stained for T cell antigens using monoclonal antibodies that cross-reacted with rhesus macaque receptors: CD2 (Leu 5b: pan-T cell), CD4 (Leu 3a: T-helper cell), and CD8 (Leu 2a: T-cytotoxic/suppressor) (Becton Dickenson, Mountain View, CA). The proportion of T-helper cells to T-suppressor cells (CD4:CD8 ratio) is often used clinically to evaluate and diagnose immunodeficiency and autoimmune disorders, and we employed this ratio as an index of the normalcy of lymphocyte development in the MR and NR infants. In the second assay, to evaluate proliferative potential, other mononuclear cells (MNC) from the same samples were resuspended in culture medium and stimulated in triplicate with one of three plant lectins: concanavalin-A (Con A), phytohemagglutinin (PHA), and pokeweed mitogen (PWM). In the third assay, done to determine lymphocyte cytolytic activity, we employed the traditional natural killer (NK) cell assay used in human research. This assay quantifies the ability of MNC to lyse radiolabeled K562 target cells (human erythroleukemia cells). Cells were cocultured in quadruplicate at three effector : target ratios: 100 : 1, 33 : 1, and 11 : 1. Lymphocyte killing capacity is often considered a later-maturing system, although in many ways it should also be thought of as an aspect of innate immunity because it is an important first line of defense. We, and others, have found that cytolytic activity is remarkably functional in young monkeys (Coe and Erickson, 1997). For statistical analysis, we used repeated-measures analysis of variance to compare responses averaged at 6-month intervals across the first 24 months of life.

To further assess one aspect of humoral immunity, at 8–15 months of age all infants received three separate vaccines: tetanus toxoid (TT) (Wyeth Laboratories, Philadelphia, PA), diphtheria toxoid (DT) (Connaught Laboratories, Swiftwater, PA), and trivalent influenza (TF) (A/Taiwan, A/Leningrad, B/Ann Arbor) (Connaught Laboratories). Primary antibody responses were assessed at 4 weeks postimmunization, and a secondary response to TF was assessed at 8 weeks. Antibody levels were quantified by enzyme-linked immunosorbent assay (ELISA).

Our findings of immune dysregulation generally concurred with earlier studies showing elevated proliferative activity and provided further insight into the extent of differences between NR and MR infants. The most dramatic new finding was a difference in the ratio of the T cell subsets. The ratios of $CD4^+$ to $CD8^+$ cells were significantly higher in the NR infants at all ages ($F[1,14] = 14.69$, $p < 0.002$) (Fig. 9–1). This shift was due primarily to a reduced percentage of $CD8^+$ cells found in the NR infants. In contrast, the overall numbers and percentage of total T cells (CD2) were similar in infants from both rearing conditions. Compared with other evaluations of rhesus macaque lymphocytes—in adults and in other laboratories—our NR infants appeared to have an abnormally high ratio (Murayama *et al.*, 1986; Ershler *et al.*, 1988; Gust *et al.*, 1991).

Recent studies have made our results more provocative, and extended the implications for assignment of animals to certain types of research on infectious illness. High $CD4^+/CD8^+$ ratios in NR rhesus macaques have also been found at the New England Primate Research Center (DeMaria *et al.*, 2000), in a project undertaken to establish a database of cell subset norms for use in the study of infant retroviral infection. The ratio was 3.3 : 1 at birth, but had decreased to 1.7 : 1 by 10 months of age. That is comparable to our data on NR infants, but very different from the 1 : 1 ratio found in our MR rhesus macaques. However, there is very little longitudinal information on MR infants, except for a report by Baroncelli *et al.*, (1997) and two cross-sectional reports (Nam *et al.*, 1998; Dykhuizen *et al.*, 2000) in which decreases in CD4/CD8 ratios were found with increasing infant age. We recently collected new cell

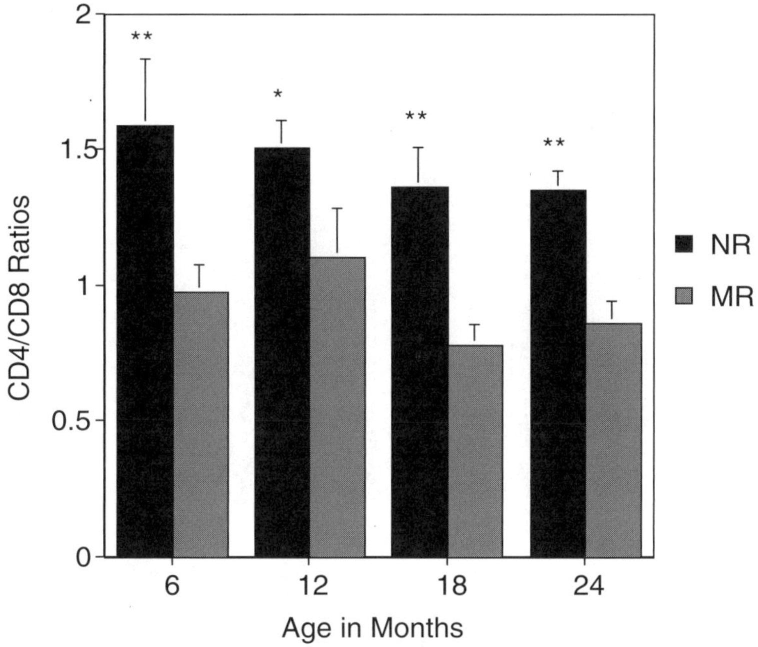

Figure 9–1. Ratios of CD4 to CD8 cells for NR and MR monkeys over the first 2 years of life. Mean (±SE) of two to three samples per subject portrayed at 6-month intervals. NR monkeys had significantly higher ratios than MR monkeys at all ages ($*p < 0.05$, $**p < 0.01$).

subset data on another cohort of MR rhesus infants, using whole blood instead of just the mononuclear cell fraction that had been separated from the rest of the blood by density gradient centrifugation. Using the whole-blood approach and newer monoclonal antibodies (CD4, CD8; BD Biosciences, San Diego, CA) we obtained values that differ from the prior numbers. Based on an analysis of 27 rhesus macaque infants, this approach yields a higher percentage of $CD4^+$ than $CD8^+$ cells, resulting in ratios of 1.55 and 1.45 to 1 at 3 and 5 months of age, respectively (Fig. 9–2). These ratios are higher than we had found in the earlier study, but are still below those for NR rhesus macaques published by DeMaria *et al.* (2000) and Van Rompay *et al.* (2003). Our newer ratios are also in keeping with the results for MR monkeys reported by Baroncelli *et al.* (1997) and by Dykhuizen *et al.* (2000). In human children, ratios begin

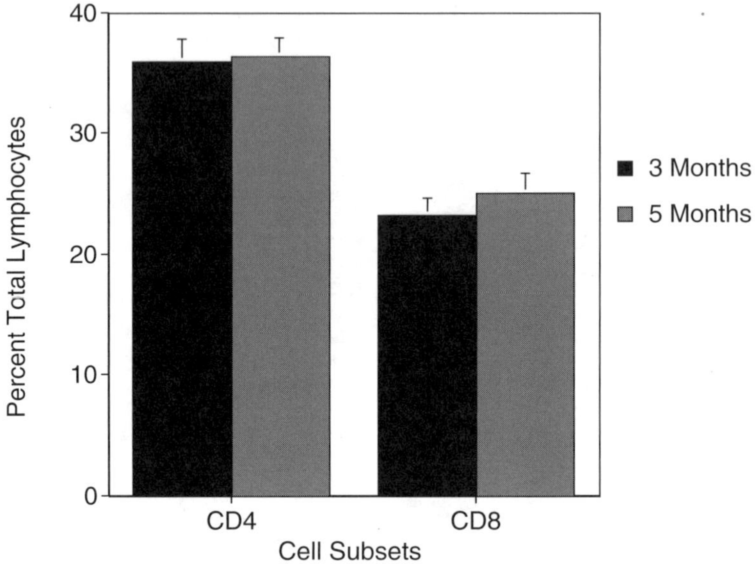

Figure 9–2. Percentages of $CD4^+$ and $CD8^+$ lymphocytes in MR infants at 3 and 5 months of age.

high, at ~3.0 to 1 at <1 year of age, and decline to a ratio of 1.5 to 1 by 7–10 years of age (Hannet *et al.*, 1992; Foerster *et al.*, 1997; Neubert *et al.*, 1998; de Vries, 2000).

In addition to the differences in T cell subsets, lymphocyte responses to stimulation *in vitro* with Con A, PHA, and PWM showed a trend toward higher proliferation in NR than MR infants, although this reached significance at only two age points (6 and 18 months) for PHA and three age points (6, 18, and 24 months) for PWM ($F[1,14] = 13.01$, $p < 0.003$) (Fig. 9–3). The reason for the higher proliferation remains unclear; it could be due to differences in the number and type of cell subsets, variation in the activation state of the cells, or perhaps altered release of soluble mediators that promote proliferation [e.g., interleukin-2 (IL-2)]. However, the numbers do correspond to findings in breast-fed (BF) and formula-fed (FF) human infants (Juto *et al.*, 1982; Pabst *et al.*, 1997). Lymphocyte proliferation responses to PHA stimulation were significantly higher in the FF babies, and intermediate responses were found in children who had received both formula and breast milk.

Pabst *et al.* (1997) also found a higher expression of activation markers on the lymphocytes from FF infants, indicating that the cells are in a state of chronic activation. The significance of these differences for long-term health is unknown, but an overactive immune response could lead to inappropriate and ineffective responses to immune challenges. There are also suggestions from epidemiological surveys of children that FF babies may be at greater risk for allergies and asthma (Hanson, 1998; Bernt and Walker, 1999).

In contrast to exaggerated proliferative responses, NK cells from NR infants had significantly decreased cytotoxic responses over the first 2 years of life. These assays are typically set up with three ratios of effector cells to target cells, and the differences were evident at all E:T ratios, i.e., 100:1 ($F[1,14] = 9.42$, $p < 0.008$), 33:1 ($F[1,14] = 6.68$, $p < 0.02$), and 11:1 ($F[1,14] = 9.44$, $p < 0.008$) (Fig. 9–4). Despite the contin-

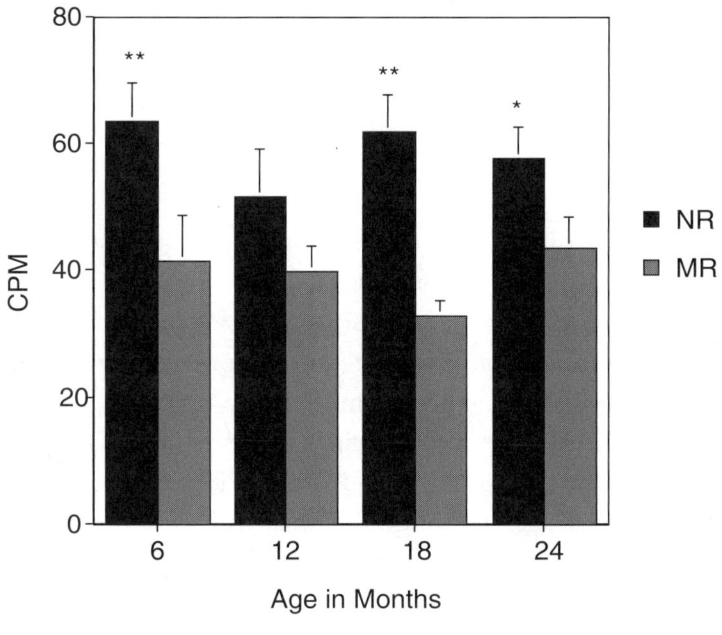

Figure 9–3. Lymphocyte proliferation responses (mean ±SE) to pokeweed mitogen (PWM) (1 μg/ml) for NR and MR monkeys over the first 2 years of life. NR monkeys had significantly higher ratios at 6, 18, and 24 months (*$p < 0.05$, **$p < 0.01$). (Adapted from Lubach *et al.*, 1995.)

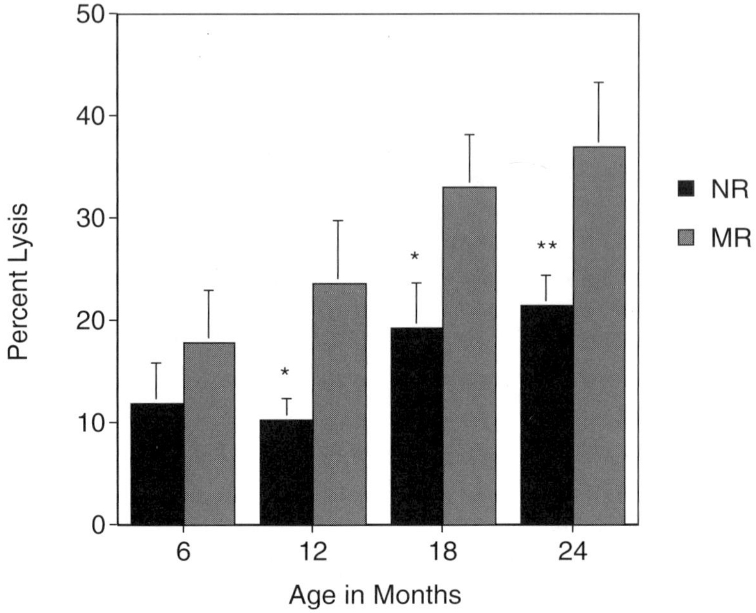

Figure 9–4. Natural killer (NK) cell cytotoxicity (mean ±SE) in NR and MR monkeys over the first 2 years of life. NR monkeys had significantly lower lytic activity at 12, 18, and 24 months, at effector: target ratios of 33:1 ($*p < 0.05$, $**p < 0.01$). (Adapted from Lubach *et al.*, 1995.)

ued effect of the two rearing conditions, both groups showed a comparable age-related increase in NK activity, indicating that immune responses in both groups were maturing.

Although we found that these aspects of immunity were distinctive, it should be noted that others were similar in the two groups of monkeys. Antibody responses to the three vaccinations were not different between groups. In the case of diphtheria and tetanus, it may reflect the efficacy of both immunizations, which are designed to elicit antibody responses even in children that may be somewhat immune compromised. But both MR and NR infant monkeys also showed comparable responses to flu vaccinations, which were not very potent in the young infant. Antibody responses to flu were low in both groups of infants following the initial, primary vaccination, and then both groups responded more robustly to a secondary, booster vaccination for influenza.

The heightened proliferation responses, lower NK activity, and increased $CD4^+/CD8^+$ ratios in the NR infants would suggest a greater susceptibility to certain types of infectious disease, or perhaps an altered immunological way of responding to a disease challenge. It is known that primate nurseries frequently have high incidences of *Campylobacter* and *Shigella* infections (Hird *et al.*, 1984; Russell *et al.*, 1987). The NR, but none of the MR, infants in our study were also more susceptible to this type of GI infection with gram-negative bacterial pathogens. Over the 2 years of the study, all eight NR monkeys had to receive antibiotic treatments at least once for GI pathogens. While it is possible that the immune dysregulation contributed to this increased risk, we also have to remain open to the alternative hypothesis, which is that bacterial exposure altered the developing immune system. Pathogens, along with a lack of maternally derived immune factors in breast milk, may have contributed to shifting the monkey off a normal immunological trajectory. As noted earlier, animals raised in germ-free environments have disturbed developmental patterns, and conversely it is known that both animals and humans raised in environments with high allergen or parasite loads have distinctive immune profiles (Holt and Macaubas, 1997; Le Souef *et al.*, 2000).

The possible influence of nursery rearing on disease susceptibility was addressed in a recent survey of pigtailed macaques (*M. nemestrina*) between 2 and 10 years of age (Sackett *et al.*, 2002). NR monkeys were shown to be just as healthy as their MR counterparts, based on colony data records. The variables examined included survival, clinical treatments, and pregnancy outcomes. The authors concluded that if there were significant immune differences, they did not result in increased disease risk, at least through 10 years of age. They argued further that breast milk may not be essential for producing healthy monkeys. Their data are certainly encouraging, but in the absence of direct assessment, it still leaves open the possibility of different set points for specific immune parameters, and perhaps a different response when challenged in a study of responses to pathogens. Although surveys of clinical records do provide valuable information, certain conditions may be overlooked by veterinary and care staff. For example, in a recent study investigating the development of GI flora in infant rhesus macaques, we found that many infants had asymptomatic infections with *Shigella*, which became evident only when fecal cultures were obtained (Bailey *et al.*, 2004). In a further examination of this issue, subclinical *Shigella* became manifest and symptomatic only during later periods of challenge, such as during our routine husbandry practice of weaning older infants from their mothers at 6–8 months of age (Bailey and Coe, 1999). Three days after separation from the mother, and rehousing with peers, infant rhesus macaques show a marked decrease in lactobacilli, which normally provide a protective function, and experience the onset of diarrheic symptoms, revealing the inherent risk of this common laboratory practice.

4. IMMUNE MODULATORS IN BREAST MILK

Much research has been done on the benefits of breast milk beyond just supplying nutrients (Losonsky and Ogra, 1981; Woodward and Draper, 2001). Breast milk contains many bioactive agents, of which most are active in the GI tract (Duffy, 2000; Kelleher *et al.*, 2003). These agents can affect the motility and pH of the gut as well as the overall integrity of the GI tract. Goldman (2000, 2002) has organized these bioactive agents into five groups: hormones (e.g., insulin, cortisol, growth hormone); growth factors (e.g., epithelial growth factor, lactoferrin); neuropeptides (vasoactive peptide, substance P); antiinflammatory agents (e.g., antioxidants, antiinflammatory cytokines); and immunomodulatory agents (e.g., prolactin, cytokines, nucleotides).

Based on evidence from human and animal studies, Goldman suggests three possible targets for these agents in breast milk: the epithelium, which controls the absorption of nutrients and acts as a barrier to foreign substances; the enteric nervous system, because the GI tract is heavily innervated by both sympathetic and parasympathetic nerves; and the mucosal immune system, which is extensive and can act either independently or in a coordinated manner with systemic immunity.

As mentioned at the beginning of this chapter, breast feeding allows the immune relationship between mothers and infants to continue past the moment of birth. Thus, the newborn infant's delay in immune function is compensated in part by supportive immunity provided through breast milk (Goldman, 2000). Antibodies in breast milk (secretory IgA) provide protection against pathogens endemic in nature or in a laboratory colony (Cleary *et al.*, 1989; Dewey *et al.*, 1995).

Breast milk also contains a variety of immune modulators that appear to initiate or mediate the infant's own immune responses. These substances include several cytokines that may help to guide inflammatory reactions, as well as a number of different leukocytes whose functions remain to be discerned (Goldman, 2000; Jones and Warner, 2000). It has been speculated that the memory T cells from the mother may contribute to the infant's cellular repertoire, perhaps even transferring across the gut and reaching the infant's thymus, but this topic needs further exploration in primates. There have been reports of larger thymuses in BF human infants, but the significance is unknown (Hasselbalch *et al.*, 1996, 1999; Jeppesen *et al.*, 2003). In fact, young monkeys and children have exraordinarily large thymuses relative to the adult; the thymus decreases in size with age, so the meaning of a difference in size would have to be considered carefully with respect to the normal maturational pattern. For example, a difference in size could also be interpreted as a reflection of a different rate of thymic regression, rather than as an improvement in T cell function. Moreover, a number of stress-related processes, especially the adrenal hormone cortisol, can cause marked and rapid decreases in thymic size, even within a few days (Kelley *et al.*, 1984).

5. BREAST MILK AND GUT MATURATION

From birth on, both helpful and harmful bacteria affect gut maturation and associated immune responses (Hooper *et al.*, 2001). Intestinal flora mature in fairly specific patterns in most mammals, including human infants (Lejeune *et al.*, 1984), and this can even influence patterns of gene expression in the intestine (Hooper *et al.*, 1999). It is amazing to think about the fact that an infant is born from a sterile environment with no bacteria in its gut, but within a day will have millions of bacteria (many entering even during the first swallows as the infant is being delivered). Within a few days after the infant starts to nurse, *Lactobacillus*, a strain of acid-preferring bacteria in breast milk that competes against other harmful bacteria can comprise up to 99% of the total intestinal microflora, and remains predominant in the intestine as long as breast feeding continues (Mata and Wyatt, 1971; Packard, 1982).

The pattern of microfloral colonization of the rhesus infant gut has been studied in detail in our laboratory (Bailey *et al.*, 2004). Measures of intestinal bacteria were determined from stool cultures, generated from rectal swabs, over the first 4 months of life. As in human infants, the initial intestinal colonization by aerobic bacteria is gradually replaced by anaerobic bacteria. Two types of protective bacteria, lactobacilli and bifidobacteria, predominate during the nursing phase, and then shift toward the adult profile after the infant is weaned and begins to include more solid food in its diet (Fig. 9–5).

An interesting aspect to this study was the finding that infants born to mothers that had been part of a study investigating the influence of

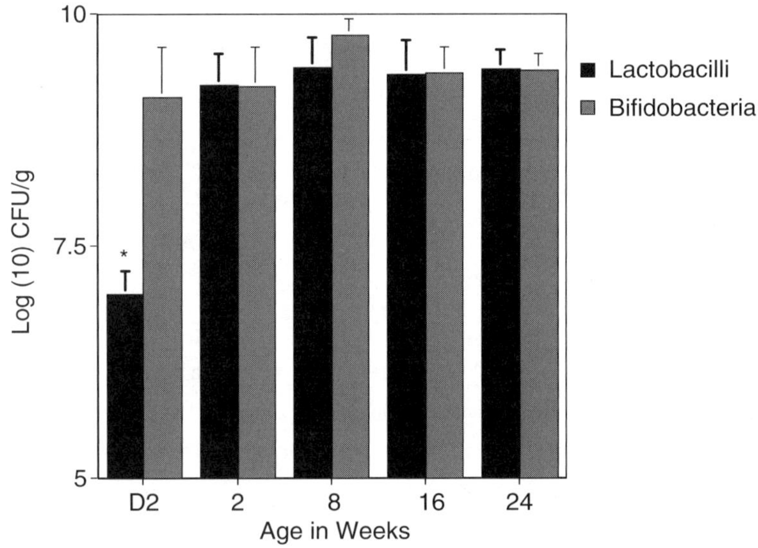

Figure 9–5. Anaerobic *Lactobacillus* spp. and *Bifidobacterium* spp. during the first 24 weeks of life. Data are the mean (±SE) of $\log_{10}$ transformed number of colony-forming units per gram of fecal matter (CFU/g). There was a developmental trend for increasing concentrations of lactobacilli, with day 2 of life significantly lower (*$p < 0.05$). (Adapted from Bailey *et al.*, 2004.)

gestational disturbance were slower to develop high levels of lactobacilli and bifidobacteria at 2 and 8 weeks of age, and never attained the same concentrations as the infants from undisturbed pregnancies. These infants also appeared to be more vulnerable to colonization by *Shigella flexneri*, a gram-negative pathogen. About one-third of the infants from the prenatal-disturbance condition exhibited asymptomatic *S. flexneri*, whereas none of the control infants did across the first 6 months of life. This finding is in agreement with other studies showing that disruption of the normal pattern of intestinal colonization and maturation can affect the risk for infection with GI pathogens (Stevens *et al.*, 1984; Schiffrin *et al.*, 1993; Bailey and Coe, 1999). Other studies in rodents also have shown that periods of acute stress, as well as corticosteroid hormones, can influence the capacity of bacterial pathogens to translocate across the gut wall and cause more serious infections within the body compartment. This type of vulnerability could be a potential concern in NR monkeys as well as in monkeys that may become immunocompromised in the course of an experimental protocol.

We believe that these findings in monkeys may have some bearing on children. It has been shown that FF human infants are slower to develop the microfloral colonization normally seen in BF infants, and likely never maintain the same levels of *Bifidobacterium*, which is promoted by substances in breast milk (Kleessen *et al.*, 1995). Thus, not only are the patterns different, but the number of beneficial species are also reduced. Furthermore, the fecal pH of FF infants is significantly higher, and therefore the shift in acidity may create an internal milieu more favorable to the growth of pathogenic organisms. Some attempts have been made to improve infant formulas by adding substances found only in breast milk (Heird, 2001). In one such study, formula was supplemented with lactobacilli and then tested in rhesus macaque infants that were challenged with *Escherichia coli* (Kelleher *et al.*, 2002). Infants given the lactobacilli had less severe diarrhea and resolved the infection much more quickly. Given the information above, it would not be surprising to find that intestinal maturation was delayed or altered in NR primate infants. A study specifically focused on gut pathogens and mucosal immunity in MR and NR monkeys is warranted.

6. TH1/TH2

Episodes of GI illness and exposure to other environmental antigens could predispose infants to accelerate maturation of their immune system (Granot *et al.*, 1999). Responses to antigens are mediated through cytokines, or hormone-like immune modulators that can affect the activities of several types of cells (Glimcher and Murphy, 2000). Helper T cells (Th), also known as $CD4^+$ cells, produce a specific profile of cytokines, which is also dependent on the type of stimulus. This realization has led to division of T cell responses into two types, labeled either Th1 or Th2 (reflecting both the predominant types of cytokines and the effect on other aspects of immunity). The Th1 cytokines [IL-2, IL-12, interferon-γ (IFN-γ)] promote cellular immunity, macrophage activation, IgG synthesis, and antiviral activity. In contrast, the Th2 cytokines (IL-4, IL-5, IL-10) preferentially predispose the individual toward a humoral immune response, promoting B cell replication, IgE production, and an eosinophilia (Folkerts *et al.*, 2000). It is of interest that a bias toward Th2 predominance is found in children with allergies and asthma (Folkerts *et al.*, 2000; Zhang *et al.*, 2000), which also tends to be the more immature response (Prescott *et al.*, 1998).

Human children manifest a predominant Th2 response until about 12 months of age, when Th1 responses start to become more dominant (Kovarik and Siegrist, 1998; Vigano *et al.*, 1999; Ganschow *et al.*, 2001). The neonatal Th2 polarization is similar to what is found in the uterine environment, in which Th1 responses are suppressed to protect the placenta and fetus from maternal rejection (Rowe *et al.*, 2000). In fact, even the overall immune bias of the adult female is pushed somewhat in this direction toward the latter months of pregnancy. Fortunately, a predominant Th2 response does not undermine the response to immunization, which is designed to elicit a humoral or antibody response. Moreover, examination of the temporal course of the infant reaction to a vaccine reveals this predilection. In infancy, cytokine responses to DPT vaccinations are initially more Th2-like and show a delayed appearance of the Th1 response, especially in the production of IFN-γ (Rowe *et al.*, 2000). On the other hand, recurrent infections (Illi *et al.*, 2001), as well as other vaccinations in infancy, may promote a shift toward more of a

Th1 response (Johnston and Openshaw, 2001). It has been shown that even subclinical levels of some microbes stimulate cells to produce IL-12, which is important in the deviation toward the Th1 phenotype (Matricardi *et al.*, 2000). There are other instances where a Th2 phenotype can become dominant, and this has led some to postulate that the level of hygiene in modern society could be having some inadvertent and maladaptive consequences. This "hygiene hypothesis" suggests that the reduction of childhood infections, the overuse of antibiotics that wipe out normal GI tract organisms, and an antiseptic environment may in fact be contributing to the dramatic rise in the prevalence of allergies and asthma worldwide (Folkerts *et al.*, 2000; Matricardi *et al.*, 2000). A continued bias toward a Th2 response in the child could lead to atopy and then asthma. Lactoferrin, an antimicrobial agent found in breast milk, may help to down-regulate this response (Bernt and Walker, 1999). Davidson and Lönnerdal (1986) found lactoferrin concentrations to be quite similar in human and rhesus macaque milk. Additionally, a soluble cytokine receptor, CD30, which is present on activated $CD4^+$ and $CD8^+$ T cells, may also aid in the down-regulation of some types of immune responses. It is of particular interest that high levels of CD30 are found in colostrum, suggesting that they serve a role in how the neonate begins to respond to the antigenic aspects of the environment during the first day of life (Bertotto *et al.*, 1997). The significance of these immune-related products in milk and the long-term impact on the gut and on systemic immunity are important areas for further research in primates.

7. TEMPERATURE AND ENTRAINMENT

Maintaining immunity is energetically expensive, and mounting an immune response can be even more demanding. Paradoxically, during the initial response to antigen challenge and infection the host is sometimes driven by the surging cytokines to reduce food intake as well as other behavioral activities that might require energy expenditure. It is of particular interest that energy regulation in response to vaccination in infancy has been shown to differ in FF and BF human infants (Lopez-Alarcon *et al.*, 2002). FF children had significantly decreased energy intakes over the first 48 hr after immunization. The exact mechanism for this protective effect of breast milk is unknown, but it may be related to cytokines or cytokine receptors in the milk. Our laboratory has not examined NR monkeys' responses to cytokines, but we have conducted a number of studies on how monkeys generated from disturbed pregnancies respond to bacterial proteins and proinflammatory cytokines such as IL-1. In general, a history of prenatal disturbances results in a juvenile monkey that shows a smaller response than found in the ones born after undisturbed pregnancies (Reyes and Coe, 1997). By extension, it would be worthwhile to investigate the pyrogenic response of NR primate infants. It is likely that the differences in thermoregulation created by maternal versus nursery rearing right after birth set in motion processes that continue to persist through infancy. The normal pattern of development for the infant macaque, and for most other monkeys, is for the infant to remain on the mother's ventrum for the first several weeks of life. Even after it slowly starts to explore the environment, the infant returns to the mother for suckling, resting, and sleeping. Thus, for several months the infant receives not only nourishment, comfort, and security from the mother, but also warmth. These reciprocal mother–infant interactions evidently entrain the infant to the mother's activity and temperature rhythms. Most NR infants are assisted with thermal regulation during the first 7–14 days of life, either in an incubator or with a heating pad. Thereafter, the infant is thought to be capable of independent thermoregulation, although we have evidence that the cyclicity is altered from the normal pattern.

Studies of rat pups have shown the importance of the mother's body heat for many aspects of mother–infant behavior, such as grooming and suckling (Alberts and Brunjes, 1978; Alberts and May, 1984). Early separation from the dam can have long-lasting effects on the pup's sympathetic nervous system, altering energy metabolism, which, in turn, can affect temperature regulation (Hofer, 1983; Kuhn, 1990). Such thermal challenges early in development can affect other aspects of physiology, including the immune and GI systems (Ackerman *et al.*, 1978; Kelley, 1980; Stone *et al.*, 1976).

Our data on rhesus macaques were obtained noninvasively via small biotelemetry transmitters. Between 2 and 4 weeks of age, all NR and MR infants were surgically implanted with a small biotelemetry transmitter

(2 cm × 8 mm disk, weight: 5 g) (Mini-Mitter, Inc., Sunriver, OR). These transmitters measured body core temperature and locomotor activity for 24 hr/day, up to 6 months in some infants. The antennas were enclosed in PVC pipes that were mounted on the inside of the home cages. Room temperature was monitored, and all activity in the rooms was recorded. Both groups were on the same light/dark schedule. Group means were compared for 7-day periods at 6, 10, and 14 weeks of age.

The data quickly revealed that NR infants had lower nocturnal temperatures than MR infants even with a supplemental heat source (Lubach *et al.*, 1992). More importantly, there were differences in the phase and shape of the diurnal temperature rhythm. Although all infants had well-developed circadian rhythms in temperature, the pattern of those rhythms differed significantly throughout the 14 weeks. NR infants reached their peak daytime temperatures in the afternoon, whereas the MR infants peaked in the early morning and maintained that temperature during the day. Activity levels did not differ between groups, but the phasing of the activity levels reflected the phasing of the body temperatures in the two groups (Table 9–1). Additional data from a subset

Table 9–1. Body Temperature and Activity of NR and MR Monkeys at 6, 10, and 14 Weeks of Age

	Temperature[a]		Activity[a]	
	PH	PL	PH	PL
6 Weeks				
NR	149[b]	260	134[b]	258
MR	86[b]	223	97[b]	243
10 Weeks				
NR	128[b]	248	130[b]	266[c]
MR	75[b]	237	75[b]	242[c]
14 Weeks				
NR	139[b]	254	131[b]	261
MR	59[b]	244	78[b]	240

[a] Data were computed from 12 samples per hour for 24 hr. PH, phase of the highest hourly mean; PL, phase of the lowest hourly mean; (0° = lights on and 180° = lights off).
[b] $p < 0.01$.
[c] $p < 0.05$.

of infants indicated that the phase shift continued to 8.5 months of age in the NR infants.

The data suggest that the monkey mother does have an important influence on the infant's body temperature regulation and phasing. NR infants may experience a thermal challenge during the night, especially during the first 10 weeks of life, as indicated by their lower body temperatures. Studies with human infants have shown that FF infants have a higher metabolic rate at night, along with increased energy expenditure (Butte *et al.*, 1990). In this case, what is operational is not the effect of direct maternal contact, but the way the milk is being used by the body. With a higher metabolic demand, it may be that NR infants divert energy away from the maturation or maintenance of other systems that require large amounts of energy, such as the immune system.

Differences in body temperature were not as great as the differences in the phasing of temperature and activity. One reason for the phase shift in the NR infants may be the loss of cues that the mother would normally provide (Reite *et al.*, 1982). Feeding schedules and light:dark cycles are also known to entrain rhythms (Sulzman *et al.*, 1977; Aschoff and von Goetz, 1986; Hiroshige and Honma, 1987). Since the lights were on a similar cycle for the MR and NR monkeys, it is most likely that the NR monkeys became entrained to the bottle-feeding schedules, and that this entrainment persisted well beyond the age when they were getting bottles. Whether the differences presented here affected any aspects of immunity is unknown, but the possibility exists because the immune system also has a diurnal rhythm (Smolensky *et al.*, 1980; Abo *et al.*, 1981; Terao *et al.*, 2002).

8. CONCLUSIONS

It is clear that raising infant primates in a nursery, in the absence of breast milk and the warmth and biorhythms of the mother, has multiple impacts on physiological development. Although surveys may indicate that there are no long-term differences in health outcomes, at least in the pigtail macaque (Sackett *et al.*, 2002), it is very evident that there are numerous differences in the baseline physiology of monkeys raised this way. The differences span stress-related hormones of the endocrine system (growth hormone and cortisol), neurotransmitter systems of the brain (including the important monoamines), and a number of significant aspects of the immune system.

The differences we have found in primate infants raised under different rearing conditions concur with observations on laboratory rodents as well as farm animals. Both age at weaning and acute disturbances such as of mother–infant separations have been found to affect many immune responses. Thus, it is not surprising that a drastic deviation from the normal rearing provided by the biological mother should have a sustained effect on immune responses. Indeed, it is possible to express it in the converse manner and comment on the resiliency of the primate infant that it fares so well when reared in a way that deviates so far from the species-typical and "biologically expected" mode of care. In part, this speaks to the care and creativity of the human caregivers in recreating the essential aspects of maternal rearing, including high-quality formula, thermal supplementation, and environmental stimulation. In current studies of nursery rearing, in contrast to early investigations designed to assess the effects of minimal stimulation, we carefully avoid any types of environmental deprivation known to have adverse consequences for animal health and well-being. In the study reported here, for example, infants were routinely provided with peer interactions to help promote behavioral normalcy.

These findings from monkeys raised in different conditions may also have some relevance for children. In human infants, there is evidence of differences between BF and FF infants in several aspects of immunity. We have also looked at some types of immune responses in children raised in orphanages, and the data suggest both more infections and altered immune responses. Even with more benign variations in rearing, there appear to be important consequences of environmental hygiene and allergen exposure. Indeed, many immune responses are designed to handle a heavy parasite load, and in the absence of this "normal" environmental challenge, the body develops differently. We believe that the NR environment creates its own unique influence and challenge for the NR infant, which in turn influences some aspects of immune maturation at least temporarily and often permanently. We have found that lymphocytes from NR rhesus macaques are hyperresponsive to stimulation *in vitro*, whereas NK cells seem relatively deficient. The long-lasting effects of these responses, and the differences in T cell subsets, suggest there may be other, more permanent, alterations in immunity, especially if the monkey was evaluated in the context of a virulent infection (Capitanio and Lerche, 1991). At a minimum, some of these effects seem to indicate a strong role for breast milk in the development of infant immunity, including the many antimicrobial and antiinflammatory factors, and components for energy regulation. Differences in gut maturation, pathogen exposure, temperature regulation, biorhythms, and energetics may synergize to have cumulative effects on infant immunity.

If there is a resurgence of nursery rearing to meet the needs of certain research endeavors, it will be of value to explore these physiological processes in greater detail. It will be important to understand the underlying mechanisms, in order to devise appropriate strategies to minimize the dysfunction and to better mimic the normalcy afforded by maternal care. Tracking immunological development has become even easier with new methods and new "primate-appropriate" reagents, decreasing the need to rely on products originally developed for human cells (Sopper *et al.*, 1997; Bleavins and de la Iglesia, 2000; Rosenberg and Reimann, 2000; Yoshino *et al.*, 2000) (also see the Appendix). It is also possible to use smaller amounts of blood, both to follow the development of T cell subsets and to assess the response to antigen challenge and cytokine biology. Cytokines are especially important for investigators conducting viral studies, and there are still very few data on the development of these immune modulators in primates (Abel *et al.*, 2001). In fact, given that there are more than 200 species of primates, it is remarkable how little is known about most parameters in either NR or MR animals, beyond the macaque species. Indeed, we know almost nothing about the immunity of primates reared in nature, where they would be exposed to a much higher pathogen load than any animal maintained in captivity. To truly appreciate the impact of nursery rearing on immune development, ideally we need to know much more about the "natural" immune trajectory of infant primates (Le Souef *et al.*, 2000).

Infancy is an especially vulnerable period, when the developing physiological systems rely on the environment for stimulation and guidance. This is as true for immunity as it is for the brain. If the development of the immune system depends on extraimmune influences, both from the mother and the physical environment, the nature of an infant's rearing experience will de facto have important consequences for health and well-being (Ader, 1983).

ACKNOWLEDGMENTS

This research was supported in part by NSF grant BNS-8713940 (G.R.L.), NIMH grant MH41659, and NIH grant AI46521 (C.L.C.), ONR Contract N00014-87-0227 (C.L.C., W.B.E.), NIMH grant MH30915 (E.M.W.K.), NRSA fellowship MH12458 (M.T.B.), and the Wisconsin Primate Research Center (5P51-RR000167).

9. APPENDIX

The following companies represent a partial list of sources for primate-specific, or cross-reacting, reagents.

These and other companies usually have methods and cross-reactivity charts available on-line and in their catalogs.

We have also included, below, a website that is not company specific. It contains a database of monoclonal antibodies that cross-react with a variety of Old-World and New-World primate species: http://nhprimate.bidmc.harvard.edu/NHP/

Companies

BD Biosciences
Pharmingen
10975 Torreyana Rd.
San Diego, CA 92121-1106
877-232-8995
www.bdbiosciences.com

Biosource
542 Flynn Road
Camarillo, CA 93012

800-242-0607
www.biosource.com

Caltag Laboratories
1849 Bayshore Boulevard, #200
Burlingame, CA 94010
800-874-4007
www.caltag.com

Mabtech, Inc.
Mariemont Executive Building
Suite 112
3814 West Street
Mariemont, OH 45227
866-354-7768
www.mabtech.com
(ELISA and ELISPOT)

R&D Systems, Inc.
614 McKinley Place N.E.
Minneapolis, MN 55413
800-343-7475
www.rndsystems.com
(ELISA and ELISPOT)

Serotec Inc.
3200 Atlantic Ave.
Suite 105
Raleigh, NC 27604
800-265-7376
www.serotec.com

U-Cytech, Biosciences
Yalelaan 48
3584 CM Utrecht
The Netherlands
31.30.253.5960
www.ucytech.com
(ELISA and ELISPOT)

REFERENCES

Abel, K., Alegeria-Hartman, M.J., Zanotto, K., McChesney, M.B., Marthas, M.L., and Miller, C.J. (2001). Anatomic site and immune function correlate with relative cytokine mRNA expression levels in lymphoid tissues of normal macaques. *Cytokine* 16:191–204.

Abo, T., Kawate, T., Itoh, K., and Kumagai, K. (1981). Circadian rhythms of human T, B, and K cell traffic in the peripheral blood. *J. Immunol.* 126:1360–1363.

Ackerman, S.H., Hofer, M.A., and Wiener, H. (1978). Early maternal separation increases gastric ulcer risk in rats by producing a latent thermoregulatory disturbance. *Science* 201:373–376.

Ader, R. (1983). Developmental psychoneuroimmunology. *Dev. Psychobiol.* 16:251–267.

Alberts, J.R., and Brunjes, P.C. (1978). Ontogeny of thermal and olfactory determinants of huddling in the rat. *J. Comp. Physiol. Psychol.* 92:897–906.

Alberts, J.R., and May, B. (1984). Non-nutritive, thermotactile induction of olfactory preferences for huddling in rat pups. *Dev. Psychobiol.* 17:161–181.

Anderson, J.H. (1986). Rearing and intensive care of neonatal and infant nonhuman primates. In: Benirschke, K. (ed.), *Primates: The Road to Self-Sustaining Populations.* Springer-Verlag, New York, pp. 747–762.

Aschoff, J., and von Goetz, C. (1986). Effects of feeding cycles on circadian rhythms in squirrel monkeys. *J. Biol. Rhythms* 1:267–276.

Bailey, M.T., and Coe, C.L. (1999). Maternal separation disrupts the integrity of the intestinal microflora in infant rhesus monkeys. *Dev. Psychobiol.* 35:146–155.

Bailey, M.T., Lubach, G.R., and Coe, C.L. (2004). Prenatal stress alters bacterial colonization of the gut in infant monkeys. *J. Pediatr. Gastroenterol. Nutr.* 38:414–421.

Baroncelli, S., Panzini, G., Geraci, A., Pardini, S., Corrias, F., Iale, E., Varano, F., Turillazzi, P.G., Titti, F., and Verani, P. (1997). Longitudinal characterization of CD, CD8 T-cell subsets and of haematological parameters in healthy newborns of cynomolgus monkeys. *Vet. Immunol. Immunopathol.* 59:141–150.

Bernt, K.M., and Walker, W.A. (1999). Human milk as a carrier of biochemical messages. *Acta Paediatr. Suppl.* 430:27–41.

Bertotto, A., Vagliasindi, C., Gerli, R., Spinozzi, F., Castellucci, G., Fabietti, G., Crupi, S., Radicioni, M., Cozzali, R., Ferraro, L., Niccoli, A., De Rosa, O., Lupi, C., Brunelli, R., Merluzzi, A., Guadalupi, D., Prente, C., Pertici, L.,

Serra, M., and Vaccaro, R. (1997). Soluble CD30 antigen in human colostrum. *Biol. Neonate* 71:69–74.

Bleavins, M., and de la Iglesia, F.A. (2000). Preclinical immune function testing. An approach using the cynomolgus monkey (*Macaca fascicularis*). In: Korte, R., and Weinbauer, G.F. (eds.), *Towards New Horizons in Primate Toxicology*. Waxmann Munster, New York, pp. 127–141.

Blomquist, A.J., and Harlow, H.F. (1961). The infant rhesus monkey program at the University of Wisconsin Primate Laboratory. *Proc. Anim. Care Panel* 11:57–64.

Burbacher, T.M., and Grant, K.S. (2000). Methods for studying nonhuman primates in neurobehavioral toxicology and teratology. *Nerotoxicol. Teratol.* 22:475–486.

Buse, E., Habermann, G., Osterburg, I., Korte, R., and Weinbauer, G.F. (2003). Reproductive/developmental toxicity and immunotoxicity assessment in the nonhuman primate model. *Toxicology* 221:221–227.

Butte, N.F., Wong, W.W., Ferlic, L., Smith, E.O., Klein, P.D., and Garza, C. (1990). Energy expenditure and deposition of breast-fed and formula-fed infants during early infancy. *Pediatr. Res.* 28:631–640.

Capitanio, J.P. (1985). Early experience and social processes in rhesus macaques (Macaca mulatta): II. Complex social interaction. *J. Comp. Psychol.* 99: 133–144.

Capitanio, J.P., and Lerche, N.W. (1991). Psychosocial factors and disease progression in simian AIDS: A preliminary report. *AIDS* 5:1103–1106.

Cauchi, M.N. (1981). *Obstetric and Perinatal Immunology*. Edward Arnold Publishers, London.

Champoux, M., Coe, C.L., Schanberg, S.M., Kuhn, C.M., and Suomi, S.J. (1989). Hormonal effects of early rearing conditions in the infant rhesus monkey. *Am. J. Primatol.* 19:111–117.

Cleary, T.G., Winsor, D.K., Reich, D., Ruiz-Palacios, G., and Calva, J.J. (1989). Human milk immunoglobulin A antibodies to *Shigella* virulence determinants. *Infect. Immun.* 57:1675–1679.

Coe, C.L., and Erickson, C.M. (1997). Stress decreases lymphocyte cytolytic activity in the young monkey even after blockade of steroid and opiate hormone receptors. *Dev. Psychobiol.* 30:1–10.

Coe, C.L., Lubach, G.R., Ershler, W.B., and Klopp, R.G. (1989). Influence of early rearing on lymphocyte proliferation responses in juvenile rhesus monkeys. *Brain Behav. Immun.* 3:47–60.

Coe, C.L., Lubach, G.R., Schneider, M.L., Dierschke, D.J., and Ershler, W.B. (1992). Early rearing conditions alter immune responses in developing infant primates. *Pediatrics* 90:505–509.

Coe, C.L., Lubach, G.R., and Izard, K.M. (1994). Progressive improvement in the transfer of maternal antibody across the Order Primates. *Am. J. Primatol.* 32:51–55.

Dahlgren, U.I., Hanson, L.A., and Telemo, E. (2001). Maturation of immunocompetence in breast-fed vs. formula-fed infants. In: Woodward, B., and Draper, H.H. (eds.), *Advances in Nutritional Research*, Vol. 10. Kluwer Academic/Plenum Publishers, New York, pp. 311–325.

Davidson, L.A., and Lonnerdal, B. (1986). Isolation and characterization of rhesus monkey lactoferritin. *Pediatr. Res.* 20:197–201.

DeMaria, M.A., Casto, M., O'Connell, M., Johnson, R.P., and Rosenzweig, M. (2000). Characterization of lymphocyte subsets in rhesus macaques during the first year of life. *Eur. J. Haematol.* 65:245–257.

de Vries, E., de Bruin-Versteeg, S., Comans-Bitter, W.M., de Groot, R., Hop, W.C.J., Boerma, G.J.M., Lotgering, F.K., and van Dongen, J.J.M. (2000). Longitudinal survey of lymphocyte subpopulations in the first year of life. *Pediatr. Res.* 47:528–537.

Dewey, K., Heinig, M.J., and Nommsen-Rivers, L.A. (1995). Differences in morbidity between breast-fed and formula-fed infants. *J. Pediatr.* 126: 696–702.

Duffy, L. (2000). Interactions mediating bacterial translocation in the immature intestine. *J. Nutr.* 130:432S–436S.

Dykhuizen, M., Ceman, J., Mitchen, J., Zayas, M., MacDougall, A., Helgeland, J., Rakasz, E., and Pauza, C.D. (2000). Importance of the CD3 marker for evaluating changes in rhesus macaque CD4/CD8 T-cell ratios. *Cytometry* 40:69–75.

Ershler, W.B., Coe, C.L., Gravenstein, S., Schultz, K.T., Klopp, R.G., Meyer, M., and Houser, W.D. (1988). Aging and immunity in nonhuman primates: I. Effects of age and gender on cellular immune function in rhesus monkeys (*Macaca mulatta*). *Am. J. Primatol.* 15:181–188.

Foerster, M., Delgado, I., Abraham, K., Gerstmayr, S., and Neubert, R. (1997). Comparative study on age-dependent development of surface receptors on peripheral blood lymphocytes in children and young nonhuman primates (marmosets). *Life Sci.* 60:773–785.

Folkerts, G., Walzl, G., and Openshaw, J.M. (2000). Do common childhood infections 'teach' the immune system not to be allergic? *Immunol. Today* 21:118–120.

Ganschow, R., Broering, D.C., Nolkemper, D., Albani, J., Kemper, M.J., Rogiers, X., and Burdelski, M. (2001). Th2 cytokine profile in infants predisposed to improved graft acceptance after liver transplantation. *Transplantation* 72:929–934.

Glimcher, L.H., and Murphy, K.M. (2000). Lineage commitment in the immune system: The T helper lymphocyte grows up. *Genes Dev.* 14:1693–1711.

Goldman, A.S. (2000). Modulation of the gastrointestinal tract of infants by human milk. Interfaces and interactions. An evolutionary perspective. *J. Nutr.* 130:426S–431S.

Goldman, A.S. (2002). Evolution of the mammary gland defense system and the ontogeny of the immune system. *J. Mammary Gland Biol. Neoplasia* 7:277–289.

Goldman, A.S., Chheda, S., and Garofalo, R. (1998). Evolution of immunologic functions of the mammary gland and the postnatal development of immunity. *Pediatr. Res.* 43:155–162.

Granot, E., Rabinowitz, R., and Schlesinger, M. (1999). Lymphocyte subset profile of young healthy children residing in a rural area: Possible role of recurrent gastrointestinal infections. *J. Pediatr. Gastroenterol. Nutr.* 28:147–151.

Gust, D.A., Gordon, T.P., Wilson, M.E., Ahmed-Ansari, A., Brodie, A.R., and McClure, H.M. (1991). Formation of a new social group of unfamiliar rhesus monkeys affects the immune and pituitary adrenocortical systems. *Brain Behav. Immun.* 5:296–307.

Hannet, I., Erkeller-Yuksel, F., Lydyard, P., Denys, V., and DeBruyere, M. (1992). Developmental and maturational changes in human blood lymphocyte subpopulations. *Immunol. Today.* 13:215–218.

Hanson, L.A. (1998). Breastfeeding provides passive and likely long-lasting active immunity. *Ann. Allergy Asthma Immunol.* 81:523–533.

Harlow, H.F., and Harlow, M.K. (1969). Effects of various mother-infant relationships on rhesus monkey behaviors. In: Foss, B.M. (ed.), *Determinants of Infant Behavior*, Vol. 4. Methuen, London, pp.15–36.

Hasselbalch, H., Jeppesen, D.L., Engelmann, M.D., Michaelsen, K.F., and Nielsen, M.B. (1996). Decreased thymus size in formula-fed infants compared with breastfed infants. *Acta Paediatr.* 85:1029–1032.

Hasselbalch, H., Engelmann, M.D., Ersboll, A.K., Jeppesen, D.L., and Fleischer-Michaelsen, K. (1999). Breast-feeding influences thymic size in late infancy. *Eur. J. Pediatr.* 158:964–967.

Hayward, A.R., and Kurnick, J. (1981). Newborn T cell suppression: Early appearance, maintenance in culture, and lack of growth factor suppression. *J. Immunol.* 126:50–53.

Heird, W.C. (2001). The role of polyunsaturated fatty acids in term and preterm infants and breastfeeding mothers. *Pediatr. Clin. North Am.* 48:173–188.

Hird, D.W., Anderson, J.H., and Bielitzki, J.B. (1984). Diarrhea in nonhuman primates: A survey of primate colonies for incidence rates and clinical opinion. *Lab. Anim. Sci.* 34:465–470.

Hiroshige, T., and Honma, K. (1987). *Comparative Aspects of Circadian Clocks.* Hokkaido University Press, Sapporo, Japan.

Hofer, M.A. (1983). The mother-infant interaction as a regulator of infant physiology and behavior. In: Rosenblum, L.A., and Moltz, H. (eds.), *Symbiosis in Parent-Offspring Interactions.* Plenum Press, New York, pp. 61–75.

Holt, P.G., and Macaubas, C. (1997). Development of long term tolerance versus sensitization to environmental allergens during the perinatal period. *Curr. Opin. Immunol.* 9:782–787.

Hooper, L.V., Xu, J., Falk, P.G., Midtvedt, T., and Gordon, J.I. (1999). A molecular sensor that allows a gut commensal to control its nutrient foundation in a competitive ecosystem. *Proc. Natl. Acad. Sci. USA* 96:9833–9838.

Hooper, L.V., Wong, M.H., Thelin, A., Hansson, L., Falk, P.G., and Gordon, J.I. (2001). Molecular analysis of commensal host-microbial relationship in the intestine. *Science* 291:881–884.

Illi, S., von Mutius, E., Lau, S., Bergmann, R., Niggemann, B., Sommerfeld, C., Wahn, C., and MAS Group. (2001). Early childhood infectious diseases and the development of asthma up to school age: A birth cohort study. *BMJ* 322:390–395.

Jeppesen, D., Hasselbalch, H., Ersboll, A.K., Heilmann, C., and Valerius, N.H. (2003). Thymic size in uninfected infants born to HIV-positive mothers and fed with pasteurized human milk. *Acta Paediatr.* 92:679–683.

Johnston, S.L., and Openshaw, P.J.M. (2001). The protective effect of childhood infections. *BMJ* 322:376–377.

Jones, C.A., and Warner, J.O. (2000). Breast milk as an alternative source of cytokines for offspring. *Clin. Exp. Allergy* 30:599–601.

Juto, P., Moller, C., Engberg, S., and Bjorksten, B. (1982). Influence of type of feeding on lymphocyte function and development of infantile allergy. *Clin. Allergy* 12:409–416.

Kelleher, S., and Lonnerdal, B. (2001). Immunological activities associated with milk. In: Woodward, B., and Draper, H.H. (eds.), *Advances in Nutritional Research,* Vol. 10. Kluwer Academic/Plenum Publishers, New York, pp.39–65.

Kelleher, S., Casas, I., Carbajal, N., and Lonnerdal, B. (2002). Supplementation of infant formula with the probiotic *Lactobacillus reuteri* and zinc: Impact on enteric infection and nutrition in infant rhesus monkeys. *J. Pediatr. Gastroenterol. Nutr.* 35:162–168.

Kelleher, S.L., Chatterton, D., Nielsen, K., and Lonnerdal, B. (2003). Glycomacropeptide and alpha-lactalbumin supplementation of infant formula affects growth and nutritional status in infant rhesus monkeys. *Am. J. Clin. Nutr.* 77:1261–1268.

Kelley, K.W. (1980). Stress and immune function: A bibliographic review. *Ann. Rech. Vet.* 11:445–478.

Kelley, K.W., Mertsching, H.J., and Salmon, H. (1984). Immunity changes in confined animals: A route to disease? *Ann. Rech. Vet.* 15:201–204.

Kleessen, B., Bunke, H., Tovar, K., Noack, J., and Sawatzki, G. (1995). Influence of two infant formulas and human milk on the development of the faecal flora in newborn infants. *Acta Paediatr.* 84:1347–1356.

Klein, R.B., Fisher, T.J., Gard, S.E., Biberstien, M., Rich, K.C., and Stiehm, E.R. (1977). Decreased mononuclear and polymorphonuclear chemotaxis in human newborns, infants, and young children. *Pediatrics* 60:467–472.

Kovarik, J., and Siegrist, C-A. (1998). Immunity in early life. *Immunol. Today* 19:150–152.

Kraemer, G.W., Ebert, M.H., Schmidit, D.E., and McKinney, W.T. (1989). A longitudinal study of the effect of different rearing conditions on cerebrospinal fluid norepinephrine and biogenic amine metabolites in rhesus monkeys. *Neuropsychopharmacology* 2:175–189.

Kuhn, C.M., Park, J., and Schanberg, S.M. (1990). Endocrine responses to mother-infant separation in developing rats. *Dev. Psychobiol.* 23:395–410.

Lejeune, C., Bourrillon, A., Boussougant, Y., and de Paillerets, F. (1984). Sequential development of the intestinal flora in newborn infants: A quantitative differential analysis. *Dev. Pharmacol. Ther.* 7:138–143.

Le Souef, P.N., Goldblat, J., and Lynch, N.R. (2000). Evolutionary adaptation of inflammatory immune responses in human beings. *Lancet* 356:242–244.

Lindberg, T., Engberg, S., Jakobsson, I., and Lonnerdal, B. (1997). Digestion of proteins in human milk, human milk fortifier, and preterm formula in infant rhesus monkeys. *J. Pediatr. Gastroenterol. Nutr.* 24:537–543.

Lonnerdal, B., Keen, C., Glazier, C.E., and Anderson, J. (1984). A longitudinal study of rhesus monkey (*Macaca mulatta*) milk composition: Trace elements, minerals, protein, carbohydrate, and fat. *Pediatr. Res.* 18:911–914.

Lopez-Alarcon, M., Garza, C., Habicht, J-P., Martinez, L., Pegueros, V., and Villalpando, S. (2002). Breastfeeding attenuates reductions in energy intake induced by a mild immunologic stimulus represented by DPTH immunization: Possible roles of interleukin-1beta, tumor necrosis factor-alpha and leptin. *J. Nutr.* 132:1293–1298.

Losonsky, G.A., and Ogra, P.L. (1981). Maternal-neonatal interactions and human breast milk. In: Gleicher, N. (ed.), *Reproductive Immunology.* Alan R. Liss, Inc., New York, pp. 171–182.

Lubach, G.R., Kittrell, E.M., and Coe, C.L. (1992). Maternal influences on body temperature in the infant primate. *Physiol. Behav.* 51:987–994.

Lubach, G.R., Coe, C.L., and Ershler, W.B. (1995). Effects of early rearing environment on immune responses of infant monkeys. *Brain Behav. Immun.* 9:31–46.

Makori, N., Tarantal, A.F., Lu, F.X., Rourke, T., Marthas, M.L., McChesney, M.B., Hendrickx, A.G., and Miller, C.J. (2003). Functional and morphological development of lymphoid tissues and immune regulatory and effector function in rhesus monkeys: Cytokine-secreting cells, immunoglobulin-secreting cells, and CD5+ B-1 cells appear early in fetal development. *Clin. Diagn. Lab. Immunol.* 10:140–153.

Marthas, M.L., Van Rompay, K.K.A., Otsyula, M., Miller, C.J., Canfield, D.R., Pedersen, N.C., and McChesney, M.B. (1995). Viral factors determine progression to AIDS in simian immunodificiency virus-infected newborn rhesus monkeys. *J. Virol.* 69:4198–4205.

Martin, D.P., McGowan, M.J., and Loeb, W.F. (1973). Age related changes of hematologic values in infant *Macaca mulatta. Lab. Anim. Sci.* 23:194–200.

Mason, W.A., Mendoza, S.P., and Moberg, G.P. (1991). Persistent effects of early social experience on physiological responsiveness. In: Ehara, A., Kimura, T., Takenaka, O., and Iwamoto, M. (eds.), *Primatology Today.* Elsevier, New York, pp. 469–471.

Mata, L.J., and Wyatt, R.G. (1971). Host resistance to infection. *Am. J. Clin. Nutr.* 24:976–986.

Matricardi, P.M., Rosmini, F., Riondino, S., Fortini, M., Ferrigano, L., Rapicetta, M., and Bonini, S. (2000). Exposure to foodborn and orofecal microbes versus airborne viruses in relation to atopy and allergic asthma: Epidemiological study. *BMJ* 320:412–417.

Miller, M.E. (1980). The immunodeficiencies of immaturity. In: Steihm, E.R., and Fulginiti, V.A. (eds.), *Immunologic Disorders in Infants and Children.* W.B. Saunders Co., Philadelphia, pp. 219–238.

Murayama, Y., Fukao, K., Noguchi, A., and Takenaka, O. (1986). Epitope expression on primate lymphocyte surface antigens. *J. Med. Primatol.* 15:373–387.

Nam, K.-H., Akari, H., Terao, K., Itagaki, S., and Yoshikawa, Y. (1998). Age-related changes in major lymphocyte subsets in cynomolgus monkeys. *Exp. Anim.* 47:159–166.

Neubert, R., Delgado, I., Abraham, K., Schuster, C., and Helge, H. (1998). Evaluation of the age-dependent development of lymphocyte surface receptors in children. *Life Sci.* 62:1099–1110.

Novak, M.A., and Harlow, H.F. (1975). Social recovery of monkeys isolated for the first year of life: I. Rehabilitation and therapy. *Dev. Psychol.* 11:453–465.

Osburn, B.I., MacLachlan, N.J., and Terrel, T.G. (1982). Ontogeny of the immune system. *J. Am. Vet. Med. Assoc.* 181:1049–1052.

Pabst, H.F., Spady, D.W., Pilarski, L.M., Carson, M.M., Beeler, J.A., and Krezolek, M.P. (1997). Differential modulation of the immune response by breast- or formula-feeding of infants. *Acta Paediatr.* 86:1–7.

Packard, V.S. (1982). *Human Milk and Infant Formula.* Academic Press, New York.

Prescott, S.L., Macaubas, C., Holt, B.J., Smallacombe, T.B., Loh, R., Sly, P.D., and Holt, P.G. (1998). Transplacental priming of the human immune system to environmental allergens: Universal skewing of initial T cell responses toward the Th2 cytokine profile. *J. Immunol.* 160:4730–4737.

Reite, M., Seiler, C., Crowley, T.J., Hydinger-MacDonald, M., and Short, R. (1982). Circadian rhythm changes following maternal separation. *Chronobiologia* 9:1–11.

Reyes, T.M., and Coe, C.L. (1997). Prenatal manipulations reduce the proinflammatory response to a cytokine challenge in juvenile monkeys. *Brain Res.* 769:29–35.

Rook, G.A.W., and Stanford, J.L. (1998). Give us this day our daily germs. *Immunol. Today* 19:113–116.

Rosenberg, Y.J., and Reimann, K.A. (2000). Immunophenotypic and functional properties of lymphoid cells from nonhuman primates. In: Korte, R., and Weinbauer, G.F. (eds.), *Towards New Horizons in Primate Toxicology.* Waxmann Munster, New York, pp. 111–126.

Rowe, J., Macaubas, C., Monger, T.M., Holt, B.J., Harvey, J., Poolman, J.T., Sly, P.D., and Holt, P.G. (2000). Antigen-specific responses to diphtheria-tetanus-acellular pertussis vaccine in human infants are initially Th2 polarized. *Infect. Immun.* 68:3873–3877.

Ruppenthal, G.C. and Reese, D.J. (eds.) (1979). *Nursery Care of Nonhuman Primates.* Plenum Press, New York.

Ruppenthal, G.C., and Sackett, G.P. (1979). Experimental and husbandry procedures: Their impact on development. In: Ruppenthal, G.C., and Reese, D.J. (eds.), *Nursery Care of Nonhuman Primates.* Plenum Press, New York, pp. 269–284.

Russell, R.G., Rosenkranz, S.L., Lee, A.L., Howard, H., DiGiacomo, R.F., Bronsdon, M.A., Blakely, G.A., Tsai, C.-C., and Morton, W.R. (1987). Epidemiology and etiology of diarrhea in colony-born *Macaca nemestrina*. *Lab. Anim. Sci.* 37:309–316.

Sackett, G.P., Bowman, R.E., Meyer, J.S., Tripp, R.L., and Grady, S.S. (1973). Adrenocortical and behavioral reactions by differentially raised rhesus monkeys. *Physiol. Psychol.* 1:209–212.

Sackett, G.P., Ruppenthal, G.C., and Davis, A.E. (2002). Survival, growth, health and reproduction following nursery rearing compared with mother rearing in pigtailed monkeys (Macaca nemestrina). *Am. J. Primatol.* 56:165–183.

Schiffrin, E.J., Carter, E.A., Walker, W.A., Frieberg, E., Benjamin J., and Israel, E.J. (1993). Influence of prenatal corticosteroids on bacterial colonization in the newborn rat. *J. Pediatr. Gastroenterol. Nutr.* 17:271–275.

Smolensky, M.H., Reinberg, A., and McGovern, J.P. (eds.) (1980). *Recent Advances in the Chronobiology of Allergy and Immunology.* Pergamon Press, New York.

Sopper, S., Stahl-Hennig, C., Demuth, M., Johnston, I.C.D., Dorries, R., and ter Meulen, V. (1997). Lymphocyte subsets and expression of differentiation markers in blood and lymphoid organs of rhesus monkeys. *Cytometry* 29:351–362.

Stevens, S., Kennedy, C.R., Lakhani, P.K., and Brenner, M.K. (1984). In-vivo immune responses of breast-fed and bottle-fed infants to tetanus toxoid antigen and to normal gut flora. *Acta Paediatr. Scand.* 73:426–432.

Stone, E.A., Bonnet, K.A., and Hofer, M.A. (1976). Survival and development of maternally deprived rats: Role of body temperature. *Psychosom. Med.* 38:242–249.

Sulzman, F.M., Fuller, C.A., and Moore-Ede, M.C. (1977). Feeding time synchronizes primate circadian rhythms. *Physiol. Behav.* 18:775–779.

Suomi, S.J., and Harlow, H.F. (1972). Social rehabilitation of isolate-reared monkeys. *Dev. Psychol.* 6:487–496.

Terao, K., Suzuki, J., and Ohkura, S. (2002). Circadian rhythm in circulating CD16-positive natural killer (NK) cells in macaque monkeys, implication of plasma cortisol levels. *Primates* 43:329–338.

Van Rompay, K.K.A., Greenier, J.L., Cole, K.S., Earl, P., Moss, B., Steckbeck, J.D., Pahar, B., Rourke, T., Montelaro, R.C., Canfield, D.R., Tarara, R.P., Miller, C., McChesney, M.B., and Marthas, M.L. (2003). Immunization of newborn rhesus macaques with simian immunodeficiency virus (SIV) vaccines prolongs survival after oral challenge with virulent SIVmac251. *J. Virol.* 77:179–190.

Vigano, A., Esposito, S., Arienti, D., Zagliani, A., Massironi, E., Principi, N., and Clerici, M. (1999). Differential development of Type 1 and Type 2 cytokines and beta-chemokines in the ontogeny of healthy newborns. *Biol. Neonate* 75:1–8.

Woodward, B., and Draper, H.H. (eds.) (2001). *Advances in Nutritional Research*, Vol. 10. Kluwer Academic/Plenum Publishers, New York.

Yoshino, N., Ami, Y., Terao, K., Tashiro, F., and Honda, M. (2000). Upgrading of flow cytometric analysis for absolute counts, cytokines and other antigenic molecules of cynomolgus monkeys (*Macaca fascicularis*) by using anti-human cross-reactive antibodies. *Exp. Anim.* 49:97–110.

Zhang, X.-L., Komada, Y., Chipeta, J., Li, Q.-S., Inaba, H., Azuma, E., Yamamoto, H., and Sakurai, M. (2000). Intracellular cytokine profile of T cells from children with acute lymphoblastic leukemia. *Cancer Immunol. Immunother.* 49:165–172.

CHAPTER TEN

Special Challenges of Rearing Infant Macaques Infected with Lentivirus (SIV, HIV, SHIV)

Julie M. Worlein, James C. Ha, Christy Harris, Jennifer Leigh, Kelsey Stratton, and Rodney J.R. Ho

1. INTRODUCTION

The nonhuman primate lentivirus model is perhaps the best animal model of human immunodeficiency virus (HIV) infection. The lentiviruses (lentivirinae), a subfamily of retroviruses (family Retroviridae), include, in addition to HIV-1 and HIV-2, simian immunodeficiency virus (SIV) and a variety of simian/human immunodeficiency viruses (SHIVs). SHIVs are humanly engineered, chimeric viruses made by inserting HIV genes into an SIV. Lentivirus-infected macaques exhibit virological, immunological, pathological, and central nervous system (CNS) responses that are very similar to those seen in humans infected with HIV (Hirsch and Lifson, 2000; Sopper *et al.*, 2002). Research with these animals, and the rearing of infected infants, requires consideration of both the health of the animals and the safety of their human handlers. In this chapter we will describe lentiviruses and the symptoms associated with infection; outline biosafety 2/3 containment procedures that ensure the safety of personnel; discuss rearing and husbandry protocols that ensure the healthy growth of these inherently fragile infants; and present normative data on their growth and behavioral development.

2. LENTIVIRUSES AND VIRAL SYMPTOMOLOGY

Unlike most viruses, retroviruses (including lentiviruses) consist of RNA that serves as a template for the synthesis of complementary DNA, which is then integrated into the host DNA through the action of the enzyme reverse transcriptase. Thus, they can live within cells for years, causing persistent infection that cannot be eliminated by currently available antiretroviral drugs. Lentiviral infection is also characterized by a long latent period, CNS involvement, and a high mutation rate (thus making vaccine development extremely problematic). It is interesting to note that lentiviruses (including a nonpathogenic form of HIV) are being used experimentally as a vector for gene therapies for various diseases (Quinonez and Sutton, 2002).

It is commonly known that HIV infection in humans produces depletion of $CD4^+$ lymphocytes, resulting in profound immunosuppression, chronic diarrhea and wasting, susceptibility to a variety of opportunistic infections, and eventually death. What is less widely recognized is that HIV causes neurological symptoms in both adults and children. Approximately 30–60% of adults and 78–93% of infants and children infected with HIV exhibit symptoms denoting neurological dysfunction (Belman *et al.*, 1988; Epstein *et al.*, 1984; Belman, 1992; McArthur *et al.*, 1994; Schwarcz and Rutherford, 1989). Infants and children infected with HIV routinely show frank signs of cognitive and motor impairment. These include developmental delay and/or loss of previously acquired developmental milestones, mental retardation, and deficits of visual-spatial integrative ability and memory (Belman *et al.*, 1988, 1996; Boivin *et al.*, 1995; Coplan *et al.*, 1998; Diamond, 1989; Diamond *et al.*, 1987, 1990; Drotar *et al.*, 1997, 1999; Nozyce *et al.*, 1994). In adults, behavioral symptoms consist of impaired short-term memory, psychomotor slowing, cognitive decline, and loss of mental agility (for reviews see Clifford, 2002; Navia *et al.*, 1986; Power and Johnson, 1995; Price, 1995). Both adults and children also have been reported to have problems with attention and concentration, and both can experience apathy, depression, social withdrawal, personality changes, and sleep disturbances (Arendt *et al.*, 1990; Belman, 1990; Krovner *et al.*, 1989; Lifschitz *et al.*, 1989; Moss *et al.*, 1989, 1994, 1996; Navia *et al.*, 1986).

In recent years, better diagnostic techniques have been developed and drug therapies have become more effective, resulting in lengthened survival of HIV-infected individuals and rendering AIDS a chronic, managed disease. Unfortunately, however, the therapeutic techniques that control viral replication in peripheral blood cannot always cross the blood–brain barrier to eliminate virus from the CNS. Thus the brain continues to be vulnerable to the effects of infection (Gelbard and Epstein, 1995), and the proportion of HIV-infected individuals that experiences CNS symptoms appears unabated and may be on the rise (Masliah *et al.*, 2000; Morgello *et al.*, 2002; Neuenburg *et al.*, 2002). Once present, neurological symptoms are usually progressive and eventually debilitating, terminating in progressive encephalopathy in children and AIDS dementia complex in adults. As they increase, these neurological symptoms impinge on all realms of behavioral functioning: employment, independence, adherence to drug therapies, mood, and—foremost—quality of life. Understanding the mechanisms, as well as controlling and alleviating CNS symptoms, is the next big challenge facing AIDS researchers and will remain so until a vaccine or a cure is found.

In our AIDS research at the University of Washington, several projects address the issue of neuroAIDS. In the nursery devoted to pediatric neuroAIDS we have reared macaque infants infected with HIV-1, HIV-2, and several strains of SIV and SHIV. At present we primarily rear infants infected with HIV-2_{287}, which was derived from a human HIV-2 virus and passaged several times through macaques to increase pathogenicity (McClure *et al.*, 2000). These infant macaques infected with HIV-2_{287} present an excellent model of pediatric neuroAIDS.

3. NEURO-AIDS NURSERY PROCEDURES

3.1. Personnel Safety

Lentiviruses are extremely fragile. They can live on dry surfaces for only a short time and are destroyed by various disinfectants, including bleach and alcohol (MMWR, 1988). Just two cases of SIV seroconversion in humans have been reported, one from a needle stick and one from improper use of personal protective equipment. However, neither of these persons was ill at the time their seroconversion was reported (MMWR, 1992), and there have been no reports of workers becoming ill as a result of SIV exposure. On the other hand, SHIV and HIV-2_{287} can infect human cell lines *in vitro* (Singh *et al.*, 2003; Ho *et al.*, 1996), and although no incidents of human infection with these viruses have been reported to date, the possibility of such infection exists. Considering the ramifications of infection, every possible safety precaution must be taken to prevent exposure.

An obvious method of minimizing the chance of accidentally transmitting lentiviruses from animals to humans is to avoid handling the animals except when they are anesthetized. This is not practical in our laboratory, however, for our battery of motor and cognitive assessments requires that the infants be awake and alert during testing. Therefore, we had to develop protocols that would protect personnel from possible viral exposure and still allow us to obtain developmental measures. Most of our protocols are based on guidelines from the CDC (www.cdc.gov/od/ohs/biosfty/bmbl4/bmbl4toc.htm).

All personnel who enter the laboratory, for any purpose, are advised of the potential hazard of exposure, and personnel who deal directly with the animals are thoroughly trained on safety procedures by the Washington National Primate Research Center (NPRC) health and safety coordinator before they enter the laboratory. In addition, new personnel are accompanied by experienced staff while they gain experience working under biosafety conditions. Doors can be opened only by a key card, which is programmed by the health and safety officer and activated only when the card-holder is properly trained.

Personnel are required to work under biosafety level 2/3 (BSL2/3) containment standards (Fig. 10–1). Street clothes are not permitted in the laboratory. Instead, personnel must wear personal protection equipment (PPE), which includes hospital scrubs, moisture-resistant Tyvek coveralls, moisture-resistant protective shoe covers and surgical masks, protective hair covering, Tyvek gauntlet arm shields, and a double layer of gloves. Gloving procedure requires that a hole be punched through the arm of the Tyvek coveralls (through which the thumb is inserted)

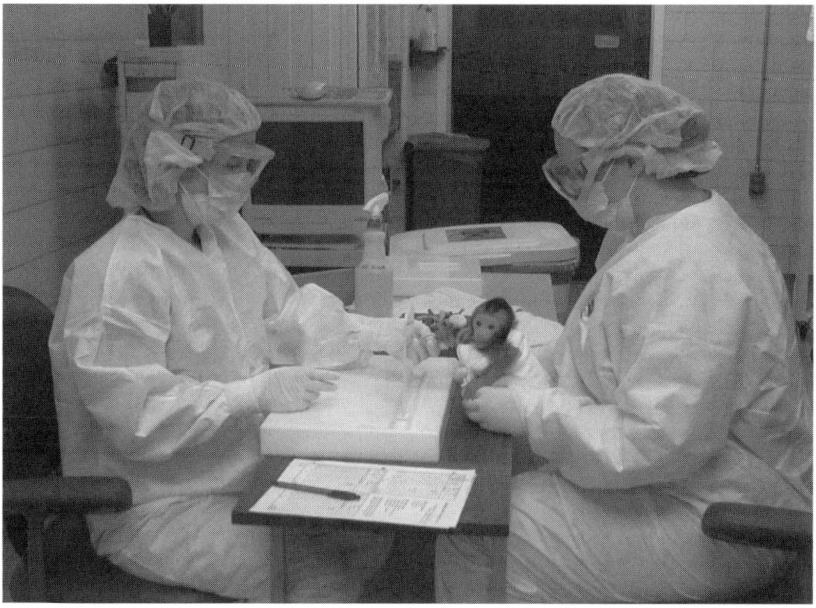

Figure 10–1. Research personnel testing an infected infant on a cognitive task (Object Concept).

after putting on the first pair of gloves. The outer pair of gloves is then put on over the Tyvek suiting. This ensures that no skin around the wrists and hands is exposed. Gowning and degowning are conducted in areas directly outside the laboratory and the PPE is removed in a specific order to prevent possible contamination.

When animals are handled they are swathed in a diaper to minimize the creation of aerosols or spatters from urine or feces. If outer gloves become contaminated with urine or feces during testing, they are removed and replaced. Equipment and work surfaces are routinely decontaminated with disinfectants. Wastes are transported in leak-proof containers and autoclaved prior to disposal in accordance with local and institutional regulations.

3.2. Rearing and Husbandry Protocols

Rearing protocols are based on those in use at the Infant Primate Research Laboratory (IPRL) at the Washington NRPC (Ruppenthal and

Sackett, 1992), tailored to the special nutritional and clinical requirements of infected infants. The nursery is staffed by professional personnel 24 hr a day, 7 days a week. Infants are housed in individual cages. For contact comfort they are supplied with a diaper surrogate or, at later ages, a hanging cloth surrogate. They are fed a formula of Enfamil. Water used to make formula and for drinking is purified through a filter that rejects substances larger than 0.5 μm. Infants initially are fed by hand every 2 hr. Once they can self-feed from a bottle on the side of the cage, they have access to formula 24 hr a day (bottles are changed every 4 hr) until 21 days of age, at which time the formula bottle is available in 4-hr increments separated by 4 hr. This protocol ("4 on 4 off") has been shown to result in superior weight gain in noninfected nursery-reared infants (Sackett and Ruppenthal, 1992). Water is supplied to the infants during the "4 off" period. In addition to formula, infants are fed high-protein monkey chow *ad libitum* from 14 days of age.

Infants are weighed at the same time each day and the amount of formula ingested at each feeding is recorded. A decrease in intake and/or lack of weight gain over a period of several days can be the harbinger of secondary infection or disease progression and is monitored carefully. Because lentivirus-infected infants (and human infants infected with HIV) often do not grow at a normal rate, we supplement the regular diet with Boost, a high-calorie, vitamin-enriched nutritional supplement. The results of this regimen in a typical infant are shown in Fig. 10–2. After infection with HIV-2_{287}, the infant's weight gain fell below that of noninfected infants reared in the IPRL. The infant began receiving Boost supplement at 88 days and, after a few days to adjust to the new taste of the formula, began to grow at an accelerated rate. At 96 days of age, its weight gain matched that of its age mates in the IPRL. Now all infants (including noninfected control infants) receive Boost supplement beginning at 18 days of age. Boost is introduced gradually, starting with a mixture of 10% Boost added to the normal formula. If, after supplementation, infants do not show an acceptable weight gain, the proportion of Boost can be increased up to 30%. During the transition from one percentage of Boost to another, a supplemental bottle is always offered at 1:00 pm (during the normal "4 off" period) so the animal can adjust to the change in flavor gradually.

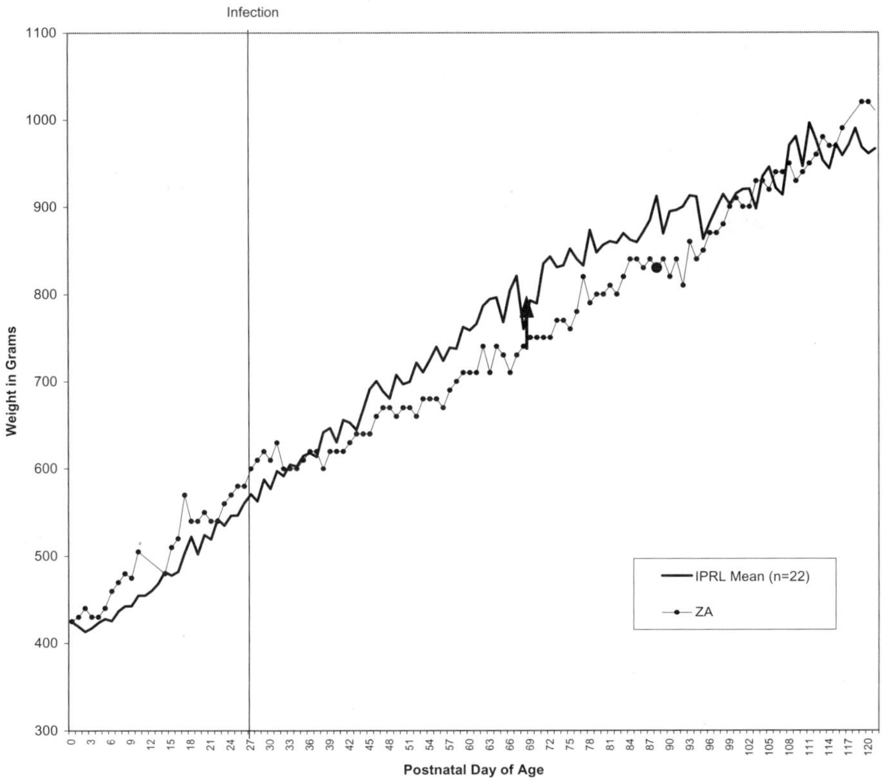

Figure 10–2. Graph of weight gain of an $HIV2_{287}$-infected infant plotted against IPRL norms ($n = 10$). The line marks the date of infection and the arrow the date of implementation of Boost supplementation.

Lentivirus-infected infants are immunosuppressed and therefore susceptible to secondary infections. To guard against these infections we take special precautions to ensure that there is minimal chance of contacting infectious agents. Bottles and enrichment toys are sanitized in a high-temperature dishwasher. Toys filled with food treats are soaked in bleach prior to dishwashing. Diapers and surrogate covers are washed at 190°F (Miele, Novotronic), which eliminates most contaminants. When resistant organisms (e.g., cryptosporidia) enter the laboratory, diapers and surrogate covers are autoclaved after washing. These measures have kept secondary infections to a minimum.

Analysis of the data on all medications administered to infants revealed that infected infants have slightly, but not significantly, more days on

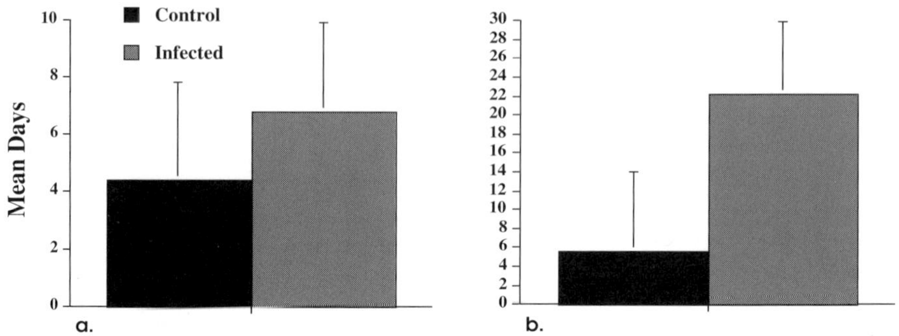

Figure 10–3. Mean days per animal over an average of 250 days that control (n = 5) and HIV2$_{287}$-infected (n = 6) infants received (**a**) antibiotic treatments and (**b**) received nonprescription remedies (Pepto Bismol and/or Kana Banana) for diarrhea.

antibiotic treatment than control infants (6.8 days versus 4.4 days per animal, $p > 0.10$) (Fig. 10–3a). However, especially at the later stages of disease, infected infants do evince a higher incidence of diarrhea, which is part of the viral syndrome. Over-the-counter remedies (Pepto Bismol, Kanna Banana) are used to control these symptoms. Infected infants receive these remedies more often than control infants (22.3 versus 5.6 days per animal). Although this is not statistically significant due to high variability among subjects ($p > 0.10$) it does reflect their proclivity for a higher incidence of diarrhea (which is also seen in humans infected with HIV).

Rectal temperature is taken with a small (1/6-inch diameter) pediatric digital probe. Use of a separate probe for each infant helps to prevent the spread of infection. Temperature is monitored every 4 hr at first, gradually decreasing to every 8 hr at 62 days of age. Taking temperatures at multiple timepoints is necessary because lentivirus-infected infants can have difficulty thermoregulating and may experience temperature spikes, sometimes on a diurnal cycle (Fig. 10–4). Rectal temperatures of ≥103°F are treated with Tylenol.

Before taking the animal's temperature, the technician records its behavioral status (asleep, passive, active, excited/disturbed, or sedated) and the consistency of any stool seen. Intakes, weights, behavioral status, and stool information are entered daily into a computer file that is accessible to clinical/veterinary and research staff. Daily review of these parameters allows clinical staff to detect any illness in its initial stages.

Infants are socialized in a large testing and socializing cage (Fig. 10–5) for 30–60 min daily with like-virus-infected infants and noninfected controls. The latter is feasible because the chance of transmitting the virus in these young animals is almost nonexistent. To date, none of our noninfected controls has become infected. Viral detection procedures are now so sophisticated that control animals can be returned to the regular colony after serving as socialization partners if they test negative twice, at monthly intervals, on polymerase chain reaction and enzyme-linked immunosorbent assays (ELISA) screens.

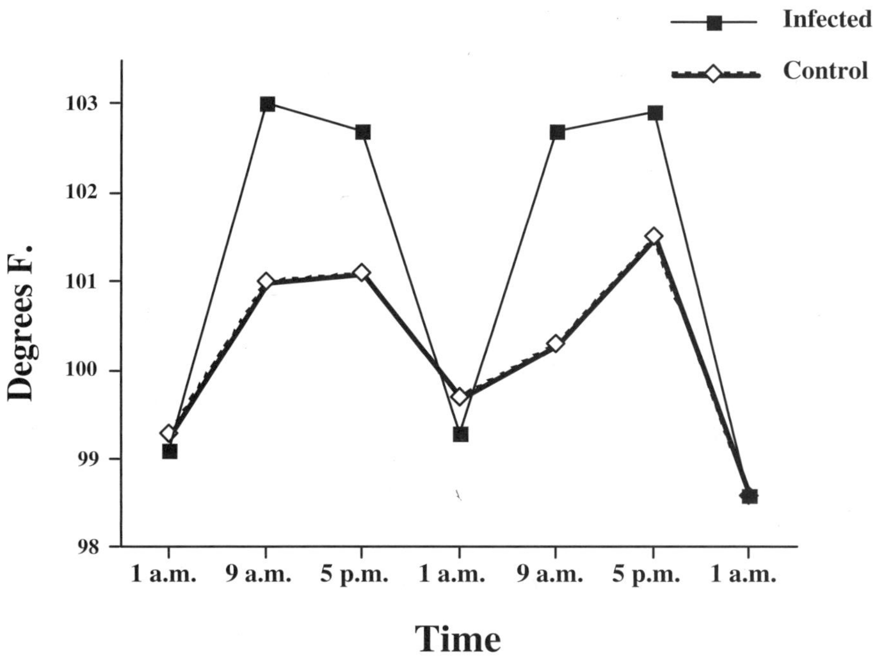

Figure 10–4. Illustration of temperature spikes for an infected infant (without a secondary infection) compared to a control infant during the same 2-day period. Both infants show diurnal cyclicity, with higher temperatures during the daytime and lower temperatures at night. However, the temperature rise during the day for the infected infant is 1 to 2 degrees higher than the noninfected infant.

Figure 10–5. Picture of infants socializing in their socialization/motor testing cage.

Socialization is extremely important in producing physiologically and behaviorally normal infants. Because nonsocialized infants have a number of behavioral, neurological, autonomic, hormonal, and immunological abnormalities (Martin *et al.*, 1988, 1991; Lewis *et al.*, 1990; Gluck *et al.*, 1989; Harlow and Harlow, 1962; Kraemer, 1985; Kraemer *et al.*, 1983; Suomi *et al.*, 1971; Sackett, 1968, 1973, 1974, 1976), which can confound experimental outcome, socialization should be a factor in all studies of lentivirus-infected infants. It is also an important clinical tool because behavioral changes during socialization are often the first signs of the onset of a secondary infection or disease progression. In our laboratory, socialization takes place in large, mesh-bottomed cages that allow urine and feces to drop to the pan below, thus decreasing the likelihood of contact with these possible vectors of contamination. This cage contains toys, shelves, and hanging chains. Each social group has its own socialization cage, which is sanitized regularly to reduce the chances of transmitting secondary infections.

4. NORMATIVE DATA

4.1. Weights and Anthropometrics

As described above (Section 3.2), we routinely give both control and infected infants Boost nutritional supplement. Figure 10–6a illustrates the weight gains of our Boost-supplemented infants compared with nonsupplemented infants, matched for birth weight, reared in the IPRL. Supplemented infants (both control and infected) gained weight at a more rapid rate than nonsupplemented IPRL infants. It is interesting to note that among the Boost-supplemented infants, initially the infected infants gained weight at a slightly greater rate than their noninfected counterparts. Longer into the infective process, however, the growth curve leveled off and noninfected controls became heavier than infected infants. Since the infected infants remained active and clinically healthy even after the growth curves leveled off, we surmise that the extra nutritional supplement supplied to them early in development buffered them somewhat from the advancing disease process.

4.2. Data Analysis

Time (age)-dependent measures of physical growth were assessed for effects of Sex and Group (Boost-supplemented noninfected, $n = 4$; Boost-supplemented infected, $n = 5$; and nonsupplemented IPRL controls, $n = 51$) using growth-curve analysis conducted by general linear modeling (GLM; Tabachnick and Fidell, 2001). In this approach, we established linear growth curves (slope and y-intercept) for each individual and physical measure, then assessed the effects of Sex and Group, and the covariate Birth Weight, using these growth curve values in a GLM. The GLM approach allowed us to simultaneously assess the effects of the categorical variables Sex and Group and the continuous variable of Birth Weight. Bonferoni adjustments were applied to the large number of statistical tests that were performed, as well as to the post hoc tests of group differences when needed. The results of this analysis of physical growth allowed us to test hypotheses about both overall differences in size at identical ages (y-intercept) and the rate of growth over time (slope).

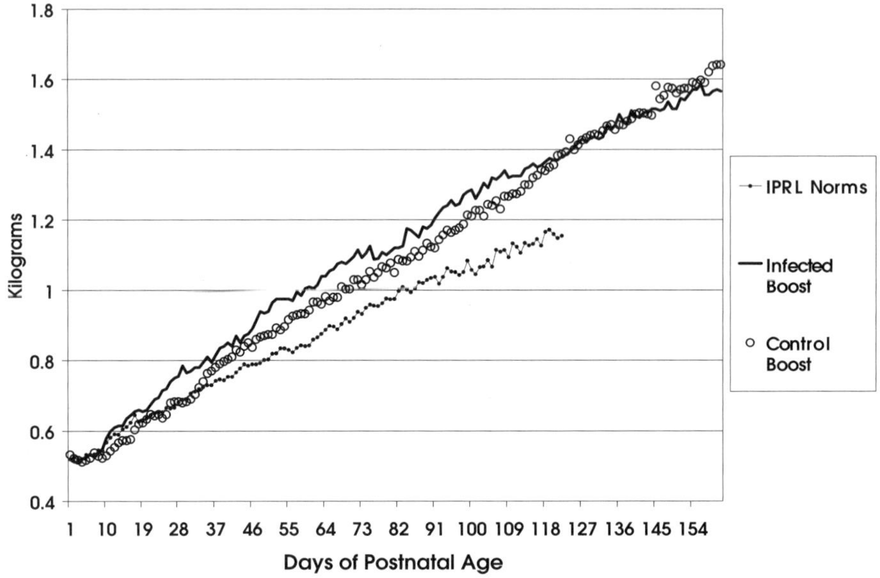

Figure 10–6. Daily weight gain in infants with comparable birthweights reared in the IPRL with no supplement (IPRL Norms, $n = 4$), control infants reared in our laboratory (noninfected Boost supplemented, $n = 3$), and infected infants (HIV-2_{287}-infected Boost supplemented, $n = 2$). Daily weights are available only until 120 days of age.

Assumptions of all statistical tests were checked, including normality and homoscedasticity. No violations of assumptions were revealed. The analysis addressed seven growth values: weight; crown–rump length; head width, length, and circumference; cephalic index; and body mass index (BMI).

No significant differences were found for Sex. However, not surprisingly, several differences in the intercept values of the covariate Birth Weight were found, indicating that birth weight should be considered in any analysis of growth. There were several significant differences in slope and intercept values among Groups. Slope values were significantly different for weight gain ($p = 0.003$), with both infected and noninfected Boost-supplemented infants gaining more weight than the nonsupplemented infants reared in the IPRL (data not shown). As with the data in Fig. 10–6, the slope levels off at later ages in infected infants. There was a trend toward statistical significance between infected and noninfected Boost-supplemented infants groups in the intercept for the BMI. Initially, infected infants had a higher BMI than noninfected Boost-supplemented controls, but this reversed at later ages. This finding suggests that early in development infected infants invest their caloric intake into fat, while noninfected infants invest their caloric intake into growth. This hypothesis will be tested when a larger cohort of Boost-supplemented infants is available.

Although deficits in head growth are often present in HIV-infected human infants (Belman *et al.*, 1988), we found no significant differences in slope values for head width, head length, head circumference, or cephalic index for HIV-2_{287}-infected infants. There was a significant difference in the intercept value for head length ($p = 0.05$), but this difference was the opposite of what we had predicted, with infected infants having a longer head length than noninfected infants (data not shown). It is possible that deficits in head growth could occur at a later stage in the infective process, but our data are silent on this issue because most infected infants in our laboratory are euthanized, according to experimental protocols, before 1 year of age.

5. COGNITIVE MEASURES

5.1. Object Concept

The Object Concept battery of tests measures the infant's ability to recognize that objects still exist after they disappear from direct view. These tests measure developmental cognitive milestones. The testing apparatus consists of a white plastic box with either a small screen or a well that can be covered by a removable lid. Infants are allowed to interact with a desirable object (a small toy or, at later ages, a piece of fruit), which is then placed on the apparatus in one of three of conditions: No-Hide (the object is placed in full view on the apparatus); Partial-Hide (the object is partially hidden behind the screen or partially covered by the lid over the well); or Full-Hide (the object is placed behind the screen or in the

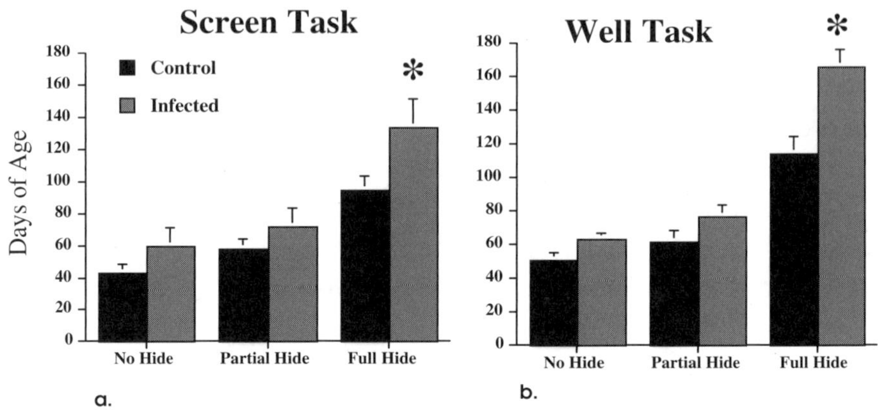

Figure 10–7. Mean days of age, and standard errors, that infected and control infants reached criterion for each stage of the Screen (**a**) and Well tasks (**b**). These differences reached statistical significance for Full Screen (infected $n = 3$, control $n = 10$, $^*p = 0.2$) and Full Well (infected $n =$ ••, control $n = 8$, $^*p = 0.03$).

well covered by the lid). Criterion is 8 retrievals in 10 attempts on 2 consecutive test days.

Infected infants reached criterion at later ages on all aspects of these cognitive tasks, showing a delay in cognitive development (Fig. 10–7). This delay reached statistical significance for Full Hide Screen ($p = 0.05$) and Full Hide Well ($p = 0.03$). Differences between infected and control infants were smaller on tasks mastered at earlier ages and greater at tasks mastered at later ages, presumably reflecting viral effects of advancing disease on the brain. These same abilities have been shown to be impaired in SIV-infected infants (Morton *et al.*, 1992) as well as infant macaques infected prenatally with HIV-2_{287} (J. Worlein *et al.*, unpublished observations). Delays in attaining cognitive milestones are also characteristic of human infants and children infected with HIV (Belman *et al.*, 1996).

5.2. Motor Development

We currently employ two tests of motor development, one for fine motor development and one for gross motor development. The Fine Motor Task assesses the ability of the infant to pick up small pieces of fruit between its thumb and index finger in a pincer grasp. Criterion is reached when the infant is able to pick up the fruit correctly on two of five trials. Infected infants were significantly older (p = 0.002) than noninfected infant when they attained this motor milestone (Fig. 10–8).

Gross motor development is assessed during socialization time in the motor testing/socialization cage 5 days a week from 1 to 6 months of age. The cage contains four items (two shelves, one mesh side and one front bar) that are scored for each infant daily. Infants are coded according to their ability to use these elements and whether they climb, jump, or fall to/from the target. Data are sampled by noting the method of attaining and exiting the target (climb or jump) for the first use of the target on that day. Proportions are calculated for each method at each target and summarized over a 6-month period (see Fig. 10–9).

A review of the data shows that infected infants are more likely to climb and less likely to jump to targets than controls. These differences reach statistical significance for the mesh side of the cage (climb, $p = 0.08$;

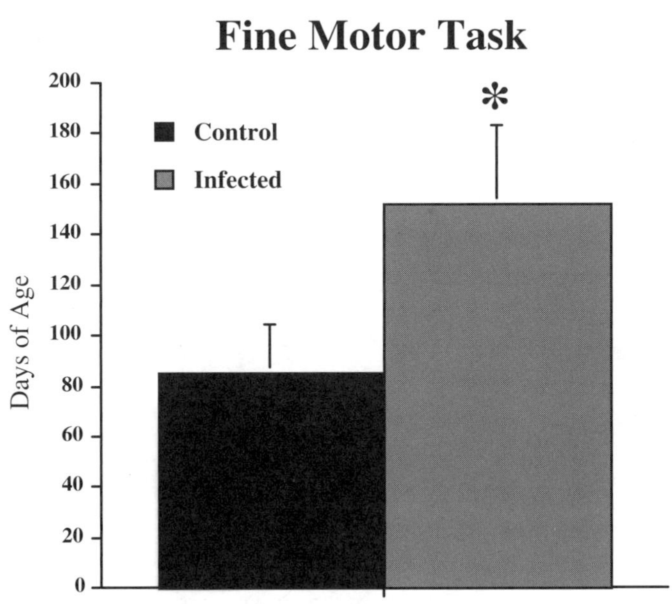

Figure 10–8. Mean days of age, and standard errors, that infected ($n = 2$) and control infants ($n = 5$) reached criterion for the Fine Motor Task.

jump, $p = 0.01$) and the back shelf, which is the most difficult target to reach (climb, $p = 0.04$; jump, $p = 0.04$). These results are indicative of delayed motor development in infected infants, as the usual developmental sequence is to climb and then jump to a target. Thus, infected infants show developmental delays on both measures of motor development. The most prevalent and severe symptom in HIV-infected human infants and children are motor abnormalities (Belman *et al.*, 1985, 1988, 1996). Sometimes these deficits are so severe that it is difficult to administer cognitive assessments (Belman, 1990) (Fig. 10–9).

5.3. Species-Typical Behaviors

Species-typical social behaviors are not often studied in lentivirus-infected macaques. Impairments of socially and emotionally expressive behavior are common in HIV-infected infants and children, including decreased sociability and emotional responsiveness, delay in acquisition of socially responsive smile, and increased irritability (Belman, 1990; Moss *et al.*, 1994, 1996). Since these behavioral symptoms are very prevalent in

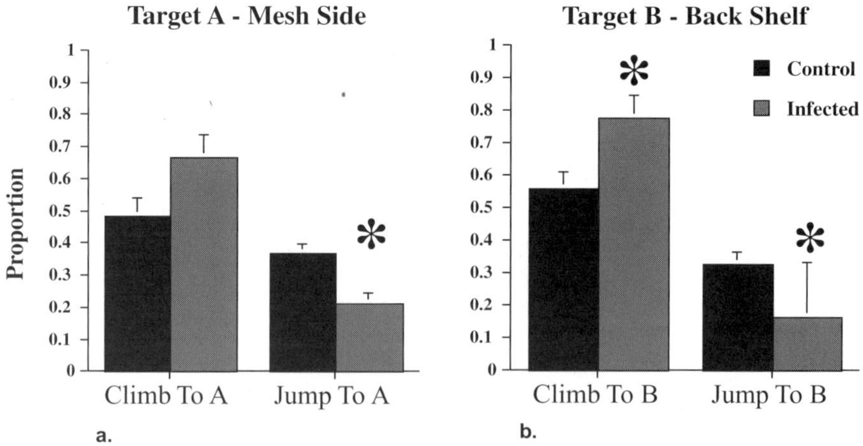

Figure 10–9. Proportion of days, with standard errors, that infected ($n = 3$) and control ($n = 5$) infants climbed and jumped to the mesh side (**a**) and back shelf (**b**) of the motor/socialization cage.

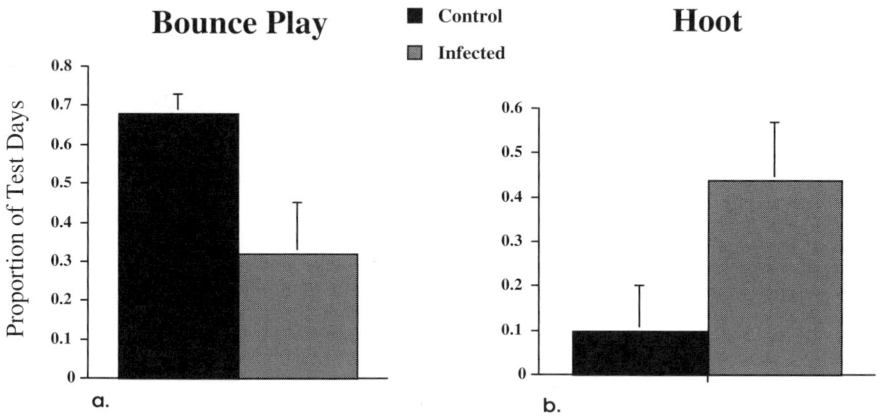

Figure 10–10. Proportion of days, with standard errors, that infected ($n = 3$) and control ($n = 5$) infants exhibited Bounce Play (**a**) and Hoot (**b**) during socialization.

HIV-infected human infants and children, they warrant further study in the nonhuman primate model.

To discover whether there are differences between control and infected infants in behavioral development, we recorded the presence or absence of species-typical behaviors (e.g., bounce play, lipsmack, len, grimace, earflip, screech, hoot, convulsive jerk, and huddle) 5 days a week during the socialization period. Data analysis revealed a trend for infected infants to exhibit bounce play fewer days ($p = 0.07$) than controls and to vocalize in hoots (a disturbance behavior) more often ($p = 0.09$) (Fig. 10–10).

6. SUMMARY

In this chapter we have shown that lentivirus-infected infant macaques can be reared and tested safely in a nursery environment. Special feeding protocols ensure that they grow at a normal weight and remain relatively free of secondary infections. Behavioral data gathered under these conditions can be attributed to effects of the virus on the brain, not to general illness or debility. Socialization is an important factor of this rearing protocol as it ensures the proper behavioral and physiological development of these infants. Socialization between infected infants and uninfected controls is feasible because vertical transmission is almost non-existent under the stringent conditions in our laboratory. Cognitive and motor delays evinced by lentivirus-infected macaques closely mirror those seen in HIV-infected infants and children, and thus provide an excellent model of pediatric neuroAIDS.

REFERENCES

Arendt, G., Hefter, H., Elsing, C., Strohmeyer, G., and Freund, H.J. (1990). Motor dysfunction in HIV-infected patients without clinically detectable central-nervous deficit. *J. Neurol.* 237:362–368.

Belman, A.L. (1990). AIDS and pediatric neurology. *Neurol. Clin.* 8:571–603.

Belman, A.L. (1992). Acquired immunodeficiency syndrome and the child's central nervous system. *Pediatr. Clin. North Am.* 39:691–714.

Belman, A.L., Diamond, G., Dickson, D., Horoupian, D., Llena, J., Lantos, G., and Rubinstein, A. (1988). Pediatric acquired immunodeficiency syndrome: Neurologic syndromes. *Am. J. Dis. Child.* 142:29–35.

Belman, A.L., Muenz, L.R., Marcus, J.C., Goedert, J.J., Landesman, S., Rubinstein, A., Goodwin, S., Durako, S., and Willoughby, A. (1996). Neurologic status of human immunodeficiency virus 1-infected infants and their controls: A prospective study from birth to 2 years. *Pediatrics* 98:1109–1118.

Boivin, M.J., Davies, A.G., Mokili, J.K.L., Green, S.D.R., Giordani, B., and Cutting, W.A.M. (1995). A preliminary evaluation of cognitive and motor effects of pediatric HIV infection in Zairian children. *Health Psychol.* 14:13–21.

Clifford, D.B. (2002). AIDS dementia. *Med. Clin. North Am.* 86:537–550.

Coplan, J., Contello, K.A., Cunningham, C.K., Weiner, L.B., Dye, T.D., Roberge, L., Wojtowycz, M.A., and Kirkwood, K. (1998). Early language development in children exposed to or infected with human immunodeficiency virus. *Pediatrics* 102(1):e8.

Diamond, G.W. (1989). Developmental problems in children with HIV infection. *Ment. Retard.* 27:213–217.

Diamond, G.W., Kaufman, J., Belman, A.L., Cohen, L., Cohen, H.J., and Rubinstein, A. (1987). Characterization of cognitive functioning in a subgroup of children with congenital HIV infection. *Arch. Clin. Neuropsychol.* 2:245–256.

Diamond, G.W., Gurdin, P., Wiznia, A.A., Bleman, A.L., Rubenstein, A., and Cohen, H.J. (1990). Effects of congenital HIV infection on neurodevelopmental status of babies in foster care. *Dev. Med. Child Neurol.* 32:999–1005.

Drotar, D., Olness, K., Wiznitzer, M., Guay, L., Marum, L., Svilar, G., Hom, D., Fagan, J.F., Ndugwa, C., and Kiziri-Mayengo, R. (1997). Neurodevelopmental outcomes of Ugandan infants with human immunodeficiency virus type 1 infection. *Pediatrics* 100(1):E5.

Drotar, D., Olness, K., Wiznitzer, M., Schatschneider, C., Marum, L., Guay, L., Fagan, J., Hom, D., Svilar, G., Ndugwa, C., and Mayengo, R.K. (1999). Neurodevelopmental outcomes of Ugandan infants with HIV infection: An application of growth curve analysis. *Health Psychol.* 18:114–121.

Epstein, L.G., Sharer, L.R., Joshi, V.V., Fojas, M.R., Koenigsberger, M.R., and Oleske, J.W. (1984). Progressive encephalopathy in children with acquired immunodeficiency syndrome. *Ann. Neurol.* 17:488–496.

Gelbard, H.A., and Epstein, L.G. (1995). HIV-1 encephalopathy in children. *Curr. Opin. Pediatr.* 7:655–662.

Gluck, J.P., Ozer, H., Hensley, L.L., Beauchamp, A., Mailman, R.B., and Lewis, M.H. (1989). Early social isolation in rhesus monkeys: Long-term effects on survival and cell mediated immunity. *Abstr. Soc. Neurosci.* 15:297.

Harlow, H.F., and Harlow, M.K. (1962). The effect of rearing conditions on behavior. *Bull. Menninger Clin.* 26:213–224.

Hirsch, V.M., and Lifson, J.D. (2000). Simian immunodeficiency virus infection of monkeys as a model system for the study of AIDS pathogenesis, treatment, and prevention. *Adv. Pharmacol.* 49:437–477.

Ho, R.J.Y., Agy, M.B., Morton, W.R., Scheibel, M., McClure, J., Watson, A., Hu, S.-L., Nosbisch, C., Dorofeeva, N., and Unadkat, J.D. (1996). Development of a chronically catheterized maternal-fetal macaque model to study in utero mother-to-fetus HIV transmission: A preliminary report. *J. Med. Primatol.* 25:218–224.

Kraemer, G.W. (1985). Effects of differences in early social experience on primate neurobiological-behavioral development. In: Reite, M., and Field, T. (eds.), *The Psychobiology of Attachment and Separation.* Academic Press, Orlando.

Kraemer, G.W., Ebert, M.H., Lake, R., and McKinney, W.T. (1983). Amphetamine challenge: Effects in previously isolated rhesus monkeys and implications for animal models of schizophrenia. *Ethopharmacology: Primate Models of Neuropsychiatric Disorders.* Alan R. Liss, Inc., New York, pp. 199–218.

Krovner, R., Perecman, E., Lazar, W., Hainline, B., Kaplan, M., and Lesser, M. (1989). Relation of personality and attentional factors to cognitive deficits in human immunodeficiency virus-infected subjects. *Arch. Neurol.* 46:274–277.

Lewis, M.H., Gluck, J.P., Beauchamp, A.J., Keresztury, M.F., and Mailman, R.B. (1990). Long-term effects of early social isolation in *Macaca mulatta*: Changes in dopamine receptor function following apomorphine challenge. *Brain Res.* 513:67–73.

Lifschitz, M., Hanson, C., Wilson, G., and Shearer, W.T. (1989). Behavioral changes in children with human immunodeficiency virus (HIV) infection. V International Conference on AIDS, June 4–9, Montreal. Abstracts 5:316.

Martin, L.J., Spicer, D.M., Lewis, M.H., Gluck, J.P., and Cork, L.C. (1991). Social deprivation of infant rhesus monkeys alters the chemoarchitecture of the brain: I. subcortical regions. *J. Neurosci.* 1:3344–3358.

Martin, R.E., Sackett, G.P., Gunderson, V.M., and Goodlin-Jones, B.L. (1988). Auditory evoked heart rate responses in pigtailed macaques (*Macaca nemestrina*) raised in isolation. *Dev. Psychobiol.* 21:251–260.

Masliah, E., DeTeresa, R.M., Mallory, M.E., and Hansen, L.A. (2000). Changes in pathological findings at autopsy in AIDS cases for the last 15 years. *AIDS* 14:69–74.

McArthur, J.C., Selnes, O.A., Glass, J.D., Hoover, D.R., and Bacellar, H. (1994). HIV dementia. Incidence and rish factors. *Res. Publ. Assoc. Res. Nerv. Ment. Dis.* 72:251–272.

McClure, J., Schmidt, A.M., Rey-Cuille, M.A., Bannink, J., Misher, L., Tsai, C.C., Anderson, D.M., Morton, W.R., and Hu, S.L. (2000). Derivation and characterization of a highly pathogenic isolate of human immunodeficiency virus type 2 that causes rapid CD+4 cell depletion in *Macaca nemenstrina*. *J. Med. Primatol.* 29(3–4):114–126.

MMWR (1988). Perspectives in disease prevention and health promotion guidelines to prevent simian immunodeficiency virus infection on laboratory workers and animal handlers. *MMWR Morb. Mortal. Wkly. Rep.* 37:693–694, 699–704.

MMWR (1992). Anonymous survey for simian immunodeficiency virus (SIV) seropositivity in SIV-laboratory researchers. *MMWR Morb. Mortal. Wkly. Rep.* 41:814–815.

Morgello, S., Mahboob, R., Yakoushina, T., Khan, S., and Hague, K. (2002). Autopsy findings in a human immunodeficiency virus-infected population over 2 decades: Influences of gender, ethnicity, risk factors, and times. *Arch. Pathol. Lab. Med.* 126:182–190.

Morton, W.R., Worlein, J.M., Kuller, L., Tsai, C.C., Benveniste, R.E., and Ochs, H.D. (1992). Maternal-fetal transmission of SIV and cognitive development. In: Bloom, F.E., and Oldstone, M.B.A. (eds.), *Discussions in Neuroscience:*

Molecular and Cellular Mechanisms of Neural-Immune Interactions. Elsevier Science Publishers, New York.

Moss, H., Wolters, P., Eddy, J., Weiner, L., Pizzo, P., and Brouwers, P. (1989). The effects of encephalopathy and AZT treatment on the social and emotional behavior in pediatric AIDS. V International Conference on AIDS, June 4–9, Montreal. Abstracts 5:328.

Moss, H.A., Brouwers, P., Wolters, P.L., Wiener, L., Hersh, S., and Pizzo, P.A. (1994). The development of a Q-sort behavioral rating procedure for pediatric HIV patients. *J. Pediatr. Psychol.* 19:27–46.

Moss, H., Wolters, P.L., Brouwers, P., Hendricks, M.L., and Pizzo, P.A. (1996). Impairment of expressive behavior in pediatric HIV-infected patients with evidence of CNS disease. *J. Pediatr. Psychol.* 21:379–400.

Navia, B.A., Jordan, B.D., and Price, R.W. (1986). The AIDS dementia complex: I. Clinical features. *Ann. Neurol.* 19:517–524.

Neuenburg, J.K., Brodt, H.R., Herndier, B.G., Bickel, M., Bacchetti, P., Price, R.W., Grant, R.M., and Scholte, W. (2002). HIV-related neuropathology, 1985 to 1999: Rising prevalence of HIV encephalopathy in the era of highly active antiretroviral therapy. *J. Acquir. Immune. Defic. Syndr.* 31:171–177.

Nozyce, M., Hittelman, J., Muenz, L., Duraki, S.J., Fischer, M.L., and Willoughby, A. (1994). Effect of perinatally acquired human immunodeficiency virus infection on neurodevelopment in children during the first two years of life. *Pediatrics* 94:883–891.

Power, C., and Johnson, R.T. (1995). HIV-1 associated dementia: Clinical features and pathogenesis. *Can. J. Neurol. Sci.* 22:92–100.

Price, R.W. (1995). AIDS dementia complex and HIV-1 brain infection: A pathogenetic framework for treatment and evaluation. *Curr. Top. Microbiol. Immunol.* 202:33–54.

Quinonez, R., and Sutton, R.E. (2000). Lentiviral vectors for gene delivery into cells. *DNA Cell Biol.* 21:937–951.

Ruppenthal, G.C., and Sackett, G.P. (1992). *Research Protocol and Technician's Manual: A Guide to the Care, Feeding and Evaluation of Infant Monkeys.* Infant Primate Research Laboratory, University of Washington, Seattle, pp. 1–83.

Sackett, G.P. (1968). Abnormal behavior in laboratory-reared rhesus monkeys. In: Fox, M.W. (ed.), *Abnormal Behavior in Animals.* W.B. Saunders Company, Philadelphia, pp. 293–333.

Sackett, G.P. (1973). Adrenocortical and behavioral reactions by differentially raised rhesus monkeys. *Physiol. Psychol.* 1:209–212.

Sackett, G.P. (1974). Sex differences in rhesus monkeys following varied rearing experiences. In: Friedman, R.C., Richart, R.M., and Vande Wiele, R.L. (eds.), *Sex Differences in Behavior.* John Wiley and Sons, New York, pp. 99–122.

Sackett, G.P. (1976). Social isolation rearing: Species differences in behavior of macaque monkeys. *Dev. Psychol.* 12:283–288.

Sackett, G.P., and Ruppenthal, G.C. (1992). Growth of nursery-raised *Macaca nemestrina* infants: Effects of feeding schedules, sex, and birth weight. *Am. J. Primatol.* 27:189–204.

Schwarcz, S.K., and Rutherford, G.W. (1989). Acquired immunodeficiency syndrome in infants, children and adolescents. *J. Drug Iss.* 19:75–92.

Singh, D.K., Chouduri, R., Pacyniak, E., Berman, N.E., and Stephens, E.B. (2003). Infection of human astrocytoma cells with simian-human immunodeficiency virus results in up-regulation of gene expression and altered growth properties. *Neurosci. Lett.* 340(3):201–204.

Sopper, S., Koutsillieri, E., Scheller, C., Czub, S., Riederer, P., and ter Muelen, V. (2002). Macaque animal model for HIV-induced neurological disease. *J. Neural Transm.* 109:747–766.

Suomi, S.J., Harlow, H.F., and Kimball, S.D. (1971). Behavioral effects of prolonged partial social isolation in the rhesus monkey. *Psychol. Rep.* 29:1171–1177.

Tabachnick, B.G., and Fidell, L.S. (2001). *Using Multivariate Statistics.* Boston, Allyn and Bacon, pp. 901–907.

CHAPTER ELEVEN

Nursery Rearing and Biobehavioral Organization

John P. Capitanio, William A. Mason, Sally P. Mendoza, Laura DelRosso, and Jeffrey A. Roberts

1. INTRODUCTION

Nursery rearing provides many benefits to the scientist, such as easy access to animals; greater uniformity in rearing conditions involving temperature, diet, social conditions, etc.; and reduced exposure to pathogens that might affect research results. A number of studies focusing on both behavioral and physiological processes, however, have suggested that nursery-raised animals differ from animals raised in a richer social and physical environment. Differences have been found for behavioral measures (Champoux *et al.*, 1991), immunological measures (Lubach *et al.*, 1995), hematological measures (Kriete *et al.*, 1995), and concentrations of pituitary–adrenal hormones (Shannon *et al.*, 1998). Many of these data are reviewed elsewhere in this volume (see Chapters 5, 9, and 12).

Other studies have also demonstrated significant biobehavioral differences between groups of macaques whose experiences differed in seemingly more minor ways. For example, Laudenslager *et al.* (1985) demonstrated persisting differences in cellular immune function between adult pigtailed macaques that had experienced a 10- to 14-day social separation during their first year of life and comparably aged animals that had served as controls for those studies. In addition, contrasts between rhesus macaques of Indian and Chinese origin have revealed differences in behavior (Champoux *et al.*, 1994), hematological measures (Champoux *et al.*, 1996), endocrine measures (Champoux *et al.*, 1989), neurochemistry (Champoux *et al.*, 1997), and even viral load following inoculation with the simian immunodeficiency virus (Marthas *et al.*, 2001). Rhesus macaques living in urban versus rural environments in India also show significant behavioral differences (e.g., Singh, 1968). Finally, even among animals of similar origin, individual differences in personality characteristics, such as sociability, are significantly related to behavioral and physiological differences in responsiveness (Capitanio, 2004; Capitanio *et al.*, 1999; Maninger *et al.*, 2003).

We believe these data have two important implications. First, they suggest that the notion of a "normal" animal is suspect. Within a species, there is substantial variation in most behavioral and physiological traits. While some of this variation can be explained by factors such as rearing history, geographic origin, or personality characteristics, it is important to remember that animals that differ in these categorical attributes may nevertheless show substantial similarities on most biobehavioral measures. The second implication is that it is imperative to understand the nature and source of these differences in biobehavioral organization. From a research perspective, knowledge of how and why animals differ from each other can improve our science and possibly reduce the number of animals needed for studies. From a colony management perspective, knowledge of these differences can inform decision making relating to, for example, which animals might be most compatible as social partners. Furthermore, knowledge of biobehavioral organization can be useful for understanding which animals might be particularly at risk for poor health outcomes following routine colony management procedures (e.g., separations, relocations, group formations). We return to these themes in Section 3 below.

2. BIOBEHAVIORAL ASSESSMENT OF INFANTS AT THE CALIFORNIA NATIONAL PRIMATE RESEARCH CENTER (CNPRC)

At the CNPRC, we have implemented a biobehavioral assessment program designed to characterize young animals from throughout the colony. The goal is to provide standardized information to colony managers and researchers alike, for use in decision making and subject selection. Animals participating in the program originate from multiple locations: half-acre outdoor field corrals; smaller outdoor corncribs; indoor cages where infants are reared with their mothers; and the primate nursery. At the CNPRC, animals are raised in the nursery for specific research projects and for specific pathogen-free (SPF) colony production. Production of rhesus macaques that are free of specific simian viruses involves early weaning from the dam and raising of the infant in the nursery environment, which eliminates the potential for infection from the dam and cagemates and allows serial testing of the offspring to ensure seronegativity.

Approximately 50% of the animals in any given birth year at the CNPRC are assigned to projects either as fetuses or as neonates and so are unavailable for testing. To date, we have assessed more than 300 rhesus macaques in each of the most recent birth seasons, 2001 and 2002. Our data analysis so far has revealed that individual variation in our measures is considerable, and that rearing history has a substantial impact on different facets of biobehavioral organization.

2.1. Subjects and Living Conditions

All subjects in the current assessments were rhesus macaques (*Macaca mulatta*) that were born during 2001. All animals were 90–120 days old at time of testing, and originated from each of the four living situations at the CNPRC.

2.1.1. Fieldcage-Raised (FR) Animals

Animals were born to mothers that resided in the 15 half-acre outdoor enclosures at the CNPRC. Each cage has chain-link sides and top and contains a natural substrate, several A-frame structures to provide sitting surfaces and protection from sun and rain, PVC-coated perches along the four inside corners and on the poles supporting the cage top, a variety of climbing devices, and several food hoppers. Purina chow is provided twice daily, fruit and vegetable supplements are provided twice weekly, and water is available *ad libitum* from several Lixit devices located along the periphery. Approximately 60–100 animals reside year-round in each field cage. We assessed 218 FR infants.

2.1.2. Corncrib-Raised (CR) Animals

CR animals are housed in small social groups, usually containing one adult male and three to six adult females and their offspring, in small "corncrib" structures. The structures, made of a variety of materials such as chain-link fencing or welded wire, typically consist of two cylindrical cages (each approximately 4 m in diameter) connected by a small cubicle (e.g., Hoffmann and Stowell, 1973). Each cage contains PVC perches, climbing structures, and food hoppers; the animals generally have free access to all parts of the two cages. Food and water are provided as for the FR animals. We assessed 32 CR infants.

2.1.3. Mother-Reared (MR) Animals

Macaque females in the CNPRC's timed-mating program are housed in standard-sized individual cages located in indoor housing rooms. Animals are fed twice daily and water is available *ad libitum*. Infants are raised with their mothers from birth, either alone in a single cage or in social contact with another mother–infant pair housed in an adjacent cage. In the latter case, the barrier between the two cages is removed daily for approximately 8 hr. We assessed 24 MR infants.

2.1.4. Nursery-Reared (NR) Animals

When research protocol or health considerations require that infants born in the timed-mating program be removed from their mothers at birth, they are housed in incubators in one of nine nursery rooms. The infants remain in incubators until they are 30 days old, whereupon they are moved to individual units in a quad cage. After 1 or 2 days of habituation, the barriers between adjacent cages are removed for 6 hr per day so that the infants can interact in pairs. At approximately 3 months of age, familiar pairs are relocated to larger cages. Beginning at birth, animals are fed a 1 : 1 solution of formula and water at 2-hr intervals, and are gradually introduced to solid foods beginning at about 14 days of age. At about 84 days of age, animals begin eating dry monkey chow and formula is discontinued. Nineteen of the 64 NR infants born during the 2001 birth season were available for our assessments and all were tested.

2.2. Assessment Procedures

2.2.1. General Procedures

Subjects were tested in cohorts of five to eight, with all members of a given cohort originating from either the indoor (MR or NR) or the outdoor (FR or CR) colonies. Beginning at approximately 8:30 am, FR and CR subjects and their mothers were captured from their living cages and taken to an outdoor holding area. Mothers were lightly immobilized with ketamine hydrochloride, and infants were removed and transported to the testing room. Mothers of MR infants were immobilized in their living cages and the infants were removed and brought to the testing room. NR infants were brought directly to the testing room. All animals arrived at the testing room by 9:30 am and were placed in individual living cages, which contained a towel and duck-shaped stuffed toy to which the infants could cling. A water bottle was attached to the front of each cage and food (standard monkey chow soaked in Tang) was available *ad libitum*. Fruit was also provided.

After an infant had been in the living cage for 15 min, it was assessed as described in the following sections. Each infant was assessed in each test situation according to a predetermined random order, which did not change from test to test. The same basic catalog of behaviors was used in all assessments, focusing on activity states, self-directed behaviors, vocalizations, facial expressions, and environmental exploration. Most assessments were conducted while the infant was in a test cage (which was cleaned after each animal was tested) located in a room adjacent to the housing room. While in this cage, the infant had no visual contact with the observer. On day 2 of testing, at approximately 10 am, FR and CR subjects were returned to their mothers in the outdoor holding area, where they remained for 1 hr before being returned, with their mothers, to their field cages. MR subjects were returned to their mothers in their living cages and NR subjects were returned to the nursery.

For this comparison, we restrict our discussion of behavioral data to measures of activity and distress vocalizations, and to positional responses to our manipulations. Duration measures are expressed as proportion of total time observed.

2.2.2. *Specific Assessments*

2.2.2a. Living Cage Observations. To assess responses to separation and relocation, we observed each infant for a 5-min period at the beginning and at the end of the 24-hr separation period. For these assessments, an observer sat in front of the rack of cages at a distance of 8 feet. Observers were instructed to minimize all direct eye contact with the animals during data collection. Focal animal sampling was used, and data were recorded with the use of the Observer program (Noldus, 1991). Contrasts were made between the day 1 and day 2 focal samples (that is, 15 min after the infants arrived in the test area, and about 2 hr before they returned to their living situation).

2.2.2b. Preferential Look Test. The goal of this test was to assess the infants' recognition memory using procedures adapted from Gunderson *et al.* (1987). Each infant was caught and carried to a small test cage that was placed in front of a large video monitor. Videotaped stimuli were presented on-screen, and the infant's responses were videotaped for later scoring. At the end of the 6-min session, the infant was returned to its living cage. Data from this assessment are not presented here.

2.2.2c. Video Playback. The purpose of this test was to assess the infants' responsiveness to a social stimulus presented via color videotape. The 10-min stimulus tape consisted of three sequences of an unfamiliar adult male rhesus macaque displaying viewer-directed aggressive behavior, interspersed with four segments of the same male displaying nonsocial (i.e., neutral) behavior. Each subject was carried to the test cage, which was placed in front of the video monitor. The infant's behaviors while it viewed the aggression and nonsocial sequences of the videotape were recorded and the infant was then returned to its living cage.

2.2.2d. Human Intruder Test. The goal of this test was to assess the infants' responsiveness under standardized and graded conditions of challenge. Each infant was carried to the test cage and experienced four 1-min trials. For the first trial, an unfamiliar human sat approximately 3 feet in front of the cage, presenting her left profile. After 1 min, the human, maintaining a profile orientation, moved to within 12 inches of the test cage. One minute later, the human returned to the more distant location and attempted to maintain direct eye contact with the infant for 1 min, after which the human again moved to within 12 inches of the test cage and attempted to maintain eye contact for 1 min. These conditions are designated, respectively, profile far, profile near, stare far, stare near.

2.2.2e. Blood Sampling. Blood was sampled on four occasions to assess infants' responses to separation and relocation, and to determine the regulatory characteristics of the hypothalamic–pituitary–adrenal (HPA) axis. Four blood samples were drawn via femoral venipuncture from each monkey. The first sample (1 ml) was taken following completion of the Living Cage observations on day 1, and was used for assessment of plasma cortisol concentrations, a complete blood count, and flow cytometry to assess numbers of $CD4^+$ and $CD8^+$ T cells. After completion of the Human Intruder testing (at approximately 4 pm), a second sample (0.5 ml) was taken, after which each subject was given 500 μg/kg dexamethasone intramuscularly (im). The third sample (0.5 ml) was taken following the Living Cage observations on day 2 (approximately 8:30 am), after which each subject was injected with 2.5 IU adrenocorticotropic hormone (ACTH) im. The fourth sample (0.5 ml) was taken 30 min after the ACTH injection. Because of a nationwide shortage of ACTH for several weeks during our testing, sample sizes for the analysis of the blood measures are somewhat reduced. We present data only for subjects for which we have results from all four blood samples. A slightly different version of this analysis appears in Capitanio *et al.* (2005).

2.2.2f. Novel Objects. The goal of this test was to assess the infants' responsiveness to unfamiliar objects in their living cages. The first object (black hollow cylinder with ridges on the exterior surface, 9 cm long × 4 cm in diameter, weighing 0.090 kg) was present when the animal was first placed in the living cage, and remained in the cage until after the second blood sample was drawn. The second object (white, crayon-shaped object with ridges on the outer surface, same dimensions and weight) was then placed in the cage and remained until the assessments ended on day 2. Each object contained an Actiwatch activity monitor that recorded the force exerted on the object. These data are not reported here.

2.2.2g. Temperament Ratings. The goal of the rating was to provide an overall "thumbnail" picture of each infant's behavioral characteristics during the testing period. At the end of the 24-hr assessment period, just before the infant was returned to its housing situation, it was rated by the technician who had performed the assessments. The rating instrument consisted of 20 adjectives (e.g., calm, bold, excitable, playful, tense) and their definitions. The technician rated each animal on each adjective using a seven-point Likert-type scale, reflecting "total absence" to "extremely large amount" of the characteristic displayed. Factor analysis was performed on the ratings for the FR animals, and a three-factor solution, explaining 53.1% of the total variance, was suggested. The factors were (negatively loaded items are italicized): engaged (curious, active, bold, confident, playful, flexible, *depressed*, *slow*, *tense*, *timid*), equable (gentle, calm, *aggressive*, *irritable*), and excitable (impulsive, excitable).

2.3. Results

2.3.1. Living Cage Observations

Reactions by members of the four rearing groups to being removed from their home cages and placed in a novel cage and in novel surroundings differed considerably. This was clearly evident in the principal behavioral measures of modes of reaction: activity (locomotion, hanging from the ceiling/side of the cage, sitting, and lying) and distress vocalizations.

Mean scores for duration of locomotion and frequency of distress vocalizations differentiated groups [main effects: locomotion, $F(3,286) = 5.21$, $p < 0.01$; distress vocalizations, $F(3,286) = 5.17$, $p < 0.01$], and values across both days were highest for the MR infants. Post hoc analyses revealed that means for MR infants were significantly different from those of the other three groups, whose means did not differ significantly from each other. Duration of hanging from the ceiling/side of the living cage also differentiated groups [main effect: $F(3,286) = 15.57$, $p < 0.001$]. NR infants displayed significantly shorter durations than the other three groups, and CR infants displayed significantly longer durations than did FR infants.

The passive responses of sitting and lying also varied significantly with rearing condition. On both observation days, the NR group had the longest durations of the four groups, while the mean scores were lowest for the CR group [main effects: sitting, $F(3,286) = 9.18$, $p < 0.001$; lying, $F(3,286) = 33.38$, $p < 0.001$, respectively]. Mean values (across the 2 days) of the four activity categories are shown in Fig. 11–1.

The outdoor-raised infants (FR and CR) engaged in significantly less hanging behavior on day 2 than on day 1 [interaction: $F(3,286) = 6.53$, $p < 0.001$], and FR animals showed a significant increase in locomotion [interaction: $F(3,286) = 7.57$, $p < 0.001$].

2.3.2. *Video Playback*

As described earlier, this test assessed responses to a videotape of an unfamiliar adult male macaque alternately showing neutral behaviors and threatening behaviors that seemed to be directed toward the viewing infant. Whereas the CR, FR, and MR infants were familiar with adult male macaques, the NR infants were not. Although members of the different rearing groups did not show differences in the amount of time spent watching the video displays, they differed in both activity levels and position in the viewing cage, with NR infants showing more of a "sit and stare" pattern of response. Across both conditions, activity levels were highest for CR infants, with significantly lower levels for NR (which had the lowest levels) and FR infants. MR infants showed intermediate levels, which were not significantly different from levels in the other groups [main effect for rearing group: $F(3,279) = 5.21$, $p < 0.01$]. Regarding position, although all subjects showed slight increases in the amount of time spent in the back of the viewing cage (i.e., away from the playback

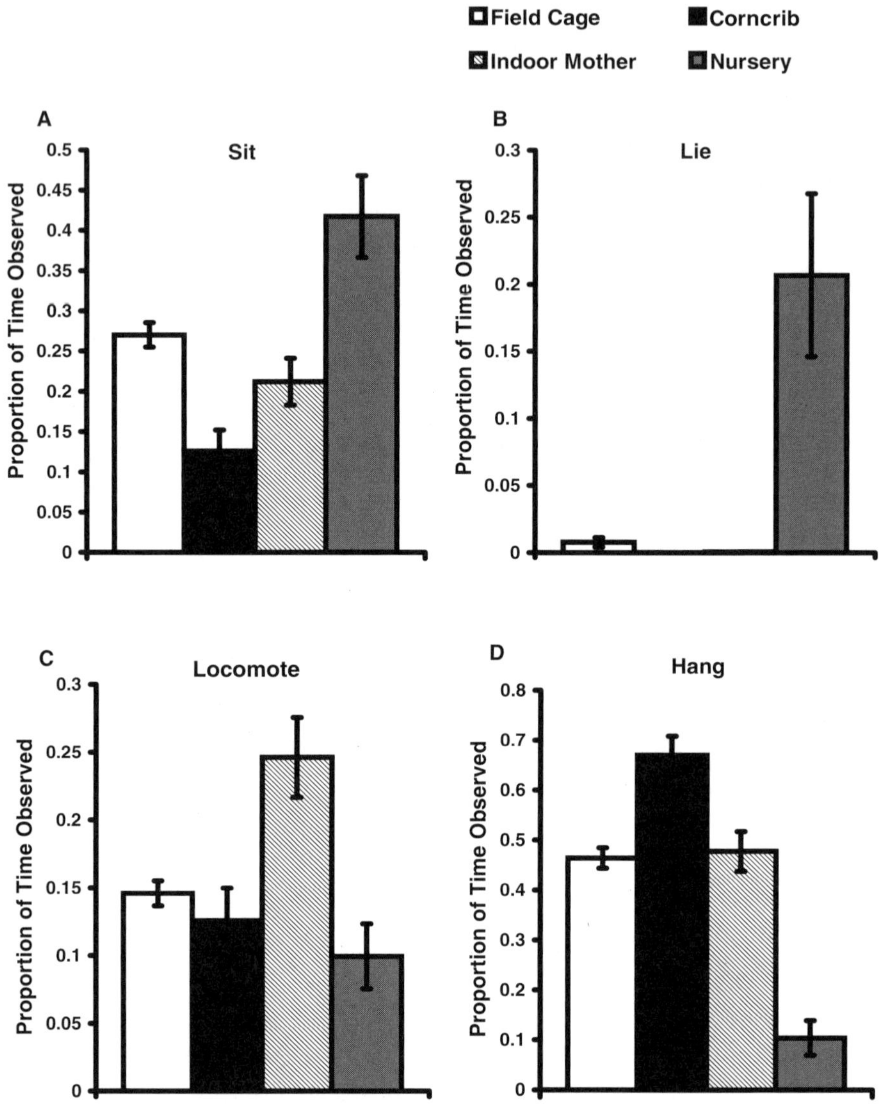

Figure 11–1. (**A–D**) Proportion of time observed for activity measures in Living Cage

monitor) during the aggression compared with the nonsocial segments [significant effect of condition: $F(1,279) = 5.59$, $p < 0.05$], NR infants had longer durations overall compared with MR, FR, and CR infants [main effect of group: $F(3,279) = 6.89$, $p < 0.001$], which did not differ from each other. Finally, groups differed significantly in frequency of distress vocalizing. NR infants had the highest frequencies, though they were significantly higher only when compared with frequencies for FR infants, which displayed the lowest frequencies [$F(3,279) = 4.08$, $p < 0.01$]. Frequencies of distress vocalization for CR and MR infants were intermediate and not significantly different from those of other groups (Fig. 11–2).

2.3.3. *Human Intruder*

Different patterns of response to the human intruder were evident based on rearing condition, as well as distance and orientation of the intruder. For all three measures, we found significant three-way interactions of rearing condition by orientation (profile versus stare) by distance (near

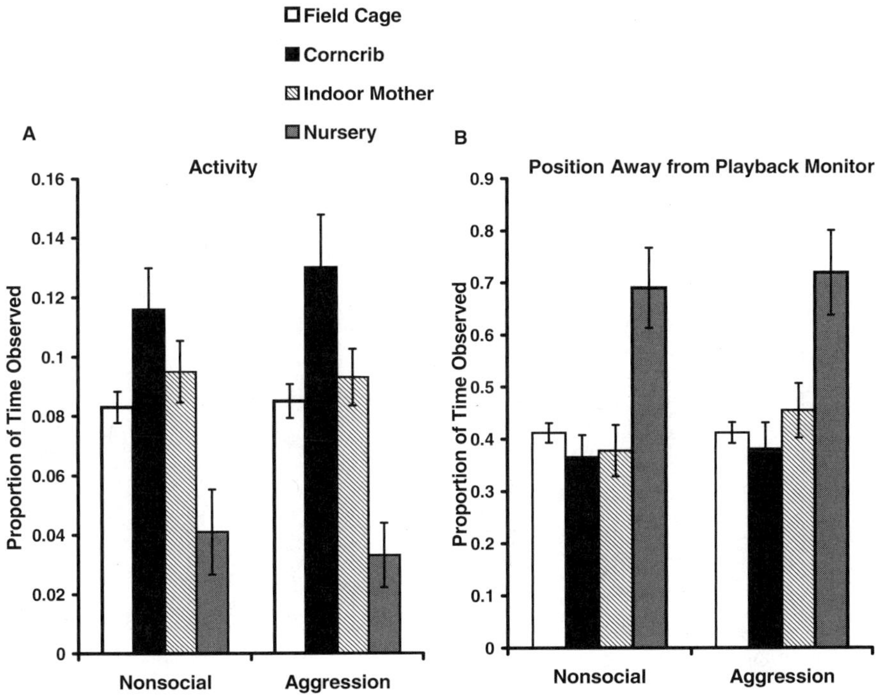

Figure 11–2. Proportion of time observed for activity (**A**) and position (**B**) during Video Playback assessment

versus far): activity: [$F(3,281) = 4.88$, $p < 0.01$]; position: [$F(3,281) = 5.12$, $p < 0.01$]; distress vocalizations [$F(3,281) = 2.70$, $p < 0.05$]. For all interactions, we performed simple effects analyses to determine the nature of the differences. For activity, outdoor-raised infants (FR and CR) showed sharper differentiation among the four conditions, compared with indoor-reared infants. Whereas NR infants showed consistently low levels of activity across all four conditions, and MR infants' activity levels were consistently high across all four conditions, the outdoor-raised infants showed the highest activity only in the most intense condition, Stare Near. During the other three conditions, activity was more consistent, at about one-half to one-third the levels shown during Stare Near (Fig. 11–3A).

Regarding position, two rearing groups showed simple interaction effects. FR infants had shortest durations away from the intruder in the

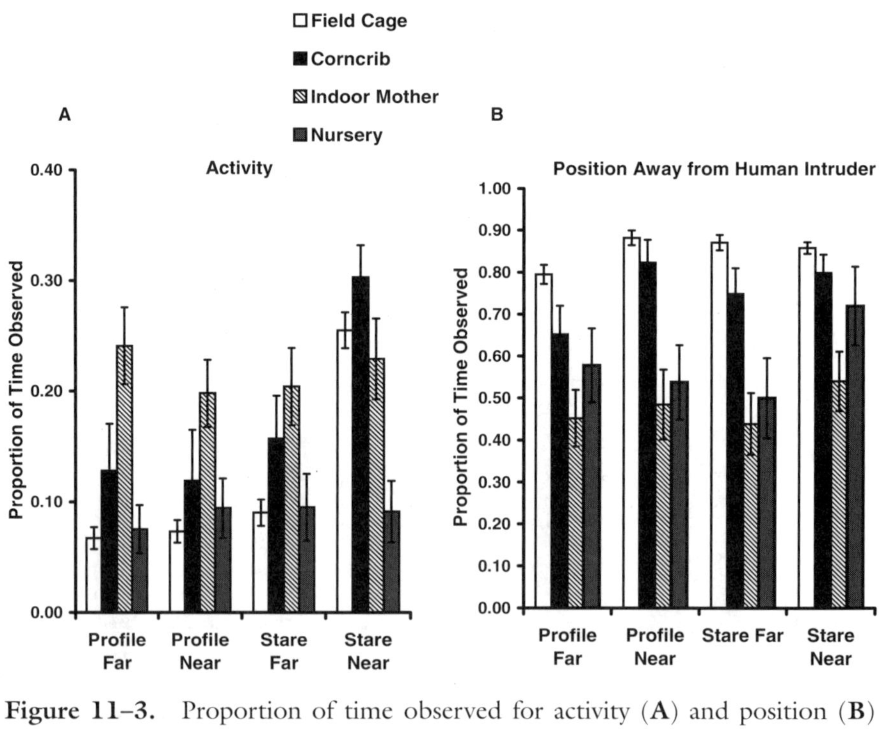

Figure 11–3. Proportion of time observed for activity (**A**) and position (**B**) during Human Intruder assessment

least intense condition, Profile Far. For the other three conditions, distances were high and consistent. NR infants, in contrast, showed relatively short durations away from the human in all but the most intense condition, Stare Near. Distances shown by CR and MR infants did not differ by condition, but overall the former had relatively long durations and the latter had relatively short durations (Fig. 11–3B).

Finally, decomposition of the significant interaction for distress vocalizations revealed significant simple effects for all rearing groups except the NR infants, which showed consistently high frequencies across all four experimental conditions. In contrast, members of the other three rearing groups had low frequencies of distress vocalizations for all conditions except for the Stare Near condition, during which frequencies were at least double the frequencies of the other three trials.

2.3.4. *Hypothalamic Pituitary-Adrenal Regulation*

Overall, NR infants had significantly lower concentrations of plasma cortisol than NR, FR, and CR infants, whose cortisol levels did not differ from each other. However, this effect varied somewhat by sample. For the first two samples, taken on day 1 approximately 1.0 and 6.5 hr after the subjects were brought to the testing room, NR infants had lower cortisol levels than all other animals; members of the other three rearing groups did not differ from each other. In fact, only NR infants showed the expected circadian decline. For the first day 2 sample, cortisol levels were expected to be lower owing to suppression of the HPA system by the high dose of dexamethasone the animals received the previous afternoon. This was the case for FR and CR infants, but not MR and NR infants. The second sample on day 2 showed increased cortisol in all groups in response to ACTH stimulation, as expected. It should be noted, however, that cortisol levels for the NR infants on day 2 remained substantially lower than values for the FR and MR infants, but not significantly below those for CR infants. These effects were indicated by follow-up tests for significant main effects of rearing group [$F(3,215) = 8.88$, $p < 0.001$], and for the interaction of rearing group by sample [$F(9, 645) = 4.23$, $p < 0.001$] (Fig. 11–4).

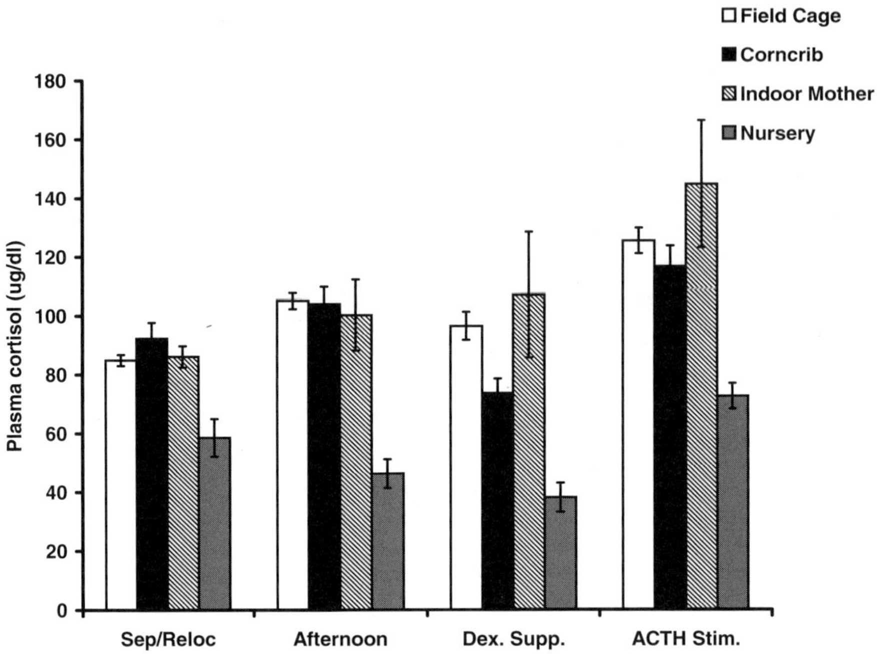

Figure 11–4. Plasma cortisol concentrations after initial separation/relocation, after completion of day 1 testing, after overnight dexamethasone suppression, and 30 min after ACTH injection

2.3.5. Hematology

Overall, we found significant differences between rearing conditions for all measures except the CD4/CD8 ratio. Among leukocyte subsets, NR subjects had significantly lower neutrophil numbers than did FR subjects. Moreover, indoor-raised animals (NR and MR) had significantly elevated total lymphocyte numbers, as well as $CD4^+$ and $CD8^+$ subset numbers, compared with levels in subjects that were reared outdoors (Fig. 11–5). Finally, outdoor-raised subjects had significantly lower red blood cell counts and hematocrit than did MR monkeys, and significantly lower hemoglobin concentrations than did NR monkeys (see Fig. 11–6). Statistical results are as follows: total white blood cells [$F(3,288) = 3.45$, $p < 0.05$], neutrophils [$F(3,288) = 8.10$, $p < 0.001$], lymphocytes [$F(3,288) = 10.39$, $p < 0.001$], $CD4^+$ [$F(3,287) = 46.30$, $p < 0.001$], $CD8^+$ [$F(3,287) = 24.21$, $p < 0.001$], red blood cell numbers [$F(3,288) = 5.20$, $p < 0.01$], hematocrit [$F(3,288) = 5.46$, $p = 0.001$], and hemoglobin [$F(3,288) = 6.43$, $p < 0.001$].

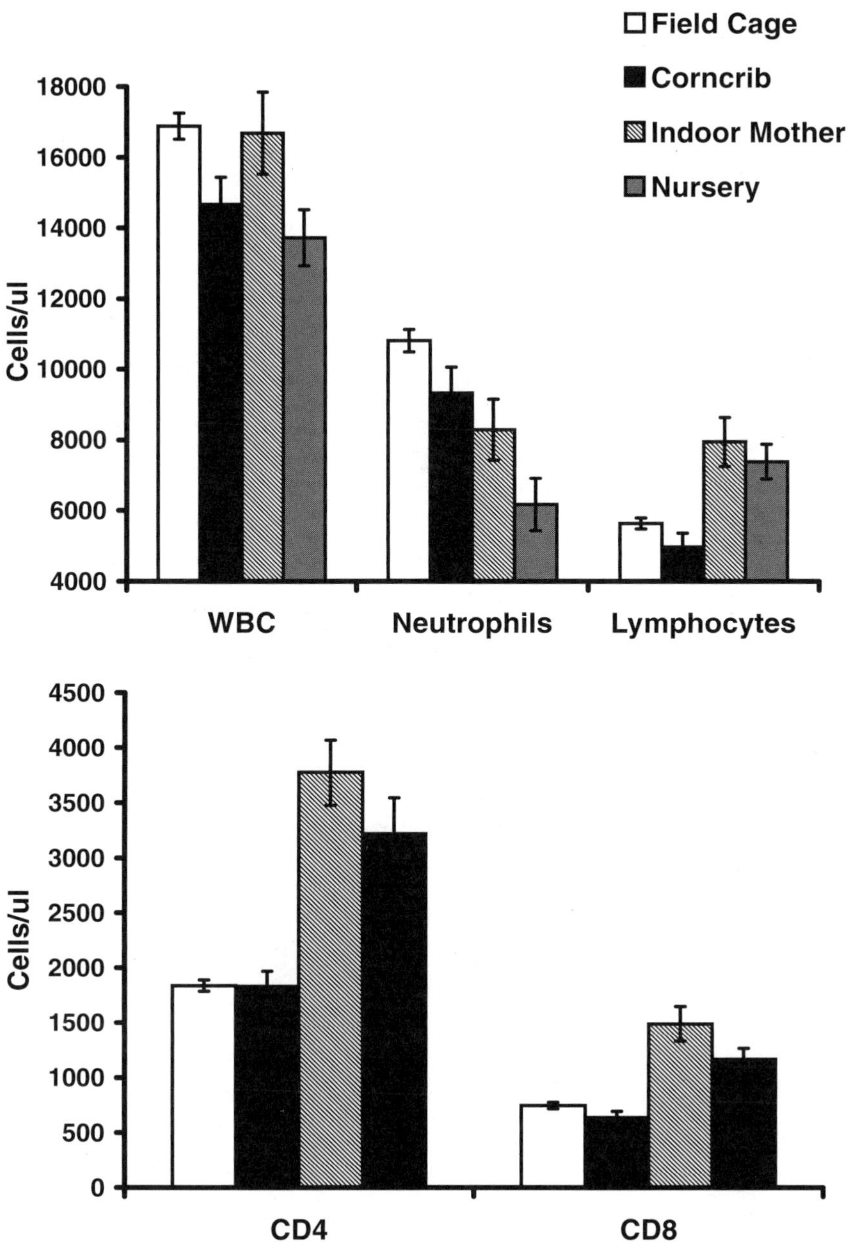

Figure 11–5. Numbers of leukocytes and lymphocytes

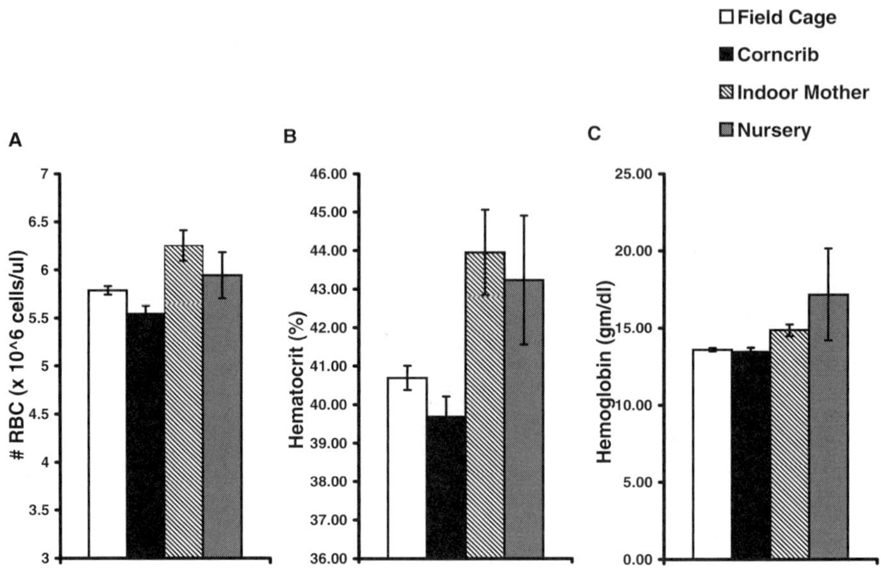

Figure 11–6. Red blood cell numbers (**A**), hematocrit (**B**), and hemoglobin (**C**) values

2.3.6. Temperament Ratings

A significant effect of rearing condition was found for the scale Excitable [$F(3,274) = 6.54$, $p < 0.05$]. Post hoc comparisons indicated that CR subjects were significantly more excitable than were FR subjects.

3. DISCUSSION

Nursery rearing has a substantial impact on biobehavioral organization. The NR infants in this study were generally less active than animals from the other rearing groups (Living Cage Observations, Video Playback, Human Intruder), and they seemed to differentiate between experimental conditions somewhat less than did members of the other rearing groups (day 1 versus day 2 of Living Cage Observations, Human Intruder). Moreover, NR infants displayed a very different pattern of HPA regulation and substantial differences in hematological parameters compared with those shown by infants reared in the other conditions.

The plasma cortisol results were particularly striking. Others (e.g., Shannon *et al.*, 1998) have found differences in basal and stress cortisol concentrations in NR infants, but to our knowledge our data are the first to demonstrate that the regulation of the HPA axis is different: standard doses of dexamethasone and ACTH resulted in cortisol levels that were significantly lower in the NR subjects than in the other monkeys. Had we measured only the day 1 cortisol concentrations, we might have concluded that the lower concentrations in NR infants were merely a reflection of lower stress. On the face of it, this would seem plausible. Unlike subjects in the other three rearing conditions, NR infants did not undergo a maternal separation, and the housing room and the cages in which they were placed for testing were much more similar to those they routinely experienced, particularly compared with conditions experienced by the outdoor-raised infants. The pharmacological challenges, however, suggest that the cortisol differences reflect a more fundamental difference in HPA regulation.

In terms of absolute values, the plasma cortisol concentrations in the NR infants are more similar to those we have found for adult rhesus macaques that were born and raised in the field cages (e.g., Capitanio *et al.*, 1998). This could mean that nursery rearing accelerates HPA maturation. Alternatively, it is possible that NR animals never experience the heightened cortisol concentrations that animals from the other conditions experience. The latter interpretation is consistent with data by Shannon *et al.* (1998), showing that surrogate-peer-reared animals (whose rearing experiences were similar to those of our NR animals) had consistently lower cortisol concentrations compared with MR animals at multiple time points both earlier and later than the age range (90–120 days) examined in our study. Thus, nursery rearing appears to alter, from an early age, the regulatory characteristics of the HPA axis. It is unclear whether nursery rearing might also influence development of other physiological systems, and to what aspect of nursery rearing the HPA effects might be attributed—lack of a mother, intermittent social experience, nutrition (e.g., formula feeding), etc.

NR subjects also showed differences in a variety of hematological measures. Interestingly, for many of these measures, the differences were between indoor-raised (NR and MR) and outdoor-raised (CR and FR) animals. Whereas group differences in plasma cortisol concentrations seem to reflect nursery rearing per se, group differences in hematological values seem to reflect more whether animals were raised indoors or outdoors. The presence of a mother may be a less important influence on development of the physiological systems associated with these measures (e.g., Bicknese *et al.*, 1993). Indoor and outdoor rearing differ in many ways. Indoor-raised animals, at least in our colony, live in small laboratory cages and their opportunities for physical activity and social interaction are limited. These animals are typically not exposed to seasonal fluctuations in weather and lighting; are exposed to a variety of sights and sounds that outdoor-raised animals largely do not experience; and are much more familiar with humans. The latter was evident during the Human Intruder study, in which indoor-raised animals spent less time distant from the human intruder than did outdoor-raised animals (Fig. 11–3B). Thus, our comparisons reveal that some of the differences that might be attributed to nursery rearing may simply reflect indoor housing. Given that NR animals are fed on formula, and MR animals are fed breast milk (just like both sets of outdoor-raised animals), their similarities on our hematological measures suggest that nutritional differences were not the critical factor responsible for the indoor versus outdoor differences.

Inasmuch as both the FR and CR animals were reared outdoors in relatively rich social environments, it was surprising that some significant differences were evident. For example, during the Living Cage observations, CR monkeys showed the longest durations of hanging behavior, and did not show the significant increase in locomotion on day 2 that the FR monkeys did; CR monkeys had the longest durations of activity in the Video Playback assessment; and overall, CR monkeys were rated as more Excitable than were FR monkeys. CR animals also showed a greater response to dexamethasone than did FR animals (CR animals showed a 30-μg/dl drop in cortisol, versus an 8.5-μg/dl decline for FR animals). What might be the source of these differences? One factor may be the much higher proportion of primiparous mothers in the CR versus the FR groups. In our facility, many of the animals living in the corncribs are young animals that will ultimately be placed together in a new field cage. In addition, many of the corncrib mothers were themselves reared indoors, either with their mothers or in the nursery. The field cages, in contrast, comprise animals representing a wide variety of ages and backgrounds. Data from subsequent birth cohorts may help elucidate whether the CR effects we have seen relate to maternal parity, mothers' early history, or some aspect of the housing condition itself, all of which, for the present data set, are confounded.

A final result worth noting reflects the important role that context plays during any type of assessment. All subjects in our study were relocated from their familiar living environments to an unfamiliar room, and animals in three of the four rearing conditions experienced a separation from mother, the usual response to which involves increased activity and distress vocalizing. In fact, this response pattern was seen in the indoor MR subjects, but not in the FR or CR subjects, which were also raised with their mothers. We suspect that the greater change in physical surroundings for the outdoor-raised animals inhibited both the activity and distress vocalization responses that usually accompany maternal separation. Indoor MR animals did show elevations in these responses, probably because the novelty of the physical surroundings was much lower. These results indicate that physical conditions that appear to the human experimenters to be identical for all subjects may not be identical psychologically for animals that have had very similar early experiences, and may result in substantial variability in outcome measures of interest.

4. IMPLICATIONS

There is no doubt that nursery rearing, as carried out at the CNPRC, produces rhesus macaques that show substantial differences in biobehavioral organization compared with animals raised in a more species-typical environment, such as the half-acre field cages. As previously discussed, however, some of the effects of nursery rearing may be superimposed on effects deriving from the fact that the animals are raised indoors, which by itself imposes different opportunities for social and nonsocial behavior and for interaction with humans, and which involves substantial differences in sensory input and environmental consistency. While our data suggest that NR animals differ in important ways from animals with other rearing histories, the data also suggest that animals reared in richer social environments can differ from each other as well. Our analysis contrasted animals raised outdoors in small (corncrib) and large (field cage) enclosures, but others have reported differences based on other factors, such as separation history, geographic origin, and personality factors. We believe that differences between animals raised in nurseries and those raised elsewhere should be considered within this broader context of individual and group differences, and this view entails several implications.

One implication, as described earlier, is conceptual, and relates to what constitutes a "normal" rhesus macaque. Clearly, there is a range of variation on almost all measures one can make, and this applies as well to those measures that we (and others) have found to differ significantly among different types of animals. For example, we found strong group differences in numbers of $CD4^{+}$ T cells (Fig. 11–5B). Inspection of individual data points, however, reveals that some NR and MR monkeys had values that were lower than the mean values found among FR and CR monkeys. Knowing the differences in central tendency (means) among rearing groups can help in predicting that an NR monkey is likely to have higher $CD4^{+}$ cell numbers than an FR monkey, but it does not ensure that prediction. Rather than focusing on whether a set of animals is normal with respect to some absolute standard, it may make more sense to focus on the characteristics of specific animals relative to the norms for their particular experience. Thus, all FR animals are not by definition normal and all NR animals are not by definition abnormal, but rather there are extreme individuals in each rearing group. This approach reminds us that an organism develops within an environment—biotic, abiotic, and conspecific—and, to a large extent, what matters is the animal's ability to adapt or "fit" the circumstances of its life. In both nursery and field cage contexts, some animals "fit" better than others.

Other implications of the view that rearing differences might best be considered in the broader context of individual differences concern methodology. It is, of course, unclear whether statistically significant differences in test results, whether stemming from rearing differences or other sources, translate necessarily into consequences that have biological significance regarding health or other outcomes. This is a fertile area of research in its own right. Nevertheless, caution would suggest that careful attention be paid to subject selection for studies focusing on behavioral or physiological outcomes that have been shown to be influenced by rearing condition. For example, while it is very unclear how circulating numbers of lymphocytes translate into functional immune outcomes (Westermann and Pabst, 1990), the substantial differences in $CD4^+$ and $CD8^+$ cell numbers from indoor-raised animals in general, compared with outdoor-raised animals, should cause investigators to be careful when selecting subjects for studies of immune function. Similarly, differences in HPA regulation have been found to affect immune function (e.g., DeRijk *et al.*, 1996); the fact that NR animals differ substantially from others might be a concern for immunological studies, or for studies focusing on glucose metabolism.

The potential problems that differences arising from rearing history and other sources pose for experimental studies are complex. On the one hand, as just suggested, attention must be paid to subject selection. For example, it is possible to use only animals from the same rearing condition (e.g., all FR or all CR, but not a mix of FR and CR animals), the net effect of which is to increase the homogeneity of the sample. By reducing variability in this way, smaller samples can probably be utilized, since statistical power is inversely related to variability. On the other hand, this strategy limits generalizability. That is, it may not be possible to assume necessarily that results obtained on a homogeneous set of animals (e.g., FR animals) will generalize well to other conspecifics (e.g., CR animals), or even across species.

While using only animals from one rearing condition is probably wise in any event, it does not eliminate other sources of variability, ranging from personality factors (e.g., Maninger *et al.*, 2003), to differential expression of genes based on rearing history (Bennett *et al.*, 2002; P.J. Pierre *et al.*, Chapter 22, this volume). Another (or additional) solution that addresses the issue of variation is to utilize measures of individual differences as covariates in statistical analyses. However, such a strategy requires quantifying the relevant variables, a process that could be time consuming. Broad-based assessments of biobehavioral organization, such as the one we have begun at the CNPRC, may provide investigators with relevant information that could be useful either as selection criteria or as covariates in analyses.

A final implication of our results and approach to these issues pertains specifically to colony management. During the course of its life in a primate colony, an individual animal experiences a variety of procedures, ranging from the routine (e.g., daily cleaning and feeding) to the rare (e.g., relocation). Not all animals respond in the same way to such procedures, and animals that adapt poorly may be at increased risk for health problems, such as recurring diarrhea. Having quantified data on biobehavioral characteristics of animals from every rearing circumstance can help identify at-risk animals and inform decision making. In our colony, for example, a small but consistent percentage of NR animals typically develops health problems in response to relocation and group formation. Previously, these health problems were believed to be associated with the lack of exposure to breast feeding and associated maternal immunoglobulins in NR infants. This association may, in fact, reflect a more complex interaction of nutrition, temperament, and early experience. Knowing what the biobehavioral risk factors are that might be associated with that outcome can help with designing management strategies that reduce distress and ease the transition to group living for those that are at risk. Such an issue cuts across categories based on rearing condition, and focuses attention instead on the more fundamental issue of adaptation: some animals from any rearing condition will be less adaptable than others from the same condition.

In conclusion, we believe that a focus on group and individual differences in biobehavioral organization—their origin as well as their consequences—can contribute to the more efficient utilization of increasingly scarce captive primate resources. In terms of research such a focus can help with issues associated with experimental design that might help reduce animal numbers in experiments. Having quantified data may also be useful for husbandry purposes, in helping, for example, to find compatible social partners or identifying risk factors associated with compromised health. At an important conceptual level, a focus on individual differences can reveal the wide variability that exists among animals, and cause us to reexamine the notion of what a normal animal is. Nursery rearing definitely produces animals with particular characteristics—but then so does rearing in corncribs, in the natural habitats of China, and in the forests of India. We believe that an understanding of the causes and consequences of variation in biobehavioral organization, from whatever sources, will become an increasingly important factor in research involving nonhuman primates.

ACKNOWLEDGMENTS

Supported by National Institutes of Health grant RR00169 to the California National Primate Research Center. We thank C. Brennan, J. Horrocks, E. Lowe, C. Stanko, G. Vicino, R. Vogt, and the animal care staff (especially A. Cabrera, L. Dorsey, and R. Walgenbach) at CNPRC for their expert help and cooperation.

REFERENCES

Bennett, A.J., Lesch, K.P., Heils, A., Long, J.C., Lorenz, J.G., Shoaf, S.E., Champoux, M., Suomi, S.J., Linnoila, M.V., and Higley, J.D. (2002). Early experience and serotonin transporter gene variation interact to influence primate CNS function. *Mol. Psychiatry* 7:118–122.

Bicknese, E.J., George, J.W., Hird, D.W., Paul-Murphy, J., Anderson, J.A., and Roberts, J.R. (1993). Prevalence and risk factors for iron deficiency anemia in weanling rhesus macaques. *Lab. Anim. Sci.* 43:434–438.

Capitanio, J.P. (2004). Intra- and interspecific variation in personality. In: Thierry, B., Singh, M., and Kaumanns, W. (eds.), *Macaque Societies: A Model for the Study of Social Organization*. Cambridge University Press, Cambridge.

Capitanio, J.P., Mendoza, S.P., and Lerche, N.W. (1998). Individual differences in peripheral blood immunological and hormonal measures in adult rhesus macaques (*Macaca mulatta*): Evidence for temporal and situational consistency. *Am. J. Primatol.* 44:29–41.

Capitanio, J.P., Mendoza, S.P., and Baroncelli, S. (1999). The relationship of personality dimensions in adult male rhesus macaques to progression of simian immunodeficiency virus disease. *Brain Behav. Immun.* 13:138–154.

Capitanio, J.P., Mendoza, S.P., Mason, W.A., and Maninger, N. (2005). Rearing environment and hypothalamic-pituitary-adrenal regulation in young rhesus monkeys (Macaca mulatta). *Dev. Psychobiol.* 46:318–330.

Champoux, M., Coe, C.L., Schanberg, S.M., Kuhn, C.M., and Suomi, S.J. (1989). Hormonal effects of early rearing conditions in the infant rhesus monkey. *Am. J. Primatol.* 19:111–117.

Champoux, M., Metz, B., and Suomi, S. (1991). Behavior of nursery/peer-reared and mother-reared rhesus monkeys from birth through 2 years of age. *Primates* 32:509–514.

Champoux, M., Suomi, S.J., and Schneider, M.L. (1994). Temperament differences between captive Indian and Chinese-Indian hybrid rhesus macaque neonates. *Lab. Anim. Sci.* 44:351–357.

Champoux, M., Kriete, M.F., Higley J. Dee, and Suomi, S.J. (1996). CBC and serum chemistry differences between Indian-derived and Chinese-Indian hybrid rhesus monkey infants. *Am. J. Primatol.* 39:79–84.

Champoux, M., Higley, J.D., and Suomi, S.J. (1997). Behavioral and physiological characteristics of Indian and Chinese-Indian hybrid rhesus macaque infants. *Dev. Psychobiol.* 31:49–63.

DeRijk, R.H., Petrides, J., Deuster, P., Gold, P.W., and Sternberg, E. (1996). Changes in corticosteroid sensitivity of peripheral blood lymphocytes after strenuous exercise in humans. *J. Clin. Endocrinol. Metab.* 81:228–235.

Gunderson, V., Grant-Webster, K., and Fagan, J.F. (1987). Visual recognition memory in high- and low-risk infant pigtailed macaques (Macaca nemestrina). *Dev. Psychol.* 23:671–675.

Hoffmann, R.A., and Stowell, R.E. (1973). Outdoor housing of nonhuman primates. *Lab. Anim. Sci.* 23:74–83.

Kriete, M.F., Champoux, M., and Suomi, S.J. (1995). Development of iron deficiency anemia in infant rhesus macaques. *Lab. Anim. Sci.* 45:15–21.

Laudenslager, M., Capitanio, J.P., and Reite, M. (1985). Possible effects of early separation experiences on subsequent immune function in adult macaque monkeys. *Am. J. Psychiatry* 142:862–864.

Lubach, G.R., Coe, C.L., and Ershler, W.B. (1995). Effects of early rearing environment on immune responses of infant rhesus monkeys. *Brain Behav. Immun.* 9:31–46.

Maninger, N., Capitanio, J.P., Mendoza, S.P., and Mason, W.A. (2003). Personality influences tetanus-specific antibody response in adult male rhesus macaques after removal from natal group and housing relocation. *Am. J. Primatol.* 61: 73–83.

Marthas, M.L., Lu, D., Penedo, M. C.T., Hendrickx, A.G., and Miller, C.J. (2001). Titration of an SIV mac251 stock by vaginal inoculation of Indian and Chinese origin rhesus macaques: Transmission efficiency, viral loads, and antibody responses. *AIDS Res. Hum. Retrovir.* 17:1455–1466.

Noldus, L.P.J.J. (1991). The observer: A software system for collection and analysis of observational data. *Behav. Res. Methods Instrum. Comput.* 23:415–429.

Shannon, C., Champoux, M., and Suomi, S.J. (1998). Rearing condition and plasma cortisol in rhesus monkey infants. *Am. J. Primatol.* 46:311–321.

Singh, S.D. (1968). Social interactions between the rural and urban monkeys, Macaca mulatta. *Primates* 9:69–74.

Westermann, J., and Pabst, R. (1990). Lymphocyte subsets in the blood: A diagnostic window on the lymphoid system? *Immunol. Today* 11:406–410.

CHAPTER TWELVE

Neurobehavioral Assessment of Nonhuman Primate Neonates

Mary L. Schneider, Maribeth Champoux, and Colleen F. Moore

1. INTRODUCTION

There are a number of reasons for assessing the neurobehavior of neonatal nonhuman primates. A central premise is that neurobehavior—characterized by rapid neuromotor development, increasing state regulation, and an emerging attentional system (Als, 1986)—is presumed to be linked to the growth of neural systems in the brain. Although the state of knowledge regarding brain–behavior relations in early development is rudimentary, new knowledge is emerging regarding the relation between early brain functioning and behavior (Nelson *et al.*, 2002). For instance, there is evidence that in large areas of the cerebral cortex in rhesus macaques, synapses grow and are pruned in approximately parallel cycles that seem to be closely related to broad cognitive changes (Goldman-Rakic, 1987; Rakic *et al.*, 1986). Thus, the study of neurobehavioral development during the neonatal period of life may provide some insight into the development of neural networks that support behavioral systems and may provide meaningful inferences related to central nervous system (CNS) integrity relevant to certain human disorders.

Because of the close link between brain development and neurobehavior, early neurobehavioral assessments have been used in studies of a variety of substances and conditions that either mimic or replicate disorders found in humans. For example, prenatal exposure to alcohol has been modeled in nonhuman primates (Clarren and Bowden, 1982; Schneider *et al.*, 1997, 2001), as has pre- and postnatal exposure to environmental toxins such as lead and PCBs (Levin *et al.*, 1988; Schantz *et al.*, 1991; Rice and Hayward, 1997). Naturally occurring conditions such as low birth weight and prenatal stress, which can compromise neurodevelopment in the human fetus, have been studied in nonhuman primates (Gunderson *et al.*, 1989; Schneider, 1992a,b,c; Schneider *et al.*, 1999; Schneider and Moore, 2000). Behavioral studies have also been done to characterize similarities and differences between human and nonhuman primate neonates (Hallock *et al.*, 1989; Bard *et al.*, 1992).

To facilitate our studies of neurobehavioral development in rhesus macaques, we developed a multiitem tool for measuring the rapid growth and development during the earliest postnatal period. Called the Primate Neonatal Neurobehavioral Assessment (PNNA), this panel of tests measures characteristics viewed as critical to early development, including the dimensions of state modulation or arousal, orienting or attention, and neuromotor maturity, all of which have important ramifications for later cognitive and social–emotional functioning. The primate assessment panel was based primarily on the best-known tool for assessing human neonates, the Brazelton Newborn Assessment Scale (NBAS; Brazelton, 1984), which itself was derived from other research on neonatal functioning, including the seminal work of Frances Graham and her colleagues on high-risk infants (Graham, 1956; Graham *et al.*, 1956, 1962). The PNNA also contains items adapted from existing nonhuman primate assessments (Mowbray and Caddell, 1962; Goodlin *et al.*, 1982) and from other human tests (Bayley, 1969; Ayres, 1976; Rothbart, 1981).

The impetus for developing the PNNA was twofold. First was the observation that process-oriented instruments designed to reflect dynamic aspects of behavior, such as Brazelton's NBAS, may be more useful than assessments of developmental milestones (such as motor milestones). What was needed was a panel of tests that could measure important, but generally neglected, early developmental outcomes, such as the complex components of self-regulatory capacities.

Tests for human infants were first developed and standardized in the early 1900s. However, most of the items involved visually guided manipulations of objects and gross and fine motor behavior of infants 3 months or older. In tests for older infants (18–24 months of age), the main developmental component was verbal in character (McCall *et al.*, 1972). A number of longitudinal studies initiated in the 1930s and 1940s were designed to predict an individual's developmental outcome from these types of infant measures (Gessell and Armatruda, 1941). But these efforts were characterized by many reviewers as having only a limited success. For the most part, correlations of infant test scores with later measures of IQ did not reach a level of strength that was clinically useful or theoretically interesting (Kopp and McCall, 1982). As Graham *et al.* remarked in 1956, "similar data have been interpreted by different standards. A correlation of a given size does not arouse the same response in all psychologists" (p. 30).

The first tests for nonhuman primate infants focused primarily on delineating milestones in motor development, such as the appearance and disappearance of reflexes (Mowbray and Caddell, 1962; King and King, 1970; Castell and Sackett, 1973; Ehrlich, 1974; King *et al.*, 1974; Sackett *et al.*, 1982). Other tests have included Apgar-type assessments of respiration rate, muzzle color, initial state, and muscle tone, along with observations of activities (locomotion and exploration), visual exploration, and scanning (Golub and Gershwin, 1984). Thus, there was a need for a process-oriented neonatal test for primates, one that could measure the developing motor system as well as self-regulation, sensory processing, and the emerging attention system.

The second impetus for developing the PNNA was interest, both historical and revitalized, in the influence of early life experiences on the development of the brain and behavior (Irwin, 1942; Shonkoff and Phillips, 2000). Animal studies are needed to study the dynamic and continuous interaction between biology and experience, in order to delineate the biobehavioral mechanisms for the effects of adverse early life experiences. For example, elegant studies of rodents have provided evidence that experiences during early development are assimilated into the individual's brain by altering the neural processes of the developing organism (Greenough and Black, 1992). This incorporation of experience in the form of altered neural substrates then influences the way the individual experiences new events (Boyce *et al.*, 1998; Cicchetti and Tucker, 1994; Kraemer, 1992). Animal studies have shown that what happens early in life is not only critical for later development, but also has potential for altering the developmental trajectory throughout life. Thus, we designed the PNNA to capture the nonhuman primate's earliest characteristics and capabilities and to investigate the vulnerability or resiliency of these characteristics to pre- and early postnatal insults.

In this chapter we will describe our test procedure and review data from experimental studies conducted with the PNNA at the University of Wisconsin-Madison Harlow Center for Biological Psychology and at the Laboratory of Comparative Ethology, National Institute of Child Health and Human Development. The initial study in which the PNNA was developed was the dissertation research of M.L. Schneider, conducted under the direction of Stephen J. Suomi and Colleen F. Moore at the University of Wisconsin-Madison (Schneider, 1987; Schneider and Suomi, 1992).

2. DESCRIPTION OF THE PNNA

The PNNA originally was designed to assess aspects of neurobehavior that show developmental changes during the first month of life. Subsequently, we combined some items into categories modeled closely on the Brazelton NBAS, namely, Orientation, Motor Maturity, Motor Activity, and State Control (Schneider *et al.*, 1991; Schneider and Suomi, 1992). The items in each category are defined in Table 12–1. Orientation items consist of neonatal orienting and visual following responses elicited by a visual stimulus (three-dimensional toy with Mickey-Mouse face) and lip-smacking sound made by the examiner. The approximate percentage of time during which the infant is attentive is rated, as well as the duration of gaze while orienting to the toy or sound. Neuromotor items in the Motor Maturity category include ratings of muscle tonus, coordination, labyrinthine righting, response speed, and spontaneous motor activity.

Table 12–1. Definition of Items on the Primate Neonatal Neurobehavioral Assessment

Items	Definitions
Orientation cluster	
Visual Orient	Eyes oriented toward plastic Mickey Mouse face held in four positions in infant's periphery (0 = no orient; 1 = brief direct contact; 2 = prolonged direct contact)
Visual Follow	Eyes follow moving toy in both horizontal and vertical directions (0 = eyes contact target but don't follow; 1 = eyes start to follow; 2 = eyes follow target completely)
Duration of Looking	Duration of gaze toward orienting items (0 = brief gaze; 1 = 1-sec gaze; 2 = ≥2-sec gaze)
Attention Span	Percentage of trials infant attends toward orienting items (0 = lack of attention on all items; 1 = infant attends in 25% of trials; 2 = infant attends in 75% of trials)
Motor Maturity cluster	
Head Posture Prone	Infant's ability to hold head up when held in prone position (0 = flaccid tone with head hanging down; 1 = head lifted but not maintained for 3 sec; 2 = head lifted and maintained for ≥3 sec)
Head Posture Supine	Infant's ability to hold head up when held in supine position (rated identical to Head posture prone)
Response Speed	Speed of response to stimuli (0 = 25% of responses quick; 1 = 75% of responses quick; 2 = all responses quick)
Coordination	Quality of movement (0 = clumsy movements; 1 = adequate movements; 2 = agile movements)
Labyrinthine Righting	Realignment of head when body is tilted 45° sideways (0 = no righting, head and body in same plane; 1 = head partially rights; 2 = head rights and lines up with the vertical plane)
Activity cluster	
Motor Activity	Amount of motor activity (0 = in motion 25% of the time; 1 = in motion 50% of the time; 2 = continuous motion)
Coordination	Quality of movement (0 = clumsy movements; 1 = adequate movements; 2 = agile movements)
Spontaneous Crawl	Maturity of locomotor pattern (0 = no reciprocal locomotor pattern; 1 = weak attempt to locomote; 2 = coordinated locomotor pattern)
Passive	Duration of time inactive (0 = no inactivity; 1 = inactive 50% of the time; 2 = inactive ≥75% of the time)
State Control cluster	
Irritability	Amount of distress noted (0 = distress minimal during testing; 1 = distress apparent 50% of time; 2 = distress observed continuously)
Consolability	Ease of consoling infant (0 = infant quickly consoled when picked up; 1 = infant consoled only after prolonged holding, swaddling, rocking, and/or stroking; 2 = infant inconsolable)
Struggle during Testing	Amount of squirming noted (0 = infant squirmed 25% of the time; 1 = infant squirmed 50% of the time; 2 = infant squirmed continuously)
Predominant State	Infant's behavioral state during examination (0 = alert, awake, aware; 1 = alert but somewhat agitated; 2 = extremely agitated throughout examination)

Reflex items include righting reactions, grasp reflex, rotation or tonic deviation of the head and eyes, Moro reflex, and postrotatory nystagmus, a measure of vestibular-ocular function. Muscle tonus is assessed in three conditions: (1) while the infant tries to maintain its head in a position against gravity as it is held in prone and supine conditions in the air; (2) while the infant's relaxed limbs are flexed and extended; and (3) while the infant actively contracts its limb muscles. Temperament ratings in the State Control category are based on behaviors observed during the administration of the orienting and neuromotor items. As the goal is to maintain the infant in a quiet, alert state throughout the examination, the infant's temperament is monitored constantly. Several temperament dimensions are rated, including consolability, irritability, and fearfulness during the testing session.

We administer the 20-min battery of developmental tests throughout the first 30 days postpartum. Although newborn primates do not show the pronounced diurnal activity patterns older primates do, testing at a standard time each day obviates the possibility of confounding effects relating to time of day. In our laboratories, testing occurs between 1000 and 1200 hr, about midway between feedings for nursery-reared infants.

The test begins with the examiner wrapping the infant in a diaper from the waist down, leaving the arms free to move. The Orientation items are administered first, followed by the Neuromotor, Motor Maturity, and State Control items in an invariant sequence. Each item is rated on a scale ranging from 0 to 2, with half-point scores allowed as in some human neonatal assessments. Observers are trained to reliability and are blind to the infants' pregnancy and rearing conditions.

After testing, four composite scores are computed for each infant based partly on prior factor analyses (Schneider *et al.*, 1991). The composite cluster scores are based on a joint consideration of factor analysis results as well as the strength of the Brazelton scale. Because we intended the PNNA results to serve as an animal analogue of human development, we based the cluster scores on items that not only loaded together in the factor analyses but also corresponded closely to items in the Brazelton scale for human infants. Thus, the cluster scores we use are not "factor scores," because they do not strictly follow the factor analysis. Rather, they parallel composite scores of the Brazelton NBAS.

3. STUDIES AT THE UNIVERSITY OF WISCONSIN

3.1. Comparisons of Nursery-Reared and Mother-Reared Infants

In the initial studies using the PNNA, we compared 36 (16 female, 20 male) nursery-reared infants with 28 (18 female, 10 male) mother-reared infants. The nursery-reared infants were separated from their mothers at birth and reared for the first 30 days of life in a primate nursery (Schneider and Suomi, 1992). Nursery-reared infants were tested with the PNNA on postnatal days 4, 6, 8, 11, 15, 18, 22, 25, and 29. The frequent assessment allows detailed mapping of the developmental course of behavior on each item. Mother-reared infants were assessed only twice (postnatal days 15 and 29) in order to minimize the stress of separation and decrease the risk of maternal rejection of the infant.

As expected, the nursery-reared infants displayed significant developmental changes across the first 30 days of life. Infant scores increased with age on Orientation, Motor Maturity, and Motor Activity. Similarly, for individual items within the categories, scores increased across age for visual orienting, visual following, reach and grasp, duration of looking, and attention span, as well as for muscle tonus, body righting, response speed, motor activity, coordination, locomotion, and balance. At the same time, primitive reflexes (Moro, rooting, grasp reflexes) decreased significantly across testing sessions. Items in the State Control category—i.e., response intensity, fearfulness, and distress to limitation—increased with age (see Schneider and Suomi, 1992, for details). The correlations of scores across test days both for individual items and for cluster scores were moderately stable, with the exception of the State Control cluster, which was unstable until approximately postnatal day 12. The cluster score stability correlations accounted for 10–49% of the variance. In fact, composite scores for Orientation, Motor Maturity, and Motor Activity were significantly correlated across the nine testing days from the first day of testing.

Comparisons of nursery-reared and mother-reared infants on days 15 and 29 revealed a number of interesting differences between the two groups. First, nursery-reared infants demonstrated higher sensory responsiveness, such as greater auditory startle and a stronger aversive reaction to tactile stimuli, than did mother-reared monkeys. This finding is very interesting when considered within the context of Denenberg's (1964) theory of monotonicity, which postulates that low stimulus input in infancy produces a highly reactive organism later in life. Others have noted that animals reared without species-typical stimulation exhaust the available sensory stimuli of their surroundings in early exploratory activity and therefore lack the opportunity to habituate to higher levels of stimulation (Mason, 1968; Harlow and Harlow, 1969; Rosenblum, 1971). When these animals are exposed to higher levels of input they quickly become overaroused and tend to demonstrate extreme reactions. Nursery-reared infants experience less species-typical tactile, auditory, and vestibular-proprioceptive stimulation than their mother-reared counterparts. For instance, mother-reared infants experience tactile input from maternal body contact and grooming, and they receive vestibular-proprioceptive input as the mother moves about, shifts positions, and rhythmically expands and contracts her ventrum while breathing. Mother-reared infants also are required to make adaptive movements in response to maternal movement, leading to meaningful proprioceptive feedback as well as additional vestibular input. In addition, mother-reared infants are more likely to be exposed to more meaningful auditory stimulation, such as their mother's vocalizations and internal sounds associated with the mother's heartbeat and gastrointestinal system. Changes in auditory input, such as alarm calls or increases in maternal heart rate, also provide informational cues about the environment. In contrast, nursery-reared infants are, for the most part, stationary in their cages, with the exception of self-generated movements, and engage primarily in visual exploration of their surroundings.

Given that nursery-reared monkey infants are more likely to be relatively stationary and are exposed to the coming and going of nursery staff and activities of other infants in adjacent cages, it is not surprising that they showed a precocious emergence of visual orienting and visual following compared with mother-reared infants. Similarly, they were superior to mother-reared infants on measures of reach and grasp, not unexpected given that this ability is dependent on visual functions, which developed earlier than in mother-reared infants. On the other hand, mother-reared infants were more active, showed quicker response speeds, were more irritable and difficult to console, and vocalized more, whereas nursery-reared infants were more passive and demonstrated more self-mouthing. Taken as a whole, the picture that emerges is that nursery-reared infants show precocious development of visual orienting, visual follow, and reach and grasp, exaggerated sensory responses to tactile and auditory stimuli, and increased passivity and self-directed activities compared with mother-reared infants (Schneider, 1987).

3.2. Effects of Prenatal Stress on Early Neurobehavior

The PNNA results showing the significant influence of early rearing environment suggested that the PNNA might be useful for the study of other early influences on neurodevelopment. Although it is commonly accepted that prenatal exposure to cigarettes, alcohol, and other drugs can have negative effects on offspring, it is less well known whether environmental events, such as psychological stress in the pregnant mother's life, also have adverse effects on the offspring's neurodevelopment. Confirmation of a possible link between prenatal stress and vulnerability to neurodevelopmental maladaptations or developmental psychopathology would have important ramifications for human health: vulnerable children could be referred for services early in life. Early intervention is important because it is a time of relatively high neural plasticity and behavioral flexibility, when environmental factors can have profound effects on later development (Guralnick and Bennett, 1987).

In his pioneering studies of prenatal stress in nonhuman primates, Sackett (1981) found that prenatal stress increased fetal loss in female pigtailed macaques that, on the basis of their prior breeding history, were considered to be at low risk for fetal loss. Later, Sackett and colleagues reported that prenatally stressed pigtailed macaques showed morphological alterations, such as exaggerated dermatoglyphic asymmetry (difference in pattern and number of dermal ridges between the left and right hand), which is associated with a higher incidence of infant mortality (Newell-Morris *et al.*, 1989). The research of Coe and Lubach (2003) has also yielded important results, demonstrating that prenatal stress can alter the developmental trajectory of the infant's immune responses.

These primate studies laid the groundwork for a series of studies on prenatal stress at the University of Wisconsin-Madison Harlow Center for Biological Psychology. In these studies, we randomly assigned pregnant female monkeys to receive daily stress or to receive no stress other than normal husbandry practices. Using the PNNA, we conducted repeated observations of the offspring's neonatal development across the first month of life, and continued to follow them into adulthood, assessing social and cognitive functions as well as neurotransmitter function using positron emission tomography neuroimaging techniques. The stressor consisted of removing the pregnant female from the home cage, transporting her to a darkened room, and administering three noise bursts (115 dB sound at 1 m, 1300 Hz) randomly over a 10-min period. All females in the stress condition received the stressor at the same time of day (1600 hr), five times per week. Controls were undisturbed during pregnancy, except for normal animal husbandry.

We conceptualize the stress treatment as a model of recurrent daily episodic stress. Noise has also been used as a prenatal stressor in many rodent studies (e.g., Fride and Weinstock, 1988). Studies with humans have shown that uncontrollable noise (such as aircraft, highway traffic, and construction noise) is a source of psychological stress (Ising *et al.*, 1990; Kryter, 1990; Evans *et al.*, 1995). Our data show that the noise stressor activates the hypothalamic–pituitary–adrenal (HPA) axis in the pregnant female, significantly raising plasma cortisol levels [baseline = 25.2 ± 2.2 μg/dl; post stress = 34.8 ± 2.4 μg/dl (mean ± SEM)].

The stress treatment was administered to pregnant females on days 90–145 of a 165-day gestation period. We refer to this timing of the stressor as "mid-late gestation" stress. In our first study, we specifically avoided the early gestation period in order to minimize the risk of inducing early fetal loss. Similarly, we avoided administering the stressor very late during gestation to reduce the chance of inducing early parturition, which could complicate the interpretation of our data (see Schneider, 1992a,b,c, for details). Twenty-four infant monkeys (12 prenatally stressed and 12 controls) were born to mothers in this study. After the infants were born, they were separated from their mothers and hand-reared in the laboratory primate nursery. All infants were reared according to standard nutritional protocol, i.e., with access to Similac formula (Schneider, 1992a). To minimize any potential negative effects associated with hand rearing, each infant cage was enriched with toys and climbing devices (Schneider and Suomi, 1992). The infants were also socialized in play groups with another infant from the same experimental condition for 15 min four times each week until they were approximately 30 days of age. Hand rearing in the nursery, although labor intensive, prevents any chance of confounding the prenatal stress condition with differential maternal treatment. It also permitted continuous access to the infant for testing.

Because the human literature suggests that prenatal stress is associated with preterm birth and with low birth weight (see Wadhwa, 1998), we tested for condition differences in these variables. Gestation duration did not differ significantly across conditions, but the mean weight of the prenatally stressed infants was significantly less than that of the control offspring (495.4 g ± 13.6 g versus 527.1 g ± 11.7 g). Even so, the birth weights of the prenatally stressed infants were within 1 SD of what is considered "normal" for rhesus macaques (mean ± SD = 501 g ± 64 g for males and 478 g ± 61 g for females based on 1270 rhesus macaques at the Harlow Center for Biological Psychology from 1973 through 1997).

When tested with the PNNA, prenatally stressed infants demonstrated reduced Motor Maturity and Motor Activity, and marginally reduced Orientation compared with control infants (see Schneider, 1992, for details). On individual items, they demonstrate reduced muscle tonus, slower response speed, less motor coordination and balance, and increased distractibility.

The next logical step was to determine whether a similar profile would emerge in prenatally stressed monkeys that were reared with their mothers. Thus in our next study, which we conducted to determine whether there is a sensitive period for prenatal stress effects (Schneider *et al.*, 1999), we tested two groups of infants that were raised with their mothers so as to eliminate the confound of differential rearing experiences. One group ($n = 10$) was stressed early in gestation (days 45–90) and the other ($n = 8$) during mid to late gestation (days 90–145). Controls were 13 infants whose mothers were undisturbed during pregnancy, except for normal animal husbandry. The stressor was the three noise bursts employed in our first study. Relying on Rakic's (1985, 1988, 1995) elegant studies of fetal brain development in rhesus macaques, our early stress period (days 45–90 postconception) approximates the phase of neural migration, while the mid to late gestation stress period (days 90–145) approximates the early phase of synaptogenesis. The primate brain undergoes three broad phases of development: generation of neurons (postconception days 0–40), neuronal migration (days 40–70/100), and synaptogenesis (day 112 postconception through the third month postnatally) (Rakic, 1988, 1995; Zecevic and Rakic, 1991). Knowing what type of brain development is occurring at the time of prenatal stress is important for the ultimate goal of relating the findings to human development, and also for comparing our results with those obtained with other species.

The 10 infants that experienced prenatal stress early in gestation had slightly smaller birth weights (mean = 474 g ± 19 g) than either those that were stressed in mid-gestation (mean = 560 g ± 15.2 g) or the control infants (mean = 516 g ± 13.8 g). PNNA results showed a pattern of scores that was strikingly similar to our results with nursery-reared monkeys (Schneider *et al.*, 1999). Prenatal stress affected Motor Maturity, Motor Activity, and Orientation, with prenatally stressed infants demonstrating poorer performance than controls. The effects were also significant when birth weight was entered as a covariate. Moreover, the infants that experienced prenatal stress during early gestation appeared to be affected more markedly than those that were stressed in mid gestation. Although infants from both early and mid to late gestation stress conditions scored lower than controls on measures of Orientation and Motor Maturity, infants that experienced early gestation stress had more pervasive and pronounced developmental delays. Moreover, Condition × Day of Testing interactions were significant, indicating that the pattern of development differed across groups. Whereas controls showed rapid developmental changes across testing days, reflecting the emergence and maturation of the attention and neuromotor systems, the monkeys that were stressed in either early or mid to late gestation showed flat or variable developmental trajectories (see Schneider *et al.*, 1999, for details). These results replicate the main findings of our previous experiment with nursery-reared infants, demonstrating generalization to a mother-reared sample, and also suggest that timing of prenatal stress is important.

Applying information from Rakic's studies of CNS development in fetal rhesus macaques, it appears that a period of enhanced vulnerability to prenatal stress occurs when neurons are migrating in order to form the appropriate neural pathways. It is interesting to note that studies of radiation exposure in humans have found that a critical period for the development of mental retardation later in life occurs between approximately week 8 and week 15 of gestation (Otake and Schull, 1984). Moreover, research has demonstrated that if neural migration is disrupted, an aberration in cell position can result. Abnormal neuronal migration in humans is hypothesized to be associated with a number of syndromes, including dyslexia, autism, and mental retardation (Galaburda *et al.*, 1989; Kotrla *et al.*, 1997). Researchers have posited several mechanisms for these effects. Abnormally positioned neurons may or may not find their normal targets during synaptogenesis, and so normal functions may not be established. Alternatively, the neurons might find their targets but change their normal pathway to their targets, or fail to contact the appropriate portion of the dendrites of their target cells (Caviness and Rakic, 1978; Stanfield *et al.*, 1979; Nowakowski and Hayes, 1999), resulting in connections that may or may not support functional capacities that are normal (Drager, 1981; Nowakowski and Hayes, 1999).

One might ask what role maternal endocrine activation plays in the effects of prenatal stress on offspring. The primate HPA axis is activated by stress episodes. More specifically, information relating to stress or challenges is integrated into the paraventricular nucleus of the hypothalamus by neurons expressing corticotropin-releasing factor (CRF) (Swanson *et al.*, 1983). CRF secretion stimulates the synthesis of the precursor protein proopiomelanocortin and the release of stored adrenocorticotrophic hormone (ACTH) from the anterior pituitary. Other neuroactive peptides, such as arginine vasopressin, are also expressed in CRF neurons and act along with CRF to stimulate ACTH release (Plotsky, 1991). ACTH then stimulates the synthesis and release of glucocorticoids, which mobilize energy during stress and act as transcriptional regulators (cortisol in primates and corticosterone in rats) from the adrenal cortex. Cortisol is the major hormonal product synthesized by the HPA axis in primates.

Studies with rats have shown that ACTH treatments to the pregnant dam affect offspring much as prenatal stress does. To determine whether this effect is also obtained in nonhuman primates, we used the PNNA to assess the neonatal neurobehavior of mother-reared infants that had been exposed to ACTH for 2 weeks *in utero* (Schneider *et al.*, 1992). The results were strikingly similar to those found in infants whose mothers had received the noise stress: compared with controls, these infants had a shorter attention span, less motor maturity, increased irritability, and decreased consolability. This study suggests that activation of the maternal HPA axis is at least part of the mechanism behind the prenatal stress effects on infant functioning.

Another important stressor in primates, both human and nonhuman, is disruption of social groups as a result of relocation. Psychological studies of human victims of technological disasters that involved family relocations have found long-lasting effects in adults, including depression, anxiety, and increased stress-related hormones (Davidson *et al.*, 1982; Baum *et al.*, 1983; Bromet *et al.*, 1990; Havenaar *et al.*, 1996; Cwikel *et al.*, 1997). Some studies have shown that relocation during the prenatal period due to the Chernobyl nuclear disaster had a negative effect on children's later intellectual and emotional functioning (Nyagu *et al.*, 1998; Kolominsky *et al.*, 1999), but some have not (Litcher *et al.*, 2000). These studies of child outcomes following a stressful family relocation during the prenatal period are difficult to interpret because of the many other variables involved, such as unknown doses of radiation at Chernobyl. In animals, however, it is possible to study relocation stress per se.

In an experiment with squirrel monkeys (*Saimiri boliviensis peruviensis*), pregnant females were stressed by being removed from their stable social groups and housed in new cages with unfamiliar pregnant females (Schneider and Coe, 1993). This manipulation was based on studies indicating that changes in the composition of monkey social groups result in marked changes in behavior, accompanied by autonomic and endocrine activity that persists for several weeks (Mendoza *et al.*, 1979; Kaplan *et al.*, 1990). Moreover, naturalistic studies on monkeys have shown that dominance relations during pregnancy can influence reproductive success and infant development (Wasser and Starling, 1988). Because we used squirrel monkeys rather than rhesus macaques, this experiment also helps establish the generalizability of the results to another nonhuman primate species. The study involved three different pregnancy conditions: mid gestation stress (monkeys were moved once during mid gestation), chronic stress (monkeys were moved three times), and no stress (control). The infants remained with their mothers during rearing except for a brief separation for PNNA testing at 15 days of age. This particular age point was selected because it provides the most reliable and sensitive measure of neurobehavior in the primate infant (Schneider and Suomi, 1992).

The PNNA results showed that compared with controls, infants from the chronic stress condition had shorter attention spans during the administration of the orientation items, less motor ability, and impaired balance. This pattern was strikingly similar to that observed in our study of hand-reared rhesus macaques born to mothers that had been exposed to a noise stressor for a 6-week period during mid to late gestation. Interestingly, these effects were not evident (compared with the control group) for the offspring of the monkeys that were relocated only once during mid gestation. Thus, the results suggest that chronic stress is more deleterious to infant neurobehavioral development than a single stressful event.

In summary, across four separate experiments with two species of nonhuman primates a neurobehavioral profile emerged showing an adverse effect of prenatal stress on Motor Maturity and Orientation. Because a very similar neurobehavioral profile was observed in the offspring of monkeys that received ACTH during pregnancy (but were not otherwise stressed), these data suggest that activation of the maternal HPA axis is one important mechanism of prenatal stress effects on neonatal neurobehavior.

3.3. Studies of Fetal Alcohol or Combined Alcohol and Prenatal Stress

The impetus for these studies was the possibility that prenatal stress could render the fetus more vulnerable to the adverse effects of teratogens, such as maternal alcohol consumption during pregnancy. Some epidemiological researchers have found evidence that the effects of prenatal alcohol exposure on the offspring are mediated by other factors in the environment such as low socioeconomic status, poor prenatal care, or maternal age and/or parity (Abel, 1995; Jacobson *et al.*, 1998). Therefore, it is important to pursue primate studies in which prenatal stress and fetal alcohol exposure are systematically manipulated. Subjects were female breeders that would consistently and voluntarily consume a solution of 0.6 g/kg alcohol sweetened with Nutrasweet (300 mg/100 ml). This dosage, which is comparable to one or two drinks for an average-sized woman, yielded blood alcohol concentrations of 20–50 mg/dl. Animals were randomly assigned to one of four treatment groups: Alcohol-only, Stress-only, Alcohol + Stress, and Control. Animals in the Alcohol groups received the alcohol solution daily throughout gestation. Control females consumed an isocaloric solution. The stress treatment was identical to that described above: three short (<1-sec) noise bursts randomly administered over a 10-min period while the animal was placed in a transport cage in a darkened room.

After birth, infants remained with their mothers and were separated weekly so that we could measure their growth and weight and administer the PNNA. All animals were tested on postnatal days 4, 9, 15, and 22 (+1) by testers who were blind to the experimental conditions of the subjects. The infants from the alcohol-alone condition showed reduced Orientation and less Motor Maturity overall than controls even though they were normal in birth weight, gestation length, and apparent facial dimensions. However, males from the alcohol + stress condition had reduced birth weights, and both males and females from the alcohol + stress condition showed slower response speed and less coordination than controls. Infants from the alcohol-only condition did not differ from controls on response speed and coordination, suggesting that prenatal stress might exacerbate alcohol-induced neuromotor impairments. Finally, although all alcohol-consuming and control females produced viable offspring, alcohol accompanied by prenatal stress resulted in 23% fetal losses, i.e., abortion and stillbirth (see Schneider *et al.*, 1997, for details).

In our next study, we examined the effects of the gestational timing of alcohol exposure in neonates from four conditions: (1) early alcohol exposure, in which mothers voluntarily consumed alcohol on gestation days 0–50; (2) mid to late alcohol exposure, in which mothers voluntarily consumed alcohol on gestation days 50–135; (3) continuous alcohol exposure, in which mothers consumed alcohol on days 0–135 or

0–165 (no differences were found across these groups so they were combined); and (4) control, in which mothers consumed an isocaloric control solution during pregnancy (see Schneider *et al.*, 2001, for details). PNNA results indicated that exposure to moderate levels of alcohol early in gestation significantly reduced scores on Orientation and Motor Maturity after controlling for birth weight. Alcohol exposure in mid to late gestation also resulted in reduced Motor Maturity. PNNA score was a more sensitive marker of early-gestation moderate alcohol exposure than were the growth parameters of birth weight or crown-to-rump length. The finding that early-gestation alcohol exposure had effects on neurodevelopment assessed by the PNNA raises the possibility that subtle neurodevelopmental effects might be induced even before pregnancy is detected.

4. STUDIES AT THE LABORATORY OF COMPARATIVE ETHOLOGY

4.1. General Differences from Studies at Wisconsin

The work at the Laboratory of Comparative Ethology has followed a different thematic track than that at the University of Wisconsin, in that the PNNA has been used to explore individual differences in biobehavioral traits. Predicated on the view that the neonatal assessment captures inherent, yet potentially modifiable, traits of the organism, the main focus of the work at the Laboratory of Comparative Ethology has been the elucidation of factors contributing to individual differences in neonatal organization, with consideration of both genetic and environmental factors. Nursery rearing has become a valuable adjunct to these studies for several reasons. Initially, nursery rearing was a useful counterpart to mother rearing, as it has been shown to generate predictable behavioral and physiological outcomes (Champoux *et al.*, 1989; Higley *et al.*, 1991a,b). Later it became evident that using nursery animals to study neonatal behavior enabled the assessment of genotype–environmental effects and interactions. Finally, the nursery environment entails standardized husbandry conditions and easy accessibility to animals, allowing the possibility of intervention studies that would be difficult to perform with mother-reared infants.

The PNNA has been modified slightly at the Laboratory of Comparative Ethology. Due to the large number of infants in each birth cohort, infant assessment was conducted only on days 14 and 30 in the first 2 years of the laboratory's existence (1981–1982) and on days 7, 14, 21, and 30 (or the closest weekday thereto) in subsequent years. The postrotary nystagmus test is omitted, as is the Moro reflex. An additional observational assessment is added in which we place the infant with a blanket and a small toy in a small testing cage for 5 min following the assessment of temperament. Also, we routinely collect blood and cerebrospinal fluid (CSF) samples under anesthesia following the assessments on days 14 and 30. Testing is typically conducted between 11:00 am and 12:00 pm, although assessments may be conducted between 1:00 pm and 2:00 pm if CSF sampling is not required. Animal husbandry procedures are similar to those at the University of Wisconsin, the National Institute of Child Health and Human Development (NICHD) protocol having been derived from the University of Wisconsin protocol when Laboratory Director Stephen Suomi moved from Wisconsin to the NICHD in the early 1980s. We have collected PNNA data from more than 200 animals from 12 birth-year cohorts at the NICHD laboratory.

4.2. Genetic Influence on Behavioral Development

In recent years there has been an explosion of technologies enabling the study of the relations between specific genetic endowments and behavior. We have been able to utilize some of these methods to begin to study the contribution of genetic factors to behavior as assessed by the PNNA in the rhesus macaque. Our first study was a retrospective comparison of infants that were of Indian origin with those that had one Indian-origin parent and one Chinese-origin parent (Champoux *et al.*, 1994). Although the majority of rhesus macaques at the Laboratory of Comparative Ethology are descendants of Indian-origin monkeys (as at the University of Wisconsin colony, from which they are derived), we recently obtained Chinese-origin animals to expand the genetic diversity of the colony. Rhesus macaques from the two origins appear to differ genetically on the basis of transferrin and mitochondrial markers (Annenkov, 1972; Zhang and Shi, 1993). When we began the study, anecdotal evidence had already suggested increased aggressiveness and irritability on the part of rhesus macaques imported from China, but no definitive data were available to document this phenomenon. The nursery environment provided the ideal setting to compare infants of these two strains because there is no parental social influence on the monkeys in the nursery.

The impetus for this retrospective study came from several incidents in which even very young nursery animals attempted to bite the examiner during administration of the PNNA. This atypical behavior prompted further examination of the background of the infants. In all cases, the biting infants were Chinese–Indian hybrids. We then performed a retrospective comparison of all Indian-origin ($n = 29$; 14 males and 15 females) and Chinese–Indian hybrid ($n = 13$; 8 males and 5 females) nursery infants on the neonatal assessment. Compared with Indian-origin infants, hybrid animals obtained significantly lower scores on the Orientation cluster and significantly higher scores on the State Control cluster, indicating higher levels of distress. In addition, hybrid animals scored lower on individual tests of attention, tactile sensitivity, and fine motor ability. They were rated as more fearful and less cuddly than Indian-origin animals, and exhibited more behaviors indicative of emotional arousal such as tremors and large-body spasms. Because all animals were reared under identical conditions, the presumption is that genetic factors were the cause of these group differences. Although prenatal factors such as differential prenatal stress in Chinese and Indian dams cannot be ruled out, this explanation seems less likely, as hybrid infants were obtained from both Indian and Chinese dams.

Two other retrospective studies lent additional support to the genetic basis of variability in scores on the PNNA. The first study compared 36 mother-reared and 79 nursery-reared animals that were characterized for the serotonin transporter promoter polymorphism (5-HTTLPR) (Champoux *et al.*, 2002a). In humans, variability at this locus is associated with measures of serotonin-mediated behaviors and personality characteristics, such as neuroticism and anxiety-related traits (Heils *et al.*, 1997; Lesch and Mossner, 1998) and with neonatal orientation and emotionality (Auerbach *et al.*, 1999; Ebstein *et al.*, 1998), although conflicting data have been reported (Ebstein *et al.*, 1997). A similar functional polymorphism has been identified in rhesus macaques (Lesch *et al.*, 1997) and has been found to be significantly associated with levels of the serotonin metabolite 5-hydroxyindoleacetic acid (5-HIAA) in the cerebrospinal fluid (Bennett *et al.*, 2002).

We identified 80 animals that were homozygous for the long allele (26 mother-reared and 54 nursery-reared) and 35 animals that were heterozygous for long and short alleles of 5-HTTLPR (10 mother-reared and 25 nursery-reared) (see Champoux *et al.*, 2002a, for details). Only two infants were homozygous for the short allele. As in the studies at Wisconsin, mother-reared and nursery-reared animals differed significantly on several dimensions. However, a main effect of genotype was observed for the State Control cluster, indicating that heterozygous individuals were more distressed during the examination, regardless of rearing condition. In addition, there was a significant interaction of rearing environment and genotype for the Orientation cluster. For mother-reared animals, genotype did not affect orientation scores; however, for nursery-reared animals, heterozygotes with respect to 5-HTTLPR exhibited significantly lower orientation scores than homozygotes. This difference was most evident on test days 21 and 30. Thus, expression of this genotype difference was evident only in the nursery environment. Interestingly, Bennett *et al.* (2002) reported an identical gene–environment interaction for 5-HIAA outcomes in rhesus macaques. In other words, nursery-reared animals with low levels of serotonin were more likely to be heterozygotes for the short and long allele variants; however, this association was not found in mother-reared animals. There are several interpretations of this result. First, it is possible that maternal influence may protect against or buffer differential genotypic expression. Second, in individuals of greater or lesser susceptibility, the relative privation of the nursery environment may lead to differential expression of outcomes. And third, it may simply be that the test instrument itself is a less sensitive measure of developmental functioning for mother-reared infants than for nursery-reared infants. Nonetheless, it is evident that the early rearing environment provides a powerful venue for the assessment of how genetic influences on neonatal behavior are modified by experience.

An additional study confirmed that variability in the outcome of the PNNA is based at least partially on genetics (Champoux *et al.*, 1999).

This study included 269 animals (134 mother-reared and 135 nursery-reared) from eight birth-year cohorts; all monkeys were members of a single pedigree that extended over five generations. Identity of sire and dam were available for all but a few animals, allowing the construction of a unitary pedigree structure using the program PedSys (Southwest Foundation for Biomedical Research, San Antonio, TX). As PNNA data from some animals were available only for days 14 and 30, we used only these days for the heritability analyses. We used a variance decomposition model to estimate the heritability of PNNA cluster scores using the computer program SOLAR (Almasy and Blangero, 1998). Covariates in the analyses were sex of infant and rearing condition. With these covariates controlled, heritability values significantly greater than zero were obtained for State Control, Motor Activity, and Orientation scores on days 14 and 30 and for the Motor Maturity score on day 30. Therefore, it appears that temperament characteristics reflecting attention, activity, and affective intensity are subject to heritable influence in nonhuman primates. Although it was not possible to conduct separate analyses for mother-reared and nursery-reared infants, on the basis of our work with 5-HTTLPR, we would expect heritability scores to be higher in nursery-reared infants than in mother-reared infants. Whether this might be the result of the decreased environmental variation of the nursery environment, the increased challenge of adapting to the nursery environment, or some mitigating effects of maternal rearing is an open question.

4.3. Nutritional Influence on Development

For studies of nutritional and other environmental influences on behavioral development and expression, the primate nursery is valuable because conditions can be manipulated in a controlled fashion and the effects can be determined with the PNNA. In one study, we increased the long-chain polyunsaturated fatty acid (LC-PUFA) composition of the infant formula to a level similar to that in mother's milk (Champoux *et al.*, 2002b). It has been suggested that the higher IQ scores, educational attainment, and achievement scores in breast-fed infants are due at least partially to the presence of these essential fatty acids in breast milk (e.g., Horwood and Fergusson, 1998). It is well established that these compounds affect visual physiology and function, and possibly cognitive and behavioral functions, in human and monkey infants (e.g., Neuringer *et al.*, 1986; SanGiovanni *et al.*, 2000a,b; Birch *et al.*, 2000). In our study, 14 nursery infants received the standard formula (a blend of Primilac and Similac formulas) and 14 infants received the identical formula supplemented with docosahexaenoic acid (DHA; 22:6n-3) and arachidonic acid (AA; 20:4n6). Plasma levels of these fatty acids reflected the feeding regimen; supplemented infants exhibited levels of AA and DHA similar to those of mother-reared (breast-fed) infants. PNNA results indicated that infants in the supplemented condition obtained higher scores on the Orientation and Motor Maturity scores than the control infants. These differences were most pronounced on days 7 and 14, a result that suggests earlier maturation of specific visual and motor abilities in the supplemented infants. Although the group differences disappeared by day 30, this finding probably reflects a "ceiling effect" due to all monkeys reaching the appropriate developmental levels on the PNNA. Follow-up tests that are more sensitive to behaviors that develop after day 30 might have revealed the continuation of group differences.

5. SUMMARY AND FUTURE DIRECTIONS

The studies described in this chapter demonstrate the utility and benefits of using neurodevelopmental assessments in nursery-reared infants on several fronts.

Although characterization of infants' early neurobehavioral development is a worthy venture in and of itself, the crucial question is whether the individual differences in the behaviors assessed by the PNNA will serve as markers or indicators of characteristics, or as predictors of later life outcomes. For example, at the University of Wisconsin's Harlow Center for Biological Psychology, we have found that low PNNA scores on Orientation and Motor Maturity were correlated with more trials to criterion (inferior performance) on acquisition performance of the non-match-to-sample learning task assessed when the monkeys were adolescents (Schneider *et al.*, 2001). The challenge for our two laboratory groups, as well as for others, is to determine to what extent the neonatal behaviors represent inherent characteristics of the organism, and how those potentially inherent characteristics interact with experience throughout development to create the observed behavioral phenotype.

Further research is needed to understand the importance of early life experiences during both the pre- and postnatal periods, as well as the interactive influences of genetics and environment on the unfolding of behavior. Behavioral studies with nonhuman primates often provide the critical causal link between experimental findings in rodents, which differ quite dramatically from humans, and correlational studies of humans in epidemiological research. Primate models provide the opportunity to assess the effect of the timing of early experiences, and to investigate the balance between risk and protective factors. They also afford an opportunity to evaluate how interventions can improve the likelihood of more positive developmental outcomes.

It is clear that the PNNA is sensitive to both environmental and genetic influences on early development. Thus, another fruitful area for future research is the interaction between prenatal stress and postnatal environmental factors. For example, the PNNA could be used to document developmental outcomes when evaluating the capacity to improve outcome through planned postnatal interventions or when studying how prenatal stress interacts with teratogens such as nicotine, alcohol, and other drugs of abuse. It is important to investigate whether one particular variable potentiates the impact of another variable, so that the effect of the two variables together might surpass the sum of the variables (Rutter, 1983). Also, how such variables have different effects on certain individuals, and how one variable might buffer or reduce the impact of another, are topics that need to be investigated (Rutter and Quinton, 1977).

Another future direction is to address how performance on the PNNA relates to performance on other tests. This might be useful for determining tasks that reflect similar domains that might shed light on specific brain functions. Procedures such as positron emission tomography (PET) using radioactive ligands, although not acceptable for experimental use with children, can be employed in research with primates. For example, in our current work, we have developed a technique allowing us to examine areas of the brain that are rich in dopamine receptors. Our data so far suggest that prenatal stress alone and in combination with prenatal alcohol exposure produces an up-regulation of D2 receptor binding (Roberts *et al.*, 1999), possibly due to a reduction in availability of dopamine.

Primate studies can provide the opportunity to test possible etiological factors together and separately, and to determine neurobehavioral profiles associated with etiological factors that have relevance for human disorders. For example, using brain imaging monkeys with and without hyperactivity could be examined to investigate whether the distinction in neonatal neurobehavioral profiles is reflected in brain structure and function.

Studies of the sort proposed here might ultimately have important implications for treatment and intervention programs. Better and earlier identification of at-risk children, using sensitive neurobehavioral measures, can lead to early intervention for those who show early neurobehavioral and neuromotor effects. Early treatment is especially critical for children with unstable family environments, before these early effects translate into later behavioral problems and cognitive deficits. Intervention may be most effective when it occurs in early childhood, rather than waiting until adolescence or early adulthood (see Guralnick and Bennett, 1987, for a review).

ACKNOWLEDGMENTS

This research was supported in part by grants from the National Institute of Alcohol Abuse and Alcoholism (AA10079, AA12277) (M.L.S. and C.F.M.) and by the Intramural Research Program of the National Institute of Child Health and Human Development (M.C.).

REFERENCES

Abel, E.L. (1995). An update on incidence of FAS: FAS in not an equal opportunity birth defect. *Neurotoxicol. Teratol.* 17:437–443.

Almasy, L., and Blangero, J. (1998). Multipoint quantitative trait linkage analysis in general pedigrees. *Am. J. Hum. Genet.* 62:1198–1211.

Als, H. (1986). A synactive model of neonatal behavioral organization: Framework for the assessment of neurobehavioral development in the premature infant and for support of infants and parents in the neonatal intensive care environment. *Phys. Occup. Ther. Pediatr.* 6(3/4):3–53.

Annenkov, H.A., Mirvis, A.B., and Kortidadze, H.G. (1972). Geographical transferrin polymorphism in *Macaca mulatta*. *Primates* 13:235–242.

Auerbach, J., Geller, V., Lezer, S., Shinwell, E., Belmaker, R.H., Levine, J., and Ebstein, R. (1999). Dopamine D4 receptor (D4DR) and serotonin transporter promoter (5-HTTLPR) polymorphisms in the determination of temperament in 2-month-old infants. *Mol. Psychiatry* 4:369–373.

Ayres, A.J. (1976). *Southern California Postrotary Nystagmus Test Manual.* Western Psychological Services, Los Angeles.

Bard, K.A., Platzman, K.A., Lester, B.M., and Suomi, S.J. (1992). Orientation to social and nonsocial stimuli in neonatal chimpanzees and humans. *Infant Behav. Dev.* 15:43–56.

Baum, A., Fleming, R., and Singer, J.E. (1983). Coping with victimization by technological disaster. *J. Soc. Issues* 39:117–138.

Bayley, N. (1969). *Manual for the Bayley Scales of Infant Development.* Psychological Corporation, New York.

Bennett, A.J., Lesch, K.P., Long, J.C., Lorenz, J.G., Shoaf, S.E., Champoux, M., Suomi, S.J., Linnoila, M.V., and Higley, J.D. (2002). Early experience and serotonin transporter gene variation interact to influence primate CNS function. *Mol. Psychiatr.* 7:118–122.

Birch, E.E., Garfield, S., Hoffman, D.R., Uauy, R., and Birch, D.G. (2000). A randomized controlled trial of early dietary supplement of long-chain polyunsaturated fatty acids and mental development in term infants. *Dev. Med. Child Neurol.* 42:174–181.

Boyce, W.T., Frank, E., Jensen, P.S., Kessler, R.C., Nelson, C.A., Steinberg, L., and The MacArthur Foundation Research Network on Psychopathology and Development. (1998). Social context in developmental psychopathology: Recommendations for future research from the MacArthur Network on Psychopathology and Development. *Dev. Psychopathol.* 10:143–164.

Brazelton, T.B. (1984). *Neonatal Behavioral Assessment Scale*, 2nd ed. Lippincott, Philadelphia.

Bromet, E.J., Parkonson, D.K., and Dunn, L.O. (1990). Long-term mental health consequences of the accident at Three Mile Island. *Int. J. Ment. Health* 19:48–60.

Castell, R., and Sackett, G. (1973). Motor behaviors of neonatal rhesus monkeys: Measurement techniques and early development. *Dev. Psychobiol.* 6:191–202.

Caviness, V.S., and Rakic, P. (1978). Mechanisms of cortical development: A view from mutations in mice. *Annu. Rev. Neurosci.* 1:297–326.

Champoux, M., Coe, C.L., Schanberg, S.M., Kuhn, C.M., and Suomi, S.J. (1989). Hormonal effects of early rearing conditions in the infant rhesus monkey. *Am. J. Primatol.* 19:111–117.

Champoux, M., Suomi, S.J., and Schneider, M.L. (1994). Temperament differences between captive Indian and Chinese-Indian hybrid rhesus macaque neonates. *Lab. Anim. Sci.* 44:351–357.

Champoux, M., Jaquish, C.E., Higley, S.B., and Suomi, S.J. (1999). Heritability of standardized biobehavioral assessment scores in rhesus monkey infants. *Am. J. Primatol.* 49:42 (Abst.).

Champoux, M., Bennett, A., Shannon, C., Higley, J.D., Lesch, K.P., and Suomi, S.J. (2002a). Serotonin transporter gene polymorphism, differential early rearing, and behavior in rhesus monkey neonates. *Mol. Psychiatr.* 7:1058–1063.

Champoux, M., Hibbeln, J.R., Shannon, C., Majchrzak, S., Suomi, S.J., Salem, N., and Higley, J.D. (2002b). Essential fatty acid formula supplementation and neuromotor development in rhesus monkey neonates. *Pediat. Res.* 51:1–9.

Cicchetti, C., and Tucker, D. (1994). Development and self-regulatory structures of the mind. *Dev. Psychopathol.* 6:533–549.

Clarren, S.K., and Bowden, D.M. (1982). Fetal alcohol syndrome: A new primate model for binge drinking and its relevance to human ethanol teratogenesis. *Pediatrics* 101:819–824.

Coe, C.L., and Lubach, G.R. (2003). Critical periods of special health relevance for psychoneuroimmunology. *Brain Behav. Immun.* 17:3–12.

Coe, C.L., Lubach, G.R., and Schneider, M.L. (1999). Neuromotor and socioemotional behavior in the young monkey is presaged by prenatal conditions. In: Lewis, M., and Ramsay, D. (eds.), *Soothing and Stress.* Erlbaum, Hillsdale, NJ, pp. 19–38.

Cwikel, J., Abdelgani, A., Goldsmith, J.R., Quastel, M., and Yevelson, I.I. (1997). Two-year follow-up study of stress-related disorders among immigrants to Israel from the Chernobyl area. *Env. Health Perspect.* 105(Suppl. 6):1545–1550.

Davidson, L.M., Baum, A., and Collins, D.L. (1982). Stress and control-related problems at Three Mile Island. *J. Appl. Soc. Psychol.* 12:349–359.

Drager, U.C. (1981). Observations on the organization of the visual cortex in the reeler mouse. *J. Comp. Neurol.* 201:555–570.

Ebstein, R.P., Gritsenko, I., Nemanov, L., Frisch A., Osher, Y., and Belmaker, R.H. (1997). No association between the serotonin transporter gene regulatory region polymorphism and the Tridimensional Personality Questionnaire (TPQ) temperament of harm avoidance. *Mol. Psychiatr.* 2:224–226.

Ebstein, R.P., Levine, J., Geller, V., Auerbach, J., Gritsenko, I., and Belmaker, R.H. (1998). Dopamine D4 receptor and serotonin receptor promoter in the determination of neonatal temperament. *Mol. Psychiatry* 3:238–246.

Ehrlich, A. (1974). Infant development in two prosimian species: Greater galago and slow loris. *Dev. Psychol.* 7:439–454.

Evans, G.W., Hygge, S., and Bullinger, M. (1995). Chronic noise and psychological stress. *Psychol. Sci.* 6:333–338.

Fride, E., and Weinstock, M. (1988). Prenatal stress increases anxiety related behavior and alters cerebral lateralization of dopamine activity. *Life Sci.* 42:1059–1065.

Galaburda, A.M., Rosen, G.D., and Sherman, G.F. (1989). The neural origin of developmental dyslexia: Implications for medicine, neurology, and cognition. In: Galaburda, A.M. (ed.), *From Reading to Neurons.* MIT Press, Cambridge, pp. 377–404.

Gessell, A., and Armatruda, C. (1941). *Developmental Diagnosis.* Hoeber, New York.

Goldman-Rakic, P.S. (1987). Circuitry of the primate prefrontal cortex and regulation of behavior by representational memory. In: Plum, F., and Mountcastle, V. (eds.), *Handbook of Physiology*, Vol. 5. American Physiological Society, Bethesda, MD, pp. 373–418.

Golub, M.S., and Gershwin, M.E. (1984). Standardized neonatal assessment in the rhesus monkey. In: Nathanielsz, P.W., and Parer, J.T. (eds.), *Research in Perinatal Medicine.* Perinatology Press, New York, pp. 55–81.

Goodlin, B.L., Caffery, S.A., and Sackett, G.P. (1982). Neonate behavioral assessment of pigtail macaques (*Macaca nemestrina*). *Int. J. Primatol.* 3:286.

Graham, F.K. (1956). Behavioral differences between normal and traumatized newborns. I. The test procedures. *Psychol. Monogr.* 70(427). (Entire issue.)

Graham, F.K., Matarazzo, R.G., and Caldwell, B.M. (1956). Behavioral differences between normal and traumatized newborns: II. Standardization, reliability, and validity. *Psychol. Monogr.* 70(428). (Entire issue.)

Graham, F.K., Ernhart, C.B., Thurson, D., and Craft, M. (1962). Development three years after perinatal anoxia and other potentially damaging newborn experiences. *Psychol. Monogr.* 76(522). (Entire issue.)

Greenough, W.T., and Black, J.E. (1992). Induction of brain structure by experience: Substrates for cognitive development. In: Gunnar, M.R., and Nelson, C.A. (eds.), *Developmental Behavioral Neuroscience*, Vol. 24. Erlbaum, Hillsdale, NJ.

Gunderson, V.M., Grant-Webster, K.S., and Sackett, G.P. (1989). Deficits in visual recognition in low birth weight infant pigtailed monkeys (*Macaca nemestrina*). *Child Dev.* 52:389–392.

Guralnick, M.J., and Bennett, F.C. (1987). A framework for early intervention. In: Guralnick, M.J., and Bennett, F.C. (eds.), *The Effectiveness of Early Intervention for At-Risk and Handicapped Children.* Academic Press, New York, pp. 3–29.

Hallock, M.B., Worobey, J., and Self, P.A. (1989). Behavioral development in chimpanzee (*Pan troglodytes*) and human newborn across the first month of life. *Int. J. Behav. Dev.* 12:527–540.

Harlow, H.F., and Harlow, M.K. (1969). Effects of various mother-infant relationships on rhesus monkey behaviors. In: Foss, B. (ed.), *Determinants of Infant Behavior*, Vol. 4. Methuen, London.

Havenaar, J.M., Van den Brink, W., Van den Bout, J., Kasyanenko, A.P., *et al.* (1996). Mental health problems in the Gomel region (Belarus): An analysis of risk factors in an area affected by Chernobyl disaster. *Psychol. Med.* 26:845–855.

Heils, A., Mossner, R., and Lesch, K.P. (1997). The human serotonin gene polymorphism: Basic research and clinical implications. *J. Neural Transm.* 104:1005–1014.

Higley, J.D., Hasert, M.F., Suomi, S.J., and Linnoila, M. (1991a). Nonhuman primate model of alcohol abuse: Effects of early experience, personality, and stress on alcohol consumption. *Proc. Natl. Acad. Sci. USA* 88:7261–7265.

Higley, J.D., Suomi, S.J., and Linnoila, M. (1991b). CSF monoamine metabolite concentrations vary according to age, rearing, and sex, and are influenced by the stressor of social separation in rhesus monkeys. *Psychopharmacology* 103:551–556.

Horwood, L.J., and Fergusson, D.M. (1998). Breastfeeding and later cognitive and academic outcomes. *Pediatrics* 101:9–20.

Irwin, D.C. (1942). Can infants have IQ's? *Psychol. Rev.* 49:69–79.

Ising, H., Rebentisch, E., Poustka, F., and Curio, I. (1990). Annoyance and health risk caused by military low-altitude flight noise. *Int. Arch. Occup. Environ. Health* 62:357–363.

Jacobson, S.W., Jacobson, J.L., Sokol, R.J., and Ager, J.W. (1998). Relation of maternal age and pattern of pregnancy drinking to functionally significant cognitive deficit in infancy. *Alcohol. Clin. Exp. Res.* 22:345–351.

Kaplan, J.R., Manuck, S.B., and Gatsonis, C. (1990). Heart rate and social status among male cynomolgus monkeys (*Macaca fascicularis*) housed in disrupted social groupings. *Am. J. Primatol.* 21:175–187.

King, J.E., and King, P.A. (1970). Early behaviors in hand-reared squirrel monkeys (*Saimiri sciureus*). *Dev. Psychobiol.* 2:251–256.

King, J.E., Fobes, J.T., and Fobes, J.L. (1974). Development of early behaviors in neonatal squirrel monkeys and cotton-top tamarins. *Dev. Psychobiol.* 7:97–109.

Kolominsky, Y., Igumnov, S., and Drozdovitch, V. (1999). The psychological development of children from Belarus exposed in the prenatal period to radiation from the Chernobyl atomic power plant. *J. Child Psychol. Psychiatry* 40:299–305.

Kopp, C.B., and McCall, R.B. (1982). Predicting later mental performance for normal, at-risk, and handicapped infants. In: Baltes, P.B., and Brim, O.G. (eds.), *Life-Span Development and Behavior*, Vol. 4. Academic Press, New York, pp. 33–61.

Kotrla, K.J., Sater, A.K., and Weinberger, D.R. (1997). Neuropathology, neurodevelopment and schizophrenia. In: Keshavan, M.S., and Murray, R.B. (eds.), *Neurodevelopment and Adult Psychopathology*. Cambridge University Press, Cambridge, UK, pp. 187–198.

Kraemer, G.W. (1992). A psychobiological theory of attachment. *Behav. Brain Sci.* 15:493–511.

Kryter, K.D. (1990). Aircraft noise and social factors in psychiatric hospital admission rates: A re-examination of some data. *Psychol. Med.* 20:395–411.

Lesch, K.P., and Mössner, R. (1998). Genetically driven variation in serotonin uptake: Is there a link to affective spectrum, neurodevelopmental, and neurodegenerative disorders? *Biol. Psychiatry* 44:179–192.

Lesch, K.P., Meyer, J., Glatz, K., Flügge, G., Hinney, A., Hebebrand, J., Klauck, S.M., Poustka, A., Poustka, F., Bengel, D., Mössner, R., Riederer, P., and Heils, A. (1997). The 5-HT transporter gene-linked polymorphic region (5-HTTLPR) in evolutionary perspective: Alternative biallelic variation in rhesus monkeys. *J. Neural Transm.* 104:1259–1266.

Levin, E.D., Schneider, M.L., Ferguson, S.A., Schantz, S.L., and Bowman, R.E. (1988). Behavioral effects of developmental lead exposure in rhesus monkeys. *Dev. Psychobiol.* 21:371–382.

Litcher, L., Bromet, E.J., Carlson, G., *et al.* (2000). School and neuropsychological performance of evacuated children in Kyiv 11 years after the Chernobyl disaster. *J. Child Psychol. Psychiatry* 41:291–299.

Mason, W. (1968). Naturalistic and experimental investigations of the social behavior of monkeys and apes. In: Jay, P. (ed.), *Primates: Studies in Adaptation and Variability*. Rinehart and Winston, New York, pp. 398–420.

McCall, R.B., Hogarty, P.S., and Hurlburt, N. (1972). Transitions in infant sensori-motor development and the prediction of childhood IQ. *Am. Psychol.* 27:728–748.

Mendoza, S., Coe, C.L., and Levine, S. (1979). Physiological response to group formation in the squirrel monkey. *Psychoendocrinology* 3:221–229.

Mowbray, J.B., and Caddell, T.E. (1962). Early behavior patterns in rhesus monkeys. *J. Comp. Physiol. Psychol.* 55:350–357.

Nelson, C.A., Bloom, F.E., Cameron, J.L., Amaral, D., Dahl, R.E., and Pine, D. (2002). An integrative, multidisciplinary approach to the study of brain-behavior relations in the context of typical and atypical development. *Dev. Psychopathol.* 14:499–520.

Neuringer, M., Connor, W.E., Lin, D.S., Barstad, L., and Luck, S. (1986). Biochemical and functional effects of prenatal and postnatal omega-3 fatty acid deficiency on retina and brain in rhesus monkeys. *Proc. Natl. Acad. Sci. USA* 83:4021–4025.

Newell-Morris, L.L., Fahrenbruch, C.E., and Sackett, G.P. (1989). Prenatal psychological stress, dermatoglyphic asymmetry and pregnancy outcome in the pigtailed macaque (*Macaca nemestrina*). *Biol. Neonate* 56(2):61–75.

Nowakowski, R.S., and Hayes, N.L. (1999). CNS development: An overview. *Dev. Psychopathol.* 11:395–417.

Nyagu, A.I., Loganovsky, K.N., and Loanovskaja, T.K. (1998). Psychophysiologic after-effects of prenatal irradiation. *Int. J. Psychophysiol.* 30:303–311.

Otake, M., and Schull, W.J. (1984). In utero exposure to A-bomb radiation and mental retardation. *Br. J. Radiol.* 57:409–414.

Plotsky, P.M. (1991). Pathways to the secretion of adrenocorticotropin: A view from the portal. *J. Neuroendocrinol.* 3:1–9.

Rakic, P. (1985). Limits of neurogenesis in primates. *Science* 227:154–156.

Rakic, P. (1988). Defects of neuronal migration and pathogenesis of cortical malformations. *Progr. Brain Res.* 73:15–37.

Rakic, P. (1995). Development of cerebral cortex in human and nonhuman primates. In: Lewis, M. (ed.), *Child and Adolescent Psychiatry*, 2nd ed. Williams & Wilkins, Baltimore, pp. 9–29.

Rakic, P., Bourgeois, J.P., Eckenhoff, M.F., Zecevic, N., and Goldman-Rakic, P.S. (1986). Concurrent overproduction of synapses in diverse regions of the primate cerebral cortex. *Science* 232:232–235.

Rice, D.C., and Hayward, S. (1997). Effects of postnatal exposure to a PCB mixture in monkeys on nonspatial discrimination reversal and delayed alternation performance. *Neurotoxicology* 18:479–494.

Roberts, A.D., DeJesus, O.J., Schneider, M.L., Schueller, M.J., Shelton, S., and Nickles, R.J. (1999). Dopamine system characterization of rhesus monkeys exposed to moderate dose alcohol in utero. Proceedings of the Society of Nuclear Medicine 46th Annual Meeting, Los Angeles, CA.

Rosenblum, L. (1971). The ontogeny of mother-infant relations in macaques. In: Moltz, H. (ed.), *Ontogeny of Vertebrate Behavior.* Academic Press, New York, pp. 315–367.

Rothbart, M.K. (1981). Measurement of temperament in infancy. *Child Dev.* 52:569–578.

Roughton, E.C., Schneider, M.L., Bromley, L.J., and Coe, C.L. (1998). Maternal endocrine activation during pregnancy alters neurobehavioral state in primate infants. *Am. J. Occup. Ther.* 52:90–98.

Rutter, M. (1983). Statistical and personal interactions: Facets and perspectives. In: Magnusson, D., and Allen, V. (eds.), *Human Development: An Interactional Perspective.* Academic Press, New York, pp. 295–319.

Rutter, M., and Quinton, D. (1977). Psychiatric disorder—Ecological factors and concepts of causation. In: McGurk, H. (ed.), *Ecological Factors in Human Development.* North-Holland, Amsterdam.

Sackett, G.P. (1981). A nonhuman primate model for studying causes and effects of poor pregnancy outcomes. In: Friedman, S., and Sigman, M. (eds.), *Preterm Birth and Psychological Development.* Academic Press, New York, pp. 41–63.

Sackett, G.P., Gunderson, V., and Baldwin, D. (1982). Studying the ontogeny of primate behavior. In: Fobes, J.L., and King, J.E. (eds.), *Primate Behavior.* Academic Press, New York.

SanGiovanni, J.P., Berkey, C.S., Dwyer, J.T., and Colditz, G.A. (2000a). Dietary essential fatty acids, long-chain polyunsaturated fatty acids, and visual resolution acuity in healthy fullterm infants: A systematic review. *Early Hum. Dev.* 57:165–188.

SanGiovanni, J.P., Parra-Cabrera, S., Colditz, G.A., Berkey, C.S., and Dwyer, J.T. (2000b). Meta-analysis of dietary essential fatty acids and long-chain polyunsaturated fatty acids as they relate to visual resolution acuity in healthy preterm infants. *Pediatrics* 105:1292–1298.

Schantz, S.L., Levin, E.D., and Bowman, R.E. (1991). Long-term neurobehavioral effects of perinatal polychlorinated biphenyl (PCB) exposure in monkeys. *Environ. Toxicol. Chem.* 10:747–756.

Schneider, M.L. (1987). *A Rhesus Monkey Model of Human Individual Differences.* Unpublished doctoral dissertation. University of Wisconsin-Madison.

Schneider, M.L. (1992a). The effect of mild stress during pregnancy on birth weight and neuromotor maturation in rhesus monkey infants (*Macaca mulatta*). *Infant Behav. Dev.* 15:389–403.

Schneider, M.L. (1992b). Delayed object permanence development in prenatally stressed rhesus monkey infants (*Macaca mulatta*). *Occup. Ther. J. Res.* 12(2):96–110.

Schneider, M.L. (1992c). Prenatal stress exposure alters postnatal behavioral expression under conditions of novelty challenge in rhesus monkey infants. *Dev. Psychobiol.* 25(7):141–152.

Schneider, M.L., and Coe, C.L. (1993). Repeated social stress during pregnancy impairs neuromotor development of the primate infant. *J. Dev. Behav. Pediatr.* 14(2):81–87.

Schneider, M.L., and Moore, C.F. (2000). Effect of prenatal stress on development: A nonhuman primate model. In: Nelson, C. (ed.), *Minnesota Symposium on Child Psychology.* Erlbaum, Hillsdale, NJ, pp. 201–243.

Schneider, M.L., and Suomi, S.J. (1992). Neurobehavioral assessment in rhesus monkey neonates (*Macaca mulatta*): Developmental changes, behavioral stability, and early experience. *Infant Behav. Dev.* 15(2):155–177.

Schneider, M.L., Moore, C., Suomi, S.J., and Champoux, M. (1991). Laboratory assessment of temperament and environmental enrichment in rhesus monkey infants (*Macaca mulatta*). *Am. J. Primatol.* 25:137–155.

Schneider, M.L., Coe, C.L., and Lubach, G.R. (1992). Endocrine activation mimics the adverse effects of prenatal stress on the neuromotor development of the infant primate. *Dev. Psychobiol.* 25(6):427–439.

Schneider, M.L., Roughton, E.C., and Lubach, G.R. (1997). Moderate alcohol consumption and psychological stress during pregnancy induces attention and neuromotor impairments in primate infants. *Child Dev.* 68:747–759.

Schneider, M.L., Roughton, E.C., Koehler, A., and Lubach, G.R. (1999). Growth and development following prenatal stress in primates: An examination of ontogenetic vulnerability. *Child Dev.* 70:263–274.

Schneider, M.L., Moore, C.F., and Becker, E.F. (2001a). Timing of moderate alcohol exposure during pregnancy and neonatal outcome in rhesus monkeys (*Macaca mulatta*). *Alcohol. Clin. Exp. Res.* 25:1238–1246.

Schneider, M.L., Moore, C.F., and Kraemer, G.W. (2001b). Moderate alcohol during pregnancy: Learning and behavior in adolescent rhesus monkeys. *Alcohol. Clin. Exp. Res.* 25:1383–1392.

Shonkoff, J.P., and Phillips, D.A. (eds.) (2000). *From Neurons to Neighborhoods: The Science of Early Childhood Development.* National Academy Press, Washington, D.C.

Stanfield, B.B., Caviness, V.S.J., and Cowan, W.M. (1979). The organization of certain afferents to the hippocampus and dentate gyrus in normal and reeler mice. *J. Comp. Neurol.* 185:461–483.

Swanson, L.W., Sawchenko, P.E., Rivier, J., and Vale, W. (1983). Organization of ovine corticotropin-releasing factor (CRF)-immunoreactive cells and fibers in the rat brain: An immunocytochemical study. *Neuroendocrinology* 36:165–186.

Wadhwa, P.D. (1998). Prenatal stress and life-span development. In: *Encyclopedia of Mental Health*, Vol. 3. Academic Press, Orlando, pp. 265–280.

Wasser, S., and Starling, A.K. (1988). Proximate and ultimate causes of reproductive suppression among female yellow baboons at Mikumi National Park, Tanzania. *Am. J. Primatol.* 16:97–121.

Zecevic, N., and Rakic, P. (1991). Synaptogenesis in monkey somatosensory cortex. *Cerebral Cortex* 1:510–523.

Zhang, Y., and Shi, L. (1993). Phylogeny of rhesus monkeys (*Macaca mulatta*) as revealed by mitochondrial DNA restriction enzyme analysis. *Int. J. Primatol.* 14:587–605.

CHAPTER THIRTEEN

Is It Nutrients or Nurturing? Comparison of the Growth and Development of Mother-Reared and Laboratory-Reared Macaque Infants (*Macaca nemestrina*)

Debra L. Durham and Laura L. Newell-Morris

1. INTRODUCTION

Many aspects of early nutrition have important implications for later biological programming and health outcome (Henry and Ulijaszek, 1996; Lucas, 1992; McGill *et al.*, 1996). For example, the positive role of breast milk in the development of infant immunocompetence is widely recognized (Arke, 1992; Kelleher and Lonnerdal, 2001; Newburg, 1999). The relation between feeding regimen and growth is more complex, and studies comparing breast-fed infants with formula-fed infants have yielded an inconsistent picture. Some report no difference in weight between the two feeding categories from birth up to 4 or 6 months of age (Auestad *et al.*, 1997; Innis *et al.*, 1996; Pomerance, 1987); others have found that formula-fed infants are heavier than their breast-fed counterparts before this time, starting at 6 weeks (Fomon, 2004; Nelson *et al.*, 1989) or 3 months of age (Dewey *et al.*, 1992). Results for body composition are also conflicting. Some studies report that formula-fed infants are fatter than their breast-fed counterparts (Dewey *et al.*, 1993; Yeung *et al.*, 1981), while others show overall fatness and the proportion of lower-body fat distribution are greater in breast-fed infants (Stuff and Mueller, 1997).

The relation between infant feeding and development is equally complex. While some investigators have reported significant differences in cognitive development and intelligence between breast-fed and formula-fed infants (Bier *et al.*, 2002; Horwood and Fergusson, 1998; Quinn *et al.*, 2001), others point to confounding factors, such as duration of breast feeding or maternal traits, as etiological variables (Drane and Logemann, 2000; Feldman and Eidelman, 2003; Gale and Martyn, 1996). Studies targeting the consequences of specific nutritive compounds for motor and cognitive development have yielded variable results (e.g., Auestad *et al.*, 2003; Marin *et al.*, 2000).

To control for some of the confounding conditions that characterize the human studies (Glassman and Coelho, 1988), the use of nonhuman primate (NHP) models of growth and development has proven to be a productive approach. For example, human mothers can be involved in both formula/bottle- and breast-feeding regimens, whereas NHP mothers participate only in breast feeding, and bottle-fed infants are housed without their mothers, often in single cages. The term *rearing* is used here to reflect this linked combination of diet and housing that yields two discrete groups of NHP infants, viz., mother-reared (breast-fed) infants and laboratory-reared (formula/bottle-fed) infants without mothers. In studies of infant baboons (*Papio cynocephalus*) and capuchins (*Cebus apella*), weight and rate of weight gain did not differ by rearing regimen during early infancy (Glassman and Coelho, 1988; Patino *et al.*, 1997). Differences did emerge for female baboons after 15 weeks of age, when breast-fed animals outweighed their formula-fed counterparts

(Glassman and Coelho, 1988). Studies of experimental formulas (e.g. Kelleher *et al.*, 2003) and feeding protocols (Sackett and Ruppenthal, 1992) have also shown associations between rearing category and outcome of growth.

Infant body size and composition are not the only variables that may influence later health, reproduction, and longevity; infant development also plays an important role (Bastian *et al.*, 2003; Capitanio, 1985). During infancy, even subtle variations in micronutrient content can lead to significant differences not only in weight gain but also in neuromotor development (Champoux *et al.*, 2002). Yet, to the best of our knowledge, no studies of NHP infants under different rearing regimens have integrated body size and growth with indices of physical development. One such index is the status of skeletal maturation, which, in turn, is correlated with many other indices of development and has been used for decades to assess the status of physical development in both humans (Tanner *et al.*, 1983) and NHPs (Newell-Morris *et al.*, 1991).

The present study explores the association of age, sex, and rearing regimen with weight, weight gain, linear growth and rate of growth, and skeletal development during the first 6–7 months postpartum in pigtailed macaque (*Macaca nemestrina*) infants. We test three null hypotheses regarding the growth and development of infants raised under either laboratory-reared (LR) or mother-reared (MR) regimens:

1. There is no significant rearing effect on weight from birth through 6 months of age.
2. There is no significant rearing effect on crown–rump length (CRL) from birth through 6 months of age.
3. There is no significant rearing effect on skeletal development (DEV) from birth through 6 months of age.

The null hypothesis is particularly appropriate here, because modern animal husbandry practices seek to replicate in LR infants the normative infant growth and developmental patterns seen in MR infants (e.g., Sackett *et al.*, 2002).

2. MATERIALS AND METHODS

2.1. Samples

The study sample consisted of 144 pigtailed macaque infants, 73 LR (35 female, 38 male) and 71 MR (36 female, 35 male). The infants were followed through the first 6–7 months of life. Only infants and dams judged to be clinically normal were included. Infants with birth weights <5th centile or >95th centile of sex-specific colony standards were excluded. The average parity of LR dams was 2.6 ± 1.6 and that of MR dams, 1.6 ± 0.72.

2.1.1. LR Sample

The LR infants were housed in individual cages at the Infant Primate Research Laboratory (IPRL) of the University of Washington National Primate Research Center. Of the 73 subjects, 26 had been removed from their mothers per study protocol, 26 were sent to the IPRL because their mothers were unable to provide adequate care, and five had been assigned as the result of injury or illness. The remaining 16 animals came in under miscellaneous conditions, such as transfer from another facility. The housing and husbandry procedures at the IPRL are described in detail elsewhere (Sackett *et al.*, 2002). Briefly, infants are bottle-fed a 10% dextrose solution every 2 hr, beginning 2 r after birth; on days 2 and 3 they are fed a 50/50 mixture of 10% dextrose and SMA or Enfamil; beginning with day 4, they receive SMA or Enfamil formula on a 4-hr on/off feeding schedule. Weaning is initiated at 3 months (day 91) with a 40% decrease in formula concentration and is completed by 4 months (day 119). Infants have access to Purina® high-protein monkey chow beginning at 75–80 days. They are weighed daily.

To promote psychological well-being and encourage species-typical behavior, the infants receive social contact through daily half-hour play sessions with two or three peers.

2.1.2. MR Sample

The 71 MR infants were housed with their mothers at the previous Primate Field Station (PFS), Medical Lake, WA, in social groups of eight lactating dams matched for parity. Infants were allowed to suckle on demand until they were separated from their mothers at 6 ± 0.5 months of age. They also had *ad libitum* access to Purina® high-protein monkey chow and water while still with their mothers.

2.2. Data Collection

Somatometric and radiographic data were collected on LR and MR subjects between birth and 6–7 months of age postpartum The first assessment was made within 1 week of birth or upon arrival at the IPRL. Subsequently, infants were measured at 90-day intervals (±14 days) for a minimum of three records per subject. In addition, the weights of LR infants taken at 4 months of age during bottle weaning were used in data analysis. Body weight was measured with a digital scale calibrated to the nearest gram. CRL was measured on a sliding scale with the subject in a supine position. The infant's rump was positioned against the base of the device and a sliding scale was brought to the crown. The measurement was read to the nearest millimeter after the subject was removed from the scale. At the IPRL, two observers measured each subject and the average of the two measures was used in the analysis. Interobserver error was <5%. At the PFS, one technician took all measurements over the duration of the study. The technician was trained by an experienced observer (L.N.M.); interobserver error during training was <2%.

A PA-view radiograph of the left hand and wrist, i.e., palm flat against the table surface, was taken at a standardized film-tube distance of 39 inches. An ossification rating was assigned to each of 13 epiphyseal centers in the hand/wrist using the TW2-RUS Method (Tanner *et al.*, 1983). The ratings were summed to yield a total DEV score for each animal at each age. Two experienced assessors read all films independently. Interobserver error was <2%; in the few cases of disagreement, an average of the two rankings was used.

2.3. Statistical Analysis

The outcome variables of weight, CRL, and DEV were analyzed using a Mixed Model that included Age, Sex, and Rearing as well as the Age: Sex and Age: Rearing interactions. For subject i at time j, each of the three outcome variables was fit to the model:

$$y_{ij} - \mu + Age_{ij} + Sex_i + Rearing_i + Sex{:}Age_{ij} + Rearing{:}Age_{ij} + \tau_i + \varepsilon_{ij}$$

where $\tau_i \sim N(0, \Sigma_\tau)$, $\varepsilon_{ij} \sim N(0, \sigma^\varepsilon)$, and it is assumed that the subject-specific random effect, $\tau_{I,}$ is independent of the error ε_{ij}. Age was treated as a continuous, fixed variable. The subject-specific random effect accounted over the study period for the baseline values of weight, CRL, and DEV of the individual animal. The factor Sex was estimated relative to males. Rearing was estimated relative to MR infants. In our analysis of CRL, we raised Age to the $\frac{3}{4}$ power to linearize its relation to CRL. We tested our three hypotheses by determining whether coefficients in the Mixed Model were statistically significant ($\alpha \leq 0.05$).

Table 13–1 shows the cross-tabulation of subjects by sex, age in 1-month blocks, and rearing group. Sample numbers within each 1-month

Table 13–1. Cross-Tabulation of Laboratory-Reared (LR) and Mother-Reared (MR) Infant Pigtailed Macaques (*M. nemestrina*) by Sex and Age in 1-Month Blocks at Time of Measurement

	Rearing group		
Sex	Laboratory	Mother	Total
Female			
Month			
1.00	31	36	67
2.00	5	0	5
3.00	8	13	21
4.00	33	23	56
5.00	21	0	21
6.00	8	19	27
7.00	18	17	35
8.00	2	0	2
Total	126	108	234
Male			
Month			
1.00	29	35	64
2.00	6	0	6
3.00	9	10	19
4.00	34	25	59
5.00	30	0	30
6.00	8	19	27
7.00	19	16	35
8.00	2	0	2
Total	137	105	242

block varied, especially in blocks on or near the scheduled month of assessment. For example, measurements taken after an infant was exactly 6 months (180 days) old appear in block 7 or 8. The LR sample contains eight animals in block 6, and double this number in block 7; the MR sample in block 6 is 27 animals and that in block 7 is 35 animals. For descriptive purposes, the mean ± 1 SE of each growth and developmental variable is plotted sex-specific in 1-month blocks (Figs. 13.1–13.3).

3. RESULTS

Coefficients for the three Mixed Models are given in Table 13–2. For all three variables there was a highly significant association with Age ($p < 0.001$), as would be expected for growing animals. The factor Sex was significant only for DEV ($p < 0.001$). Rearing was significant only for CRL ($p = 0.007$).

Sex : Age interaction was significantly positive for DEV ($p < 0.001$) and negative for weight ($p < 0.001$), evidence that female infants matured skeletally at a faster rate than did male infants but gained weight at a slower rate. The Sex : Age interaction term was not significant for CRL, indicating similar linear growth rates for male and female subjects. The

Table 13–2. Mixed Model Coefficients for the Outcomes of Weight, DEV, and CRL in Laboratory-Reared (LR) and Mother-Reared (MR) Pigtailed Macaque Infants (*M. nemestrina*)

	Outcome[a]								
	Weight (g)			DEV (TW2-RUS)			CRL (mm)		
	Coefficient	SE	*p*-value	Coefficient	SE	*p*-value	Coefficient	SE	*p*-value
Intercept	498.2	47.1	<0.001	69.8	13.3	<0.001	185.8	4.20	<0.001
Age	4.49	0.22	<0.001	0.61	0.07	<0.001	1.75	0.080	<0.001
Sex	−27.5	21.7	0.21	52.5	6.15	<0.001	−3.26	1.91	0.09
Rearing	23.0	21.8	0.29	−1.69	6.16	0.78	5.25	1.95	0.007
Sex:Age	−0.53	0.12	<0.001	0.35	0.03	<0.001	−0.042	0.038	0.25
Rearing: Age	0.68	0.12	<0.001	−0.086	0.03	0.01	0.21	0.039	<0.001

[a] DEV, developmental rating; CRL, crown–rump length.

Rearing:Age interaction was significantly positive for weight and CRL, and negative for DEV ($p < 0.01$). Thus, LR infants grew more rapidly than did their MR counterparts (Figs. 13–1 and 13–2), but matured more slowly (Fig. 13–3). Although it might be predicted that LR infants would slow in their rate of weight gain during the 1-month interval of bottle-weaning at 3–4 months of age, the model supports a higher rate of weight gain for LR infants than for the MR group that was not weaned until 6–7 months Daily weight records for LR infants throughout bottle-weaning revealed no consistent pattern of weight loss nor slowing of the rate of weight gain (data not shown). A paired t-test on the rate of weight gain before and after bottle weaning indicated no significant change (male: $t = -1.27$, $p = 0.27$; female: $t = 1.67$, $p = 0.11$).

Although the association of the Sex effect with weight was not significant, within each rearing group male infants tended to be, on an average, heavier than female infants (Fig. 13–1). In addition, male and female body weights diverged progressively over the study period, supporting the finding from the Model that female infants gain weight more slowly than their male counterparts.

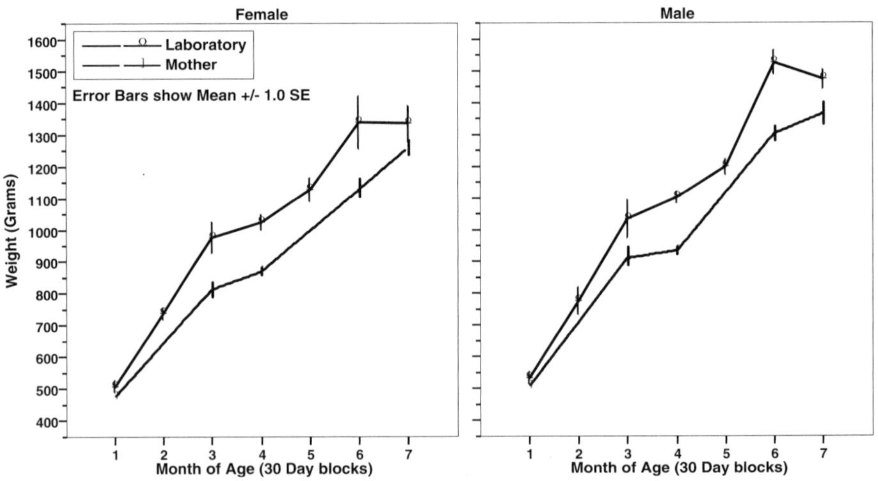

Figure 13–1. Mean and SE of weight for laboratory-reared (LR) and mother-reared (MR) female (left) and male (right) pigtailed macaque infants (*M. nemestrina*), birth to 7 months of age.

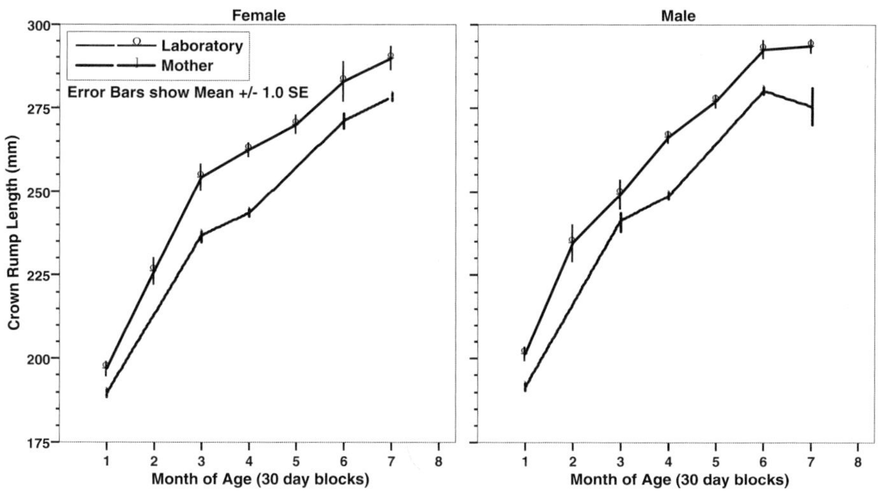

Figure 13–2. Mean and SE of crown–rump length (CRL) of laboratory-reared (LR) and mother-reared (MR) female (left) and male (right) pigtailed macaque infants (*M. nemestrina*), birth to 7 months of age.

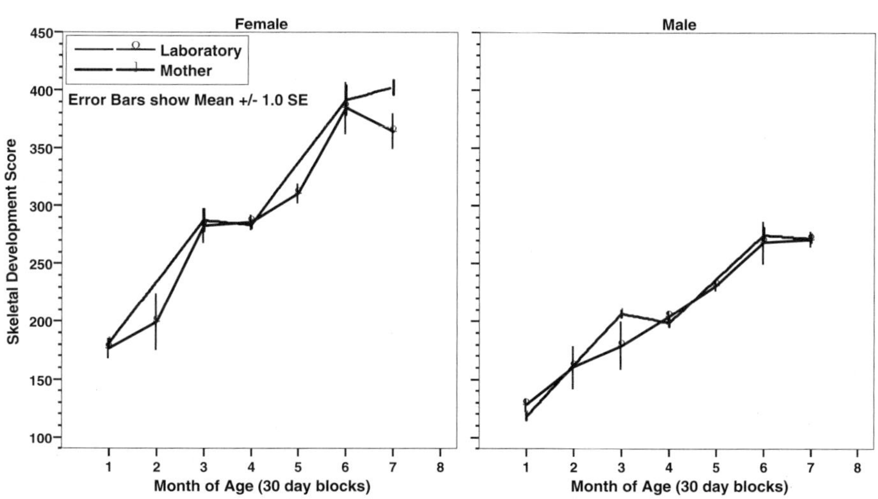

Figure 13–3. Mean and SE of skeletal developmental rating (DEV) of laboratory-reared (LR) and mother-reared (MR) female (left) and male (right) pigtailed macaque infants (*M. nemestrina*), birth to 7 months of age.

4. DISCUSSION

The results for these 144 pigtailed macaque infants show that rearing is significantly associated with growth and development in this species. Thus, the three null hypotheses are rejected.

4.1. Sex Effects

4.1.1. Growth

The Sex effect was not significantly associated with weight, although within each rearing category male infants gained weight faster than female infants and were generally heavier. The lack of association may be attributed to a number of factors. First, sample sizes were small. Significant sex differences in weight both at birth and throughout infancy have been reported for a larger sample from the same research facilities from which our subjects were drawn (Sackett and Ruppenthal, 1992). Second, it is possible that the Model was not sensitive to Sex as main effect because of the subject-specific random effect, τ_i. The latter was incorporated to account for differences between individuals in their starting values. As such, the random effect may already effectively capture the association between Sex and weight where a sex-determined difference in weight is present at birth. Finally, because our subjects were followed for a relatively short time during early infancy, detectable sexual dimorphism in weight may not yet have developed. Analyses of the growth of other macaques suggest that weight differences by sex do not always exist in early infancy (Bowman and Lee, 1995; Willner and Martin, 1985), but rather, appear more during juvenile development in sexually dimorphic species (Smith and Leigh, 1998).

The significant Sex:Age interaction supports the interpretation that sex-specific rates of growth during infancy are responsible for the progressive divergence between male and female body weights over time observed in both rearing regimens. Though not all macaques demonstrate an early divergence between the sexes in weight gain (cf. *M. fuscata*, Hiraiwa, 1981; *M. fasicularis*, Willner and Martin, 1985), a sex difference in the rate of gain has been reported for other macaque species (Bowman and Lee, 1995; Sackett and Ruppenthal, 1992). Where a sex difference does exist, it could be the result of higher food intake or lower expenditure of energy on the part of male primates, but these factors were not measured in our study. We found little published evidence to substantiate sex differences in feeding duration or frequency in NHP infants (Brown, 2001; Gomendio, 1990), and reports of sex differences in energy expenditure are inconclusive. Activity levels of male and female human infants can be similar (Wells and Davies, 1995); when differences do exist, male infants are the more active gender (Campbell and Eaton, 1999). The possibility exists that male macaque infants therefore expend more, not less, energy, as might be assumed from their greater weight gain; we found no studies suggesting that males are the energetically conservative sex among primates.

Unlike weight, linear growth (CRL) was not significantly associated with Sex or the Sex:Age interaction. Consequently, at any given age female infants were lighter than male infants of the same length. Given the similar length-for-age values, the sex differences in weight are probably not a function of skeletal growth, but rather may reflect a sex-specific differential accumulation of other lean tissues. If sex differences in energy intake and expenditure cannot be shown, it is possible that male pigtailed macaque infants translate calories into lean tissue mass differently than do female infants, as has been reported for baboons (e.g., Lewis *et al.*, 1984). The measures collected in our study did not allow us to address this question. We suggest that future studies of hormonal mechanisms controlling the deposition of fat and lean tissues might provide insight into the mechanisms that underlie the sexual dimorphism observed in body composition during infancy and later.

4.1.2. Skeletal Development

Skeletal development in the pigtailed infants followed the normative trend for primates. DEV was associated with Sex, differing by sex in both absolute value and velocity consistently across rearing categories (Fig. 13–3). Accelerated skeletal development characterizes human females, who also tend to show advanced physiological and cognitive development relative to their male peers (Ulijaszek *et al.*, 1998). Relatively earlier female maturation is normal among sexually dimorphic NHPs that live in mixed-sex groups (Leigh, 1995). Because sexual dimorphism in DEV is already evident at birth, it appears that skeletal development is influenced by sex-specific endocrinological and/or physiological mechanisms that operate both pre- and postnatally.

4.2. Rearing Effects

4.2.1. Growth

LR infants gained weight significantly faster than did MR infants. The differences in weight gain for infants between the two rearing regimens may be attributable to greater caloric intake by LR infants, more limited energy expenditure, or some combination thereof. If the higher rate of weight gain in LR infants were attributable to basic caloric or nutritional differences, it might be expected that the 1-month interval of weaning would depress weight gain, but this effect was not observed; in fact, rates of gain for weight and linear growth were higher in LR than in MR subjects. Thus, despite a major change in nutrition, LR infants were able to sustain their relatively faster rates of gain. In addition, bottle-weaning also seems not to have affected linear growth, as the positive association between Rearing and CRL implies that LR infants were also longer than MR infants.

The differences observed between rearing regimens in rates of weight gain and linear growth could ultimately lead to differing weight-for-length ratios between LR and MR infants. The positive coefficients for Rearing on CRL and the Rearing:Age interactions on weight and CRL strongly suggest that LR infants could have a greater weight-for-length ratio than do MR infants. Potential differences in weight-for-length ratios between the two rearing regimens are partially dependent upon growth velocities between birth and 6 months of age, but here the number of intervals between assessments was not sufficient to permit the calculation of changes in weight-for-length over time.

The vast literature on the consequences of undernutrition documents the constraint that insufficient diet can impose on growth. The specific relations between diet and growth, however, are not linear, nor are they independent of the sociophysical environment or the intrinsic biological capacities of infants. In short, caloric intake cannot be directly translated into growth rates. Squirrel monkey infants showed no apparent weight gain between 6 and 8 months of age, even though diet was held constant and growth occurred before and after that period (Pereira and Leigh, 2003), suggesting that the necessary physiological mechanisms were not operative. In ring-tailed lemur infants, growth rates and metabolic rates can be manipulated by photoperiod (Pereira, 1993). Certainly there are many other factors contributing to growth that are not accounted for in the present analysis. For example, maternal weight, maternal weight gain during pregnancy, and genetic effects explain a significant proportion of variation in NHP birth weights (Bowman and Lee, 1995; Ha *et al.*, 2002; Michaelsen *et al.*, 1994).

4.2.2. Development

Though the association between Rearing and DEV was not significant, the Rearing : Age interaction was significant in the Model, and was greater for male infants than for females. Sex differences in the timing of critical periods could make males more sensitive to early rearing conditions. Within sex, LR and MR infants were at similar stages of development initially, but LR infants developed at a slower rate during later infancy. As a result, the physically larger LR infants were at the same, or earlier, state of maturation than were their smaller-bodied MR peers, i.e., for given size, LR infants were delayed in development. Or were the MR infants more precocious in their development? A more complete suite of developmental indices (neurological, physiological, behavioral) would provide a better understanding of the distinctions between LR and MR infants that emerged from the skeletal developmental data. The concentrations of certain micronutrients and lipids in infant foods have been shown to influence not only growth, but also development (Champoux *et al.*, 2002; Kelleher *et al.*, 2003; Marin *et al.*, 2000). Here, the difference in composition between breast milk and formula provides an obvious potential explanation for the significant associations found between Rearing, weight gain, and linear growth. However, the fact that formula feeding cannot be coupled with mother–infant cohousing limits the ability to distinguish between dietary and social aspects of the two infant-rearing regimens. In the absence of an obvious or documented relation to dietary regimen, therefore, it cannot be concluded that diet alone resulted in the accelerated maturation of MR infants relative to their LR counterparts.

Studies of provisioned, free-ranging, and captive primates have shown how qualitative and quantitative differences in diet affect development all the way to adulthood, at least when measured by sexual maturation (for review see McFarland, 1997). Given the growth and developmental differences between our rearing groups, it is possible that LR and MR infants go on to mature sexually at different ages or sizes. Differences in age or size at maturation ultimately depend on whether differences in growth and maturation continue at similar rates and durations for both rearing groups. The accumulation of adipose tissue that is normal during later juvenile and adolescent stages would also exert an influence on final age and size at maturity Given the many factors that influence growth and development during juvenile and adolescent development, the differences detected in our model of early infancy may not carry into adulthood. Longitudinal data from the IPRL and PFS suggest that LR and MR subjects do not differ in body weight at 12 months of age, though female subjects differed in weight after 7 years of age (LR > MR) (Sackett *et al.*, 2002). Variation in age at first reproduction (3–4 years) was also reported (Sackett *et al.*, 2002), but the data do not allow us to resolve the question of difference in age at sexual maturation as signaled by first estrus.

5. CONCLUSIONS, LIMITATIONS, AND SUGGESTIONS FOR FUTURE RESEARCH

Our study reveals the importance of including developmental measures in studies of early rearing and nutrition by highlighting that growth and development can respond differently to the same set of conditions. Specifically, the more rapid growth in weight and length of LR infants was not accompanied by accelerated development. By comparison with the challenges that investigators encounter in human studies of infant nutrition, where variations between households are difficult to describe and take into account, the rearing environments in both of our experimental regimens were well controlled. Yet, even in this context, some of our findings are difficult to explain because we were unable to tease apart the effects of feeding from the effects due to the presence/absence of mothers. Thus, we are left to consider the implications of both nutrients and nurturing for the patterns of infant growth and development that were observed. In short, infant formula does not equal mother's milk, nor does the absence of mother equal the presence of mother.

Since it is not possible to assign the mode of feeding without also affecting the presence of the mother, the effects of maternal care must be considered when rearing differences in growth and development are found. The classic studies of early isolation in infant NHPs demonstrated the critical nature of maternal–infant attachment for normal development and outcomes later in life (Harlow and Mears, 1979). Later suitable alternatives to mother–infant housing appear to mitigate many of the negative effects originally induced (e.g., Sackett *et al.*, 2002), but still cannot replicate the maternal–infant dyad. In a recent study, infants raised without their mothers had different social interactions and lower dominance ranks when introduced to peer groups than subjects raised with their mothers, and rank, in turn, affected weight gain between 1 and 3 years of age (Bastian *et al.*, 2003). If, as these studies imply, the social components of the rearing environment influence growth and development, either in concert with or independently of nutrition, specific effects will be difficult to quantify in NHP model systems. Given the complexity of the relationship, and the number of unknowns involved, researchers using NHP models of growth and development are advised to interpret results conservatively.

One of the limitations of our study was that infants were not measured frequently enough to establish serial rates of change directly. The assumption of constant gain between 3-month intervals is not necessarily realistic for infants between 0 and 6 months of age. Thus, we were unable to identify critical periods, growth spurts, or catch-up growth or to address weight-for-length differences. We also lacked data on energy intake and expenditure. In the absence of reliable values for metabolic rates and caloric intake and expenditure for infant and juvenile NHPs, we are of the opinion that future studies of infant metabolic rates and energetics will be essential. Similarly, data on the composition of NHP breast milk and the changes in composition and quantity over the period from birth to weaning are essential for the documentation of infant nutritional needs and metabolic strategies. A more complete understanding of the differences between breast milk and formula and of how infants use the nutritive and nonnutritive components of food is basic to the elucidation of the relations between early nutrition, growth, and development. Ideally, experimental studies involving the manipulation of the ingredients of infant formula may allow us to isolate the effects of individual components or interactions of interest. We emphasize that questions about the consequences of early rearing conditions for later health outcome require longitudinal data. Here, NHPs with their faster rates of growth and development compared with the human species will continue to be valuable in such research and stand to make significant contributions toward understanding and improving infant nutrition for both humans and NHPs.

Our results emphasize the need to consider potential implications of early nutrition and rearing conditions for a broad range of traits that are potentially subject to biological programming. The effects of early biological programming for growth, physiology, and reproduction are of interest, not only because they are central to our understanding of life history, health, and biology, but also because they have potential implications for biomedical research. It has been suggested elsewhere that rearing conditions lead to long-term differences in metabolism, development, and health outcome (Lucas, 1992). If early differences between LR and MR infants persist or reemerge during adulthood, the possibility exists that early rearing conditions could translate into differential responses to later research protocols (National Research Council, 2002). Retrospective and prospective studies of NHP infants from LR and MR environments will provide valuable tools to further explore such issues.

REFERENCES

Arke, J. (ed.) (1992). *Infant Feeding: The Physiological Basis.* World Health Organization, Geneva.

Auestad, N., Montalto, M., Hall, R., Fitzgerald, K., Wheeler, R., Connor, W., Neuringer, M., Connor, S., Taylor, J., and Hartmann, E. (1997). Visual acuity, erythrocyte fatty acid composition, and growth in term infants fed formulas with long chain polyunsaturated fatty acids for one year. *Pediatr. Res.* 41:1–10.

Auestad, N., Scott, D.T., Janowsky, J.S., Jacobsen, C., Carroll, R.E., Montalto, M.B., Halter, R., Qiu, W., Jacobs, J.R., Connor, W.E., Connor, S.L., Taylor, J.A., Neuringer, M., Fitzgerald, K.M., and Hall, R.T. (2003). Visual, cognitive, and language assessments at 39 months: A follow-up study of children fed formulas containing long-chain polyunsaturated fatty acids to 1 year of age. *Pediatrics* 112:177–183.

Bastian, M.L., Sponberg, A.C., Suomi, S.J., and Higley, J.D. (2003). Long-term effects of infant rearing condition on the acquisition of dominance rank in juvenile and adult rhesus macaques (*Macaca mulatta*). *Dev. Psychobiol.* 42:44–51.

Bier, J.A., Oliver, T., Ferguson, A.E., and Vohr, B.R. (2002). Human milk improves cognitive and motor development of premature infants during infancy. *J. Hum. Lact.* 18:361–367.

Bowman, J.E., and Lee, P.C. (1995). Growth and threshold weaning weights among captive rhesus macaques. *Am. J. Phys. Anthropol.* 96:159–175.

Brown, G. (2001). Sex-biased investment in nonhuman primates: Can Trivers & Willard's theory be tested? *Anim. Behav.* 61:683–694.

Campbell, D., and Eaton, W. (1999). Sex differences in the activity levels of infants. *Infant Child Dev.* 8:1–17.

Capitanio, J.P. (1985). Early experience and social processes in rhesus macaques (*Macaca mulatta*): II. Complex social interaction. *J. Comp. Psychol.* 99:133–144.

Champoux, M., Hibbeln, J.R., Shannon, C., Majchrzak, S., Suomi, S.J., Salem, N., Jr., and Higley, J.D. (2002). Fatty acid formula supplementation and neuromotor development in rhesus monkey neonates. *Pediatr. Res.* 51:273–281.

Dewey, K.G., Heinig, M.J., Nommsen, L.A., Peerson, J.M., and Lonnerdal, B. (1992). Growth of breast-fed and formula-fed infants from 0 to 18 months: The DARLING Study. *Pediatrics* 89:1035–1041.

Dewey, K.G., Heinig, M.J., Nommsen, L.A., Peerson, J.M., and Lonnerdal, B. (1993). Breast-fed infants are leaner than formula-fed infants at 1 y of age: The DARLING study. *Am. J. Clin. Nutr.* 57:140–145.

Drane, D.L., and Logemann, J.A. (2000). A critical evaluation of the evidence on the association between type of infant feeding and cognitive development. *Paediatr. Perinat. Epidemiol.* 14:349–356.

Feldman, R., and Eidelman, A.I. (2003). Direct and indirect effects of breast milk on the neurobehavioral and cognitive development of premature infants. *Dev. Psychobiol.* 43:109–119.

Fomon, S.J. (2004). Assessment of growth of formula-fed infants: Evolutionary considerations. *Pediatrics* 113:389–393.

Gale, C.R., and Martyn, C.N. (1996). Breastfeeding, dummy use, and adult intelligence. *Lancet* 347:1072–1075.

Glassman, D., and Coelho, A, Jr. (1988). Formula-fed and breast-fed baboons: Weight growth from birth to adulthood. *Am. J. Primatol.* 16:131–142.

Gomendio, M. (1990). The influence of maternal rank and infant sex on maternal investment trends in rhesus macaques: Birth sex ratios, inter-birth intervals and suckling patterns. *Behav. Ecol. Sociobiol.* 27:365–375.

Ha, J.C., Ha, R.R., Almasy, L., and Dyke, B. (2002). Genetics and caging type affect birth weight in captive pigtailed macaques (*Macaca nemestrina*). *Am. J. Primatol.* 56:207–213.

Harlow, H., and Mears, C. (1979). *The Human Model: Primate Perspectives.* John Wiley & Sons, New York.

Henry, C., and Ulijaszek, S. (eds.) (1996). *Long-Term Consequences of Early Environment: Growth, Development and the Lifespan Developmental Perspective.* Cambridge University Press, Cambridge.

Hiraiwa, M. (1981). Maternal and alloparental care in a troop of free-ranging Japanese monkeys. *Primates* 22:309–329.

Horwood, L.J., and Fergusson, D.M. (1998). Breastfeeding and later cognitive and academic outcomes. *Pediatrics* 101:E9.

Innis, S.M., Auestad, N., and Siegman, J.S. (1996). Blood lipid docosahexaenoic and arachidonic acid in term gestation infants fed formulas with high docosahexaenoic acid, low eicosapentaenoic acid fish oil. *Lipids* 31:617–625.

Kelleher, S.L., and Lonnerdal, B. (2001). Immunological activities associated with milk. *Adv. Nutr. Res.* 10:39–65.

Kelleher, S.L., Chatterton, D., Nielsen, K., and Lonnerdal, B. (2003). Glycomacropeptide and alpha-lactalbumin supplementation of infant formula affects growth and nutritional status in infant rhesus monkeys. *Am. J. Clin. Nutr.* 77:1261–1268.

Leigh, S.R. (1995). Socioecology and the ontogeny of sexual size dimorphism in anthropoid primates. *Am. J. Phys. Anthropol.* 97:339–356.

Lewis, D.S., Bertrand, H.A., Masoro, E.J., McGill, H.C., Jr., Carey, K.D., and McMahan, C.A. (1984). Effect of interaction of gender and energy intake on lean body mass and fat mass gain in infant baboons. *J. Nutr.* 114:2021–2026.

Lucas, A. (1992). Early nutrition and later outcome. In: Widdowson, E., and Mathers, J. (eds.), *The Contribution of Nutrition to Human and Animal Health*. Cambridge University Press, Cambridge, pp. 266–277.

Marin, M.C., Rey, G.E., Pedersoli, L.C., Rodrigo, M.A., and de Alaniz, M.J. (2000). Dietary long-chain fatty acids and visual response in malnourished nursing infants. *Prostaglandins Leukot. Essent. Fatty Acids* 63:385–390.

McFarland, R. (1997). Female primate: Fit or fat? In: Morbeck, M., Galloway, A., and Zihlman, A. (eds.), *The Evolving Female: A Life-History Perspective*. Princeton University Press, Princeton, pp. 163–175.

McGill, H.C., Jr., Mott, G.E., Lewis, D.S., McMahan, C.A., and Jackson, E.M. (1996). Early determinants of adult metabolic regulation: Effects of infant nutrition on adult lipid and lipoprotein metabolism. *Nutr. Rev.* 54:S31–S40.

Michaelsen, K.F., Larsen, P.S., Thomsen, B.L., and Samuelson, G. (1994). The Copenhagen Cohort Study on infant nutrition and growth: Breast-milk intake, human milk macronutrient content, and influencing factors. *Am. J. Clin. Nutr.* 59:600–611.

National Research Council (ed.) (2002). *Nutrient Requirements of Nonhuman Primates*. National Academies Press, Washington, D.C.

Nelson, S.E., Rogers, R.R., Ziegler, E.E., and Fomon, S.J. (1989). Gain in weight and length during early infancy. *Early Hum. Dev.* 19:223–239.

Newburg, D.S. (1999). Human milk glycoconjugates that inhibit pathogens. *Curr. Med. Chem.* 6:117–127.

Newell-Morris, L., Carrol, B., Covey, A., Medley, S., and Sackett, G.P. (1991). Postnatal growth and skeletal maturation of experimental preterm macaques (*Macaca nemestrina*). *J. Med. Primatol.* 20:17–22.

Patino, E., de Olivera, C., Zetterman, C., Bordal, J., and Ruiz, J. (1997). Hand-rearing *Cebus apella*. *Lab. Prim. Newsl.* 36:8–10.

Pereira, M. (1993). Agonistic interactions, dominance relation, and ontogenetic trajectories in ringtailed lemurs. In: Pereira, M., and Fairbanks, L. (eds.), *Juvenile Primates: Life History, Development, and Behavior*. Oxford University Press, New York, pp. 285–305.

Pereira, M., and Leigh, S. (2003). Modes of primate development. In: Kappeler, P., and Pereira, M. (eds.), *Primate Life Histories and Socioecology*. University of Chicago Press, Chicago, pp. 149–176.

Pomerance, H.H. (1987). Growth in breast-fed children. *Hum. Biol.* 59:687–693.

Quinn, P.J., O'Callaghan, M., Williams, G.M., Najman, J.M., Andersen, M.J., and Bor, W. (2001). The effect of breastfeeding on child development at 5 years: A cohort study. *J. Paediatr. Child Health* 37:465–469.

Sackett, G.P., and Ruppenthal, G.C. (1992). Growth of nursery-reared Macaca nemestrina infants: Effects of feeding schedules, sex, and birth weight. *Am. J. Primatol.* 27:189–204.

Sackett, G.P., Ruppenthal, G.C., and Davis, A.E. (2002). Survival, growth, health, and reproduction following nursery rearing compared with mother rearing in pigtailed monkeys (*Macaca nemestrina*). *Am. J. Primatol.* 56:165–183.

Smith, R.J., and Leigh, S.R. (1998). Sexual dimorphism in primate neonatal body mass. *J. Hum. Evol.* 34:173–201.

Stuff, J., and Mueller, W. (1997). Role of infant feeding practice, sex and age on fatness and subcutaneous fat distribution in infancy. *Am. J. Hum. Biol.* 9:179–190.

Tanner, J., Whitehouse, R., Cameron, N., Marshall, W., Healy, M., and Goldstein, H. (1983). *Assessment of Skeletal Maturity and Prediction of Adult Height (TW2 Method).* Academic Press, London.

Ulijaszek, S., Johnston, F., and Preece, M. (eds.) (1998). *The Cambridge Encyclopedia of Human Growth and Development.* Cambridge University Press, Cambridge.

Wells, J.C., and Davies, P.S. (1995). Diet and behavioural activity in 12-week-old infants. *Ann. Hum. Biol.* 22:207–215.

Willner, L., and Martin, R. (1985). Some basic principles of mammalian sexual dimorphism. In: Ghesquire, J., Martin, R., and Newcombe, F. (eds.), *Human Sexual Dimorphism.* Taylor and Francis, London, pp. 1–42.

Yeung, D.L., Pennell, M.D., Leung, M., and Hall, J. (1981). Infant fatness and feeding practices: A longitudinal assessment. *J. Am. Diet. Assoc.* 79:531–535.

CHAPTER FOURTEEN

Baboon Nursery Rearing Practices and Comparisons between Nursery-Reared and Mother-Reared Individuals

Linda Brent and Anne Bode

1. INTRODUCTION

The size, hardiness, and adaptability of baboons have influenced their ability to populate vast regions of Africa, from sub-Saharan deserts to lush tropical forests. These factors are also key reasons they are chosen for many types of biomedical and behavioral research, including studies of cardiovascular disease, infectious disease, reproduction, osteoporosis, genetics, and social organization.

The Southwest Foundation for Biomedical Research (SFBR), formerly known as the Southwest Foundation for Research and Education, maintains the largest population of captive baboons in the world. The name "baboon" usually refers to a number of closely related subspecies of *Papio hamadryas*, including olive or anubis (*P. h. anubis*), sacred or hamadryas (*P. h. hamadryas*), yellow or cynocephalus (*P. h. cynocephalus*), red or guinea (*P. h. papio*), and chacma baboons (*P. h. ursinus*). The baboon breeding and research program at the SFBR was initiated in 1957 and included anubis, guinea, and cynocephalus baboons (Vice and Rodriguez, 1965; Hendrickx and Kriewaldt, 1965). Published information on nursery rearing of these animals dates back to 1966 (Vice *et al.*, 1966). The SFBR currently maintains over 3900 baboons, with more than 95% of the colony housed in social groups in 6-acre corrals or large outdoor cages. More than 2500 individuals are pedigreed and about 1000 are genotyped, with a few animals selectively bred for particular physiological characteristics. Currently, about 500 infants are born each year (mainly *P. h. anubis*), with most raised by their mothers in mixed-age and mixed-sex groups.

The purpose of this chapter is to review data on physiological and behavioral differences between mother-reared and nursery-reared baboons, to outline data on the impact of variation in the nursery environment on behavior and development, and to summarize changes in baboon nursery-rearing practices over time. This information stems primarily from studies at the SFBR, supplemented from other sources when available.

2. NURSERY-REARED VERSUS MOTHER-REARED BABOONS

As is common with captive rearing strategies for most nonhuman primates, infant baboons are sometimes removed from their mothers and reared in a nursery setting. The most common reasons are rejection or abuse by the mother or cagemates, injury or illness of the mother or infant, or research protocol. Occasionally an infant is "kidnapped" by another female (Luft and Altmann, 1982), becomes dehydrated if the kidnapper is not lactating, and ends up in the nursery if the mother is unable to reclaim and nourish the infant.

Although nursery rearing may be required to save the life of the infant, it does not come without potential costs in terms of altered growth, physiology, and behavior of the primate and increased effort and expense for the staff. The necessity of removing an infant from the mother must be determined objectively, recognizing that such a decision can have negative consequences for the development of appropriate maternal behavior in the mother as well as normative species-typical behavior, development, and survival in the infant. The degree of risk to the infant must be weighed against the value of the infant as an individual and the benefits of future mother-reared infants (Bielitzki and Ruppenthal, 1989). The best way to produce healthy nonhuman primates that are able to interact socially, cope with environmental challenges, reproduce, and successfully rear offspring is to support a program of mother rearing. At a minimum, the rearing history of each individual research or breeding animal should be considered for proper individual lifetime management and for consideration of any effect on research variables.

2.1. Behavior

The SFBR population provides a ready pool of subjects for research on the development of social and other behavior in nursery-reared versus mother-reared baboons. In a series of studies, baboon were raised either in a nursery setting, in which they had several hours of peer socialization each day, or with their mothers in a mixed-age group for the first 3 months of life. They were then placed together in peer groups. Extensive observation of these individuals revealed surprisingly few differences in social behavior (Table 14–1). Some positive social behaviors varied for young infants (Young and Bramblett, 1977), but affiliative behaviors, including play, approach, and contact, did not differ between groups after the first 3 months of age (Coelho and Bramblett, 1982, 1984; Bramblett and Coelho, 1985; Erhart, 1995). Rearing condition did have an effect on aggressive and dominance-related behaviors, which were higher in mother-reared baboons (Coelho and Bramblett, 1981). The authors concluded that affiliative behaviors might be less responsive to environmental changes than aggression-related behaviors, which are often situation specific (Bramblett and Coelho, 1985). During the first 3 months of the infant's life, nursery-reared baboons showed more environmental exploration, stereotyped behavior, tension-related behavior, and vocalizations than did mother-reared baboons (Young and Bramblett, 1977).

While the patterns of social behavior shown by the nursery-reared baboons were almost indistinguishable from those of the mother-reared baboons by adolescence, the nursery-reared baboons did appear to be

Table 14–1. Behavioral Effects of Nursery Rearing on Baboons

Subject	Findings	Reference
Social and general behavior measuredin infants (0–3 months) $n = 33$, 20 nursery-reared, 13 mother-reared	Nursery-reared infants displayed higher levels of nonaggressive social behavior, exploration of environment, stereotyped behavior, threats, avoidance, tension, rough-and-tumble play, and vocalizations	Young and Bramblett, 1977
Social behavior measured in infants (0–24 months) $n = 77$, 58 nursery-reared, 19 mother-reared	Nursery-reared males received more bites than mother-reared males, but no other differences in social behavior, including grooming, play, aggression, sociosexual behavior	Young *et al.*, 1982
Social behavior measured in infants (6–12 months) and juveniles (36–42 months) $n = 87$, 64 nursery-reared, 23 mother-reared	No rearing effect on social play; aggressive behavior and dominance behavior lower in nursery-reared infants; inappropriate submissive behavior in nursery-reared infants;allogrooming higher in mother-reared infants; no other effects on approach or friendly behaviors	Coelho and Bramblett, 1981, 1982, 1984
Social behavior measured from 0 to 84 months $n = 96$, 72 nursery-reared, 24 mother-reared	No rearing effect on friendly behaviors: embrace, groom, muzzle, contact, hold, touch	Bramblett and Coelho, 1985
Social behavior measured in adolescents (43–96 months) $n = 22$	Interactions of gender and rearing for avoidance behavior; no rearing effects on other social behaviors, including groom, present, approach, lipsmack	Erhart, 1995 (abstract)
Abnormal behavior in adults $n = 142$	Severe behavioral disturbances more frequent in nursery-reared individuals than in mother-reared	Veira and Brent, 2000

more prone to stress-related abnormal behavior. More recent study of the colony indicated that the two groups differ in terms of both type and frequency of abnormal behavior. The records of our Behavioral Intervention Program show that 142 baboons have exhibited high levels of behavioral disturbance (e.g., self-aggression, hair pulling, pacing, rocking). We found that significantly more of these individuals were raised in the nursery than would be expected, given the proportion of nursery-reared and mother-reared individuals in the colony ($\chi^2 = 77.13$, $p < 0.001$) (Veira and Brent, 2000). Similarly, Bellanca and Crockett

(2002) found that nursery rearing was related to higher levels of self-aggression and self-stimulatory abnormal behaviors in male pigtailed macaques (*Macaca nemestrina*).

In our experience, nursery-reared baboons may seem to function fairly well in normal social settings, but when faced with stress (e.g., single caging or changes in group composition), they revert to maladaptive abnormal behaviors, probably developed during infancy as self-comforting mechanisms. The reaction to stress in nursery infants may result from the lack of a normal attachment figure (mother) or altered stress physiology due to a deprivation in social and environmental stimulation during early rearing. Support for the second idea comes from an early study by Sackett *et al.* (1973), who found that rhesus macaques (*M. mulatta*) reared in social isolation had abnormal adrenal response systems and higher basal cortisol levels than peer-reared rhesus macaques. Consistent with our findings in baboons, macaque infants that were removed from their mothers and raised in nursery peer groups were more reactive to environmental change and stressors than mother-reared monkeys, and this may be due to alterations in adrenocortical, noradrenergic, and/or serotonergic systems related to the early rearing environment (Higley and Suomi, 1989; Suomi, 1991).

2.2. Growth and Development

Baboons show a high degree of sexual dimorphism in weight and growth, with males experiencing a growth spurt around 4 years of age and reaching almost twice the weight of females by 8 years of age (Coehlo, 1985). Comparison of some developmental trends for mother- versus nursery-reared baboons indicated that there were no differences in weight measurements before weaning, but mother-reared baboons weighed more after weaning and had accelerated developmental trends (Glassman and Coelho, 1986) (Table 14–2). The timing and reason for entry into the nursery may have affected weight parameters (Levin and Carey, 1988). Although detailed information on the development of neurobehavioral and circadian rhythm is available for nursery-reared baboons (Mirmiran *et al.*, 2001), comparable information for mother-reared individuals is lacking.

Table 14–2. Growth and Development of Nursery-Reared Baboons

Subjects	Findings	Reference
Weight and developmental measures from 0 to 7 years $n = 87$, 65 nursery-reared, 22 mother-reared for first 15 months, then all peer-reared	No difference in weight before weaning; weight of mother-reared infants higher after 15 weeks of age; adolescent growth spurt earlier and more intense in mother-reared infants; mean age of menarche 8 weeks earlier for mother-reared infants	Glassman and Coelho, 1986 (abstract)
Weight measured from 0 to 22 months $n = 93$, all nursery-reared	Age of entry into the nursery was related to increased body weight; orphaned infants weighed less than mother- or healthy nursery-reared infants	Levin and Carey, 1988 (abstract)

2.3. Physiology

Studies of physiological measures in mother- versus nursery-reared baboons stem from an interest in the dietary difference between mother's milk and infant formula. Rearing condition and diet have a direct effect on the metabolism of bile acids (Jackson *et al.*, 1993) and plasma concentrations of thyroid hormone (Lewis *et al.*, 1993) (Table 14–3). Studies of breast-fed versus formula-fed infants demonstrated significant differences in cholesterol levels and atherosclerotic lesions related to the higher ratios of low-density lipoprotein cholesterol to high-density lipoprotein cholesterol (Lewis *et al.*, 1988; Mott *et al.*, 1990, 1993). An ongoing investigation of hematological values of nursery- and mother-reared baboon infants suggests that the albumin : globulin ratio is higher in nursery-reared infants, while the number of white blood cells and glucose levels are lower in nursery-reared baboons (L.M. Havill, L. Brent, C.L. Snider, M.L. Leland, G.B. Hubbard, S.R. Theriot, and M.C. Mahaney, unpublished observations). Because rearing and diet are confounded, it is unclear whether the environment per se had an influence over physiological measurements. Regardless of the source of variation, such differences in physiological processes and measurements between mother- and nursery-reared baboons are important considerations when interpreting data and selecting subjects for research projects. Of interest would be comparative data on the stress physiology of nursery-reared versus mother-reared baboons (see Section 2.1).

2.4. Mortality and Morbidity

Infancy and the juvenile period are particularly critical times in terms of survival. Captive primates have a higher mortality rate than wild-born primates before 1 month of age and trauma is a more likely cause of death than infectious disease (Debyser, 1995). In a captive colony of hamadryas baboons, 14% of pregnancies resulted in abortions; of 100 live births,

Table 14–3. Physiological Effects of Nursery Rearing in Baboons

Subjects	Findings	Reference
Plasma thyroid hormones $n = 30$, 27 nursery-reared (13 high-fat formula; 14 low-fat formula), 17 mother-reared	Plasma total triiodothyronine (T_3) and free T_3 levels were higher in nursery infants in late pre-weaning period (4–14 weeks); plasma total thyroxine (T_4) and free T_4 were not different	Lewis *et al.*, 1993
Bile acid metabolism in 14-week-old infants $n = 43$, 26 nursery-reared (12 high-fat formula; 14 low-fat formula), 17 mother-reared	Cholic acid metabolism significantly affected by rearing, with little effect of formula fat content; 50% lower biliary cholic acid in mother-reared infants; little effect on chenodeoxycholic acid metabolism	Jackson *et al.*, 1993
Hematology and blood chemistry in infants $n = 159$, 41–56 nursery reared, 82–103 mother reared	Albumin/globulin ratio higher, white blood cells and glucose lower in nursery-reared animals	L.M. Havill, L. Brent, C.L. Snider, M.L.Leland, G.B. Hubbard, S.R. Theriot, and M.C. Mahaney, unpublished observations
Cholesterol metabolism measured through young adulthood $n = 45$–80, 32–58 nursery-reared (various infant formulas), 13–22 mother-reared	Cholesterol levels and lipoprotein metabolism affected by rearing and variation in infant diet; mother-reared infant and young adult baboons had higher serum cholesterol and lower VLDL cholesterol + LDL cholesterol : HDL cholesterol ratios;[a] they also responded more strongly to dietary changes in fat content and had more atherosclerotic lesions; fat ratio in infant formula had little effect on group differences	Lewis *et al.*, 1988; Mott *et al.*, 1990, 1993

[a] VLDL, very low-density lipoprotein; LDL, low-density lipoprotein; HDL, high-density lipoprotein.

only 46% survived past the first day of life (Zinner *et al.*, 1993). In the early years of the SFBR's breeding program, about 4% of 167 pregnancies ended in abortion or stillbirth (Kraemer and Vera Cruz, 1972).

To determine some possible differences between nursery-reared and mother-reared individuals, we conducted a survey of mortality and morbidity indicators for baboons that were born at the SFBR and died between 1985 and 2002. Mother-reared individuals were defined as infants that were raised by their mothers for ≥30 days, and nursery-reared individuals were defined as infants that were with their mothers <30 days before being placed in the nursery. The data set included only baboons that died of natural causes or for management reasons. To determine differences in infant mortality between mother- and nursery-reared baboons, we calculated the rate of death in the first year of life for the two groups. The mortality rate of the nursery-reared individuals ($n = 256$) during this time period was 38.3%, compared with the 33.9% mortality rate of the mother-reared individuals ($n = 578$).

To gauge the lifelong impact of nursery rearing on baboons, we used only animals that lived to >1 year of age, so as to eliminate infants that may have died or become ill as a result of birth trauma, congenital problems, or infectious disease. A total of 527 baboons were used in the analysis. The variables considered were age at death and the number of admittances to the clinic for trauma and diarrhea (corrected for lifespan by dividing the number of clinical visits by the number of years alive). For analyses, scores were square root transformed to result in more normal distributions.

The average age at death was significantly lower for nursery-reared infants (4.7 years) than for mother-reared infants (5.7 years) [$F(1,525) = 7.76$, $p < 0.006$]. The incidence of trauma during the lifetime was not significantly different between groups (average = 0.26 incidence/year for nursery-reared infants, 0.30 incidence/year for mother-reared infants). Diarrhea, commonly caused by bacterial infection in baboons, occurred significantly more often in nursery-reared baboons (0.34 incidence/year) than in mother-reared baboons (0.21 incidence/year) [$F(1,525) = 9.38$, $p < 0.002$] (see Fig. 14–1).

Our results indicate that for baboons at this facility, nursery rearing has a long-term negative consequence on their health and longevity, but

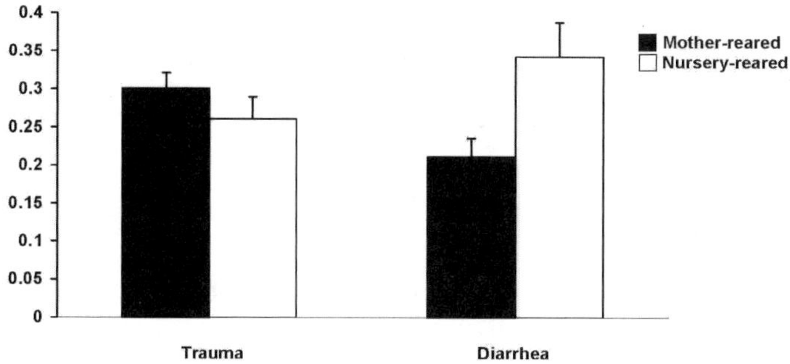

Figure 14–1. Incidence of trauma and diarrhea per year for nursery-reared and mother-reared baboons. The nursery-reared infants had significantly higher rates of treatment for diarrhea.

not on the incidence of trauma. Trauma usually is induced socially through aggressive wounding and our results provide indirect evidence that the nursery-reared baboons did not experience more aggression. Diarrhea is often infectious in nature (resulting from bacterial infections such as *Campylobacter*, *Shigella*, and *Salmonella*). The higher rate of diarrhea for nursery-reared baboons may be related to reduced development of normal antibodies resulting from living in the sterile nursery setting and lack of protective immunity gained from mother's milk. Susceptibility to intestinal pathogens has been related to nursery rearing and the lack of mother's milk, resulting in altered immune responses in rhesus macaques (Lubach *et al.*, 1995). Diarrhea also accompanies a stress response, and the nursery-reared baboons may experience increased stress owing to environmental and social changes.

Nursery-reared infants experience a number of confounds, including lack of protective immune factors usually present in breast milk and exposure to environmental variables such as human contact and social stimulation. From our relational survey it is not possible to clearly identify causative factors in the long-term negative consequences of nursery rearing. However, in a comprehensive study of 1187 mother- and 506 nursery-reared pigtailed macaques between 2 and 10 years of age, Sackett *et al.* (2002) found no differences in survival, growth, health, or reproduction between the two groups. The nursery-reared monkeys in that study received intensive human handling, and daily playroom socialization with peers, and were exposed to learning and cognitive tests. The high level of physical, social, and cognitive stimulation provided in their nursery setting may have had a positive influence on health and survival.

3. VARIATION IN THE NURSERY ENVIRONMENT

The type of nursery-rearing environment is also important when reviewing outcome variables. Total social isolation is known to induce severe behavioral distortions, beginning with a redirection of normal behaviors toward one's own body (self-clinging, self-sucking), and with prolonged isolation, deficiencies in social behavior that may last a lifetime (Capitanio, 1986; Harlow *et al.*, 1965; Mitchell *et al.*, 1966). Less isolated forms of rearing usually produce less severe problems (e.g., Suomi *et al.*, 1971). Gauging the effects of rearing environment can lead to improved methods so that nursery-reared individuals more closely resemble their mother-reared counterparts (Ruppenthal and Sackett, 1979).

Numerous studies on different methods of rearing nonhuman primates that had to be removed from their mother have identified several important factors. Most studies have been conducted with macaques. The presence of an inanimate surrogate mother provides an attachment figure and surface to which the infant can cling (Suomi, 1973). A movable surrogate provides vestibular stimulation and reduces the incidence of stereotyped self-stimulation (e.g., rocking, swaying) (Mason and Berkson, 1975), and also serves to reduce emotional responsiveness (Eastman and Mason, 1975). Social interaction with peers is also important, although continuous peer pairing may result in high levels of clinging behavior and low levels of exploration and other normal behavior (Ruppenthal *et al.*, 1991). An enriched environment can help to provide novel stimulation and encourage exploratory behavior and the ability to cope with environmental changes. In a study by Clarke *et al.* (1989), one group of nursery-reared rhesus macaques was given moving surrogates, rubber toys, and hanging diapers in their cages, while the control group received no cage enrichment. The enriched animals showed less stress-related and stereotyped behaviors than the nonenriched control group.

Unfortunately, there is not as much experimental information about the impact of variation in the nursery environment and care program for baboons as there is for the more commonly studied macaques.

3.1. Impact of Infant Formula Composition on Baboon Growth

Variations in the feeding regimen for nursery-reared infant nonhuman primates can have important consequences for growth and survival. For example, infant pigtailed macaques that had *ad libitum* access to formula consumed 20–25% more calories than did infants on a distributed schedule (4 hr access/4 hr no access). Owing to the presence of bacteria-causing diarrhea and an inability to metabolize the additional nutrients, however, they did not exceed the body weight of infants on the distributed schedule (Sackett and Ruppenthal, 1992).

Three different infant formulas were provided to 48 baboons from birth to 16 weeks of age (Rutenberg and Coehlo, 1988). The low- and high-calorie diets were 40% lower and higher, respectively, than the standard diet. The subjects' weight, crown–rump length, and triceps circumference were measured during the first 5 years of life. Measurements early in life indicated group differences based on diet, but after the dietary treatment had stopped, both low- and high-calorie subjects returned to average measurements by 26 weeks of age. Long-term differences in growth measurements were not found for males, but the effects were evident for females throughout the study.

3.2. Impact of Human Handling on Behavior

To determine if increased human handling has an impact on later behavior, 16 infant baboons were observed between 5 and 10 months of age for 30 min twice per week. While normal procedures in the nursery involved minimal human contact (usually just for biweekly cage changing), half of the subjects had received additional human handling (twice per day) from the age of about 3–42 days as part of an experimental protocol to take physiological measurements of unsedated animals. All other rearing conditions were similar.

Preliminary analyses considered the effect of group (low versus high amount of early human handling) and housing condition (single versus pair) on common behaviors, including locomotion, environmental exploration, inactivity, abnormal behavior, and social behavior. Results revealed little influence of early handling. Social behavior was higher during the paired condition [$F(1,13) = 57.77$, $p < 0.001$], as would be expected due to increased social opportunities. An interaction between group and housing condition was significant for environmental exploration [$F(1,13) = 13.31$, $p < 0.003$], with infants that received little human handling exhibiting lower levels of environmental manipulation in the paired condition than the animals that received a lot of human handling.

From these results, it appears that at least for baboons in this setting, increased amounts of human handling early in life did not have much influence over later behavior. The incidence of abnormal behaviors was elevated (average 8.6 min/hr) and did not decrease over the observational period. More dramatic changes in nursery practices are probably required to influence later behavior significantly.

3.3. Impact of Socialization Program

An infant socialization program was started at SFBR in 1999 to provide more social stimulation in the hope of reducing the high levels of abnormal behavior exhibited by many infants. Socialization included transfer of nursery infants from typical single housing to a playroom with 5–14 peers for 1–1.5 hr, 3 days per week. The playroom contained several toys, structures, and brightly painted murals on the wall (Fig. 14–2). An evaluation of the behavior of 28 infants using the playroom indicated that it was effective in reducing abnormal behavior development in the nursery-reared baboons (J. Stein and L. Brent, unpublished observations). Ten infants began having access to the playroom at >7 weeks of age, while the other 18 had access beginning at 2–7 weeks of age. Infants that were placed in the playroom at a younger age had a lower incidence of abnormal behavior (67%) than those that started at an older age (80%). Although most infants initially showed some form of abnormal behavior (head clutching, rocking, clinging, self-biting), by 8 weeks in the playroom the incidence was near zero. Eight infants never displayed any abnormal behavior, and nine others exhibited no abnormal behavior after 2 weeks in the playroom. Of six infants that did not show improvement, four were in the group that was older when it began having access to the playroom. These results demonstrate that even a few hours of access to peers in an enriched playroom is effective at reducing levels of abnormal behaviors in the nursery-reared baboons, and the sooner the infants have access to the playroom, the greater the improvement. Similarly for pig-

Figure 14–2. Infant socialization playroom, with colored murals, climbing structures, toys, and towels.

tailed macaques, access to a social playroom for 30 min/day, 5 days/week, was associated with generally normal species-typical behavior patterns by 8 months of age (Worlein and Sackett, 1997).

4. CHANGES IN NURSERY-REARING PRACTICES OVER TIME

Because removing infant baboons from their mothers and raising them in a nursery environment has a potential impact on survival, behavior, development, and physiology, an emphasis on improving nursery-rearing protocols is an important component of an animal care plan. Considering the long history of nursery-rearing baboons at the SFBR, a review of nursery care protocols provides an interesting insight into changes in philosophy and methods over time.

4.1. Early Published Reports

Information on the care of nursery reared baboons at SFBR dates back to the 1960s. Vice *et al.* (1966, 1968) described the nursery rearing of 21 infants that were removed from their mothers and placed in incubators. To reduce the incidence of respiratory disease, which accounted for the deaths of seven infants, incubator temperature was increased to 88–92°F. Infants were given daily physical examinations and their temperature was taken every 12 hr. For the first 2 weeks they were hand-fed infant formula. Comparing two different formula concentrations, the authors found that baboons had a higher caloric requirement than is provided by standard infant formula (26 kcal/oz), probably because baboon breast milk has a high fat content of 7.2–18%. Baby food was provided at 14 days of age and alternated with formula, and solid food was provided as soon as incisor teeth erupted (21–30 days of age). Infants were weaned from formula at 30–40 days of age.

In an update in 1979, Moore and Cummins described similar practices. By then, infant formula of 20 kcal/oz was considered optimal for growth and health. Infants were hand fed for the first month to reduce dribbling and wastage of formula. Baby food was not mentioned, and solid food (biscuits soaked in formula) was provided much later than previously (98–112 days of age). Infants were housed in an isolette incubator for the first 2 weeks of life, and then alone in a wire cage. The authors noted that the necessity of housing the infants singly for nutrition studies resulted in abnormal characteristics, including huddling on the bottom of the cage, reduced activity, and self-sucking. Changes in protocol to reduce these abnormal behaviors included the provision of a towel hung from the top of the cage so that instead of huddling, the infants would cling to the towel, thereby engaging in more species-typical behavior. Also, beginning at 1 week of age, infants were placed in an activity cage for socialization with peers for 2 hr daily. After these changes, no behavioral problems were noted. In this report, more emphasis is placed on maintaining sanitary conditions, including protective clothing for staff (gloves, masks, foot covers, etc.) and reducing the number of staff allowed in the nursery area. A thorough cage sanitation program was also implemented. The nursery program had a death rate of less than 5%. The main cause of death was respiratory disease and birth anomalies.

4.2. Changes over Time

These procedures remained the same during the late 1970s to early 1980s, the time period during which Bramblett, Coelho, and colleagues studied the nursery-reared subjects (Section 2.1). Infants were provided with a towel surrogate, handled five times per day for feeding, and provided with peer socialization (small or large group) for 2–4 hr daily. At 3–6 months of age, the infants were placed in social groups in large cages in the nursery. Once weaned to solid food, they were moved outside to peer social groups (Coelho and Bramblett, 1981).

By the late 1980s, the baboon nursery at SFBR had grown, housing up to 60 infants at any one time. Probably due to the increased number of infants, the emphasis on sterility was increased to reduce disease transmission. Staff effort to maintain such a nursery also must have influenced procedures. Towels were provided on the floor of the cage but not hung from the top, and no peer socialization in an activity area was provided. Changes extended to housing and feeding schedules as well. The infants were housed in incubators at a temperature of 85°F for 2–4 weeks and then housed singly in wire cages until 5–10 months of age, whereupon they were paired or placed in small groups and then released to outdoor peer groups. The 20-kcal/oz formula was standard, with 24-kcal/oz formula available as needed. Formula-soaked biscuits were provided at 4 months and the infants were weaned from formula at 6–9 months.

With the emphasis on psychological well-being of nonhuman primates in the early 1990s, nursery personnel began to recognize the underlying reasons for behavioral problems among nursery-reared individuals and instituted a number of changes. The hanging towel surrogate for the infant baboons was reinstated and manipulable toys were provided to encourage species-typical behavior. In 1999, an infant socialization program was developed in which the infants were provided with access to a playroom with compatible peers three times per week. The socialization program has been helpful in reducing the incidence of abnormal behavior (Section 3.2). Feeding schedules were also modified, so that infants now receive solid food at 3 months of age. Infants are paired at about 3–4 months of age. Sanitary conditions and a reduced number of people entering the nursery are still important, but during fair weather the infants are also placed outside near the entrance to the nursery in a cage with peers in an effort to ease the transition to the outside environment and to assist in priming their immune system.

4.3. Learning from History

When wild nonhuman primates were captured and brought into the captive environment, there was little experience on which to base management decisions. Early animal care professionals looked to the natural history of the species, using the natural environment, social structure, and eating habits as the basis for care (e.g., Yerkes and Yerkes, 1929). As captive colonies increased in size and captive breeding became important, it became critical to reduce mortality and increase efficiency. In keeping with new knowledge on the identification and treatment of particular diseases, the emphasis shifted toward sanitary conditions and physical health.

With the more recent emphasis on the connection between physical and psychological health, and public pressure on improving living conditions for captive animals, consideration of the behavioral health of the individual has become more widespread. Although it is not often realized, many of the "new" improvements from environmental enrichment and behavioral management studies can be traced to basics of care from the early years of captive propagation of nonhuman primates. The history of the nursery care of baboons at SFBR is just one example of this. What we thought was a new program of nursery infant socialization and enrichment turns out to have been common practice in the 1960s to 1980s. Thus we should not only continue to study rearing practices and make improvements as new information, problems, and solutions appear, but also learn from the past.

5. CONCLUSIONS

The information presented here demonstrates that nursery-reared baboons differ significantly from mother-reared individuals in behavior, growth, physiology, and long-term survival. Minor modifications in human handling in the nursery were found to make little difference to behavior, but access to peers in a playroom setting helped to reduce or avoid the development of abnormal behavior patterns. The levels of social and environmental stimulation available in the nursery setting certainly affect infant outcome, and determining the optimal nursery protocol is a worthy effort. Equally important is an emphasis on mother rearing to produce the most species-normal nonhuman primates.

ACKNOWLEDGMENTS

We thank Dr. Karen Rice, Dr. Michelle Leland, Michelle Foley, Elaine Windhorst, and Dr. Lorena Havill for their assistance with the preparation of this review. This work was made possible in part by a grant from the National Institutes of Health HL028972 to J.L. VandeBerg.

REFERENCES

Bellanca, R.U., and Crockett, C.M. (2002). Factors predicting increased incidence of abnormal behavior in male pigtailed macaques. *Am. J. Primatol.* 58:57–69.

Bielitzki, J.T., and Ruppenthal, G.C. (1989). Nursery rearing of old world monkeys. *Proceedings of the Annual Meeting of the American Association of Zoo Veterinarians*, pp. 193–196.

Bramblett, C.A., and Coelho, A.M., Jr. (1985). Age changes in affinitive behaviors of baboons. *Am. J. Primatol.* 9:259–271.

Capitanio, J.P. (1986). Behavioral pathology. In: Mitchell, G., and Erwin J. (eds.), *Comparative Primate Biology:* Vol. 2, Part A: *Behavior, Conservation, and Ecology*. Alan R. Liss, New York, pp. 411–454.

Clarke, M.R., Koritnik, D.R., Martin, L.N., and Baskin, G.B. (1989). Cage enrichment, physiology, and behavior in nursery-reared rhesus monkeys. *Am. J. Primatol.* Suppl. 1:53–57.

Coelho, A.M. (1985). Baboon dimorphism: Growth in weight, length, and adiposity from birth to 8 years of age. In: Watts, E.S. (ed.), *Nonhuman Primate Models for Growth and Development, Monographs in Primatology*, Vol. 6. Alan R. Liss, New York, pp. 125–159.

Coelho, A.M., Jr., and Bramblett, C.A. (1981). Effects of rearing on aggression and subordination in *Papio* monkeys. *Am. J. Primatol.* 1:401–412.

Coelho, A.M., Jr., and Bramblett, C.A. (1982). Social play in differentially reared infant and juvenile baboons (*Papio* sp.). *Am. J. Primatol.* 3:153–160.

Coelho, A.M., Jr., and Bramblett, C.A. (1984). Early rearing experiences and the performance of affinitive and approach behavior in infant and juvenile baboons. *Primates* 25:218–224.

Debyser, I.W.J. (1995). Catarrhine juvenile mortality in captivity, under semi-natural conditions, and in the wild. *Int. J. Primatol.* 16:935–969.

Eastman, R.F., and Mason, W.A. (1975). Looking behavior in monkeys raised with mobile and stationary artificial mothers. *Dev. Psychobiol.* 8:213–221.

Erhart, E.M. (1995). Savanna baboons (*Papio cynocephalus*): Early rearing experiences and adolescent behavioral development. *Am. J. Phys. Anthropol. Suppl.* 20:89 (abstract 100).

Glassman, D.M., and Coelho, A.M. (1986). Comparisons of longitudinal weight growth in nursery and mother-reared baboons. *Am. J. Phys. Anthropol.* 69:205 (abstract).

Harlow, H.F., Dodsworth, R.O., and Harlow, M.K. (1965). Total social isolation in monkeys. *Proc. Natl. Acad. Sci. USA* 54:90–97.

Hendrickx, A.G., and Kriewaldt, F.H. (1965). Observations on a controlled breeding colony of baboons. In: Vagtborg, H. (ed.), *The Baboon in Medical Research, Volume II: Proceedings of the Second International Symposium on the Baboon and Its Use as an Experimental Model.* University of Texas Press, Austin, pp. 69–83.

Higley, J.D., and Suomi, S.J. (1989). Temperamental reactivity in non-human primates. In: Kohnstamm, G.A., Bates, J.E., and Rothbart, M.K. (eds.), *Temperment in Childhood.* John Wiley & Sons Ltd., New York, pp. 153–167.

Jackson, E.M., Lewis, D.S., McMahan, C.A., and Mott, G.E. (1993). Preweaning diet affects bile lipid composition and bile acid kinetics in infant baboons. *J. Nutr.* 123:1471–1479.

Kraemer, D.C., and Vera Cruz, N.C. (1972). Breeding baboons for laboratory use. In: Beveridge, W.I.G. (ed.), *Breeding Primates. Proceedings of the International Symposium on Breeding Non-Human Primates for Laboratory Use,* Berne, 28–30 June 1971. S. Karger, New York, pp. 42–47.

Levin, J.L., and Carey, K.D. (1988). Factors influencing body weight in nursery reared baboons. *Lab. Anim. Sci.* 34:506 (abstract 65).

Lewis, D.S., Mott, G.E., McMahan, C.A., Masoro, E.J., Carey, K.D., and McGill, H.C., Jr. (1988). Deferred effects of preweaning diet on atherosclerosis in adolescent baboons. *Arteriosclerosis* 8:274–280.

Lewis, D.S., McMahan, C.A., and Mott, G.E. (1993). Breast feeding and formula feeding affect differently plasma thyroid hormone concentrations in infant baboons. *Biol. Neonate* 63:327–335.

Lubach, G.R., Coe, C.L., and Ershler, W.B. (1995). Effects of early rearing environment on immune responses of infant rhesus monkeys. *Brain Behav. Immun.* 9:31–46.

Luft, J., and Altmann, J. (1982). Mother baboon. *Nat. Hist.* 91:30–39.

Mason, W.A., and Berkson, G. (1975). Effects of maternal mobility on the development of rocking and other behaviors in rhesus monkeys: A study with artificial mothers. *Dev. Psychobiol.* 8:197–211.

Mirmiran, M., Bernardo, L., Jenkins, S.L., Ma, X.H., Brenna, J.T., and Nathanielsz, P.W. (2001). Growth, neurobehavioral and circadian rhythm development in newborn baboons. *Pediatr. Res.* 49:673–677.

Mitchell, G.D., Raymond, E.J., Ruppenthal, G.C., and Harlow, H.F. (1966). Long-term effects of total social isolation upon behavior of rhesus monkeys. *Psychol. Rep.* 18:567–580.

Moore, G.T., and Cummins, L.B. (1979). Nursery rearing of infant baboons. In: Ruppenthal, G.C. (ed.), *Nursery Care of Nonhuman Primates.* Plenum Press, New York, pp. 145–151.

Mott, G.E., Jackson, E.M., McMahan, D.A., and McGill, H.C., Jr. (1990) Cholesterol metabolism in adult baboons is influenced by infant diet. *J. Nutr.* 120:243–251.

Mott, G.E., Lewis, D.S., and McMahan, C.A. (1993). Infant diet affects serum lipoprotein concentrations and cholesterol esterifying enzymes in baboons. *J. Nutr.* 123:155–163.

Ruppenthal, G.C., and Sackett, G.P. (1979). Experimental and husbandry procedures: Their impact on development. In: Ruppenthal, G.C. (ed.), *Nursery Care of Nonhuman Primates.* Plenum Press, New York, pp. 269–284.

Ruppenthal, G.C., Walker, C.G., and Sackett, G.P. (1991). Rearing infant monkeys (*Macacanemestrina*) in pairs produces deficient social development compared with rearing in single cages. *Am. J. Primatol.* 25:103–113.

Rutenberg, G.W., and Coelho, A.M., Jr. (1988). Neonatal nutrition and longitudinal growth from birth to adolescence in baboons. *Am. J. Phys. Anthropol.* 75:529–539.

Sackett, G.P., and Ruppenthal, G.C. (1992). Growth of nursery-raised *Macaca nemestrina* infants: Effects of feeding schedules, sex, and birth weight. *Am. J. Primatol.* 27:189–204.

Sackett, G.P., Bowman, R.E., Meyer, J.S., Tripp, R.L., and Grady, S.S. (1973). Adrenocortical and behavioral reactions by differentially raised rhesus monkeys. *Physiol. Psychol.* 1:209–212.

Sackett, G.P., Ruppenthal, G.C., and Davis, A.E. (2002). Survival, growth, health, and reproduction following nursery rearing compared with mother rearing in pigtailed monkeys (*Macaca nemestrina*). *Am. J. Primatol.* 56:165–183.

Suomi, S.J. (1973). Surrogate rehabilitation of monkeys reared in total social isolation. *J. Child Psychol. Psychiatry* 14:71–77.

Suomi, S.J. (1991). Early stress and adult emotional reactivity in rhesus monkeys. *CIBA Found. Symp.* 156:171–188.

Suomi, S.J., Harlow, H.F., and Kimball, S.D. (1971). Behavioral effects of prolonged partial social isolation in the rhesus monkey. *Psychol. Rep.* 29: 1171–1177.

Veira, Y., and Brent, L. (2000). Behavioral intervention program: Enriching the lives of captive nonhuman primates. *Am. J. Primatol.* 51(Suppl. 1):97 (abstract).

Vice, T.E., and Rodriguez, A.R. (1965). Care and maintenance of baboon colonies in San Antonio, Texas, and Darajani, Kenya, East Africa. In: Vagtborg, H. (ed.), *The Baboon in Medical Research: Proceedings of the First International Symposium on the Baboon and Its Use as an Experimental Animal.* University of Texas Press, Austin, pp. 125–136.

Vice, T.E., Britton, H.A., Ratner, I.A., and Kalter, S.S. (1966). Care and raising of newborn baboons. *Lab. Anim. Care* 16:12–22.

Vice, T.E., Britton, H.A., Ratner, I.A., and Kalter, S.S. (1968). Care and raising of newborn baboons. *J. Small Anim. Pract.* 9:124–129.

Worlein, J.M., and Sackett, G.P. (1997). Social development in nursery-reared pigtailed macaques (*Macaca nemestrina*). *Am. J. Primatol.* 41:23–35.

Yerkes, R.M., and Yerkes, A.W. (1929). *The Great Apes: A Study of Anthropoid Life.* Yale University Press, New Haven.

Young, G.H., and Bramblett, C.A. (1977). Gender and environment as determinants of behavior in infant common baboons (*Papio cynocephalus*). *Arch. Sex. Behav.* 6:365–385.

Young, G.H., Coelho, A.M., Jr., and Bramblett, C.A. (1982). The development of grooming, sociosexual behavior, play and aggression in captive baboons in their first two years. *Primates* 23:511–519.

Zinner, D., Kaumanns, W., and Rohrhuber, B. (1993). Infant mortality in captive hamadryas baboons (*Papio hamadryas*). *Prim. Rep.* 36:97–113.

CHAPTER FIFTEEN

Early Rearing Conditions and Captive Chimpanzee Behavior: Some Surprising Findings

Mollie A. Bloomsmith, Kate C. Baker, Stephen R. Ross, and Susan P. Lambeth

1. INTRODUCTION

While it might seem that one of the first tasks in the formal study of behavior in captive nonhuman primates would be to determine the conditions required for normal behavioral development, this task has not yet been accomplished nearly four decades after the National Institutes of Health first established its network of primate research centers. It is widely thought that normal behavioral development depends in part on the bond between an infant primate and its mother, but not all maternally reared infants develop "normal" behavior or physiology (Sackett *et al.*, 2002). Since captive nonhuman primates are reared in social and physical conditions that are not identical to the conditions in the wild, we need to understand how these dissimilar aspects of the rearing environment can influence behavioral development, to determine whether alternative social or physical settings can approximate the circumstances for normal development, and to define exactly what constitutes "normal" behavior and the desired developmental outcome for captive nonhuman primates.

To date, behavioral development has been studied most thoroughly in macaques. Rearing experience has persistent consequences on later sociality and social competency in macaques (Mason, 1960, 1961a; Harlow and Harlow, 1969; Ruppenthal *et al.*, 1976; Suomi and Ripp, 1983; Champoux *et al.*, 1992). These effects have been demonstrated in studies on dominance and aggressive behavior (Mason, 1961b; Mitchell *et al.*, 1966; Suomi, 1974), reconciliation (Ljungberg and Westlund, 2000), sexual behavior (Chamove *et al.*, 1973; Goldfoot 1977), maternal behavior (Arling and Harlow, 1967; Suomi, 1978; Champoux *et al.*, 1992; Lindell *et al.*, 1995; Goin and Gust, 1998), and attachment, responsiveness, and social sophistication (Capitanio, 1984, 1985; Anderson and Chamove, 1986; Mason and Capitanio, 1988; Ruppenthal *et al.*, 1991).

Behavioral development in captivity has been studied much less extensively in chimpanzees (*Pan troglodytes*) and other apes. In contrast to the numerous rearing environments that have been investigated for macaques, only two situations other than natural mother rearing have been studied with chimpanzees. Early studies examined the effects of severe sensory and social isolation during nursery rearing. Such experiences resulted in high levels of abnormal behavior and dramatic deficits in social behavior and cognitive ability (Davenport and Menzel, 1963; Berkson and Mason, 1964; Davenport and Rogers, 1968, 1970; Davenport *et al.*, 1969, 1973; Turner *et al.*, 1969; Menzel *et al.*, 1970; Rogers and Davenport, 1970, 1971; Reisen, 1971; Walsh *et al.*, 1982). The pronounced behavioral consequences diminished the chimpanzees' ability to live in social groups, to copulate, and to raise infants. In more recent work, investigators examining the effects of rearing in socially and environmentally complex nursery environments have found that even these enriched conditions affect behavioral development in a manner that is qualitatively similar to, albeit more moderate than, the behavioral effects of rearing in deprivation. Nursery-reared chimpanzees display higher levels of abnormal behavior than their mother-reared counterparts (Maki *et al.*, 1993; Bloomsmith and Haberstroh, 1995; Spijkerman, 1997; Ross 2003), reduced sexual and maternal abilities (King and Mellen, 1994; Brent *et al.*, 1996; M.A. Bloomsmith *et al.*, unpublished observations), and reduced problem-solving skills (Brent *et al.*, 1995). It bears repeating, however, that their behavioral problems are not as severe and debilitating as those described in the earlier years for chimpanzees raised in deprived circumstances.

2. EVOLUTION OF NURSERY-REARING PRACTICES FOR CHIMPANZEES

The shift in research focus has been paralleled by a change in nursery-rearing practices over the past 20 years. In the 1960s and early 1970s the goal of nursery care of apes was to keep the young animals physically healthy and available to investigators for scientific study. To facilitate these goals, infants often were removed from their mothers at birth (Maple, 1980) and housed for long periods of time alone in incubators or other fairly sterile enclosures to protect their health. More recent nursery-rearing practices have been aimed at fostering species-appropriate behavioral development (Maple, 1980; Fulk and Garland, 1992) by emphasizing social stimulation and physical activity in more varied and complex enclosures. In general, nursery-reared chimpanzees are now housed with peers as a matter of course, often have extensive interaction with human caregivers, and enjoy access to the outdoors. Some also have visual access or other exposure to adult chimpanzees. In addition, human caregivers have attempted to model interactions between mother and infant chimpanzees, and to use species-appropriate gestures and facial expressions in these interactions (Bard, 1995; Bard and Gardner, 1996).

This type of enriched environment was in place at the M.D. Anderson Cancer Center chimpanzee facility during the course of many of the studies described in this chapter. Infants were housed in crib-type enclosures made of wire mesh, allowing them to stand and climb once they were able, rather than in barred cribs that do not facilitate such movement. When the infants grew somewhat older, they were placed in larger, enclosed wire mesh pens so they could locomote over all of the interior surfaces. These cages contained platforms, swinging ladders, and arboreal cage furnishings to increase locomotor opportunities. Fleece pads, toys and other manipulable objects, as well as feeding enrichment, were provided. Typically, infants were placed with other infants in peer groups at very young ages, as soon as a suitable partner was available. These groups varied from two to six animals and the chimpanzees were usually together 24 hr/day. Additional nursery personnel provided infants with considerable human contact. Each youngster received about 2 hr of attention, handling, playing, and carrying per day. Most of the infants also had free interaction with a canine companion that lived with the groups of chimpanzees for a large portion of the day and night (Thompson *et al.*, 1991; Pazol and Bloomsmith, 1993). This practice has been shown to reduce the incidence of stereotyped body rocking in captive infant chimpanzees (Pazol and Bloomsmith, 1993).

3. EFFECTS OF CURRENT NURSERY PRACTICES AND OTHER EARLY REARING EXPERIENCES ON BEHAVIORAL DEVELOPMENT OF CHIMPANZEES

The goal of current nursery practices in zoos and breeding facilities is to produce animals that display intact behavioral repertoires, social competency, flexibility, and adaptability (Fulk and Garland, 1992), but the degree to which this goal has been met has not been subjected to sufficient experimental assessment. In this chapter we will describe a series of studies conducted over the past 15 years, examining a variety of behavioral characteristics of chimpanzees that were raised by humans in a nursery and comparing their behavior with that of chimpanzees that experienced other social histories. In general, these nursery-reared chimpanzees do not seem to be showing the devastating behavioral deficits that were present among chimpanzees reared in restrictive settings in the past.

This is not to say that there are no behavioral manifestations of early rearing by humans—there are many. But the behavioral contrasts between mother- and nursery-reared chimpanzees are not expressed across animals as reliably as they once were, nor are they as debilitating as they once were, or were believed to be. The procedural changes in nursery rearing may have fundamentally altered the development of chimpanzees reared in the nursery to the extent that the general behavioral consequences of nursery rearing have now been altered. A number of our findings are somewhat surprising, given that they conflict with "conventional wisdom" about the behavioral characteristics of nursery-reared chimpanzees. Even though the sample sizes described in this chapter are modest and the dependent measures are strictly behavioral, our questioning of conventional wisdom is similar to that expressed by Sackett *et al.* (2002), who used a huge and diverse database to contest many time-worn conventional expectations related to the growth, morbidity, mortality, and reproduction of nursery-reared pigtailed macaques. In this chapter, we describe a similar examination for chimpanzees.

3.1. Abnormal Behavior

One of the most recognizable ways that the behavior of nursery-reared chimpanzees often varies from that of chimpanzees raised by their mothers is the greater expression of stereotyped and apparently non-functional behaviors. Chimpanzees display a wide array of such behaviors with extensive individual variations. These behaviors include repetitive rocking, stereotyped body movements, unusual posturing, self-orality, depilitation, regurgitation, coprophagy, urophagy, and others (Davenport and Menzel, 1963; Berkson and Mason, 1964; Davenport and Rogers, 1970; Walsh *et al.*, 1982; Capitanio, 1986). In early studies of the effects of severe sensory and social isolation (from both conspecific and human interaction) during development, deprived chimpanzees displayed high levels of abnormal behavior (Davenport and Menzel, 1963; Berkson and Mason, 1964; Davenport and Rogers, 1970; Walsh *et al.*, 1982). Even with more recent nursery practices, nursery-reared chimpanzees at a number of facilities display higher levels of abnormal behaviors than do their mother-reared counterparts (Maki *et al.*, 1993; Bloomsmith and Haberstroh, 1995; Spijkerman *et al.*, 1997; Ross, 2003).

An unanticipated finding of recent studies is that even mother-reared chimpanzees display behaviors typically thought to be abnormal. In one study, we investigated the effect of early social experience on the quality and quantity of abnormal behaviors expressed by 41 juvenile chimpanzees when they were 2–6 years old. Twenty-two subjects had continuous mother rearing, seven were reared by their mothers until they were 2 years old and then lived in an alternative social setting with peers or with peers and an adolescent female, and 12 were nursery reared by humans and lived in a peer group. The mother-reared youngsters lived from birth in large, complex social groups in fairly spacious indoor/outdoor housing with a strong environmental enrichment program. Yet they still exhibited some abnormal behaviors. Examining more than 1800 hr of observational data, we identified 10 different types of abnormal behavior. Nursery-reared subjects showed the highest levels of abnormal behavior overall, devoting a mean of 4.5% of their time to it, versus 0.4% for mother-reared subjects and 0.2% for subjects that were mother reared for 2 years before moving to an alternative social setting. This distinction was particularly strong in the case of stereotyped rocking, which was expressed by most nursery-reared subjects (mean of 3.5% of their time) and was never recorded for chimpanzees with the other two types of social backgrounds. Thus a 2-year period of maternal rearing is apparently sufficient to forestall the development of stereotyped rocking, but not other abnormal behaviors, in young chimpanzees.

While the expression of abnormal behavior among young chimpanzees varied across social settings, it is interesting that none of the social settings we studied was sufficient to prevent the development of some atypical behavior. It is somewhat surprising that the subjects with the lowest amount of time devoted to abnormal behavior were those that lived with their mothers for 2 years and then lived in an alternate social grouping, rather than the subjects that lived with their mothers continually. Apparently, a 2-year period of maternal rearing is sufficient to buffer against the development of a variety of forms of abnormal behavior, at least through the juvenile years.

One type of abnormal behavior, coprophagy, occurred in an unexpected pattern: mother-reared youngsters displayed this behavior more often (0.3% of the time) than did nursery-reared youngsters. In fact, coprophagy accounted for the vast majority of abnormal behavior recorded for mother-reared youngsters. The subjects that were mother reared for 2 years and then lived in alternative social groups showed intermediate levels of coprophagy (0.14%), and the nursery-reared subjects showed the lowest levels (0.04%). Hypotheses that call on stimulus deprivation to explain abnormal behavior cannot account for this finding, as the behavior was expressed most often in the most stimulus-rich environment that we studied. It is possible that coprophagy may be learned socially, or may be facilitated by the presence of adult chimpanzees. Such a possibility is corroborated by the findings of Hook *et al.* (2002), who identified significant differences in the expression of a variety of abnormal behaviors across captive social groups of chimpanzees and rhesus macaques. Group membership, independent of relatedness, significantly affected the expression of these behaviors, leading Hook and colleagues to suggest that social learning processes may be involved in the propagation of these behaviors.

The general point from this analysis is that even extensive maternal rearing does not guarantee so-called "normal" development in chimpanzees. Within this one type of social setting there was variation among individuals (some expressed abnormal patterns of behavior and some did not), and even those subjects that lived in the most naturalistic social and physical settings of our study expressed some atypical behavior.

3.2. Response to Novelty

The response of nonhuman primates to novelty is a valuable measure in the study of early rearing and behavioral development. In some situations, behavioral changes associated with differential early rearing conditions may not be detected under routine circumstances, but are manifested during exposure to novel objects or novel settings (Suomi, 1974; Andrews and Rosenblum, 1993). In general, restrictive rearing produces neophobia (Menzel, 1964; Sackett 1972; Timmermans *et al.*, 1994). For example, when compared with mother-reared individuals, chimpanzees and rhesus macaques that were socially isolated early in life showed more stress vocalizations and stereotyped behaviors, and less exploratory behavior when exposed to novel objects (Mason and Green, 1962; Mason, 1963; Menzel *et al.*, 1963; Menzel, 1963, 1964; Berkson and Mason, 1964). However, wild-caught rhesus macaques were less responsive to novel objects and novel situations than were urban (city living) counterparts (Singh, 1966).

To better understand the effects of early rearing environments on response to novelty, we evaluated the behaviors of 15 young chimpanzees (3–6 years old) as they were introduced to a novel enclosure (M.A.

Bloomsmith and colleagues, unpublished observations). Seven of the subjects were reared with their mothers in a social group whereas eight were reared with peers in a nursery environment. The mother-reared subjects lived from birth in large, complex social groups (8–15 group members) in fairly spacious indoor/outdoor housing with a strong environmental enrichment program. The nursery setting included peers, a moderate amount of caregiver attention, a dog surrogate that lived with the chimpanzees, enclosures designed to elicit locomotion and play, exposure to a variety of objects, access to the outdoors, and visual access to adult chimpanzees. In addition to their rearing differences during infancy, the subjects were exposed to different levels of social complexity as juveniles. At approximately 2 years of age, both the mother-reared and nursery-reared subjects were placed in indoor/outdoor run-type housing and introduced into small groups with one to three other young chimpanzees from the same rearing background. Nine of the 15 subjects lived in groups with only these like-aged individuals. Six subjects lived with peers and were introduced to an adolescent female chimpanzee as an additional social partner. These adolescent females had been mother reared for most of their lives and had experience with infants in their natal groups.

For the challenge test, subjects were exposed, in pairs, to a new outdoor enclosure that differed in construction from that of any enclosure they had experienced previously. The adolescent social partners were not introduced to the novel enclosure. On the basis of published literature, we expected that mother-reared chimpanzees, and those housed in the more complex social setting (with peers and an adolescent), would show less distress and would be quicker to explore the novel environment than would nursery-reared chimpanzees or those housed in less complex social settings. However, our expectations were incorrect. While the experience did seem to be a mildly stressful challenge (within the first hour 12 of the 15 subjects displayed some form of distress-related behaviors, and 14 of the 15 sought physical contact with their partners), early nursery rearing did not lead to any evidence of neophobia. Latency to enter the novel enclosure was uninfluenced by rearing condition. Nursery-reared subjects showed more exploratory behavior, and spent less time within 1 m of their partners than did the mother-reared subjects. Rates of distress-related behaviors over the entire test period were similar in the two rearing groups. The chimpanzees that had been housed in the more complex social setting (with peers and an adolescent) actually took longer to enter the novel environment than those housed only with peers, and spent less time in the novel enclosure during the testing period. Thus both of our hypotheses were disproved. The consequences of deprived rearing environments documented by others (e.g., Davenport and Menzel, 1963; Berkson and Mason, 1964; Turner *et al.*, 1969; Davenport and Rogers, 1970) were not reflected in the behavior of these young chimpanzees. This finding may be due to the more socially and physically enriched nursery environment in which these youngsters were raised compared with the rearing conditions experienced by the chimpanzee subjects in the 1960s. Clearly, current nursery practices seem to prepare chimpanzees to cope with at least some of the changes they are likely to encounter.

3.3. Sexual Competence

Another common finding is that chimpanzees reared in social settings other than with their mothers may not develop normal sexual competence. Severe sensory and social isolation during early development leads to disturbed sexual behavior (Turner *et al.*, 1969; Rogers and Davenport, 1970; Reisen, 1971). Even within the zoo environment, one of the major behavioral deficits associated with nursery-reared chimpanzees has been inadequate copulation skill when they grow to adult age (King and Mellen, 1994). These animals, which are sometimes referred to as "behavioral nonbreeders," typically show no interest in copulating or, if they do, cannot achieve the correct position.

In the study described in Section 3.1, we also examined the expression of sexual behavior in young subjects raised in the social contexts described above. Not surprisingly, the nursery-reared chimpanzees engaged in sexual behav-ior less often than subjects in the other two groups (Baker *et al.*, 2000; Bloomsmith *et al.*, 2002; Ross, 2003). Mother-reared subjects solicited sex, copulated, and explored others' genitals more often than did nursery-reared subjects. However, the extent of this contrast diminished as subjects matured, particularly for the subjects that lived with their mothers for 2 years before living in peer groups. At 2–4 years of age, this group resembled nursery-reared chimpanzees in their levels of sexual behavior, whereas at 4–6 years of age they behaved more like subjects that lived continually with their mothers.

While we did find differences in the rates of expression of sexual behavior, the more surprising finding is that *all* of the subjects engaged in copulation during the period of study. Each of the 12 nursery-reared subjects displayed appropriate copulatory behavior, as did each of the seven subjects that had 2 years of maternal rearing followed by other social housing. These chimpanzees are now in their juvenile, adolescent, and early adult years. All of them live socially and continue to display competent sexual behavior.

Obviously, at least minimally competent sexual behavior is crucial to maintaining self-sustaining populations of captive chimpanzees in zoological parks and laboratory colonies. It is important to demonstrate that this behavior can be promoted even among chimpanzees that must be reared by humans. This is not to say that every nursery-reared chimpanzee will become an adequate breeder. The numbers of study subjects here are low and the exact features of nursery rearing that facilitated the development of adequate sexual behavior have not been identified. Nevertheless, in future population planning efforts the breeding potential of nursery-reared chimpanzees should not be discounted.

3.4. Maternal Competence

Another constellation of behaviors often altered by early rearing conditions is later maternal behavior and competence. Most of the published studies in this area have involved macaques, which show severe deficits in mothering behavior if they have been reared in social isolation. As many as 75% of female rhesus macaques that are reared without mothers and peers are entirely incompetent with their first infants (Ruppenthal *et al.*, 1976; Suomi, 1978; Suomi and Ripp, 1983). In contrast, peer rearing results in competent maternal performance for most female macaques (Ruppenthal *et al.*, 1976; Suomi, 1978; Suomi and Ripp, 1983; Timmermans and Vossen, 1996). Even visual access to normal mother–infant behavior improves the ability to mother (Dienske *et al.*,

1980), as does parity and duration of exposure to previous infants before removal (Ruppenthal *et al.*, 1976).

Early studies of chimpanzees showed that rearing without a mother produced maternally incompetent adults whose infants had to be hand reared in a nursery to ensure survival (Yerkes and Tomilin, 1935; Rogers and Davenport, 1970). Three recent evaluations of the effects of early rearing on maternal competency in mother- and nursery-reared chimpanzees have yielded conflicting results. One survey found no difference in maternal performance between wild-born females and those that had been nursery reared or reared with their mothers in captivity (Toback *et al.*, 1992). Correspondingly, Bard (1995) reported that mothers' rearing histories were not related to maternal competence based on an analysis of archival data. However, in a similar analysis from a different facility, Brent *et al.* (1996) found that early rearing history and subsequent maternal performance were related. All of these analyses included both primiparous and multiparous chimpanzees as study subjects. Since parity influences maternal performance in chimpanzees (Rogers and Davenport, 1970), the lack of statistical control for parity may have contributed to conflicting findings in these studies.

To address this issue, we ran an archival analysis of 99 chimpanzees, focusing only on primiparous females (Bloomsmith *et al.*, 2003a). For each of these females we examined the effects of duration of time raised by their own mothers, and the opportunity for interacting with infants during their developing years, on the likelihood that they would care for their own first infants. The data were drawn from three biomedical institutions and 15 zoos over a period of nearly four decades (1954–1989), and therefore represent a variety of nursery experiences.

Neither origin (captive- versus wild-born), type of facility (biomedical institution versus zoo), nor year of birth contributed to the prediction of an animal's maternal competence. However, we did find that older first-time mothers were less likely to be competent. Statistically controlling for the mother's age at first parturition, we found that the odds of maternal competence were 3.9 times greater when mothers had more than 1 year of mother rearing than when they had less. Given the background literature described above, this compromised maternal ability of nursery-reared chimpanzee mothers is not surprising. What did surprise us was that a significant portion of the nursery-reared chimpanzees did care for their first offspring. Of the mothers that experienced less than 1 month of mother rearing themselves, 35% competently cared for their first infants. This is a substantial proportion of competence among this group, not simply an anomaly of an occasional nursery-reared mother that was competent, as we might have predicted. Another surprise was that 25% of the subjects with more than 2 years of maternal rearing were incompetent with their first infants. Even some primiparous females that were still housed with their mothers when they gave birth did not care for their first infants. It will be important to continue this work if we wish to understand the mechanisms underlying this variation.

In the same study we examined the effect of a young female's opportunity to interact with infants (<3 years old) for more than 6 months during juvenility and adolescence on her later primiparous maternal behavior. When we controlled for the females' ages when they first gave birth, the odds of maternal competency were 1.66 times greater for animals that had had an opportunity for allomaternal experience. The demonstrated effect of allomaternal exposure lends support to the hypothesis that the phenomenon of allomothering functions to improve later maternal performance in chimpanzees (Nishida, 1983; Nicolson, 1991). This idea has been termed the "learning-to-mother" hypothesis, suggesting that observational learning and interactions with infants are critical in the acquisition of maternal behavior (Spencer-Booth, 1968; Lancaster, 1971; Hinde 1974; Hrdy, 1976).

This study helps delineate the conditions for rearing maternally competent chimpanzees, and for maintaining future maternal competence. Nursery rearing did not reduce maternal competence as dramatically as we might have predicted. In future population planning efforts we should not discount the mothering potential of nursery-reared chimpanzees; instead, we can predict that about one-third of the females might care for their first infants. These data included chimpanzee mothers that were raised in a nursery as early as 1954. It is interesting that we did not detect any effect of the mothers' year of birth on their maternal performance. We had expected that with advancements in nursery-rearing techniques across the decades of this study, we might have measured some improvement in competence related to the year of the subjects' birth. That we found no such effect runs counter to the reduced degree of deficits in other aspects of development as nursery practices improved, and may call into question some of our other statements about changes in nursery rearing perhaps being responsible for improvements in behavior seen over the years. On the other hand, it is not surprising that some domains of behavioral competence would be differentially sensitive to varying degrees of perturbation from naturalistic mother rearing. For example, the social and environmental complexities associated with recent nursery practices may moderate the cognitive deficits, neophobia, and qualitatively abnormal behaviors shown by chimpanzees reared in the deprived nursery settings of the past. However, these improvements may provide relatively little in terms of compensation for the absence of the mother.

3.5. Maternal Response to Separation from Offspring

For a variety of management and/or research reasons, chimpanzee mothers and their offspring are sometimes separated from one another at an earlier age than they would be in the wild. When this occurs, the potential for distress in both the youngster and the mother is of great concern. The distress of chimpanzee infants under these conditions is addressed by Howell *et al.* (chapter 16) in this volume. To complement this information, we conducted a study to assess the behavioral evidence of distress exhibited by chimpanzee mothers when their offspring were removed (Bloomsmith *et al.*, 2003b). The few studies that have examined nonhuman primate mothers' responses to such separations suggest that they are somewhat different from those of infants, with mothers displaying signs of protest, particularly vocalizing, but relatively few or no signs of despair (Jensen, 1968; Kaplan, 1970; Hinde and Davies, 1972; Vogt and Levine, 1980; Champoux and Suomi, 1994).

We documented the behavior of 12 mothers during 15 instances of mother/offspring separation (three of the 12 mothers were separated from two different offspring over the course of study). Three male and 12 female offspring were 1.8–5.4 years old when they were permanently separated from their mothers. Most of the mothers were sedated for the removal of their infants, and we studied their behavior once they had recovered and returned to their social groups later in the day. Comparison of their behaviors in the 3 weeks before and after separation revealed no statistically significant changes in behavior after separation. Eight of these mothers were studied more intensively on the first day of separation and their data showed a significant increase in inactivity on the day of separation. No other behavioral changes were evident, and the inactivity measure returned to baseline levels after the first day of separation. Even if the temporary increase in inactivity was a symptom of despair, as it is in infants (Bowlby, 1960), it was a minor symptom because the difference was not numerically large (64.8% during preseparation to 77.3% during postseparation). These 15 mothers gave no evidence of protest or agitation in response to the removal of their offspring. Maternal age, infant age, presence of other offspring, and past experience with mother/offspring separation had no effect on response to separation.

These results contradict those of mother/infant separation in monkeys, and show that most behavioral indicators of well-being are stable in chimpanzee mothers that remain in their familiar environment and social groups after their offspring are removed. These results were corroborated by Tarou and colleagues (2000), who found that an orangutan mother showed no evidence of protest, and only a slight indication of despair, following separation from her offspring. The findings suggest that mother/offspring separation in apes may be less stressful for mothers than might be anticipated. However, it is important to note that the reaction might be different if younger offspring are removed, particularly if the infants still spend a good deal of time clinging to and being carried by their mothers. Monitoring for individual mothers that may be inactive or show another undesirable response is appropriate in the postseparation period, but since the 15 mothers in this study generally showed minimal responses, the majority of personnel effort probably should be devoted to managing the welfare of offspring when a mother/offspring separation is planned.

4. CONCLUSIONS

It is indisputable that great strides have been made in the management of captive chimpanzee infants during development, whether the infants are reared by their mothers or in a nursery. Taken as a whole, findings from these recent studies of chimpanzee behavioral development indicate that we may need to reexamine ideas about the long-term behavioral consequences of early rearing conditions. These findings can be summarized as follows: (1) some atypical behavior was present even in the rearing condition that most closely modeled the wild situation; (2) nursery-reared subjects showed no evidence of neophobia when they were introduced to a novel environment; (3) nursery-reared chimpanzees displayed normal copulation skills; (4) about one-third of nursery-reared mothers adequately cared for their first infants; and (5) mothers showed very little response to the permanent removal of their offspring. Each of these findings challenges conventional wisdom concerning the behavior of captive chimpanzees.

Some of the unexpected findings may be due to improvements in nursery-rearing practices, emphasizing socialization, sensory stimulation, and physical activity, over the past 15–20 years. While we propose that modern advances in nursery practices may account for the apparent changes in behavior characteristic of nursery-reared chimpanzees, one piece of our own evidence contradicts that premise. In our study of the effects of early rearing on primiparous maternal competence, we found no effect of changes in such practices over the four decades of data we analyzed. It is possible that maternal behavior of primipara does not show the same kind of sensitivity to improvements in nursery-rearing practices as other behaviors studied. Since the latest date of nursery rearing of the primiparous mothers studied was 1989, it is also possible that we do not have a sufficient sample of females reared under more modern nursery practices to measure any influence.

There are a number of unanswered questions and difficulties in studying the influences of early rearing on chimpanzees. The length of time required to generate data into adolescence and early adulthood is a major constraint on our ability to determine the developmental outcomes of different aspects of behavioral management. Answering some questions about effects of particular early rearing variables would run counter to today's ethical mandate to provide the most appropriate nursery-rearing environment possible to the few young chimpanzees that are being raised in nurseries. Also problematic is the transmission of at least some behavioral problems across generations. This phenomenon hampers our ability to differentiate between the effects of current conditions on immature chimpanzees versus the effect of the behavioral profile of other group members (which, in turn, could have been fostered by past conditions). Clearly, however, the context in which even currently maturing chimpanzees began their lives is not sufficient for a consistent and thoroughly desirable behavioral outcome in all respects.

These findings should not be misinterpreted to indicate that we advocate nursery rearing in situations where mother rearing is possible. We do not. Simply put, maternal rearing of young chimpanzees leads to the most species-appropriate behavioral development (Bloomsmith *et al.*, 1994). Our main point in this chapter is, instead, that nursery rearing may not be as behaviorally devastating as was once thought, or as it once was. This is important for colony managers, veterinarians, researchers, and population managers to recognize.

Even though a number of our findings illustrate less severe behavioral effects than we had expected, there is reason to be cautious. Many aspects of behavioral development have not yet been examined experimentally to determine if nursery rearing is adequate to promote their expression. As juveniles, nursery-reared chimpanzees show less dominance behavior and display less coalitionary support than do their mother-reared peers (Spijkerman *et al.*, 1997; Bloomsmith *et al.*, 1991). Since social play functions in establishing and maintaining dominance rank in adolescent chimpanzees, nursery-reared adolescents may be hampered in the process of learning to control aggression and negotiate rank relationships because they tend to play less and show species-inappropriate sex differences in play and teasing (Paquette, 1994; Spijkerman, 1995, 1996; Spijkerman *et al.*, 1997). Delays in the development of sex differences in the initiation of conflict also have been observed among nursery-reared adolescents (Spijkerman, 1995; Spijkerman *et al.*, 1997). Adult nursery-reared females experience more wounding in complex social settings, but not in small social groups, suggesting that some of these effects of rearing on juveniles and adolescent chimpanzees persist, thereby impacting the frequency or consequences of conflict (Baker *et al.*, 2000). Such social deficits, and others yet unidentified, may bear on how well nursery-reared chimpanzees can be housed in large naturalistic and socially enriched groups. Also, beyond two studies of tool use (see Brent *et al.*, 1995; Menzel *et al.*, 1970), no studies have assessed the cognitive capacities of nursery-reared chimpanzees, or whether specific deficits may present in nursery-reared subjects.

Beyond deepening our understanding of behavioral development in chimpanzees, the value of the research described in this chapter lies in improving our care and management of captive chimpanzee populations. This research helps to identify the preferred conditions for raising young chimpanzees, and to predict behavioral problems that may surface under particular rearing conditions. Being able to anticipate behavioral problems improves our ability to prevent or ameliorate them. For example, since we have found that coprophagy is common among mother-reared chimpanzees, we can implement environmental enrichment techniques shown to reduce the incidence of this behavior. Applied behavioral research also can be of use when particular roles are planned for individual animals. For example, a female that is selected early in life for future breeding could be reared by her mother for at least a year and exposed to infants during her juvenile and adolescent years to maximize the likelihood that she would rear her first infant successfully.

Information gained from behavioral research also can be applied to the long-term management of captive chimpanzee populations in laboratory or zoo settings. Projections for the desired demographics of a population should include consideration of animals' social histories to improve the accuracy of calculations. For example, colony managers may safely expect that one-third of nursery-reared chimpanzee females will care for their first infants, rather than assuming that none will do so. In addition, most young chimpanzees raised with the more enlightened nursery techniques described in this chapter can be expected to be adequate breeders. This type of informed analysis is especially important now, since little reproduction has been allowed in recent years among either the laboratory population of chimpanzees (see report from the Committee on Long-Term Care of Chimpanzees, 1997) or the zoological population. Ramifications of this change in housing conditions (such as few infants in groups with developing females) can be ameliorated by the findings of applied behavioral research.

ACKNOWLEDGMENTS

We thank Dr. Patricia Alford, Dr. Michale Keeling, Kimberly Neu, and Caroline Griffis for their contributions to this work. The project was supported by the National Institutes of Health/National Center for Research Resources grants RR03578, RR03591, RR00165, and U42-RR03589. Animals are maintained in facilities approved by the American Association for Accreditation of Laboratory Animal Care and in accordance with current regulations and standards of the United States Department of Agriculture, Department of Health and Human Services, and National Institutes of Health.

REFERENCES

Anderson, J.R., and Chamove, A.S. (1986). Infant stumptailed macaques reared with mirrors or peers: Social responsiveness, attachment, and adjustment. *Primates* 27:63–82.

Andrews, M.W., and Rosenblum, L.A. (1993). Assessment of attachment in differentially reared infant monkeys (*Macaca radiata*): Response to separation and a novel environment. *J. Comp. Psychol.* 107:84–90.

Arling, G.L., and Harlow, H.F. (1967). Effects of social deprivation on the maternal behavior of rhesus monkeys. *J. Comp. Physiol. Psychol.* 64:371–377.

Baker, K.C., Seres, M., Aureli, F., and de Waal, F.B. M. (2000). Evaluating social enrichment of chimpanzees: Injury risks under three housing conditions. *Am. J. Primatol.* 51:161–175.

Bard, K. (1995). Parenting in primates. In: Bornstein, M.H. (ed.), *Parenting in Primates. Handbook of Parenting: Biology and Ecology of Parenting*, Vol. 2. Lawrence Erlbaum, Mahwah, NJ, pp. 27–58.

Bard, K.A., and Gardner, K.H. (1996). Influences on development in infant chimpanzees: Enculturation, temperament, and cognition. In: Russon, A., Bard, K.A., and Parker, S.T., (eds.), *Reaching into Thought: The Minds of the Great Apes.* Cambridge University Press, Cambridge, pp. 235–256.

Berkson, G., and Mason, W.A. (1964). Stereotyped behaviors of chimpanzees: Relation to general arousal and alternative activities. *Percept. Motor Skills* 19:635–652.

Bloomsmith, M.A., and Baker, K.C. (2001). Social management of captive chimpanzees. In: Brent, L. (ed.), *The Care and Management of Captive Chim-*

panzees. American Society of Primatologists: Special Topics in Primatology Series (J. Wallis, series ed). Wiley-Liss, New York, pp. 204–241.

Bloomsmith, M.A., and Haberstroh, M.D. (1995). Effect of early social experience on the expression of abnormal behavior among juvenile chimpanzees. *Am. J. Primatol.* 36:110.

Bloomsmith, M.A., Alford, P.L., and Pazol, K.A. (1991). Juvenile chimpanzee behavioral development in four social settings. *Am. J. Primatol.* 24:90.

Bloomsmith, M.A., Pazol, K.A., and Alford, P.L. (1994). Juvenile and adolescent chimpanzee behavioral development in complex groups. *Appl. Anim. Behav. Sci.* 39:73–87.

Bloomsmith, M., Baker, K.C., Ross, S.R., and Pazol, K.A. (2002). The behavioral effects of early rearing experiences on captive chimpanzee behavioral development:The juvenile years. *Am. J. Primatol.* 57(Suppl. 1):54–55.

Bloomsmith, M.A., Baker, K.C., Ross, S.R., Lambeth, S.P., Brent, L., and Toback, E. (2003a). Social rearing conditions and later maternal performance of primiparous chimpanzees. *Am. J. Primatol.* 60(Suppl. 1):38–39.

Bloomsmith, M.A., Lambeth, S.P., Tarou, L.R., and Haberstroh, M.D. (2003b). Maternal response to mother—offspring separation in the chimpanzee (*Pan troglodytes*). *Anim. Welfare* 12:359–368.

Bowlby, J. (1960). Grief and mourning in infancy and early childhood. *Psychoanal. Study Child* 15:9–52.

Brent, L., Bloomsmith, M.A., and Fisher, S.D. (1995) Factors determining tool-using ability in two captive chimpanzee (*Pan troglodytes*) populations. *Primates* 36(2):265–274.

Brent, L., Williams-Blangero, S., and Stone, A.M. (1996). Evaluation of the chimpanzee breeding program at the Southwest Foundation for Biomedical Research. *Lab. Anim. Sci.* 46:405–409.

Capitanio, J.P. (1984). Early experience and social processes in rhesus macaques (*Macaca mulatta*): I. Dyadic social interactions. *J. Comp. Psychol.* 98:35–44.

Capitanio, J.P. (1985). Early experience and social processes in rhesus macaques (*Macaca mulatta*): II. Complex social interactions. *J. Comp. Psychol.* 99:133–144.

Capitanio, J.P. (1986). Behavioral pathology. In: Mitchell, G., and Erwin, J. (eds.), *Comparative Primate Biology*, Vol. 2, Part A: *Behavior, Conservation, and Ecology*. Alan R. Liss, New York, pp. 411–454.

Chamove, A.S., Rosenblum, L.A., and Harlow, H.F. (1973). Monkeys (*Macaca mulatta*) raised only with peers: A pilot study. *Anim. Behav.* 21:316–324.

Champoux, M., and Suomi, S.J. (1994). Behavioral and adrenocortical responses of rhesus macaque mothers to infant separation in an unfamiliar environment. *Primates* 35:191–202.

Champoux, M., Byrne, E., DeLizio, R., and Suomi, S.J. (1992). Motherless mothers revisited:Rhesus maternal behavior and rearing history. *Primates* 33:251–255.

Committee on Long-Term Care of Chimpanzees. (1997). *Chimpanzees in Research: Strategies for Their Ethical Care, Management and Use.* National Academy Press, Washington, D.C.

Davenport, R.K., Jr., and Menzel, F.W., Jr. (1963). Stereotyped behavior of the infant chimpanzee. *Arch. Gen. Psychol.* 8:99–104.

Davenport R.K., and Rogers, C.M. (1968). Intellectual performance of differentially reared chimpanzees. I. Delayed response. *Am. J. Ment. Defic.* 72:674–680.

Davenport, R.K., and Rogers, C.M. (1970). Differential rearing of the chimpanzee: A project survey. In: Bourne, G.H. (ed.), *The Chimpanzee*: Vol. 3. *Immunology, Infections, Hormones, Anatomy, and Behavior of Chimpanzees.* University Park Press, Baltimore, pp. 337–360.

Davenport, R.K., Rogers, C.M., and Menzel, E.W. (1969). Intellectual performance of differentially reared chimpanzees. II. Discrimination-learning set. *Am. J. Ment. Defic.* 73:963–969.

Davenport, R.K., Rogers, C.M., and Rumbaugh, D.M. (1973). Long-term cognitive deficits in chimpanzees associated with early impoverished rearing. *Dev. Psychol.* 9:343–347.

Dienske, H., van Vreeswijk, W., and Konig, H. (1980). Adequate mothering by partially isolated rhesus monkeys after observation of maternal care. *J. Abnorm. Psychol.* 89:489–492.

Fulk, R., and Garland, C. (1992). *The Care and Management of Chimpanzees* (Pan troglodytes) *in Captive Environments.* North Carolina Zoological Park, North Carolina Zoological Society.

Goin, D.L., and Gust, D.A. (1998). Peer-rearing influences subsequent maternal behavior and infant survival in a newly formed herpes B-virus negative rhesus macaque group. *Primates* 39:539–543.

Goldfoot, D. (1977). Rearing conditions which support or inhibit later sexual potential of laboratory-born rhesus monkes: Hypotheses and diagnostic behaviors. *Lab. Anim. Sci.* 27:548–556.

Harlow, H.F., and Harlow, M.K. (1969). Effects of various mother-infant relationships on rhesus monkey behaviors. In: Foss, B.M. (ed.), *Determinants of Infant Behaviour*, Vol. IV. Methuen, London, pp. 15–36.

Hinde, R.A. (1974). Mother/infant relations in rhesus monkeys. In: White, N.F. (ed.), *Ethology and Psychiatry*. University of Toronto Press, Toronto, pp. 29–46.

Hinde, R.A., and Davies, L. (1972). Removing infant rhesus from mother for 13 days compared with removing mother from infant. *J. Child Psychol. Psychiatry* 13:227–237.

Hook, M.A., Lambeth, S.P., Perlman, J.E., Stavisky, R., Bloomsmith, M.A., and Schapiro, S.J. (2002). Inter-group variation in abnormal behavior in chimpanzees (*Pan troglodytes*) and rhesus macaques (*Macaca mulatta*). *Appl. Anim. Behav. Sci.* 76:165–176.

Hrdy, S.B. (1976). Care and exploitation of nonhuman primate infants by conspecifics other than the mother. *Adv. Study Behav.* 6:101–158.

Jensen, G.D. (1968). Reaction of monkey mothers to long-term separation from their infants. *Psychonom. Sci.* 11:171–172.

Kaplan, J. (1970). The effects of separation and reunion on the behavior of mother and infant squirrel monkeys. *Dev. Psychobiol.* 3:43–52.

King, N.E., and Mellen, J.D. (1994). The effects of early experience on adult copulatory behavior in zoo-born chimpanzees (*Pan troglodytes*). *Zoo Biol.* 13:51–59.

Lancaster, J. (1971). Play-mothering: The relations between juvenile females and young infants among free-ranging vervet monkeys (*Cercopithecus aethiops sabaeus*). *Folia Primatol.* 15:161–182.

Lindell, S., Dodson, A., and Champoux, M. (1995). Early rearing experiences of adult females affect their mothering behaviors and their offspring's behaviors. *Am. J. Primatol.* 36:139.

Ljungberg, T., and Westlund, K. (2000). Impaired reconciliation in rhesus macaques with a history of early weaning and disturbed socialization. *Primates* 41:79–88.

Maki, S., Fritz, J., and England, N. (1993). An assessment of early differential rearing conditions on later behavioral development in captive chimpanzees. *Infant Behav. Dev.* 16:373–381.

Maple, T.L. (1980). Chimpanzee reproduction, rearing and rehabilitation in captivity. Technical Report Presented to the Ad Hoc Task Force National Chimpanzee Breeding Program, Tanglewood, North Carolina.

Mason, W.A. (1960). The effects of social restriction on the behavior of rhesus monkeys: I. Free social behavior. *J. Comp. Physiol. Psychol.* 53:582–589.

Mason, W.A. (1961a). The effects of social restriction on the behavior of rhesus monkeys: II. Tests of gregariousness. *J. Comp. Physiol. Psychol.* 54:287–290.

Mason, W.A. (1961b). The effects of social restriction on the behavior of rhesus monkeys: III. Dominance behavior. *J. Comp. Physiol. Psychol.* 54:694–699.

Mason, W.A. (1963). Social development of rhesus monkeys with restricted social experience. *Percept. Motor Skills* 16:263–270.

Mason, W.A., and Capitanio, J.P. (1988). Formation and expression of filial attachment in rhesus monkeys raised with living and inanimate mother substitutes. *Dev. Psychobiol.* 21:401–430.

Mason, W.A., and Green, P.C. (1962). The effects of social restriction on the behavior of rhesus monkeys. IV. Responses to a novel environment and to an alien species. *J. Comp. Physiol. Psychol.* 55:127–132.

Menzel, E.W. (1963). The effects of cumulative experience on responses to novel objects in young isolation-reared chimpanzees. *Behaviour* 21:147–160.

Menzel, E.W. (1964). Responsiveness to object-movement in young chimpanzees. *Behaviour* 24:147–160.

Menzel, E.W., Jr., Davenport, R.K., and Rogers, C.M. (1963). The effects of environmental restriction upon the chimpanzee's responsiveness to objects. *J. Comp. Physiol. Psychol.* 56:78–85.

Menzel, E.W., Jr., Davenport, R.K., and Rogers, C.M. (1970). The development of tool using in wild-born and restriction-reared chimpanzees. *Folia Primatol.* 12:273–283.

Mitchell, G.D., Raymond, E.J., Ruppenthal, G.C., and Harlow, H.F. (1966). Long-term effects of total social isolation upon behavior of rhesus monkeys. *Psychol. Rel.* 18:567–580.

Nicolson, N.A. (1991). In: Loy, J.D., and Peters, C.B. (eds.), *Understanding Behavior: What Primate Studies Tell Us About Human Behavior.* Oxford University Press, New York, pp. 17–50.

Nishida, T. (1983). Alloparental behavior in wild chimpanzees of the Mahale Mountains, Tanzania. *Folia Primatol.* 41(1–2):1–33.

Paquette, D. (1994). Fighting and playfighting in captive adolescent chimpanzees. *Aggress. Behav.* 20:49–65.

Pazol, K.A., and Bloomsmith, M.A. (1993). The development of stereotyped body rocking in chimpanzees (*Pan troglodytes*) reared in a variety of nursery setting. *Anim. Welfare* 2:113–129.

Pazol, K.A., and Bloomsmith, M.A. (1997). The effect of a short-term biomedical research protocol on stereotyped rocking in juvenile chimpanzees. *Anim. Welfare* 6:133–138.

Reisen, A.H. (1971). Nissen's observations on the development of sexual behavior in captive-born, nursery-reared chimpanzees. In: Bourne, G. (ed.), *The Chimpanzee*, Vol. 4. S. Karger, Basel, pp. 1–18.

Rogers, C.M., and Davenport, R.K. (1970). Chimpanzee maternal behavior. In:Bourne, G. (ed.), *The Chimpanzee*, Vol. 3: *Immunology, Infections, Hormones, Anatomy, and Behavior of Chimpanzees.* S. Karger, Basel, pp. 361–368.

Rogers, C.M., and Davenport, R.K. (1971). Intellectual performance of differentially reared chimpanzees. 3. Oddity. *Am. J. Ment. Defic.* 75:526–530.

Ross, S.R. (2003). The Effects of Early Rearing Condition on the Social Behavioral Development of Captive Chimpanzees (*Pan troglodytes*). Master's Thesis, Department of Social Sciences, University of Chicago.

Ruppenthal, G.C., Arling, G.L., Harlow, H.F., Sackett, G.P., and Suomi, S.J. (1976). A ten year perspective of motherless mother monkey behavior. *J. Abnorm. Psychol.* 85:341–349.

Ruppenthal, G.C., Walker, C.G., and Sackett, G.P. (1991). Rearing infant monkeys (*Macaca nemestrina*) in pairs produces deficient social development compared with rearing in single cages. *Am. J. Primatol.* 25:103–113.

Sackett, G.P. (1972). Exploratory behavior of rhesus monkeys as a function of rearing experience and sex. *Dev. Psychol.* 6:260–270.

Sackett, G.P., Ruppenthal, G.C., and Davis, A.E. (2002). Survival, growth, health, and reproduction following nursery rearing with mother rearing in pigtail monkeys (*Macaca nemestrina*). *Am. J. Primatol.* 56:165–183.

Singh, S.D. (1966). The effects of human environment upon the reactions to novel situations in the rhesus. *Behaviour* 26:243–250.

Spencer-Booth, Y. (1968). The behaviour of group companions towards rhesus monkey infants. *Anim. Behav.* 16:541–557.

Spijkerman, R. (1995). Differences in play development of young chimpanzees reared in family groups and in peer groups. *Neth. J. Zool.* 45:402–421.

Spijkerman, R. (1996). Behavioural development of infant chimpanzees with the mother and in peer groups. *Annual Report of the Primate Center TNO*, 1987, 309–312.

Spijkerman, R.P., Van Hooff, J.A.R.A.M., Dienske, H., and Jens, W. (1997). Differences in subadult behaviours of chimpanzees living in peer groups and in family groups. *Int. J. Primatol.* 18:439–454.

Suomi, S.J. (1974). Social interactions of monkeys reared in a nuclear family environment vs. monkeys reared with mothers and peers. *Primates* 15:311–320.

Suomi, S.J. (1978). Maternal behavior by socially incompetent monkeys:Neglect and abuse of offspring. *J. Pediatr. Psychol.* 3:28–34.

Suomi, S.J., and Ripp, C. (1983). A history of motherless mother monkey mothering at the University of Wisconsin Primate Laboratory. In: Reite, M., and Caine, N. (eds.), *Child Abuse: The Nonhuman Primate Data*. Alan R. Liss, New York, pp. 49–78.

Tarou, L.R., Bashaw, M.J., and Maple, T.L. (2000). Empty nest: A case study of maternal response to infant separation in a captive Sumatran orangutan (*Pongo pygmaeus abelii*). *J. Appl. Anim. Welfare Sci.* 3:203–213.

Thompson, M.A., Bloomsmith, M.A., and Taylor, L.L. (1991). A canine companion for a nursery-reared infant chimpanzee. *Lab. Primate Newsl.* 30(2): 1–4.

Timmermans, P.J.A., and Vossen, J.M.H. (1996). The influence of rearing conditions on maternal behavior in cynomolgus macaques (*Macaca fascicularis*). *Int. J. Primatol.* 17:259–276.

Timmermans, P.J.A., Vochteloo, J.D., Vossen, J.M.H., Röder, E.L., and Duijghuisen, J.A.H. (1994). Persistent neophobic behavior in monkeys: A habit or a trait? *Behav. Processes* 31:177–196.

Toback, E., Granholm, C., and McNary, J. (1992). Assessment of maternal behavior in captive chimpanzees. *Am. J. Primatol.* 27:61.

Turner, C.H., Davenport, R.K., Jr., and Rogers, C.M. (1969). The effect of early deprivation on the social behavior of adolescent chimpanzees. *Am. J. Psychiatry* 125:1531–1536.

Vogt, J.L., and Levine, S. (1980). Response of mother and infant squirrel monkeys to separation and disturbance. *Physiol. Behav.* 24:829–832.

Walsh, S., Bramblett, C., and Alford, P. (1982). A vocabulary of abnormal behavior in restrictively reared chimpanzees. *Am. J. Primatol.* 3:313–319.

Yerkes, R.M., and Tomilin, M.I. (1935). Mother-infant relations in chimpanzees. *Comp. Psychol.* 20:321–358.

CHAPTER SIXTEEN

Effects of Early Rearing History on Growth and Behavioral Development in Captive Chimpanzees (*Pan troglodytes*)

Susan M. Howell, Melanie Schwandt, Jo Fritz, Mary W. Marzke, James Murphy, and Dennis Young

1. INTRODUCTION

Chimpanzees have one of the longest developmental periods of any mammal (Pereira and Fairbanks, 1993). Development includes a lengthy period of physical maturation (Riopelle, 1963; Doran, 1997), cognitive development (Plooij, 1987; Tomasello *et al.*, 1993, 1994), and behavioral changes as the animal advances from infant (0–3.9 years) to juvenile (4–6.9 years), adolescent (7–9.9 years), and adult (10+ years). Weaning from their mothers or from their nursery environment is a long, gradual process (Horvat *et al.*, 1980; Horvat and Kraemer, 1981; Fritz *et al.*, 1991; Howell *et al.*, 1993). Juveniles and adolescents develop social skills and relationships with others and gain new behaviors including foraging, grooming, and sexual behavior (Riopelle, 1963; Mason, 1967; Russon, 1990; Tomasello *et al.*, 1994). While the purpose of the long developmental period is debated, it likely functions to promote physical growth, the acquisition of survival skills, and development into a socially and sexually competent adult (Walters, 1987; Pereira and Fairbanks, 1993).

Nonhuman primates that are reared in a deprived environment exhibit long-term negative effects on their physical growth (Frasier and Rallinson, 1972), brain development (Prescott, 1970), and sociosexual behaviors (Goldfoot, 1977a,b). These negative effects are detailed in an extensive literature (e.g., Davenport *et al.*, 1961, 1973; Harlow and Harlow, 1962; Denenberg, 1969; Randolph and Mason, 1969; Goosen, 1986; Dienske, 1988; Kolter, 1995). While relatively few studies have focused on rearing deficits in chimpanzees (Davenport *et al.*, 1961), Jo and Paul Fritz had a first-hand view of the effects of impoverished rearing and housing conditions on the behavior of young chimpanzees. In 1970 the Fritzes incorporated the Primate Foundation of Arizona (PFA) to provide a home for chimpanzees that had become retired performers, "unmanageable" pets, "undesirable" zoo exhibits, or "finished" laboratory subjects. Because there were few long-term housing facilities for these great apes, chimpanzees came from a variety of backgrounds across the nation, most of them lacking the social skills necessary to adapt to life in a social group (Fritz and Fritz, 1979; Fritz, 1986). To accommodate these animals, the Fritzes established resocialization and rehabilitation methods (Fritz and Fritz, 1979; Nash and Fritz, 1982; Fritz, 1986) and designed nursery-rearing methods to ameliorate developmental deficits in their offspring (Fritz and Fritz, 1982, 1985).

Jo Fritz harkened back to earlier work that suggested the importance of sensory stimulation (Davenport *et al.*, 1961; Sackett, 1972), compassionate humans as surrogates for maternal care (Gardner and Gardner, 1970; Hill *et al.*, 1973), and contact with peers (Chamove *et al.*, 1973; Suomi *et al.*, 1973a,b) and other companion animals (Mason and Kenney, 1974). Zoo facilities followed suit (Hannah and Brotman, 1990; Warner *et al.*, 1997) and intensive nursery rearing by compassionate caregivers (Cooper and Markowitz, 1979; Kraemer, 1995) has gradually become the normal rearing situation for many captive chimpanzees in zoos and research facilities. Improved nursery-rearing techniques (Hannah and Brotman, 1990) have included environmental enrichment, round-the-clock care by human caregivers (Warner *et al.*, 1997), and the provision of conspecific peers, alloparents, and companion animals (Thompson *et al.*, 1991; Kraemer, 1995). Whether the intensive, enriched nursery rearing of infant chimpanzees produces adult animals that are similar in all respects to chimpanzees that are reared with their mothers is an important question. There have been a number of studies of growth and development in captive chimpanzees (Grether and Yerkes, 1940; Nissen and Riesen, 1949a,b; Gavan, 1953, 1971; Davenport *et al.*, 1961; Smith *et al.*, 1975; Watts and Gavan, 1982; Lawrence and Gorzitze, 1985; Kimura and Hamada, 1996; Marzke *et al.*, 1996a,b; Winkler, 1996; Hamada *et al.*, 1996, 1998; Vancatova *et al.*, 1999; Bolter and Zihlman, 2002; Buchanan, 2002; Hamada and Udono, 2002), but only a few have directly compared growth for nursery-reared and mother-reared individuals, and these have considered growth only during the first 2 years of life (Davenport *et al.*, 1961; Smith *et al.*, 1975; Marzke *et al.*, 1996a,b).

An important measure of growth and development is positional behavior (Prost, 1965). Comprising locomotion and posture, positional behavior represents a complex set of behaviors that is influenced by a number of factors including, but not limited to, age, sex, body size, musculoskeletal anatomy, and environment. It is well documented that positional behavior goes through stages of development in which the relative frequencies of specific behaviors change as primates grow and mature (Rawlins, 1976; Crompton, 1983; Sugardjito and van Hooff, 1986; Boinski, 1989; Vilensky and Gankiewicz, 1989; Gebo, 1992; Hunt, 1992a; Doran, 1997; Remis, 1995, 1998; Wells and Turnquist, 2001; Schwandt, 2002). However, only a few studies have focused on the effects of the early rearing environment on the development of positional behavior. While two studies document postural development for nursery-reared chimpanzees (Riesen and Kinder, 1952; Brakke and Savage-Rumbaugh, 1991), no comparative data are available for mother-reared captive subjects.

A substantial number of studies have considered the effects of nursery rearing on the behavior of monkeys (Erwin *et al.*, 1973; Sackett *et al.*,

1973; Ruppenthal *et al.*, 1976; Capitanio, 1986; Higley *et al.*, 1990; Mason *et al.*, 1991; Schneider and Suomi, 1992) or compared the behavior of mother-reared versus nursery- or peer-reared monkeys (Boccia *et al.*, 1991; Clarke and Snipes, 1998; Sackett *et al.*, 1999). However, relatively few studies have compared the behaviors of nursery- and mother-reared chimpanzees (Gabriel *et al.*, 1986; Maki *et al.*, 1993) or considered the effect of an enriched nursery and responsive early rearing environment on chimpanzee behavior (Pazol and Bloomsmith, 1993; King and Mellen, 1994).

Besides the effects of nursery rearing versus maternal rearing on the behavior of nonhuman primates, the age at which an animal is separated from its mother can have a significant effect (Bernstein and Dobrofsky, 1981; Fritz *et al.*, 1990). It has been commonplace for captive chimpanzees to be reared by their mothers for the first year of life and then separated into a peer social group. However, over the past 15 years, mother-reared chimpanzees at the PFA increasingly have been allowed to remain in their natal groups to ensure normal sociosexual development. This practice is in response to considerable literature suggesting that primate infants, whether wild (Berman *et al.*, 1994) or captive (Harlow and Suomi, 1974), can become depressed following maternal separation. Their bereavement can negatively impact behavior (Codner and Nadler, 1984; Spijkerman, 1987; Hoff *et al.*, 1994) and physiology (Champoux and Suomi, 1984; Coe *et al.*, 1985; Laudenslager *et al.*, 1993; McKinney, 1985) and result in considerable psychological stress (Suomi *et al.*, 1973a, 1976; McGinnis, 1980; Kraemer *et al.*, 1991; Boccia *et al.*, 1994; Biondi and Picardi, 1996). Although 1 year of maternal rearing is considered by some to be sufficient to overcome severe agitation and depression in captive chimpanzees upon separation from their mothers, there have been few empirical studies to support this as an optimal age for maternal separation (Fritz *et al.*, 1990).

In this chapter we will attempt to fill the gaps in the existing literature by presenting the results of a series of analyses of data collected on socially housed captive chimpanzees (*Pan troglodytes*) at the PFA between 1981 and 2000. To assess the effect of an enriched early rearing environment on growth and on positional and behavioral development, we compared behaviors before and after maternal separation for short-term mother-reared infants (1–1.3 years), long-term mother-reared juveniles (3.3–5 years), and long-term mother-reared juvenile and adolescent chimpanzees separated after their mothers had weaned them from nipple contact nursing (4.8–9 years). We predicted that mother rearing and the management practice of leaving infants to grow up with their mothers to at least 3.3 years of age would eliminate behavioral indications of maternal separation stress. We also predicted that intensive hand rearing in an enriched nursery environment would result in normative growth and behaviors comparable to those displayed by mother-reared chimpanzees.

2. METHODS

2.1. Housing and Husbandry

2.1.1. Nursery Housing

The PFA nursery environment was developed to provide 24-hr care in an enriched environment (Fig. 16–1). The nursery contained 424 square feet of floor space and was furnished with standard infant-care equipment (isolette, crib, refrigerator, microwave oven, storage areas). Floors were covered with washable carpet and walls were coated with epoxy and decorated with items in high-contrast colors. Plush and manipulable toys, climbing structures, fleece blankets, and other bedding materials were available. The nursery was located within the colony environment to allow infants to hear chimpanzee vocalizations. Soft music was played during quiet times for auditory stimulation. A built-in play cage (71.25 square feet) was furnished with benches, ropes, nest-building materials, and manipulable objects to allow infants to acclimate to a colony setting and provide an area to encourage the development of species-typical nesting, foraging, and locomotor activities.

2.1.2. Nursery Husbandry

PFA nursery-rearing methods were responsive and intensive (see Fritz and Fritz, 1982, 1985 for details). Infants were separated from their

Figure 16–1. Playing in the nursery environment.

mothers only due to illness, injury, or severe maternal neglect. The infant, if removed immediately after birth, was placed in an isolette until all vital signs stabilized. Appropriate attentive care was provided while the infant was in the isolette and caregiver body contact and vertical position were maintained in the nursery after the infant was removed from the isolette. The infant was fed in the vertical position to simulate natural chimpanzee nursing. Infants wore cotton shirts or similar clothing for tactile stimulation. Infants were carried constantly except when they slept. Daily, while in bodily contact with their caregivers, infants were carried through the adult housing areas of the colony so that they could become accustomed to the presence and vocalizations of other chimpanzees. Continuous human tactile stimulation was available to infants until they began to sleep through the night. When infants awoke during the night, they were immediately picked up, fed if hungry, and held until they fell asleep again.

At 18–24 months, infants were gradually introduced to the main colony environment. Accompanied by a familiar caregiver, they spent increasing amounts of time in a cage adjacent to other chimpanzees. Stimuli in the cage included manipulable items, forage materials, climbing apparatuses, and toys of all kinds. Behavior was closely monitored and recorded on daily tally sheets for each individual. If an animal began to show abnormal behavior, such as excessive aggression or timidity, caregivers intervened with increased attention.

The feeding regimen and nutritional guidelines are described in detail by Fritz and Fritz (1985). For the first 16 weeks of life, infants received a low-iron human infant formula on demand at 1- to 3-hr intervals. At 17 weeks, the formula was changed to an iron-fortified formula and rice cereal (Pablum®), mixed with formula, was fed with a spoon. At 27 weeks, fresh fruits and vegetables, cut into medium-sized pieces, were offered twice daily. Cereal, fruits, and vegetables were increased according to the demands of the infant. Between 52 and 78 weeks, infants were gradually weaned from formula and cereal, and pellet monkey chow biscuits (Purina Lab Diet #5045) were introduced. By 78 weeks of age, chimpanzees received a varied diet of seasonal fruits and vegetables along with the pellet monkey chow biscuits.

At 3 weeks of age the infants began receiving vitamin C supplements, 125 mg/day in four to six equal doses. The dose was gradually increased to a single daily amount of 1000 mg/day by 78 weeks of age. Between 37 and 52 weeks, infants were conditioned to take a children's chewable multivitamin once a day. Infants continued to be fed individually, and to receive a bottle at least once a day, until they were 3 years old.

2.1.3. Colony Housing

For a detailed description of the PFA colony environment and environmental enrichment program, see Fritz and Howell (1993). Mother-reared chimpanzees were housed continuously in compatible social groups of three to seven animals in indoor cages (Fig. 16–2). The composition of most groups was mixed with regard to age and sex and included at least one peer. Indoor cages included a series of three interconnected cages (~17.5 m^2 of floor space and 2.8 m in height). After 1993, chimpanzees additionally had rotating access to adjoining outdoor cages (67.9–249.7 m^2 of floor space and 4.8–6.3 m in height) either every other week (1993–1999) or every other day (2000 to present) (Fig. 16–3). Cages were constructed of woven wire mesh and furnished with multiple benches and climbing devices (ropes, vertical and horizontal poles, fire hoses, and swinging tires). Destructible and indestructible objects and forage materials were provided to encourage species-typical object manipulation and play behavior.

2.1.4. Colony Husbandry

Chimpanzees at the PFA are fed an omnivorous diet that includes standard and seasonal fruits and vegetables along with commercial monkey chow (Born *et al.*, 1997). Forage (e.g., chicken scratch, sweet feed, air-popped popcorn, sunflower seeds) and browse (leaves, branches) are broadcast into bedding (straw or shredded paper). Females with infants 6 months of age or older are fed more than other chimpanzees to make certain there is ample food for the infant. By 78 weeks of age, the mother-

Figure 16–2. Interconnected indoor colony caging.

Figure 16–3. Outdoor area adjoining the indoor cages.

reared chimpanzees in this survey, like their nursery-reared counterparts, were provided a varied diet of pellet monkey chow biscuits (Purina Lab Diet #5045) along with seasonal fruits and vegetables. When infants consistently broke contact with their mothers and approached human caregivers (between 6 and 9 months), they were conditioned to suck from a soda straw to receive vitamin C supplements. Doses were gradually increased from 500 mg/day to 1000 mg/day in a manner similar to that for their nursery-reared counterparts. In addition, infants were conditioned to take a children's chewable multivitamin each day.

2.2. Data Collection and Analysis

2.2.1. Growth

Body weights were collected on a sample of 72 chimpanzees (37 females, 35 males) across a 15-year period between 1981 and 1996. Of the 37 females, 17 were nursery reared and 20 were mother reared. Of the 35 males, 10 were nursery reared and 25 were mother reared. All data were collected on healthy individuals while they were anesthetized for routine medical examinations every 6 months (Marzke *et al.*, 1996b).

Weight measurements from pregnant females were not included in this analysis.

Using weight measurements collected on subjects between 1 and 150 months of age, growth curves were fit to the data by means of "locally weighted regression scatterplot smoothing" (LOWESS). LOWESS fits a nonlinear regression surface to data via local smoothing (Cleveland, 1979; Cleveland and Devlin, 1988). A "window," or a portion of the surrounding data, was used to construct a line or curve at each data point. LOWESS fits were generated for nursery-reared and mother-reared subjects of each sex with the use of SAS macros as described by Friendly (1991), with modifications. First, a range of "windows" (*f*-values) was tested to identify the optimal *f* used to produce the final fits. The optimal *f*-value of 0.4 was identified using the M plot method of Cleveland and Devlin (1988). The final fits for nursery-reared and mother-reared subjects were then compared with the jackknife method used by Moses *et al.* (1992), which in effect deals with the problem of dependent data points that result from repeated measurements on each animal (see Marzke *et al.*, 1996b, for more details). Because this method requires data for at least two animals in the lowest and highest ranges, the full age range of 1–150 months had to be narrowed. The age range included in the jackknife analysis was 12–147 months for females and 14–134 months for males.

2.2.2. *Positional Behavior*

The subject sample included 31 chimpanzees (13 females, 18 males). Seven of the subjects were observed in 1995 as part of an earlier study concerning the effects of housing types on positional behavior. The complete sample (including those seven observed in 1995) was observed for some or all of the period between July 1996 and February 2000. Of the 31 subjects, 8 were nursery reared and 23 were maternally reared. The sample included individuals in three age categories: juveniles (4–6.9 years of age), adolescents (7–9 years), and young adults (10–19.9 years). Infants were not included because, although some infants were observed as part of a larger study on the ontogeny of positional behavior (Schwandt, 2002), there were not enough nursery-reared individuals for this comparison of nursery- and mother-reared subjects. As the majority

Table 16–1. Subject Sample for Observations on Positional Behavior[a]

Age category	Age range (years)	Rearing	Number of subjects	Number of observations
Juvenile	4.0–6.99	Nursery	4	684
		Mother	16	5577
Adolescent	7.0–9.99	Nursery	3	959
		Mother	12	3873
Young adult	10.0–19.99	Nursery	3	1381
		Mother	8	1805

[a] Age categories are based on the PFA age scale, which in turn is based on the International Species Information System age scale for chimpanzees (Fritz *et al.*, 1999).

of data were collected over a 3.5-year period, some subjects were observed in more than one age category. Table 16–1 includes information on the number of subjects and the number of observations for each age category.

All nursery-reared subjects were observed only after they had been introduced into the colony and placed within social groups. Consequently, all subjects were observed in groups of two to seven individuals. Several of the mother-reared subjects were separated from their maternal social group during the course of the study. However, a comparison of positional behavior pre- and postseparation revealed no significant differences for these subjects (Schwandt, 2002).

Subjects were observed using a group-scan sampling method in 1995 and a focal-animal sampling method from 1996 to 2000 (Altmann, 1974). The resulting data from these two sampling methods were analogous because group-scan sampling, assuming that all individuals are scored within a short period of time, approaches focal animal sampling of all animals in a group (Fragaszy *et al.*, 1992). For both sampling methods, instantaneous recording (Altmann, 1974; Martin and Bateson, 1993) with an interval of 2 min was used to record behavioral states, i.e., behaviors of relatively long duration. These included quadrupedal locomotion, bipedal locomotion, climbing, suspensory locomotion, other locomotion, sitting, lying, quadrupedal standing, bipedal standing, suspensory posture, and other posture. Behavioral events of relatively short duration, or behaviors that are considered rare for chimpanzees, were recorded continuously. These included leaping, dropping, and brachiation. Brachiation is a specific type of suspensory locomotion that has been recorded as constituting only 0.1% of all positional behavior in wild chimpanzees (Hunt, 1992b). Thus in the current study it was recorded continuously. Behavioral definitions were based on those used by Hunt (1992b). For a detailed ethogram, see Schwandt (2002).

Behaviors were summarized either as proportions of overall positional behavior (states) or as rates per hour (events). Two-tailed *t*-tests were used to compare positional behavior means between nursery-reared and mother-reared subjects in each of the three age categories. Separate tests were run for each dependent variable (behavior), and significance was evaluated using the Bonferroni adjustment for multiple comparisons (overall, $p < 0.05$). Variances were equal for all tests. Analyses were performed using SAS version 8.0.

2.2.3. Solitary and Social Behavior

The subject sample included 23 chimpanzees (12 males, 11 females). Eleven subjects (5 males, 6 females) were nursery reared (permanently separated from their mothers before 12 months of age) and 12 subjects (7 males, 5 females) were reared by their mothers. Subjects were observed longitudinally for a 5-year period between 1990 and 1995 when they were housed indoors in established social groups (most of which were mixed-sex peer groups). At the youngest ages (3–4.9 years), two mother-reared female subjects were housed with their mothers in their natal groups. Observations collected 30 days before and after final maternal separation were not used in analysis. For all other subjects and age periods, subjects were housed in compatible mixed-sex peer groups. Subject age varied at study onset (1–5 years of age), resulting in a mixed cross-sectional, longitudinal data set that included data for infants (1–3.9 years), juveniles (4–6.9 years), and adolescents (7–9.9 years).

Data were collected using a focal animal sampling method to record 12 behavioral states that occurred at 15-sec intervals across 10-min observations (Altmann, 1974). Behaviors were summarized as proportions of overall behavior (states) at 2-year intervals. Since relatively few observations were collected for younger infants (1–2.9 years), they were not included in analysis. Analysis focused on four age periods: older infants (3–3.9 years), juveniles (4–6.9 years), young adolescents (7–7.9 years), and older adolescents (8–9.9 years). Behaviors included agonistic behavior patterns (aggression, avoidance/apprehension/fear, dominance, frustration/temper tantrum), affiliative behaviors (examine genitalia, grooming, social play, and embrace/tandem walk), solitary or abnormal behaviors (solitary play, rhythmic body rocking, abnormal behavior patterns), and "other" (see Howell, 2000, for ethogram). The number of subjects and the number of observations for each age category are shown in Table 16–2. Two-tailed *t*-tests were used to compare behavior means across nursery-reared and mother-reared subjects in each of the four age categories. Separate tests were run for each dependent variable (behavior), and significance was evaluated using the Bonferroni adjustment for multiple comparisons (overall, $p < 0.05$). Variances were equal for all tests. Analyses were performed using SAS version 8.0.

2.2.4. *Behavior at Final Maternal Separation*

The subject sample included 26 mother-reared chimpanzees (16 males, 10 females). The sample included six "short-term" mother-reared chimpanzees (1 male, 5 females) permanently separated from their mothers at 1–1.3 years of age, 12 "long-term" mother-reared chimpanzees (7 males, 5 females) permanently separated from their mothers at 3.3–5 years of age, and eight "long-term/weaned" chimpanzees (4 males, 4

Table 16–2. Subject Sample for Observations of Solitary and Social Behavior[a]

Age category	Age range (years)	Rearing	Number of subjects	Number of observations
Older infant	3.0–3.99	Nursery	3	84
		Mother	7	276
Juvenile	5.0–5.99	Nursery	8	256
		Mother	10	431
Young adolescent	6.0–7.99	Nursery	11	489
		Mother	10	473
Older adolescent	9.0–9.99	Nursery	11	1163
		Mother	12	1585

[a] Age categories are based on the PFA age scale, which in turn is based on the International Species Information System age scale for chimpanzees (Fritz *et al.*, 1999).

females) permanently separated from their mothers after natural maternal weaning at 4.8–9 years of age. Subjects were considered weaned when no nipple-contact nursing was observed for two consecutive months.

A focal-animal, one/zero sampling method was used to record behavior and vocalization patterns within 30-sec intervals across 5-min samples. These included 18 behaviors or vocalizations associated with maternal separation, including agitation (rapid agitated locomotion, temper tantrum, pace, rocking), protest (scream/cry, whimper/whoo, aggressive vocalizations, glottal cramps), and despair or depression (lie/sit heavy head, sleeping) (Boccia *et al.*, 1994). We also included more-subtle indications of agitation (scratching, yawning), behaviors associated with increased coping (self-cling, attachment to an inanimate object), and normal daily behavior patterns (lie/sit alert, hang/climb, play, attention to people or chimpanzees) (for ethogram see the Appendix).

Each focal animal was observed daily (balancing am and pm observation times) for 2 weeks before permanent maternal separation ("presep"). On the morning of separation, both mother and offspring were anesthetized and a general health check was accomplished. The offspring was separated from its mother and housed in a single cage located adjacent to at least one familiar peer but out of sight of its mother. Four samples were taken at hourly intervals the afternoon of separation ("day of") after the subject had fully recovered from anesthesia. For the rest of postseparation week 1 ("postweek 1"), observations were collected four times daily, twice in the morning and twice in the afternoon. During postseparation week 2 ("postweek 2"), observations were collected once daily (balancing am and pm observation sessions). The data set included a total of 1695 5-min observation sessions (141.25 observation hours).

Scores for each behavior or vocalization were totaled across 5-min observation periods and summarized by taking the average score across observation periods (presep, day of, postweek 1, postweek 2) for each subject. Nonparametric tests were used to compensate for the relatively small and unbalanced (for males and females) subject sample ($p < 0.01$). A Wilcoxon signed rank test was used to compare behavior during "presep" and "day of" time periods and to compare "presep" with "postweek 2" time periods to provide an indication of initial and lasting reaction to final maternal separation. A Kruskal–Wallis (KW) one-way analysis of variance was used to consider the effect of rearing (short-term, long-term, and long-term/weaned) on behavior. Analyses were performed using SYSTAT version 7.0.

3. RESULTS

3.1. Growth

The LOWESS fits for females (Fig. 16–4) indicate that nursery-reared and mother-reared individuals followed essentially the same trajectory until about 80 months of age, after which nursery-reared females were larger than mother-reared females. The LOWESS fits for males (Fig. 16–5) show that nursery-reared and mother-reared males weighed about the same until about 48 months of age, after which mother-reared males were larger than nursery-reared males. However, the jackknife comparisons for both females and males were not significant (Table 16–3).

3.2. Positional Behavior

For all three age categories, *t*-tests showed that nursery-reared and mother-reared subjects did not differ significantly from each other in any of the positional behaviors tested (Table 16–4).

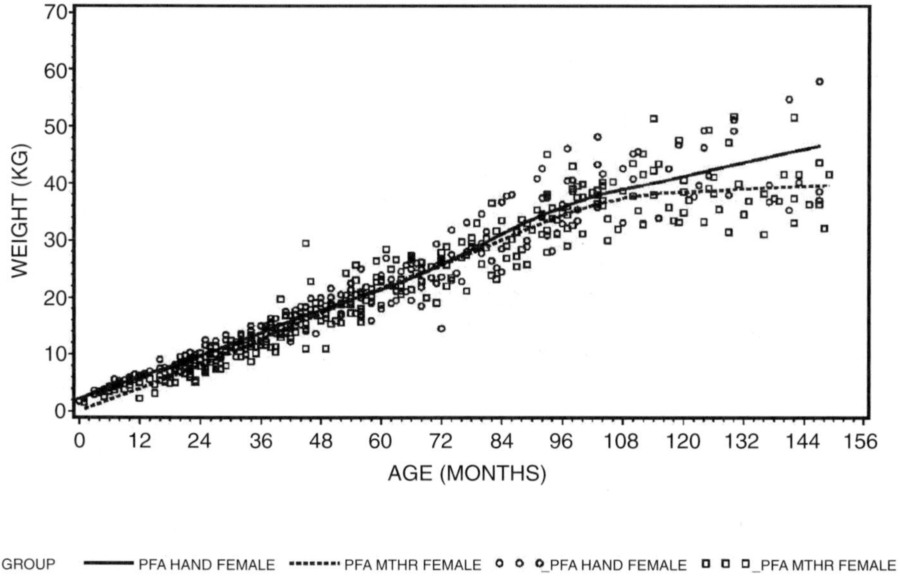

Figure 16–4. LOWESS growth curves for hand- and mother-reared females.

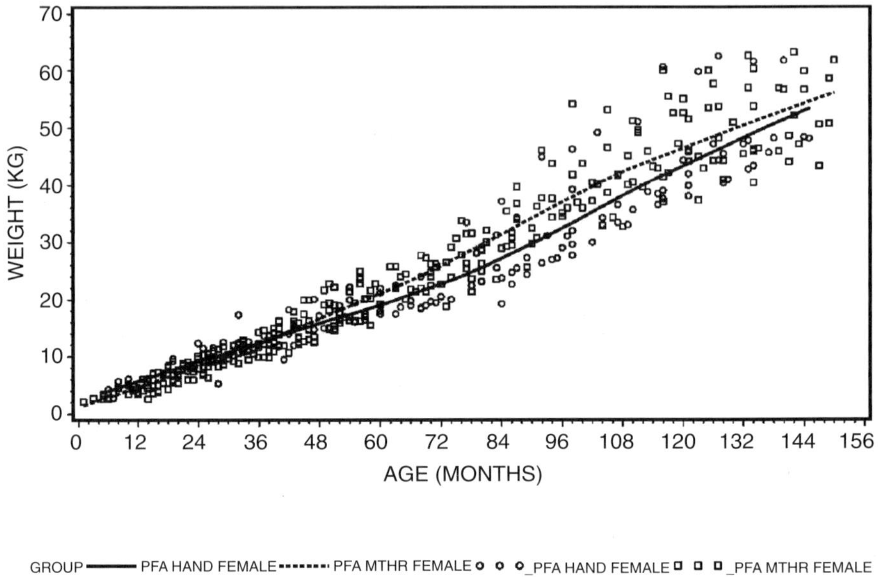

Figure 16–5. LOWESS growth curves for hand- and mother-reared males.

3.3. Solitary and Social Behavior

The *t*-test comparisons of solitary and social behavior revealed no significant differences in the behaviors of nursery- and mother-reared chimpanzees as infants, juveniles, or adolescents (Table 16–5).

Table 16–3. Jackknife Comparison of Weight versus Age in Nursery- and Mother-Reared Female and Male Chimpanzees[a]

	Age		Nursery		Mother			
Sex	Min	Max	N Subs	N Obs	N Subs	N Obs	*Z*	*p*[b]
Female	12	147	17	220	20	251	1.18	0.23
Male	14	134	10	143	25	277	−1.36	0.17

[a] Min, minimum; Max, maximum; N Subs, number of subjects; N Obs, number of observations.

[b] Two-tailed.

Table 16–4. Student's *t*-Test Results for Comparison of Positional Behavior in Nursery- and Mother-Reared Juveniles, Adolescents, and Young Adults

Age category	Behavior	*t* value[a]
Juvenile ($df = 19$)	Quadrupedal locomotion	1.08
	Bipedal locomotion	0.21
	Climbing	1.17
	Suspensory locomotion	−1.13
	Leaping	−0.11
	Brachiation	−1.33
	Dropping	0.11
	Other locomotion	0.73
	Sitting	0.99
	Lying	−0.73
	Quadrupedal standing	−0.36
	Bipedal standing	0.52
	Suspensory posture	−1.65
	Other posture	0.32
Adolescent ($df = 14$)	Quadrupedal locomotion	−0.58
	Bipedal locomotion	0.72
	Climbing	−0.39
	Suspensory locomotion	−0.39
	Leaping	−0.76
	Brachiation	−1.58
	Dropping	0.16
	Other locomotion	1.68
	Sitting	0.27
	Lying	1.25
	Quadrupedal standing	−0.08
	Bipedal standing	1.55
	Suspensory posture	−0.22
	Other posture	0.93
Young adult ($df = 9$)	Quadrupedal locomotion	−1.35
	Bipedal locomotion	3.25 ($p = 0.01$)
	Climbing	0.66
	Suspensory locomotion	−1.44
	Leaping	2.19 ($p = 0.056$)
	Brachiation	−1.38
	Dropping	2.33 ($p = 0.052$)
	Other locomotion	−0.77
	Sitting	−0.94
	Lying	1.50
	Quadrupedal standing	1.50
	Bipedal standing	1.50
	Suspensory posture	−0.13
	Other posture	−0.73

[a] All unadjusted *t*-test probabilities are >0.10 except as noted in the table. With a Bonferroni adjusted α of $0.05/14 = 0.0036$ within each age group there are no significant *t*-test values.

Table 16–5. Student's *t*-test Results for Comparison of Solitary and Social Behavior in Nursery- and Mother-Reared Subjects

Age category	Behavior	*t*-value[a]
Older infant (3 years, $df = 9$)	Aggression	0.35
	Avoidance/fear	−0.17
	Dominance	0.08
	Examine genitalia	−0.26
	Frustration/temper	−1.61
	Groom	1.23
	Other	−0.36
	Rocking	0.87
	Social play	0.88
	Solitary play	0.02
	Abnormal behavior	0.09
	Embrace/tandem walk	−1.50
Young juvenile (5 years, $df = 15$)	Aggression	0.94
	Avoidance/fear	1.56
	Dominance	−0.97
	Examine genitalia	1.31
	Frustration/temper	0.66
	Groom	−0.48
	Other	−2.17 ($p = 0.047$)
	Rocking	2.02 ($p = 0.062$)
	Social play	−0.28
	Solitary play	0.90
	Abnormal behavior	0.41
	Embrace/tandem walk	0.64
Young adolescent (7 years, $df = 19$)	Aggression	−1.36
	Avoidance/fear	0.97
	Dominance	−1.74 ($p = 0.098$)
	Examine genitalia	−2.71 ($p = 0.014$)
	Frustration/temper	−0.19
	Groom	−0.52
	Other	−1.01
	Rocking	0.81
	Social play	0.09
	Solitary play	1.35
	Abnormal behavior	1.27
	Embrace/tandem walk	1.18
Older adolescent (9 years, $df = 11$)	Aggression	0.49
	Avoidance/fear	2.27 ($p = 0.044$)
	Dominance	−0.42
	Examine genitalia	1.21
	Frustration/temper	Not observed
	Groom	−0.02
	Other	−0.26
	Rocking	1.20
	Social play	−0.33
	Solitary play	−1.24
	Abnormal behavior	0.03
	Embrace/tandem walk	Not observed

[a] All unadjusted *t*-test probabilities are >0.10 except as noted in the table. With a Bonferroni adjusted α of $0.05/12 = 0.004$ in each age group there are no significant *t*-test values.

3.4. Behavior at Final Maternal Separation

On the day of separation, Wilcoxon signed rank test results (Table 16–6) indicate a significant increase in pacing, rocking, and "whimper/whoo" vocalizations and a significant decrease in play. If we compare behaviors before and after week 2, results also indicate a significant increase in agitation (pacing, scratching) and normal behavior (lie/sit alert) and a significant decrease in play. Kruskal–Wallis results (Table 16–7) indicate that rearing category had a significant effect on agitation (pacing), protest (whimper/whoo, aggressive vocalizations), despair/depression (sleep-

Table 16–6. Wilcoxon Signed Rank Test Results (Probability Values) for Separation Effects

Behavior	Preseparation versus day of separation	Preseparation versus week 2
Agitation		
Rapid agitated locomotion	0.72	0.57
Temper tantrum	0.07	—
Pacing	<0.008[a]	<0.0028[a]
Rocking	0.06	<0.006[a]
Protest		
Scream/cry	0.33	0.27
Whimper/whoo	0.004[a]	0.03
Aggressive vocalizations	0.37	0.26
Glottal cramps	—	0.18
Despair/Depression		
Lie/sit with heavy head	0.38	0.03
Sleeping	0.11	0.05
Subtle agitation		
Scratching	0.55	<0.001[a]
Yawning	0.51	0.014
Coping activities		
Self-cling	0.06	0.09
Inanimate object attachment	0.64	0.24
Normal activities		
Lie/sit alert	0.44	<0.006[a]
Hanging/climbing	1.42	0.17
Playing	<0.0001[a]	<0.0001[a]
Attention to people	0.52	0.05
Attention to chimpanzees	0.44	0.14

[a] Significant effects ($\alpha < 0.01$).

Table 16–7. Results of Kruskal–Wallis (KW) One-Way Analysis of Variance between Short-Term, Long-Term, and Long-Term-Weaned Mother-Reared Subjects

Behavior	KW Value	Probability
Agitation		
Rapid agitated locomotion	3.58	0.17
Temper tantrum	0.00	0.99
Pacing	9.44[a]	<0.009[a]
Rocking	0.57	0.75
Protest		
Scream/cry	0.39	0.82
Whimper/whoo	21.26[a]	<0.0001[a]
Aggressive vocalizations	26.18[a]	<0.0001[a]
Glottal cramps	0.71	0.070
Despair/depression		
Lie/sit with heavy head	6.96[a]	0.03[a]
Sleeping	18.96[a]	<0.0001[a]
Subtle agitation		
Scratching	2.23	0.32
Yawning	21.50[a]	<0.0001[a]
Coping activities		
Self-cling	14.31[a]	<0.001[a]
Inanimate object attachment	0.09	0.96
Normal activities		
Lie/sit alert	54.19[a]	<0.0001[a]
Hanging/climbing	22.84[a]	<0.0001[a]
Playing	3.61	0.16
Attention to people	43.23[a]	<0.0001[a]
Attention to chimpanzees	32.04[a]	<0.0001[a]

[a] Significant tests ($\alpha < 0.01$).

ing, yawning), and normal behaviors (lie/sit alert, hang/climb, attention to people, attention to chimpanzee). Short-term mother-reared chimpanzees engaged in more protest and despair/depression and were more attentive to people than were long-term mother-reared chimpanzees (Table 16–8). While long-term/weaned chimpanzees paced more than their short-term mother-reared counterparts, they also engaged in more normal behaviors (lie/sit alert, hang/climb, and attention to other chimpanzees).

Table 16–8. Means and Standard Deviations for Each Solitary and Social Behavior by Short-Term, Long-Term, and Long-Term/Weaned Rearing Groups

	Short-Term		Long-Term		Long-term/ weaned	
Behavior	Mean	SD	Mean	SD	Mean	SD
Agitation						
Rapid agitated locomotion						
Preseparation	0.000	0.000	0.025	0.053	0.010	0.023
Day of separation	0.000	0.000	0.016	0.044	0.063	0.217
Week 1 PS[a]	0.000	0.000	0.004	0.011	0.086	0.260
Week 2 PS	0.000	0.020	0.006	0.018	0.130	0.309
Temper tantrum						
Preseparation	0.000	0.000	0.000	0.000	0.000	0.000
Day of separation	0.02	0.051	0.000	0.000	0.160	0.330
Week 1 PS	0.036	0.089	0.027	0.042	0.010	0.034
Week 2 PS	0.000	0.000	0.000	0.000	0.000	0.000
Pacing						
Pre-separation	0.000	0.000	0.106	0.178	0.118	0.267
Day of separation	0.792	1.939	0.516	1.263	1.090	1.372
Week 1 PS	0.724	1.540	1.168	1.896	1.614	1.653
Week 2 PS	0.150	0.281	0.698	0.715	1.465	1.714
Rocking						
Preseparation	0.250	0.418	0.109	0.171	0.023	0.050
Day of separation	1.458	2.240	0.016	0.044	0.493	0.898
Week 1 PS	2.245	3.671	0.196	0.367	0.838	1.108
Week 2 PS	1.367	2.404	0.275	0.438	0.509	1.070
Protest						
Scream/cry						
Preseparation	0.058	0.143	0.030	0.070	0.032	0.061
Day of separation	0.500	0.617	0.000	0.000	0.111	0.296
Week 1 PS	0.068	0.151	0.072	0.097	0.079	0.119
Week 2 PS	0.025	0.042	0.298	0.533	0.040	0.082
Whimper/whoo						
Preseparation	0.333	0.268	0.347	0.497	0.065	0.096
Day of separation	2.583	1.332	0.469	0.542	0.369	0.445
Week 1 PS	1.156	1.463	0.632	0.661	0.279	0.153
Week 2 PS	0.358	0.329	1.327	1.455	0.134	0.278
Aggressive vocalizations						
Preseparation	0.108	0.038	0.013	0.023	0.008	0.026
Day of separation	0.146	0.184	0.000	0.000	0.000	0.000
Week 1 PS	0.089	0.113	0.091	0.118	0.017	0.037
Week 2 PS	0.067	0.088	0.135	0.188	0.027	0.052
Glottal cramps						
Preseparation	0.000	0.000	0.000	0.000	0.000	0.000
Day of separation	0.000	0.000	0.000	0.000	0.000	0.000
Week 1 PS	0.000	0.000	0.000	0.000	0.000	0.000
Week 2 PS	0.000	0.000	0.008	0.024	0.008	0.026

(*Continued*)

Table 16–8. *Continued*

Behavior	Short-Term Mean	Short-Term SD	Long-Term Mean	Long-Term SD	Long-term/ weaned Mean	Long-term/ weaned SD
Despair/depression						
Lie/sit with heavy head						
Preseparation	0.067	0.000	0.094	0.135	0.182	0.390
Day of separation	0.333	0.000	0.197	0.249	0.357	0.830
Week 1 PS	1.156	1.408	0.603	0.707	0.426	0.748
Week 2 PS	1.042	1.473	0.756	0.906	0.167	0.354
Sleeping						
Preseparation	0.242	0.301	0.013	0.035	0.096	0.247
Day of separation	0.000	0.000	0.063	0.177	0.000	0.000
Week 1 PS	0.318	0.505	0.000	0.000	0.010	0.036
Week 2 PS	0.717	0.784	0.006	0.018	0.015	0.052
Subtle agitation						
Scratching						
Preseparation	0.833	0.633	0.473	0.244	0.521	0.330
Day of separation	0.500	0.737	0.278	0.277	1.188	1.374
Week 1 PS	1.328	1.496	0.933	0.336	0.908	0.454
Week 2 PS	2.017	0.392	0.958	0.430	0.683	0.517
Yawning						
Preseparation	0.158	0.111	0.019	0.026	0.023	0.050
Day of separation	0.208	0.292	0.047	0.093	0.083	0.289
Week 1 PS	0.146	0.172	0.078	0.085	0.082	0.188
Week 2 PS	0.317	0.129	0.150	0.169	0.045	0.069
Coping activities						
Self-cling						
Preseparation	0.000	0.000	0.031	0.059	0.015	0.053
Day of separation	0.000	0.000	0.413	0.704	0.642	2.070
Week 1 PS	0.000	0.000	0.264	0.408	0.401	1.083
Week 2 PS	0.000	0.000	0.183	0.395	0.059	0.158
Inanimate object attachment						
Preseparation	0.000	0.000	0.050	0.053	0.197	0.375
Day of separation	0.917	1.955	0.000	0.000	0.250	0.723
Week 1 PS	2.542	3.939	0.224	0.135	0.458	1.092
Week 2 PS	1.900	2.936	0.256	0.545	0.242	0.393
Normal activities						
Lie/sit alert						
Preseparation	1.183	1.054	2.722	1.705	6.015	1.715
Day of separation	1.396	1.901	2.350	2.732	7.068	2.060
Week 1 PS	2.677	3.170	3.091	1.662	6.242	1.355
Week 2 PS	3.558	2.724	2.948	1.740	7.079	1.436
Hanging/climbing						
Preseparation	0.133	0.189	3.548	1.668	3.643	0.727
Day of separation	2.125	1.416	2.278	0.944	4.004	1.591
Week 1 PS	3.651	1.788	3.724	1.685	4.243	1.813
Week 2 PS	1.533	0.789	3.704	1.226	3.624	1.593

Table 16–8. *Continued*

Behavior	Short-Term		Long-Term		Long-term/ weaned	
	Mean	SD	Mean	SD	Mean	SD
Playing						
Preseparation	6.758	2.673	4.631	1.330	3.822	2.430
Day of separation	0.625	0.671	0.309	0.344	0.869	0.867
Week 1 PS	1.656	1.328	0.810	0.547	0.767	0.618
Week 2 PS	4.208	3.071	1.227	1.169	1.436	0.949
Attention to people						
Preseparation	4.625	3.650	0.638	0.476	1.410	1.180
Day of separation	2.458	1.237	0.638	0.841	2.836	1.468
Week 1 PS	3.776	2.603	0.803	0.909	2.376	1.141
Week 2 PS	4.250	2.444	0.733	0.507	1.869	0.897
Attention to chimpanzees						
Preseparation	5.958	2.391	2.113	1.711	4.845	2.183
Day of separation	2.042	0.824	1.931	2.217	5.953	1.786
Week 1 PS	3.318	2.608	2.265	2.566	4.816	1.846
Week 2 PS	2.217	1.658	2.344	2.799	4.945	1.776

[a] PS, postseparation.

4. DISCUSSION

Our results suggest that chimpanzees reared in an enriched nursery with an intensive nursery-rearing protocol are similar to their mother-reared counterparts in growth and behavior. They also show that there is no significant difference in weight for nursery-reared and mother-reared chimpanzees. While the LOWESS fits suggest some variation, with nursery-reared females being larger than mother-reared females after a certain age, and mother-reared males being larger than nursery-reared males after a certain age, the differences are relatively small (in the range of 4–5 kg for both females and males). The fact that nursery-reared females seem to be larger in the long term than mother-reared females lends a strong argument against the suggestion that nursery-reared individuals are at a disadvantage when it comes to physical development.

The results also emphasize the importance of considering development not just during infancy but throughout the entire growth period. A prior study of weight gain in PFA chimpanzees aged 1–24 months also found a significant difference between nursery-reared and mother-reared females, with nursery-reared females being the heavier group (Marzke *et al.*, 1996a,b). Expansion of the analysis to a fuller age range, as the current study does, suggests that rearing differences during infancy do not persist throughout the entire growth period. This result is in accordance with other comparisons of weight growth in captive chimpanzees (Davenport *et al.*, 1961; Smith *et al.*, 1975) and macaques (Durham *et al.*, 1997; Sackett *et al.*, 2002). Nursery-reared chimpanzees at Yerkes were significantly heavier than mother-reared infants at 12–24 months (Davenport *et al.*, 1961). In a comparison of mother-reared infants at Holloman Air Force Base (HAFB) with published data on nursery-reared infants at Yerkes, Smith *et al.* (1975) found that the HAFB chimpanzees had a longer period of infancy with a slower growth rate, but by the end of the infant growth period the HAFB chimpanzees were distinctly larger then the Yerkes chimpanzees. However, the onset of what Smith *et al.* (1975) termed "juvenile growth," which generally occurred at 17–18 months of age, was not significantly different between HAFB and Yerkes chimpanzees, suggesting that differences in weight growth did not persist beyond early infancy. Durham *et al.* (1997) found that nursery-reared female pigtailed macaques (*Macaca nemestrina*) were larger at birth and gained significantly more weight during the first 3 months than their mother-reared counterparts. However, Sackett *et al.* (2002) found no overall significant differences in weight growth of nursery-reared and mother-reared individuals of either sex at 2–10 years of age.

The development of positional behavior appears to be comparable in nursery-reared and mother-reared chimpanzees. Although we were unable to test for early-rearing differences between nursery-reared and mother-reared infants, the lack of significant results for later age groups implies that the general outcome is the same under both rearing conditions. Furthermore, the development of positional behavior in PFA chimpanzees as a whole (Schwandt, 2002) is generally similar to that reported for wild chimpanzees (Doran, 1997). Had we had an adequate sample of nursery-reared and mother-reared infants, we might have found some quantitative differences in positional behavior in the earliest age group. Studies of weight gain and hand/wrist maturation in PFA infants aged 1–24 months showed that nursery-reared females were significantly heavier than their mother-reared counterparts (there was no significant difference for males), and that maturation of the hand/wrist occurred more rapidly in hand-reared females (although this difference was not significant) (Marzke *et al.*, 1996a,b).

It is possible that these differences in physical development could be correlated with differences in behavioral development. Unfortunately, not enough data were available for statistical comparison with hand/wrist maturation data. The average ages at the first appearance of several locomotor behaviors were available from unpublished PFA behavioral records for both mother-reared and nursery-reared infants. Average ages were the same in both groups for the first appearance of climbing (4 months), bipedal locomotion (6 months), and knuckle-walking (9 months). Palm-walking appeared earlier in mother-reared infants (at 6 months versus 8 months for their nursery-reared peers), while brachiation appeared earlier in nursery-reared infants (at 7 months versus 9 months). Unfortunately, not enough data were available for statistical comparison. A study of captive infant gorillas found that hand-reared infants spent more time in locomotion than did mother-reared infants in the first 6 months. However, there were no significant differences in the ages at first appearance for any particular locomotor behavior (Meder, 1989).

The behavior of hand-reared and mother-reared infants, juveniles, and adolescents also appears to be comparable. We found no significant differences in the behaviors of nursery- and mother-reared chimpanzee juveniles and adolescents. This finding is particularly interesting as there is significant literature to suggest that early rearing has a profound effect on behavior (Nicholson, 1977; Horvat *et al.*, 1980; Horvat and Kraemer, 1981; Hemelrijk and de Kogel, 1989; Plooij and van de Rijt-Plooij, 1989; Hannah and Brotman, 1990; Fritz *et al.*, 1992, 1994; Howell *et al.*, 1993; Maki *et al.*, 1993; Tomasello *et al.*, 1993; King and Mellen, 1994; Murray, 1998).

The results of our analyses suggest that improvements in nursery-rearing environment and practices for captive chimpanzees encourage the normative development of species-typical behavior and discourage the development of abnormal behavior patterns (Capitanio, 1986). Although maternal rearing is preferable whenever possible, intensive hand rearing in an enriched nursery environment seems to prevent the negative effects on behavior reported in past studies (Harlow and Harlow, 1962; Denenberg, 1969; Davenport *et al.*, 1973; Goldfoot, 1977a,b; Goosen, 1986; Kolter, 1995).

Our data show that chimpanzees that grow up with their mothers until at least 1 year of age display relatively low levels of protest and despair behaviors related to maternal separation compared with those previously reported (Brandt *et al.*, 1972; Chorazyna, 1972; Kaufman, 1973; Suomi *et al.*, 1973a, 1976; McGinnis, 1980; Kraemer *et al.*, 1991; Boccia *et al.*, 1994; Biondi and Picardi, 1996). However, short-term mother-reared chimpanzees did engage in more protest and despair/depression-related behaviors than did long-term and long-term/weaned mother-reared chimpanzees. While the latter spent more time pacing than did their younger counterparts, they also spent more time in normal everyday behaviors (lie/sit alert, hanging/climbing, and attention to other chimpanzees) after final maternal separation. We suggest that long-term maternal rearing (until at least 3.3 years of age) may be optimal for chimpanzees to prevent despair/depression-related behaviors observed in short-term mother-reared subjects.

The results presented herein suggest several areas for future research. While we included weight gain as a measure of growth, it is important to consider the effect of early rearing history on physical development. It is also important to consider growth, along with physical and behavioral development, in a matched sample of subjects observed longitudinally from infancy to adulthood. While it may not be pragmatic to replicate these studies, because few captive chimpanzee births are anticipated in the foreseeable future owing to the NRC moratorium, findings presented here should be replicated in other populations and within the zoological community. Finally, imbalances in age/sex categories should be overcome, and subjects included in this study should be followed to ascertain the effect of the early rearing environment on adult behavior and the aging process.

5. APPENDIX

5.1. Ethogram of Maternal Separation Behaviors

5.1.1. Agitation

Rapid Agitated Locomotion: Repeated, confused motion around the cage. This can follow an irregular pattern with frequent changes in direction, or a rapid, regular pattern.

Temper Tantrum: During a tantrum the chimp screams loudly and may leap up, fling its arms above its head and then slap them onto the ground, hurl itself to the ground on its face, or hug the cage mesh, an object, or itself. It may beat the ground with its hands. It may rush off, tumbling over and over, still screaming. Screaming may end in glottal cramps.

Pacing: Repeated slow locomotion (more than one complete sequence) in a regular pattern along or around the cage perimeter. Mode of locomotion may vary—quadrupedal, brachiation, bipedal, or climbing.

Rocking: Rhythmic movement of the body, either side-to-side or forward and backward. May occur while the animal is standing bipedally or quad-rupedally, sitting, crouching, or in a hanging position.

5.1.2. Protest

Scream/Cry: There is a good deal of variation in the pitch and timing of screaming. Some screaming is high pitched and long-drawn-out. Other screaming consists of a series of short, high-pitched or rasping screams. Long-drawn-out screams may also have a rasping quality.

Whimper/Whoo: Whimper: A whole series of the whoo whimpering, rising and falling in pitch, which may lead to crying and screaming. Whoo: A single-syllable soft whimper.

Aggressive Vocalizations (Barks): Clear, short sounds. Can be graded into grunts and squeaks. Soft barks are usually a mild form of threat, while waa barks are a more intensive threat.

Glottal Cramps: During loud and prolonged screaming the animal may seem almost to choke: only hoarse squeaks or rasping sounds are heard.

5.1.3. Despair/Depression

Lie/sit (Heavy Head): Individual is in either a sitting or prone position and appears to be lethargic. The head is typically drooping and may be propped against the cage or hands. The entire body is in a hunched-over posture. The infant may appear sleepy, though not asleep. There is a general lack of attention to external stimuli, though the infant is still awake.

Sleeping: Individual has eyes closed and appears to be asleep. The individual is usually lying down, though sleep occasionally occurs in a sitting position. Breathing is or becomes regular and the body generally relaxes, especially the arms and hands.

5.1.4. Subtle Agitation

Scratching: Scraping of the finger nails across the skin.

Yawning: Involuntary, wide opening of the mouth accompanied by deep inhalation.

5.1.5. Coping Behaviors

Self-Cling: Individual holds onto itself, usually in an embrace of the torso or a hand grasping the crotch.

Attachment to Inanimate Object: Individual holds an object against its body. The object may include such items as straw, toy, or blanket. The individual may also carry the object from place to place. The animal may be relaxed or exhibit a tense posture.

5.1.6. Normal Behaviors

Lie/Sit (Alert): Individual is in either a sitting or prone position, is not exhibiting signs of sleepiness, and is not involved in any active behavior such as locomotion, play, eating, rocking, or grooming.

Hanging/Climbing: Individual is hanging from some part of the cage-side, top, or shelf or is climbing from one point to another in normal locomotion.

Playing: A broad category that includes many different types of actions. Activities are nonthreatening and can be distinguished by the subject's loose body tone. Repetition and exaggeration of movements are typical of play, as well as the ease with which these activities can be interrupted. A play face frequently accompanies play. No piloerection will occur with play.

Attention to People: Individual's focus is directed toward nonchimpanzee individuals, including people, cats, or dogs. This is more than a glance toward someone. The subject's eyes may be oriented toward the individual; often the whole body, or at least the head, is turned toward the object of attention. This includes any attempt to interact with an individual, but excludes actual interactions.

Attention to Chimpanzees: Same as attention to people, except directed toward a chimpanzee.

ACKNOWLEDGMENTS

We gratefully acknowledge the PFA veterinarians and animal care staff. We also thank Caroline Robb for her assistance in manuscript preparation. This project was supported in part by the University of Texas, M.D. Anderson Cancer Research Center, as a subcontract (1U42RR15090-02) within the National Institutes of Health Biomedical Research Program. PFA is accredited by the Association for Assessment and Accreditation of Laboratory Care International. The data-collection protocol was reviewed and approved by the PFA Institutional Animal Care and Use Committee.

REFERENCES

Altmann, J. (1974). Observational study of behavior: Sampling methods. *Behaviour* 49:227–267.

Berman, C.M., Rasmussen, K.L.R., and Suomi, S. (1994). Responses of free-ranging rhesus monkeys to a natural form of social separation. *Child Dev.* 65:1028–1041.

Bernstein, I.S., and Dobrofsky, M. (1981). Compensatory social responses of older pigtailed monkeys to maternal separation. *Dev. Psychobiol.* 14(2): 163–168.

Biondi, M., and Picardi, A. (1996). Clinical and biological aspects of bereavement and loss-induced depression: A reappraisal. *Psychother. Psychosomat.* 65:229–245.

Boccia, M.I., Reite, M., and Laudenslager, M.L. (1991). Early social environment may alter the development of attachment and social support: Two case reports. *Infant Behav. Dev.* 14:253–260.

Boccia, M.I., Laudenslager, M.L., and Reite, M.L. (1994). Intrinsic and extrinsic factors that affect infant responses to maternal separation. *Psychiatry* 57:43–50.

Boinski, S. (1989). The positional behavior and substrate use of squirrel monkeys: Ecological implications. *J. Hum. Evol.* 18:659–677.

Bolter, D., and Zihlman, A. (2002). Growth and development in body tissues and proportions of African apes (*Gorilla gorilla* and *Pan troglodytes*): A preliminary report. *Am. J. Phys. Anthropol. Suppl.* 34:46.

Born, K., Fritz, J., and Trueblood, P. (1997). Chimpanzee diet at the Primate Foundation of Arizona. *The Newsletter* 8(4):1–3.

Brakke, K.E., and Savage-Rumbaugh, E.S. (1991). Early postural behavior in *Pan*: Influences on development. *Infant Behav. Dev.* 14:265–288.

Brandt, E.M., Baysinger, C., and Mitchell, G. (1972). Separation from rearing environment in mother-reared and isolation-reared rhesus monkeys (*Macaca mulatta*). *Int. J. Psychol.* 2(3):193–204.

Buchanan, L. (2002). Growth spurts in linear dimensions in *Pan troglodytes. Am. J. Phys. Anthropol. Suppl.* 34:50.

Capitanio, J.P. (1986). Behavioral pathology. In: Mitchell, G., and Erwin, J. (eds.), *Comparative Primate Biology*, Vol. 2A: *Behavior, Conservation, and Ecology*. Alan R. Liss, New York, pp. 411–454.

Chamove, A.S., Rosenblum, L.A., and Harlow, H.F. (1973). Monkeys (*Macaca mulatta*) raised only with peers. A pilot study. *Anim. Behav.* 21:316–325.

Champoux, M., and Suomi, S. (1984). Behavioral and adrenocortical responses of rhesus macaque mothers to infant separation in an unfamiliar environment. *Primates* 35(2):191–202.

Chorazyna, H. (1972). The separation of baby monkeys from their mothers. *Med. Primatol.* Part I:88–92.

Clarke, A.S., and Snipes, M. (1998). Early behavioral development and temperamental traits in mother- versus peer-reared rhesus monkeys. *Primates* 39(4):433–448.

Cleveland, W.S. (1979). Robust locally weighted regression and smoothing scatter plots. *J. Am. Statist. Assn.* 74:829–836.

Cleveland, W.S., and Devlin, S.J. (1988). Locally weighted regression: An approach to regression analysis by local fitting. *J. Am. Statist. Assn.* 83: 596–610.

Codner, M.A., and Nadler, R.D. (1984). Mother-infant separation and reunion in the great apes. *Primates* 25(2):204–217.

Coe, C.L., Wiener, S.G., Rosenberg, L.T., and Levine, S. (1985). Endocrine and immune responses to separation and maternal loss in nonhuman primates. In: Reite, M., and Field, T. (eds.), *Psychobiology of Attachment and Separation*. Academic Press, Orlando, FL, pp. 163–197.

Cooper, D.L., and Markowitz, H. (1979). Handlers' effects on contact comfort behaviors of two trios of juvenile chimpanzees in the zoo. *Psychol. Rep.* 44: 1015–1018.

Crompton, R.H. (1983). Age differences in locomotion of two subtropical galaginae. *Primates* 24:241–259.

Davenport, R.K., Menzel, E.W., and Rogers, C.M. (1961). Maternal care during infancy: Its effect on weight gain and mortality in the chimpanzee. *Am. J. Orthopsychiat.* 31:803–809.

Davenport, R.K., Rogers, C.M., and Rumbaugh, D.M. (1973). Long-term cognitive deficits in chimpanzees associated with early impoverished rearing. *Dev. Psychol.* 9:343–347.

Denenberg, V.H. (1969). Animal studies of early experience: Some principles which have implications for human development. In: Hill, J. (ed.), *Minnesota Symposium on Child Psychology.* University of Minnesota Press, Minneapolis, pp. 31–45.

Dienske, H. (1988). Ethological studies on social deprivation in monkeys, apes and humans. *Proceedings of the TNO Symposium on Biomedical Research in Primates*, pp. 63–68.

Doran, D.M. (1997). Ontogeny of locomotion in moutain gorillas and chimpanzees. *J. Hum. Evol.* 32:323–344.

Durham, D.L., Newell-Morris, L.L., Ruppenthal, G.C., and Sackett, G.P. (1997). A comparison of growth and development in mother-reared and nursery-reared pigtail macaque (*Macaca nemestrina*) infants. *Am. J. Primatol.* 42:107.

Erwin, J., Mitchell, G., and Maple, T. (1973). Abnormal behavior in non-isolate-reared rhesus monkeys. *Psychol. Rep.* 33:515–523.

Fragaszy, D.M., Boinski, S., and Whipple, J. (1992). Behavioral sampling in the field: Comparison of individual and group sampling methods. *Am. J. Primatol.* 26:259–275.

Frasier, S.D., and Rallison, M.L. (1972). Growth retardation and emotional deprivation: Relative resistance to treatment with human growth hormone. *J. Pediatr.* 80:603–609.

Friendly, M. (1991). *SAS System for Statistical Graphics*, 1st ed. SAS Institute, Inc., Cary, NC.

Fritz, J. (1986). Resocialization of asocial chimpanzees. In: Benirsckhe, K. (ed.), *Primates, the Road to Self-Sustaining Populations.* Springer-Verlag, New York, pp. 351–359.

Fritz, J., and Fritz, P. (1982). Great ape hand-rearing with a goal of normalcy and a reproductive continuum. *Proceedings of the Annual Meeting of the American Association of Zoo Veterinarians*, pp. 27–31.

Fritz, J., and Fritz, P. (1985). The hand-rearing unit: Management decisions that may affect chimpanzee development. In: Graham, C.E., and Bowen, J.A.

(eds.), *Clinical Management of Infant Great Apes*. Alan R. Liss, New York, pp. 1–34.

Fritz, J., and Howell, S.M. (1993). Psychological wellness for captive chimpanzees: An evaluative program. *Humane Intervent. Alternatives* 7:426–434.

Fritz, J., Nash, L.T., Nankivell, B., Howell, S.M., and Bloomsmith, M. (1990). Age-sex influences on maternal separation stress in chimpanzees (*Pan troglodytes*). *Am. J. Primatol.* 20:192.

Fritz, J., Howell, S.M., Hogan, H., Nankivell, B., and Nash, L. (1991). Weaning in captive chimpanzees (*Pan troglodytes*): As assessed by an easily collected measure. *Lab. Primate Newslett.* 30(4):13–16.

Fritz, J., Nash, L.T., and Howell, S.M. (1992). Effects of temporary restricted social housing on later reproductive behavior in adolescent chimpanzees. *Lab. Anim.* 21(5):21–25.

Fritz, J., Howell, S., and Selchow-Burr, C. (1994). Object attachment and stereotypic rocking in chimpanzees (*Pan troglodytes*). *Am. J. Primatol.* 33:209.

Fritz, J., Wolfle, T.L., and Howell, S. (1999). Chimpanzees. In: Poole, T. (ed.), *The United Federation for Animal Welfare Handbook on the Care and Management of Laboratory Animals*, 7th ed. Vol. 1: *Terrestrial Vertebrates.* Blackwell Science, Oxford, pp. 643–658.

Fritz, P., and Fritz, J. (1979). Resocialization of chimpanzees: Ten years of experience at the Primate Foundation of Arizona. *J. Med. Primatol.* 8:202–221.

Gabriel, S.W., Alford, P.L., and Bramblett, C.A. (1986). Differential early experience and behavior in adult male chimpanzees. *Am. J. Primatol.* 10:402.

Gardner, E.L., and Gardner, E.B. (1970). Orientation of infant macaques to facially distinct surrogate mothers. *Dev. Psychol.* 3(3):409–410.

Gavan, J.A. (1953). Growth and development of the chimpanzee: A longitudinal and comparative study. *Hum. Biol.* 25:93–143.

Gavan, J.A. (1971). Longitudinal, postnatal growth in chimpanzee. In: Bourne, G.H. (ed.), *The Chimpanzee*, Vol. 4: *Behavior, Growth, and Pathology of Chimpanzees.* Karger, Basel, pp. 47–102.

Gebo, D.L. (1992). Locomotor and postural behavior in *Alouatta palliata* and *Cebus capuchinus*. *Am. J. Primatol.* 26:277–290.

Goldfoot, D.A. (1977a). Sociosexual behaviors of nonhuman primates during development and maturity: Social and hormonal relationships. In: Schrier, A.M. (ed.), *Behavioral Primatology*, Vol. 1: *Advances in Research and Theory.* Lawrence Erlbaum Associates, Hillsdale, NJ, pp. 139–184.

Goldfoot, D.A. (1977b). Rearing conditions which support or inhibit later sexual potential of laboratory-born rhesus monkeys: Hypotheses and diagnostic behaviors. *Lab. Anim. Sci.* 27:548–556.

Goosen, C. (1986). Housing conditions and the ontogeny of stereotyped locomotor and abnormal self-directed activities in rhesus monkeys. *Annual Report of the Institute of the Division for Health Research*, pp. 287–290.

Grether, W.F., and Yerkes, R.M. (1940). Weight norms and relations for chimpanzee. *Am. J. Phys. Anthropol.* 27:181–197.

Hamada, Y., and Udono, T. (2002). Longitudinal analysis of length growth in the chimpanzee (*Pan troglodytes*). *Am. J. Phys. Anthropol.* 118:268–284.

Hamada, Y., Udono, T., Teramoto, M., and Sugawara, T. (1996). The growth pattern of chimpanzees: Somatic growth and reproductive maturation in *Pan troglodytes*. *Primates* 37:279–295.

Hamada, Y., Udono, T., Teramoto, M., and Hayasaka, I. (1998). Development of the hand and wrist bones in chimpanzees. *Primates* 39:157–169.

Hannah, A.C., and Brotman, B. (1990). Procedures for improving maternal behavior in captive chimpanzees. *Zoo Biol.* 9:233–240.

Harlow, H.F., and Harlow, M.K. (1962). The effect of rearing conditions on behavior. *Bull. Menninger Clinic* 26:213–224.

Harlow, H.F., and Suomi, S.J. (1974). Induced depression in monkeys. *Behav. Biol.* 12:273–296.

Hemelrijk, C.K., and de Kogel, C.H. (1989). What chimpanzee mothers have more sociable infants? *Behaviour* 111:305–318.

Higley, J.D., Suomi, S.J., and Linnoila, M. (1990). Parallels in aggression and serotonin: Consideration of development, rearing history, and sex differences. In: van Praag, H.M., Plutchik, R., and Apter, A. (eds.), *Violence and Suicidality: Perspectives in Clinical and Psychobiological Research*. Brunner/Mazel, New York, pp. 245–256.

Hill, S.D., McCormack, S.A., and Mason, W.A. (1973). Effects of artificial mothers and visual experience on adrenal responsiveness of infant monkeys. *Dev. Psychobiol.* 6:421–429.

Hoff, M.P., Nadler, R.D., Hoff, K.T., and Maple, T.L. (1994). Separation and depression in infant gorillas. *Dev. Psychobiol.* 27:439–452.

Horvat, J.R., and Kraemer, H.C. (1981). Infant socialization and maternal influence in chimpanzees. *Folia Primatol.* 39:99–110.

Horvat, J.R., Coe, C.L., and Levine, S. (1980). Infant development and maternal behavior in captive chimpanzees. In: Bell, R.W., and Smotherman, W.P.

(eds.), *Maternal Influences and Early Behavior*. Spectrum Publications, Inc., New York, pp. 285–309.

Howell, S.M. (2000). My friends the captive chimpanzees: Individual differences in the behavior of captive chimpanzees (*Pan troglodytes*). Ph.D. Dissertation, Arizona State University, Tempe.

Howell, S.M., Fritz, J., and Nash, L.T. (1993). Chimpanzee (*Pan troglodytes*) infant age-related changes in the dynamics of the mother-infant relationship. *The Newsletter* 5(1):1–3.

Hunt, K.D. (1992a). Social rank and body weight as determinants of positional behavior in *Pan troglodytes. Primates* 33:347–357.

Hunt, K.D. (1992b). Positional behavior of *Pan troglodytes* in the Mahale Mountains and Gombe Stream National Parks, Tanzania. *Am. J. Phys. Anthropol.* 87:83–105.

Kaufman, I.C. (1973). Mother-infant separation in monkeys: An experimental model. In: Scott, J.P., and Senay, E.C., (eds.), *Separation and Depression*. American Association for the Advancement of Science, Washington, D.C., pp. 33–52.

Kimura, T., and Hamada, Y. (1996). Growth of wild and laboratory born chimpanzees. *Primates* 37:237–251.

King, N.E., and Mellen, J.D. (1994). The effects of early experience on adult copulatory behavior in zoo-born chimpanzees (*Pan troglodytes*). *Zoo Biol.* 13:51–59.

Kolter, L. (1995). Control of behavior and the development of disturbed behavior patterns. In: Ganslober, U., Hodges, J.K., and Kaumanns, W. (eds.), *Research and Captive Propagation*. Filander Verlag, Furth, pp. 248–256.

Kraemer, G.W. (1995). Significance of social attachment in primate infants: The infant-caregiver relationship and volition. In: Pryce, C.R., Martin, R.D., and Skuse, D. (eds.), *Motherhood in Human and Nonhuman Primates*. Karger, Basel, pp. 152–161.

Kraemer, G.W., Ebert, M.H., Schmidt, D.E., and McKinney, W.T. (1991). Strangers in a strange land: A psychobiological study of infant monkeys before and after separation from real or inanimate mothers. *Child Dev.* 62:548–566.

Laudenslager, M.L., Boccia, M.L., and Reite, M.L. (1993). Biobehavioral consequences of loss in nonhuman primates: Individual differences. In: Stroebe, M.S., Stroebe, W., and Hansson, R.O. (eds.), *Handbook of Bereavement: Theory, Research, and Intervention*. Cambridge University Press, Cambridge, pp. 129–142.

Lawrence, W.A., and Gorzitze, A.B. (1985). Assessment of postnatal weight gain in nursery-rearing infant chimpanzees. In: Graham, C.E., and Bowen, J.A.

(eds.), *Clinical Management of Infant Great Apes*. Alan R. Liss, New York, pp. 157–164.

Maki, S., Fritz, J., and England, N. (1993). An assessment of early differential rearing conditions on later behavioral development in captive chimpanzees. *Infant Behav. Dev.* 16:373–381.

Martin, P., and Bateson, P. (1993). *Measuring Behavior: An Introductory Guide*. Cambridge University Press, Cambridge.

Marzke, M.W., Young, D., and Fritz, J. (1996a). Weight gain in captive chimpanzee infants: Comparisons by sex, rearing, and colony. *Am. J. Primatol.* 38:133–144.

Marzke, M.W., Young, D.L., Hawkey, D.E., Su, S.M., Fritz, J., and Alford, P.L. (1996b). Comparative analysis of weight gain, hand/wrist maturation, and dental emergence rates in chimpanzees aged 0–24 months from varying captive environments. *Am. J. Phys. Anthropol.* 99:175–190.

Mason, W.A. (1967). Motivational aspects of social responsiveness in young chimpanzees. In: Stevenson, H.W. (ed.), *Early Behavior: Comparative and Developmental Approaches*. John Wiley & Sons, New York, pp. 103–126.

Mason, W.A., and Kenney, M.D. (1974). Redirection of filial attachments in rhesus monkeys: Dogs as mother surrogates. *Science* 183:1209–1211.

Mason, W.A., Mendoza, S.P., and Moberg, G.P. (1991). Persistent effects of early social experience on physiological responsiveness. In: Ehara, A., Kimura, T., Takenaka, O., and Iwamonto, M. (eds.), *Primatology Today*. Elsevier, Amsterdam, pp. 469–471.

McGinnis, L.M. (1980). Maternal separation studies in children and nonhuman primates. In: Bell, R.W., and Smotherman, W.P. (eds.), *Maternal Influences and Early Behavior*. Spectrum Publications, Inc., New York, pp. 311–335.

McKinney, W.T. (1985). Separation and depression: Biological markers. In: Reite, M., and Fields, T. (eds.), *The Psychobiology of Attachment and Separation*. Academic Press, Orlando, FL, pp. 201–222.

Meder, A. (1989). Effects of hand-rearing on the behavioral development of infant and juvenile gorillas (*Gorilla g. gorilla*). *Dev. Psychobiol.* 22:357–376.

Moses, L.E., Gale, L.C., and Altmann, J. (1992). Methods for analysis of unbalanced, longitudinal growth data. *Am. J. Primatol.* 28:49–59.

Murray, L.E. (1998). The effects of group structure and rearing strategy on personality in chimpanzees (*Pan troglodytes*) at Chester, London ZSL and Twycross Zoos. *Int. Zool. Yearb.* 36:97–108.

Nash, L.T., and Fritz, P. (1982). Captive breeding and resocialization of chimpanzees at the Primate Foundation of Arizona. *Int. J. Primatol.* 3:316.

Nicholson, N.A. (1977). A comparison of early behavioral development in wild and captive chimpanzees. In: Chevalier-Skolnikoff, S., and Poirier, F.E. (eds.), *Primate Bio-social Development: Biological, Social, and Ecological Determinants*. Garland Publishing, New York, pp. 529–560.

Nissen, H.W., and Riesen, A.H. (1949a). Onset of ossification in the epiphyses and short bones of the extremities in chimpanzee. *Growth* 13:45–70.

Nissen, H.W., and Riesen, A.H. (1949b). Retardation in onset of ossification in chimpanzee related to various environmental and physiological factors. *Anat. Rec.* 105:665–675.

Pazol, K.A., and Bloomsmith, M.A. (1993). The development of stereotyped body rocking in chimpanzees (*Pan troglodytes*) reared in a variety of nursery settings. *Anim. Welfare* 2:113–129.

Pereira, M.E., and Fairbanks, L.A. (1993). *Juvenile Primates, Life History, Development and Behavior*. Oxford University Press, New York.

Plooij, F. (1987). Infant-ape behavioural development: The control of perception, types of learning and symbolism. In: Montangero, J., Tryphon, A., and Dionnet, S. (eds.), *Symbolism and Knowledge*. Jean Piaget Foundation Archives, Book No. 8, pp. 35–64.

Plooij, F., and van der Rijt-Plooij, H.H.C. (1989). Vulnerable periods during infancy. *Ethol. Sociobiol.* 10:279–296.

Prescott, J.W. (1970). Early somatosensory deprivation as an ontogenetic process in the abnormal development of brain and behavior. In: Goldsmith, I.E., and Moor-Jankowski, J. (eds.), *Medical Primatology*. S. Karger, Basel, pp. 356–375.

Prost, J.H. (1965). A definitional system for the classification of primate locomotion. *Am. Anthropol.* 67:135–148.

Randolph, M.C., and Mason, W.A. (1969). Effects of rearing conditions on distress vocalizations in chimpanzees. *Folia Primatol.* 10:103–112.

Rawlins, R.G. (1976). Locomotor ontogeny in *Macaca mulatta*: I. Behavioral strategies and tactics. *Am. J. Phys. Anthropol.* 44:201.

Remis, M. (1995). Effects of body size and social context on the arboreal activities of lowland gorillas in the Central African Republic. *Am. J. Phys. Anthropol.* 97:413–433.

Remis, M. (1998). The gorilla paradox: The effects of body size and habitat on the positional behavior of lowland and mountain gorillas. In: Strasser, E., Fleagle, J., Rosenberger, A., and McHenry, H. (eds.), *Primate Locomotion: Recent Advances*. Plenum Press, New York, pp. 95–106.

Riesen, A.H., and Kinder, E.F. (1952). Postural schedule for chimpanzee infants. In: Riesen, A.H., and Kinder, E.F. (eds.), *Postural Development of Infant Chimpanzees: A Comparative and Normative Study Based on The Gesell Behavior Examination*. Yale University Press, New Haven, CT.

Riopelle, A.J. (1963). Growth and behavioral changes in chimpanzees. *Z. Morph. Anthrop.* 53:53–61.

Ruppenthal, G.C., Arling, G.L., Harlow, H.F., Sackett, G.P., and Suomi, S.J. (1976). A 10-year perspective of motherless-mother monkey behavior. *J. Abnorm. Psychol.* 85:341–349.

Russon, A.E. (1990). The development of peer social interaction in infant chimpanzees: Comparative social, Piagetian, and brain perspectives. In: Parker, S.E., and Gibson, K.P. (eds.), *Language and Intelligence in Monkeys and Apes: Comparative Developmental Perspectives*. Cambridge University Press, Cambridge, pp. 379–419.

Sackett, G.P. (1972). Exploratory behavior of rhesus monkeys as a function of rearing experiences and sex. *Dev. Psychol.* 6:260–270.

Sackett, G.P., Bowman, R.E., Meyer, J.S., Tripp, R.L., and Grady, S.S. (1973). Adrenocortical and behavioral reactions by differentially raised rhesus monkeys. *Physiol. Psychol.* 1:209–212.

Sackett, G.P., Novak, M.F.S.X., and Kroeker, R. (1999). Early experience effects on adaptive behavior: Theory revisited. *Ment. Retard. Dev. Disabil. Res. Rev.* 5:30–40.

Sackett, G.P., Ruppenthal, G.C., and Davis, A.E. (2002). Survival, growth, health, and reproduction following nursery rearing compared with mother rearing in pigtailed macaques (*Macaca nemestrina*). *Am. J. Primatol.* 56:165–183.

Schneider, M.L., and Suomi, S.J. (1992). Neurobehavioral assessment in rhesus monkey neonates (*Macaca mulatta*): Developmental changes, behavioral stability, and early experience. *Infant Behav. Dev.* 15:155–177.

Schwandt, M.L. (2002). The ontogeny of positional behavior in captive chimpanzees (*Pan troglodytes*). Ph.D. Dissertation, Arizona State University, Tempe.

Smith, A.H., Butler, T.M., and Pace, N. (1975). Weight growth of colony-reared chimpanzees. *Folia Primatol.* 24:29–59.

Spijkerman, R.P. (1987). Behavior development of infant chimpanzees with the mother and in peer groups. *Annual Report of Primate Center TNO*, Netherlands, pp. 309–312.

Sugardjito, J., and van Hooff, J.A.R.A.M. (1986). Age-sex differences in the positional behavior of the Sumatran orang-utan (*Pongo pygmaeus abelii*) in the Gunung Leuser National Park, Indonesia. *Folia Primatol.* 47:14–25.

Suomi, S.J., Collins, M.L., and Harlow, H.F. (1973a). Effects of permanent separation from mother on infant monkeys. *Dev. Psychol.* 9:376–384.

Suomi, S.J., Eisele, C.D., Grady, S.A., and Tripp, R.L. (1973b). Social preferences of monkeys reared in an enriched laboratory social environment. *Child Devel.* 44:451–460.

Suomi, S.J., Collins, M.L., Harlow, H.F., and Ruppenthal, G.C. (1976). Effects of maternal and peer separations on young monkeys. *J. Child Psychol. Psychiatry* 18:101–112.

Thompson, M.A., Bloomsmith, M.A., and Taylor, L.L. (1991). A canine companion for a nursery-reared infant chimpanzee. *Lab. Primate Newslett.* 30(2):1–4.

Tomasello, M., Kruger, A.C., and Ratner, H.H. (1993). Cultural learning. *Behav. Brain Sci.* 16:495–552.

Tomasello, M., Call, J., Nagell, K., Olgain, R., and Carpenter, M. (1994). The learning and use of gestural signals by young chimpanzees: A trans-generational study. *Primates* 35(2):137–154.

Vancatova, M., Vancata, V., Jerabkova, Z., Zlamalova, H., Skrivankova, J., and Janecek, J. (1999). A longitudinal study of the growth and development of captive chimpanzees and orangutans. *Variability Evol.* 7:31–45.

Vilensky, J.A., and Gankiewicz, E. (1989). Early development of locomotor behavior in vervet monkeys. *Am. J. Primatol.* 17:11–25.

Walters, J.R. (1987). Transition to adulthood. In: Smuts, B.B., Cheney, D.L., Seyfarth, R.M., Wrangham, R.W., and Struhsaker, T.T. (eds.), *Primate Societies.* University of Chicago Press, Chicago, pp. 358–369.

Warner, A., Bitnoff, J., and McDonald, K. (1997). Responsive care update from the Oakland zoo. *Chimpanzoo Conference Proceedings*, pp. 64–67.

Watts, E.S., and Gavan, J.A. (1982). Postnatal growth of nonhuman primates: The problem of the adolescent growth spurt. *Hum. Biol.* 54:53–70.

Wells, J.P., and Turnquist, J.E. (2001). Ontogeny of locomotion in rhesus macaques (*Macaca mulatta*): II. Postural and locomotor behavior and habitat use in a free-ranging colony. *Am. J. Phys. Anthropol.* 115:80–94.

Winkler, L.A. (1996). Appearance of ossification centers of the lower arm, wrist, lower leg, and ankle in immature orangutans and chimpanzees with an assessment of the relationship of ossification to dental development. *Am. J. Phys. Anthropol.* 99:191–203.

SECTION FOUR

Introduction to Section 4: Nursery Care Methodology and Testing Techniques for the Future

This section includes eight chapters, 17–24, on husbandry and testing methodology. Chapter 17 presents basic veterinary requirements of a nonhuman primate nursery, developed for squirrel monkeys but pertinent for rearing any primate species. Chapter 18 describes methodology for rearing neonates and infants that are at high risk for medical problems. One obvious need for nursery rearing involves methods for feeding neonates and young infants. Chapters 18 and 19 detail how to perform this critical function while minimizing required human time and effort.

Until recently, most biochemical assays required blood samples, which had to be acquired by invasive techniques. For many assays it is now possible to use saliva, which can be collected noninvasively. Two chapters (20 and 21) detail collection and assay techniques for salivary studies of infant primates. Perhaps the most basic behavioral data relating to health and well-being of primate infants concern motor activity and its diurnal patterning. Chapter 22 presents a novel method for automatically collecting continuous motor activity measures under minimally invasive conditions. Chapter 23 describes noninvasive imaging techniques for use in all areas of the developmental sciences. Such techniques are becoming increasingly essential to state-of-the-art care and scientific investigation of nonhuman primates.

Chapter 24 details the methodology for performing maternal–fetal catheter blood sampling and physiological assessment. We do not present this chapter as an example of a polished, simple-to-use methodology. Rather, we present it as an example of state-of-the-art methodology for measuring longitudinal fetal development, and as a stimulus for researchers and engineers to develop less invasive and more easily instrumented techniques for studying fetal development and pregnancy processes.

CHAPTER SEVENTEEN

Squirrel Monkeys as an Example of Primate Nursery Medicine

Alan G. Brady, Susan V. Gibson, Lawrence E. Williams, and Christian R. Abee

1. INTRODUCTION

Providing nursery care for infant primates is one of the most challenging aspects of maintaining primates in captivity. Nursery infants must be protected from and treated for the many health problems to which they are vulnerable. At the same time, their developmental needs must be met through proper nutrition, environmental enrichment, and other strategies.

Primates are notoriously bad patients, and primate infants are arguably even worse. They can deteriorate rapidly when ill or injured, and they are susceptible to a variety of secondary problems such as electrolyte disturbances, malnutrition, and infection once a primary health problem occurs. For this reason, the emphasis in this chapter is on preventive medicine: creating a nursery that is conducive to preventing health problems before they start. Essentials of such a nursery include a healthy environment, qualified and motivated staff, good nutrition, and a program for the prevention of infectious disease.

2. PREVENTIVE MEDICINE CONCEPTS FOR THE NURSERY

2.1. Nursery Design

Primate nursery facilities vary widely according to amount of use, species involved, and other factors. Some amount to little more than an infrequently used part of a housing or clinical facility while others are dedicated, purpose-designed facilities that are in constant use. Certain components are useful in any primate nursery and may be developed, more or less, depending on functional requirements, budget, and other factors. These components include infant housing (which may include housing for dams and foster dams), separate storage areas for diet, medication, and supplies, hand-feeding area, space for updating and storing animal records, and a break room for staff.

Specifications for infant housing are beyond the scope of this chapter but may be found in other references (Anderson, 1986). Food and supply storage should be convenient to work areas and limited to quantities that will be used quickly. Long-term storage should be located elsewhere. Refrigerated storage for medications or chemicals should be separate from storage for infant formula and diets to avoid contamination (National Research Council, 1996). Records of each animal's growth and development, health status, treatments, and related documents are an important part of nursery operations. Space should be allowed for completion and storage of records, and for ready access to the records at all times, as required by the Animal Welfare Act of the United States (USDA, 1982). A separate break room for staff is important because the ready transmission of a variety of diseases between humans and primates makes exclusion of human food from the nursery a necessity (Jordan *et al.*, 1985; National Research Council, 1996). Staff need a break room particularly during hours when the institution's cafeteria or canteen is closed.

If infants are dependent on supplemental heating devices or other electrically powered equipment, an auxiliary power source should be available in the event of a power outage (National Research Council, 1996).

2.2. Nursery Quality Control

Record-keeping in the primate nursery should extend beyond records for individual animals. Cumulative statistics for nursery morbidity and mortality, growth measurements, reunion of infants with dams, fostering, and other measures of infant progress, as well as necropsy results for infants that die while in nursery care, can provide useful information for both husbandry and research purposes.

2.3. Nursery Staffing

The primate nursery can present special challenges to the caretaking staff. First, personnel often become attached to the infants in their care and may experience emotional difficulty when the animals sicken and die. Second, nursery staff often must work extra shifts to accommodate extended feeding and care schedules. And third, staff must take strict measures to avoid the transmission of disease to and from their infant charges. In this regard, an occupational health program is highly desirable and, in some institutions, required (National Research Council, 1996). The details of such a program are described in the National Research Council document. For animals that have a distinct delivery season, such as squirrel monkeys (*Saimiri*), an internship program using college students can be an effective way to augment the nursery staff when needed. The Squirrel Monkey Breeding and Research Resource (SMBRR) at the University of South Alabama recruits interns through contacts at regional universities and colleges. The goal of the program is to provide high-quality care for nursery infants and to educate student interns in the basics of primatology and animal care. In addition to an orientation manual and one-on-one training with supervisors, veterinary staff and a comparative psychologist, students attend weekly seminars on topics related to primatology and primate care and participate in nursery medical rounds.

2.4. Nursery Hygiene, Disinfection, and Vermin Control

The primate nursery is an environment in which infectious diseases can be easily introduced and readily propagated. Infants are immunologically naive and may enter the nursery with their health compromised by either illness or injury. Some have suggested that nursery-reared, formula-fed primate infants demonstrate altered immune function compared with maternally reared infants (Coe *et al.*, 1992; Lubach *et al.*, 1995), though this possibility is not universally accepted (Sackett *et al.*, 2002). When compromised infants are living together in close proximity and handled by the same caregivers, infectious diseases can spread rapidly (Scimeca and Brady, 1990). Thus, cleaning and disinfection are particularly important in the nursery. Any items used in caring for the infants must conform to the "3D Principle": they must be either ***disposable*** after a single use, ***duplicated*** for each infant (not shared), or ***disin-fected*** between each animal.

There are many disinfectants on the market. Classes of disinfectants include iodines and phenols, chlorine compounds, quaternary ammonia compounds, and glutaraldehydes. The use of 70% ethanol and other alcohol-based disinfectants is not recommended because they can vaporize before they have sufficient contact time to be effective (Ingraham and Fleischer, 2003). In general, disinfectants that are labeled for use in food preparation areas may be used in primate nurseries. A common and effective disinfectant is sodium hypochlorite solution (bleach), though it must be replaced regularly because its potency declines rapidly over time. It is important to note that organic material renders many disinfectants ineffective, so that soiled surfaces should be precleaned prior to disinfection. In addition, most disinfectants require set contact times to be effective. In all cases, label directions for precleaning and contact time should be followed (Huber, 1988; Ingraham and Fleischer, 2003). Finally, to prevent chemical burns and other skin problems in sensitive infants, disinfected items and surfaces should be rinsed thoroughly before infants have any contact with them.

Also important in the prevention of disease in the nursery is the control of insects and rodents, which can transmit organisms such *Yersinia enterocolitica* and *Yersinia pseudotuberculosis* to susceptible infants (Brady and Morton, 1998; Toft and Eberhard, 1998). Basic concepts of pest management include sealing of entry routes, elimination of nesting sites, and careful management of animal food and waste to eliminate sources of food for pests.

2.5. Nursery Records and Quality Control

As noted in Sections 2.1 and 2.2, thorough and accessible animal records are vitally important not only to ensure quality medical care but also to document compliance with regulatory, funding agency, and institutional requirements (USDA, 1982; National Research Council, 1996). Individual records may take the same form as adult medical records, but should also include data on feeding (Fig. 17–1) and growth (Fig. 17–2).

It is important to have procedures in place for the ongoing review of nursery progress. Cumulative statistics for measures of infant progress can provide valuable information on husbandry and management practices and primate biology (Anderson, 1986). Nursery statistics that are particularly useful include morbidity, mortality, mean daily weight gain, intake of formula, and cause of death.

Infant Weight, Feeding Chart & Log
Nursery Treatments & Comments

ID: ______ Sex: ______ Birth Date: ________ Dam: ________ Foster: ________
Birth Wt: ______ Entered Nur: ______ From Cage: ________

		Sun	Mon	Tue	Wed	Thu	Fri	Sat
	Date							
	Weight							
	Glucose							
	Temp							
Feeding Record (ml's)	7am							
	8am							
	9am							
	10am							
	11am							
	12noon							
	1pm							
	2pm							
	3pm							
	4pm							
	5pm							
	6pm							
	7pm							
	8pm							
	9pm							
	10pm							

Nursery Treatments & Comments

Does not replace Treatment Summary Sheet

Each new infant receives Rocephin @ 24hrs \ Bactrim @ 48hrs \ Complete Work up @ 72 hrs - mark on sheet below.

		Initials & Time
SUN		
MON		
TUE		
WED		
THU		
FRI		
SAT		

Strike through any unused days in both tables.

Figure 17–1. Form for recording infant feeding volumes and observations.

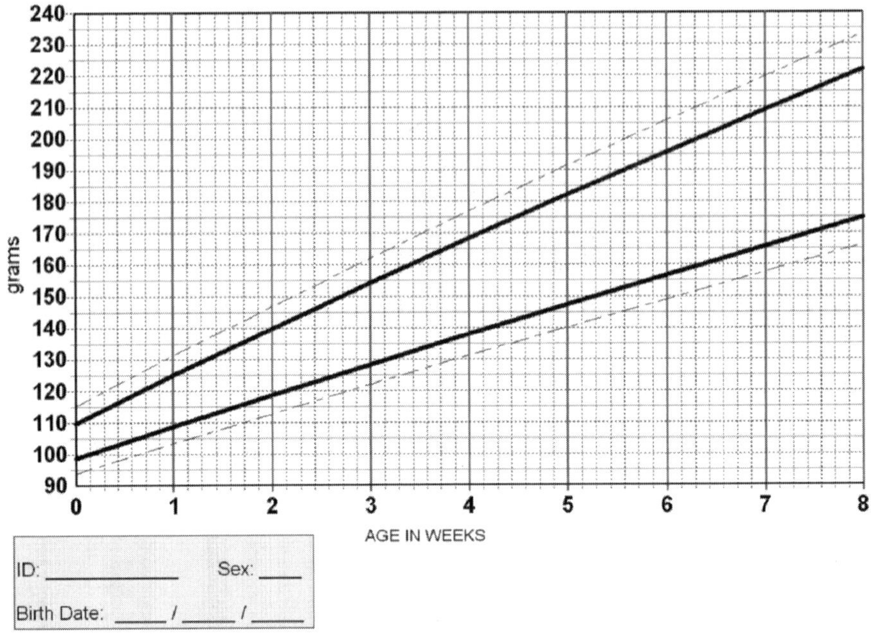

Figure 17–2. Chart for monitoring weight gain of infant *Saimiri*. Nursery staff are instructed to get veterinary consultation if infants fall outside the 10th and 90th percentile (heavy black lines) for weight gain.

3. THE SQUIRREL MONKEY NURSERY

Squirrel monkey infants present a special challenge in the nursery environment. Maternal rejection of infants is common (L.E. Willams, unpublished observations).

3.1. Admission

Upon admission to the nursery, infants should be examined for health problems and given appropriate treatment. In our experience, infants with birth weights of <90 g are less likely to survive than those of higher weight, and infants that weigh <80 g rarely survive (Williams *et al.*, 1994). Infants that are lethargic and/or hypothermic should receive supportive care and diagnostic testing (Table 17–1) in addition to therapy specific for their illness or injury. Because bacterial infections are common in the postadmission period (Scimeca and Brady, 1990; Gibson *et al.*, 1995), a 3- to 5-day course of prophylactic antibiotics (e.g., sulfamethoxazole-trimethoprim, amoxicillin-clavulanic acid, or ceftriaxone) should be considered. Selection of antibiotics should be based on sensitivity testing of common pathogenic bacterial isolates in the facility.

3.2. Feeding and Nutrition

Squirrel monkeys require a diet of high caloric density (Abee and Henrickson, 1986). To meet this need, squirrel monkey infants may be fed either milk replacers designed for nonhuman primates (Zoologic 20/14 Milk Matrix, Pet Ag Inc., Hampshire, IL) or a high calorie human formula. Whatever the choice, the formula must be highly palatable to encourage intake.

On admission to the nursery, the first feeding for a healthy neonate should consist of sterile 5% dextrose solution (Kaack *et al.*, 1979) to get the infant accustomed to handfeeding while minimizing the risk of pneumonia due to aspiration. The next feeding should consist of a 1:1 mixture of 5% dextrose and formula. Subsequent feedings should be of formula alone (Brady, 2000). Neonates should be fed hourly for at least 15 hr daily. Glucose testing (see Section 3.4.5) prior to the first feeding of the day may be used to determine if intake is adequate. For most infants, the feeding schedule may be reduced at 1–2 weeks of age. Transition to self-feeding with a bottle and rubber nipple may begin at

Table 17–1. First Aid Procedures for Hypothermic or Lethargic Squirrel Monkey Infants

1. Perform a glucose test, get a body weight, and take the animal's axillary temperature; if the temperature is <100/°F, put the animal on a hot water bottle filled with warm water; beware of scalding with hot water
2. Rapidly take note of the animal's current status: Blood in cage? Epileptic? Recently moved? Recently anesthetized? Etc.
3. Call or page the veterinarian and report information
4. If the glucose test reads <50, give 1 ml of sterile 20% dextrose by stomach tube
5. Give 5 ml lactated Ringers solution, warmed to body temperature, subcutaneously
6. If the infant is unconscious, place it on a warm (not hot) water bottle or warming pad and cover it; turn the infant over every 20 min

6–8 weeks of age. Solid food consisting of nutritionally balanced monkey biscuits softened with fruit juice may be offered at 8 weeks. (Other moistening agents should be avoided because they may neutralize vitamin C in the biscuits.) During the first 4 weeks, the infant should be weighed and its temperature checked daily. Bolivian, Guyanese, and Peruvian squirrel monkeys may be expected to lose 3–10 g during their first week of life, whereas Columbian squirrel monkeys may not experience such loss (Kaack *et al.*, 1979). Growth data for nursery-reared versus dam-reared infants in the SMBRR are shown in Fig. 17–3.

3.3. Medical Procedures

With proper technique, most basic examination procedures performed on adult primates can be performed on squirrel monkey infants.

3.3.1. Temperature Measurement

Although rectal probe may provide the most accurate reading of body temperature, in small primates the size of the probe could result in rectal

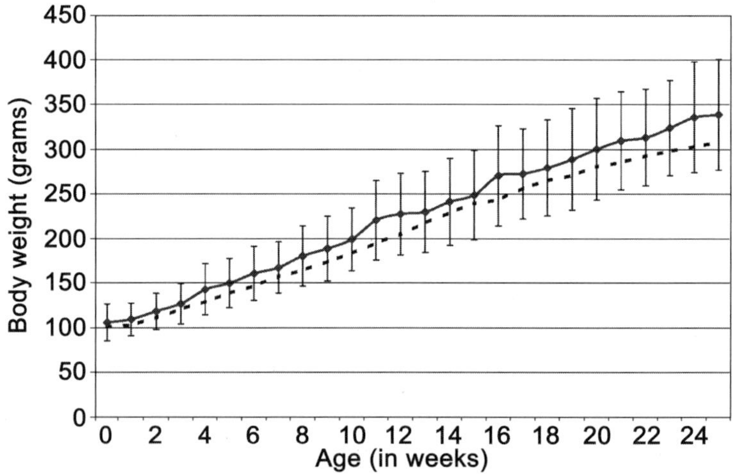

Figure 17–3. Graph showing growth of infants maintained with dams (solid line) and infants maintained in the nursery (broken line).

trauma or distress. Thus, the recommended measurement for all infants of small primate species is axillary temperature, which has been found to correlate well with rectal temperature (Brady, 2000). The temperature probe is placed in the axillary space with the arm adducted against it for the required amount of time. Tympanic thermometers, which measure temperature of the eardrum, may not be accurate in *Saimiri* infants (Brady, 2000).

3.3.2. *Auscultation*

A neonatal stethoscope (Littman Infant® stethoscope, 3M Corporation, Oakdale, MN) may be used to listen for internal body sounds. In *Saimiri* and other small infants, however, lung sounds and subtle changes in heart sounds may be difficult to detect.

3.3.3. *Blood Collection*

For an adult primate, the maximum safe volume of blood that may be sampled at one time is 1% of the animal's body weight (Fortman *et al.*, 2002). Thus, for a newborn squirrel monkey of average weight, the amount would be about 1 ml, though we recommend taking smaller volumes, especially in debilitated animals. In practical terms, the amount that can be sampled from peripheral vessels will also probably be less. Intravenous sampling and injection sites for infants include the femoral vein at the femoral triangle, the lateral tail veins at the base of the tail, and the saphenous veins on the caudal part of the lower leg.

3.3.4. *Anesthesia*

Anesthesia for adult squirrel monkeys has been described (Brady, 2000). For infants, injectable anesthesia is not recommended because it is difficult to control the depth of anesthesia. For short procedures, mask induction and maintenance with isoflurane anesthesia is recommended. Depth of anesthesia can change rapidly in infants and must be monitored closely.

3.4. Common Health Problems in Squirrel Monkey Infants

Squirrel monkey infants are admitted to the nursery with a variety of medical problems. Some dams, particularly primigravida dams, reject apparently healthy infants. Most *Saimiri* infants that are <2 months of age when presented for medical treatment have a constellation of clinical signs that include lethargy, hypothermia, hypoglycemia, and dehydration (Gibson *et al.*, 1995).

3.4.1. Prematurity

Premature delivery is relatively common in squirrel monkeys. According to one study, infants with birth weights of <90 g should be considered to be premature (Price *et al.*, 1972). Low birth weight (<100 g) is considered to be the primary predictor for infant mortality (Gibson *et al.*, 1995). In addition to low birth weight, premature infants typically have thin hair coats, are chronically hypothermic, and have limited reflexes for swallowing and grasping. Treatment is limited to supportive care. Euthanasia should be considered for markedly premature infants.

3.4.2. Bacterial, Yeast, and Viral Infections

It has been reported that almost one-third of squirrel monkey deaths in the first year are caused by infection (Gibson *et al.*, 1995). Infections may manifest as pneumonia or meningitis and may result in nursery epidemics (Scimeca and Brady, 1990). In infant squirrel monkeys, these infections often progress rapidly and there may be no sign of illness before the infant is found dead. The rapid course of these infections highlights the need for a good prevention program, including strict attention to nursery hygiene (see Section 2.3) and the use of prophylactic antibiotics upon admission to the nursery (see section 3.1). Known bacterial disease agents include *Escherichia coli*, which was implicated in a nursery epidemic that resulted in seven deaths (Scimeca and Brady, 1990). An unfortunate consequence of treatment with broad-spectrum antibiotics is *Candida* yeast infection, which can be diagnosed and treated as described by Brady and Morton (1995).

3.4.3. Herpesvirus Saimiri *Type 1 (HVS-1)*

Squirrel monkey infants are commonly infected with HVS-1, but in this genus the virus generally progresses to disease only in animals with preexisting health problems. When illness does occur, infants may develop oral ulcers that make it difficult to eat. Treatment is generally limited to supportive care to maintain fluid and electrolyte balance and caloric intake (Brady and Morton, 1995).

3.4.4. Trauma

Squirrel monkey infants are frequently admitted to the nursery with some form of trauma, including tail injury (Abee and Henrickson, 1986; Gibson *et al.*, 1995; Williams *et al.*, 1995) and skull fractures (Lee *et al.*, 1980; Abee and Henrickson, 1986; Scimeca *et al.*, 1988; Gibson *et al.*, 1995). In one study, 51 live births in a 1-year period ended in 15 neonatal deaths and of these, 5 involved skull fractures (Scimeca *et al.*, 1988). At necropsy, the skull fractures were associated with subdural hemorrhage and meningitis.

Infants with skull fractures may present with a variety of neurological deficits, but often are lethargic and demonstrate decreased consciousness. Treatment is directed to reducing damage from associated edema through the use of diuretics and steroids, but is frequently unrewarding. In general, injured infants may be treated with methods similar to those used for adults, but extra attention should be paid to infection control and supportive care. Analgesics to relieve pain associated with trauma should be administered with care. High doses of analgesics may sedate infants to a point that they are unable to swallow, making them susceptible to aspiration and hypoglycemia.

3.4.5. Hypoglycemia

Squirrel monkey infants are predisposed to hypoglycemia (Abee and Henrickson, 1986; Anderson, 1986; Brady, 2000) and neonates are especially susceptible (Brady *et al.*, 1990). Hypoglycemia is a common complication of other health problems such as maternal rejection and trauma (Brady *et al.*, 1990), and is almost always due to inadequate food intake. Clinical signs include lethargy, hypothermia, bradycardia, and, in advanced stages, seizures. The diagnosis is confirmed by measuring blood glucose concentration. This can be done by a clinical pathology laboratory, but in the interest of rapid diagnosis, which is essential for treatment to be effective, squirrel monkey nurseries should be equipped with one of the many blood glucose meters for diabetics (AccuCheck®, Boerhinger Mannheim Corp., Indianapolis, IN) or another validated cage-side rapid analyzer. If conscious, hypoglycemic infants may be treated with oral administration of glucose or sucrose solution (Gatorade®, Gatorade Corporation, Chicago, IL) or orange juice. Unconscious animals should be given 1 ml of sterile 20% dextrose solution by stomach tube. (The solution should be sterile to reduce the possibility of bacterial pneumonia if a small amount is aspirated.) The use of more concentrated glucose solutions (orally or parenterally) is not recommended. The procedure for passing a stomach tube in squirrel monkeys has been described (Brady, 2000). For infant squirrel monkeys, we recommend orogastric rather than nasogastric intubation because of the small size of the nares.

3.4.6. Hypothermia

Saimiri infants are predisposed to hypothermia, presumably because of their small size, large body surface area, and minimal body fat (Gallina and Ausman, 1979). The normal temperature range for a squirrel monkey infant is 35–37°C, with temperatures at the lower end of the range for newborns. In awake, restrained adults, normal temperature is 38–39.5°C. *Saimiri* infants typically respond to illness or other compromise by becoming hypothermic rather than feverish. Care must be taken to conserve body heat in these animals, especially with neonates and infants that are sick, injured, or anesthetized. Supplemental heating devices such as warmers (Thermocare®, Incline Village, NV) or warm-water recirculating blankets (Gaymar T Pump®, Gaymar Corporation, Orchard Park, NY) should be used with these infants. The use of electric heating pads is not recommended owing to the risk of burns, hyperthermia, and electrical shock (Brady, 2000).

3.4.7. Failure to Thrive Syndrome

A failure to thrive (FTT) syndrome, which has some parallels to human FTT, has been identified in squirrel monkey infants. Clinical signs of FTT include failure to gain weight (Fig. 17–4) and susceptibility to opportunistic infections such as *Cryptosporidia*, *Candida*, and HVS. In one squirrel monkey colony, 3.2% of liveborn infants had FTT (Gibson *et al.*, 1998). Of these, 90% had been rejected by the dam, admitted to the nursery, and/or sustained multiple episodes of trauma. At necropsy, FTT infants generally have a small or absent thymus.

3.4.8. Synthetic Fiber Gastric Foreign Bodies

Surrogate dams, made of rolled towels or fur-like fabric, are commonly used in squirrel monkey nurseries because infants are comforted by grasping an object that roughly resembles the mother's body (Ricker *et al.*, 1985). This surrogate is usually a rolled piece of cloth 4–6 cm in diameter and 12–14 cm in length. Care must be taken in selection of the fabric.

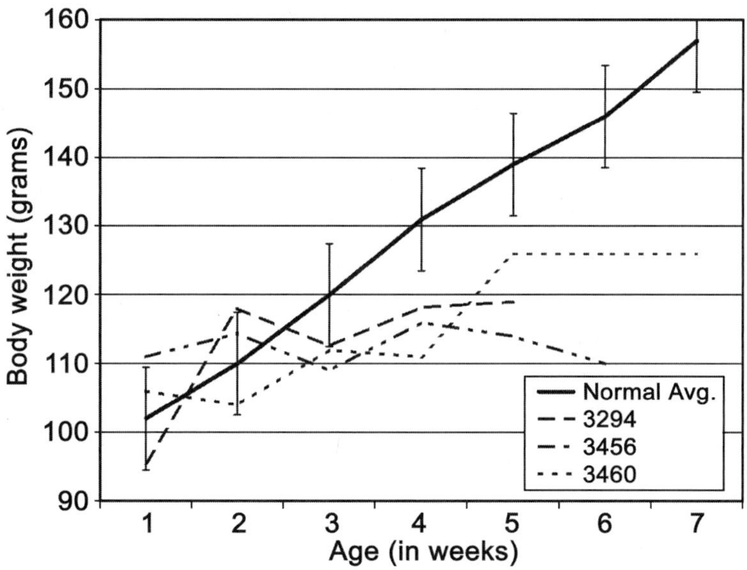

Figure 17–4. Growth of failure to thrive (FTT) infants (broken lines) versus non-FTT infants.

When the SMBRR experimented with artificial sheep skin pads that had long fibers, three infants developed the habit of ingesting the fibers, which formed gastric foreign bodies resembling trichebezoars. All three infants required gastrotomy to remove the foreign bodies and one died during surgery. The SMBRR now uses pads with short fibers. This case illustrates the importance of a preventive approach in maintaining the health of primate infants in a nursery environment.

4. CONCLUSION

Nursery care of nonhuman primates begins with a good husbandry program that addresses both basic requirements common to all primates, namely, good sanitation, housing and enrichment, and the special needs of infants, such as frequent feeding and supplemental warming. A veterinary care program that builds on good husbandry by also emphasizing disease prevention completes a high-quality program of nursery care. No primate nursery is likely to be successful without this preventive approach to care.

REFERENCES

Abee, C.R., and Henrickson, R. (eds.) (1998). *Nonhuman Primates in Biomedical Research: Diseases.* Academic Press, San Diego.

Anderson, J.H. (1986). Rearing and intensive care of neonatal and infant nonhuman primates. In: Benirschke, K. (ed.), *Primates: The Road to Self-Sustaining Populations.* Springer-Verlag, New York, pp. 747–762.

Brady, A.G. (2000). Research techniques for the squirrel monkey (*Saimiri* sp.). *ILAR J.* 41:10–18.

Brady, A.G., and Morton, D.G. (1998). In: Abee, C.R., and Henrickson, R. (eds.), *Nonhuman Primates in Biomedical Research: Diseases.* Academic Press, San Diego, pp. 363–414.

Brady, A.G., Williams, L.E., and Abee, C.R. (1990). Hypoglycemia of squirrel monkey neonates: Implications for infant survival. *Lab. Anim. Sci.* 40: 262–265.

Coe, C.L., Lubach, B.R., Schneider, M.L., Dierschke, D.J., and Ershler, W.B. (1992). Early rearing conditions alter immune responses in the developing infant primate. *Pediatrics* 90:505–509.

Fortman, J.D., Hewett, T.A., and Bennett, B.T. (2002). Experimental methodology. In: *The Laboratory Primate.* CRC Press, Boca Raton, FL, pp. 147–206.

Gallina, D.L., and Ausman, L.M. (1979). Selected aspects of the metabolic behavior of the squirrel monkey. In: Hayes, K.C. (ed.), *Primates in Nutritional Research.* Academic Press, New York, pp. 225–247.

Gibson, S.V., Williams, L.E., Brady, A.G., and Abee, C.R. (1995). Morbidity and mortality during the first year of life in colony-born squirrel monkeys (*Saimiri* spp.). *Am. J. Primatol.* 36:124 (abstract).

Gibson, S., Williams, L., Brady, A., and Abee, C. (1998). Failure to thrive syndrome in squirrel monkey infants (*Saimiri* spp.). *Am. J. Primatol.* 45: 181 (abstract).

Huber, W.G. (1988). Chemotherapy of microbial, fungal, and viral diseases. In: Booth, N.H., and McDonald, L.E. (eds.), *Veterinary Pharmacology and Therapeutics*, 6th ed. Iowa State University Press, Ames, IA, pp. 765–784.

Ingraham, A.I., and Fleischer, T.M. (2003). Disinfectants in laboratory animal science: What are they and who says they work? *Lab. Anim.* 32(1):36–40.

Jordan, H.V., van Houte, J., and Russo, J. (1995). Transmission of the oral bacterium *Streptococcus mutans* to young *Macaca fascicularis* monkeys from human nursery attendants. *Arch. Oral Biol.* 30:863–864.

Kaack, B., Walker, L., and Brizzee, K.R. (1979). The growth and development of the squirrel monkey (*Saimiri sciureus*). *Growth* 43:116–135.

Lee, K.J., Hughes, H.C., and Lang, C.M. (1980). Skull fractures in infant squirrel monkeys (*Saimiri sciureus*). *Lab. Anim. Sci.* 30:1006–1008.

Lubach, G.R., Coe, C.L., and Ershler, W.B. (1995). Effects of early rearing environment on immune responses of infant rhesus monkeys. *Brain Behav. Immun.* 9:31–46.

National Research Council. (1996). Animal environment, housing, and management. In: *Guide for the Care and Use of Laboratory Animals.* National Academy Press, Washington, D.C., p. 36.

Price, R.A., Anver, M.R., and Garcia, F.G. (1972). Simian neonatology. *Vet. Pathol.* 9:301–309.

Ricker, R.B., Brady, A.G., and Abee, C.R. (1985). Nursery rearing strategies for captive born squirrel monkeys (*Saimiri sciureus*). *Lab. Anim. Sci.* 35:529–530.

Sackett, G.P., Ruppenthal, G.C., and Davis, A.E. (2002). Survival, growth, health, and reproduction following nursery rearing compared with mother rearing in pigtailed macaques (*Macaca nemestrina*). *Am. J. Primatol.* 56:165–183.

Scimeca, J.M., and Brady, A.G. (1990). Neonatal mortality in a captive breeding squirrel monkey colony associated with an invasive *Escherichia coli*. *Lab. Anim. Sci.* 40:546–547.

Scimeca, J., Williams, L., Brady, A., and Abee, C. (1988). Skull fractures and harassment as a cause of neonatal colony mortality in squirrel monkeys (*Saimiri sciureus* and *S. boliviensis*). *Am. J. Primatol.* 14:445 (abstract).

Toft, J.D., and Eberhard, M.L. (1998). Parasitic diseases. In: Bennett, B.T., Abee, C.R., and Henrickson, R. (eds.), *Nonhuman Primates in Biomedical Research: Diseases.* Academic Press, San Diego, pp. 111–205.

United States Department of Agriculture (USDA). (1982). *Animal Welfare*, APHIS 41-35-059. Washington, D.C.

Williams, L.E., Brady, A.D., Gibson, S.V., and Abee, C.R. (1994). Reproductive performance in a breeding colony: Birth weight and perinatal mortality. *Am. J. Primatol.* 33:211 (abstract).

Williams, L.E., Brady, A.G., and Abee, C.R. (1995). A multiple year analysis of fight wounds in socially housed squirrel monkeys (*Saimiri*). *Am. J. Primatol.* 24:141 (abstract).

CHAPTER EIGHTEEN

Nursery Care of At-Risk Nonhuman Primates

Gerald C. Ruppenthal and Gene P. Sackett

1. INTRODUCTION

The Infant Primate Research Laboratory (IPRL) at the University of Washington was established in 1970 to support scientific and medical investigations that require infant nonhuman primates as subjects. In a short time an "Infant Save" program was established to care for Primate Center Colony infants with a low survival probability if they remained with their mothers in their natal social groups. Neonates qualified for the program based on low birth weight and/or prematurity, inadequate maternal care or rejection, injury, or failure to thrive. All IPRL infants, including those assigned to almost all research projects, received daily socialization in play groups. In addition, the Infant Save monkeys were utilized to develop reflex, motor, and cognitive assessment methods and to investigate aspects of feeding, care, and housing to reduce morbidity and mortality. A major goal was to develop methods that did not confound experimental studies.

The primary species used in IPRL research and in the Infant Save program is the pigtailed macaque (*Macaca nemestrina*). To a lesser extent, in order of overall numbers, the IPRL has also reared long-tailed macaques (*M. fascicularis*), baboons (*Papio* sp.), and rhesus macaques (*M. mulatta*).

Since its inception, the IPRL has undergone several changes in rearing techniques. In this chapter we will describe these changes as they relate to emerging standards for health and postnatal development, specifically the reattainment of birth weight, age at self-feeding, administration of medication, survivorship, and mortality. Data on pigtailed macaques are examined by four birth-weight percentiles: 1–10 (low birth weight), 11–50 (middle–low), 51–90 (middle–high), and 91–100 (high birth weight).

We will also present empirical data and nonexperimentally derived information that has proven useful in rearing macaques in our laboratory. For more than three decades we have expended much effort on developing and maintaining our animal records system. These records have proved invaluable for retrospective analysis of the impact on morbidity and mortality of various rearing techniques, the effects of conducting various developmental assessments, and the influence of contingent and noncontingent stimulation and human handling on behavioral development (Sackett *et al.*, 2002). Many of the developmental assessments originally were devised to support research projects and then were further developed and included in the IPRL research protocols for both husbandry and research use (Ruppenthal and Sackett, 1992). This effort generated information on normal and abnormal development, as well as clinical screening data used to avoid assigning abnormal subjects to research projects.

In addition to the Infant Save program, the IPRL has conducted a number of investigations on aspects of premature delivery. These studies have produced a significant percentage of infants in the lower percentiles of birth-weight and gestational-age distributions. Particular effort has been directed at reduction of morbidity and mortality in this class of infants, and much of the information in this chapter deals with problems arising from these at-risk individuals.

2. CRITICAL FACTORS IN CARE OF HIGH-RISK NEONATES

The IPRL maintains a permanent staff of 24-hr personnel every day of the year. This is probably the most critical aspect of successful high-risk nursery care. It is necessary for timely delivery of both routine and unscheduled clinical procedures, as well as for recognizing problems as soon as they occur. For example, high-risk neonates initially require feeding every 2 hr to avoid dehydration and hypoglycemia. Even normal perinates can become hypoglycemic rapidly. In a pilot study to examine plasma glucose levels of newborns after various fasting times, we found that glucose levels had plummeted to nearly immeasurable levels after 4 hr of fasting. Besides being life threatening, this may have long-term implications: monkeys that were made experimentally hypoglycemic for ~10 hr soon after birth had long-term deficits in learning performance and possible emotional problems (Schrier *et al.*, 1983, 1990).

The IPRL includes a maternity room, where pregnant females that will deliver infants to be reared in the nursery are housed for both clinical and experimental purposes. Females nearing term are observed over a TV system at least once every half hour from 6 pm until 6 am, when almost all deliveries occur. Newborns are usually separated from their mothers, examined, and brought to the nursery incubator within minutes of birth. This procedure also allows for collection of blood, placenta, and other tissues.

2.1. Hypothermia and Respiratory Abnormalities

A depressed body temperature for the first few hours after birth is normal in perinates and evidently has no adverse long-term sequelae (Ruppenthal *et al.*, 1983). Pregnant females lose body temperature for 60–90 min before giving birth, and this probably leads to depressed temperatures in their perinates. The infants are born at an average of 95.0°F, continue to experience temperature declines for 30–40 min, then gradually rebound in temperature to normal infant standards of 98.5–100.0°F within 2–4 hr after birth.

Normal, healthy pigtailed macaque neonates can usually maintain body temperature with incubator support. After the infants leave the incubator, usually by day 4–5, body temperature can be maintained with a diaper and heating pad at an ambient room temperature of 82–84°F. Pigtailed macaques experience more unspecified upper respiratory infections and hypothermic episodes than the other species studied in the

IPRL. The latter can maintain temperature homeostasis at lower ambient temperatures of 76–80°F.

Hypothermia after the 30-day neonatal period is an indication of at-risk health status and steps should be taken to correct the problem. This is often the result of an infection process or inadequate ambient temperature. Low-birth-weight infants and younger infants are more likely to experience late-onset hypothermia and to need temperature support.

Shivering in infants is usually an indication of a medical problem such as an infection. Infants do not use shivering thermogenesis when hypothermic, as they have a paucity of yellow fat for production of heat (Newell-Morris, 1979). Rather, they probably use blood glucose for heat production (Ruppenthal *et al.*, 1983), which underscores the need to avoid hypoglycemia by frequent feeding during the perinatal period.

2.1.1. *Incubator Methods*

In the IPRL, incubator temperature is monitored every 4 hr around the clock. The IPRL initially maintains newborns in incubators heated to 94°F, but low-birth-weight infants may require a temperature of 96–98°F. The infant is challenged by temperature reductions in 2–3°F increments on subsequent days. If it becomes hypothermic (<97.5°F), the temperature is raised to the previous setting and rectal temperature is closely monitored.

The incubator is humidified with distilled water. Failure to humidify will subject the monkey to dry heat, resulting in drying of the nasal membranes and capillary hemorrhage and often producing nasal congestion.

Healthy near- or full-term infants usually graduate from incubators without problem by day 3 or 4. If an infant again becomes hypothermic, it is returned to an incubator and the challenge steps are repeated. Low-birth-weight, often premature, infants may require incubator support for 1–3 weeks. Infants in incubators are usually quite inactive, leading to atrophy and loss of movement in joints of the arms and legs during extended stays. We recommend some form of physical therapy two or three times daily to avoid muscle problems.

2.1.2. Heating Pads

Use of a heating pad is not recommended in cases of depressed body temperature. Low-birth-weight, premature, and ill infants may be extremely lethargic, or even comatose, when experiencing hypothermia. Heating pads may heat the ventrum, which is usually in contact with the pad. This causes blood to be shunted to the ventrum, which can lead to shock. Additionally, these pads transmit heat rapidly when damp or wet from urine, and can scald or severely burn infants. Heating pads are recommended only for healthy, active infants that are experiencing mild temperature loss, and only if used on the lowest possible setting, with frequent inspection of the pad for moisture. If incubators are not available for hypothermic infants, a recirculating water heating pad, set to warm, can be an option.

2.1.3. Oxygen Support Therapy

If infants require oxygen, as indicated by clinical symptoms discussed below, it can be supplied in low concentrations of up to 30–35% using a latex supply tube placed on the side of an incubator. Higher concentrations of oxygen do not produce retinal degeneration in neonates of the species we have studied, as they do in human perinates. Bottled compressed oxygen becomes extremely dry when expanding as it leaves the bottle, so humidification may be needed. This dryness can be overcome by use of a humidifier/water heater, which can be placed in the oxygen supply line to bypass the incubator oxygen limiter.

2.1.4. Prematurity and Hyaline Membrane Disease

The average full-term delivery of pigtailed macaques in our colony is 170 days. Infants born at known gestational ages of <150 days are defined as premature. However, most neonates in our colony do not have known gestational ages unless they were conceived by timed mating or artificial insemination. Therefore, the IPRL operationally defines potential prematurity as (1) deliveries that are more than 2 standard deviations from the mean birth weight for full-term non-pigtailed macaque infants, or

(2) pigtailed macaque infants born in the lowest 10% of the birth-weight distribution. We use the term "potential" because such infants could be born at full term but be small-for-date. Without a positive pregnancy confirmation, even ultrasound estimation of gestational age will be inaccurate for either small-for-date or large-for-date fetuses and newborns.

Among other problems, prematurity may lead to hyaline membrane disease (HMD), which contributes to the greatest percentage of neonatal deaths in premature human and nonhuman primates. Among cesarian-delivered pigtailed macaques at the IPRL, severe HMD was present shortly after birth in 45% of those delivered at 135 days gestation and 27% of those delivered at 142 days gestation (Pruitt *et al.*, 1979). These monkeys had been identified from IPRL computer records as being at very high likelihood of requiring oxygen therapy during the first week of life. Observations of sternal retraction, pale coloration of the face, rapid and shallow labored breathing, puffing of the cheeks, and "yawning" or "gulping" are all useful clinical signs indicating need for oxygen therapy.

Interestingly, in our experience HMD is not always present immediately after birth. HMD began at 3–5 postnatal days in 10–15% of neonates suspected of having a mild or severe case and confirmed at a subsequent nercropsy. This suggests that surveillance for HMD should be continued after the day of birth.

Although untested experimentally, aerosol therapy (nebulizing) seems to offer relief when the HMD infant is in an incubator. Additionally, synthetic surfactant has been used with good results for infants born extremely prematurely and exhibiting obvious respiratory difficulty. This therapy is expensive, approximately $800.00 per vial, and the vial must be used within 8 hr after thawing and constituting. Consequently, the IPRL has had minimal experience with this technique. Primate caregivers that have access to a human Neonatal Intensive Care Unit may be able to obtain the surfactant when premixed vials have not been exhausted during human neonate therapy.

2.1.5. Feeding

IPRL newborns are hand fed every 2 hr, beginning 2–4 hr after birth, around the clock (Ruppenthal and Sackett, 1992). Initial fluids are 5–10%

dextrose, followed by a 50/50 mixture of dextrose/formula every 2 hr for 3–5 days, then full-strength formula every 4 hr until the infant is able to self-feed.

Many facilities routinely feed infants *ad libitum*, simply leaving the bottle attached to the cage or a surrogate device. We do not recommend this practice with neonates. Many neonates, especially those with a strong sucking response, ingest formula too rapidly and then vomit or expel contents through their noses. This can lead to aspiration, with adverse consequences leading to death if not discovered quickly. To avoid overfeeding, we limit the amount of formula or dilute it and feed more frequently. We also interrupt feedings and "burp" the infant by patting its back. For details concerning our feeding practices, see Ruppenthal and Sackett (1992). For average intake amounts and caloric intake for pigtailed macaques by birth-weight class, see Sackett and Ruppenthal (1992).

2.1.5a. Gavage Feeding. The leading cause of morbidity and mortality in premature infant monkeys, as it once was for premature human infants, is aspiration pneumonia caused by vomiting gastric fluids or inhalation of formula during bottle feeding. After experiencing a number of failures in rearing premature infants during the IPRL's first decade, we consulted with personnel from the Department of Neonatal Biology and Pediatrics at the University of Washington Medical Center. These pediatricians are invaluable as a resource for nonhuman primate nursery facilities. With their help we initiated new gavage feeding techniques. A dramatic increase in survival over decades (Fig. 18–1) is due primarily to these new feeding techniques instituted in the nursery's second decade.

In the new technique, premature and low-birth-weight pigtailed and long-tailed macaque newborns are routinely fed by gavage until they reach the gestation age cut-off (155 days for pigtailed macaques and 150 days for long-tailed macaques) or until they exceed the upper weight of the low-birth-weight category. Before these ages and/or weights, these neonates are especially susceptible to aspiration.

The percentages of infants in the four birth-weight classes that were fed by gavage are shown in Fig. 18–2. In the two lower weight classes, infants that were gavage fed for one or more days increased over the three decades of the IPRL's existence. The majority of gavage feedings for heavier infants were due to episodes of hypothermia (<97.5°F) while ill, which causes lethargy and a reduction of gut motility or diarrhea. Hypothermic infants do not feed well and require fluid support.

When drinking formula from a bottle, high-risk neonates, unlike normal neonates, often do not interrupt their bouts of sucking to pause and swallow. They also often suck fiercely when first given their bottles, especially if the fasting period between feedings is extended. In these cases, the infant may swallow a bolus of formula, then vomit and inhale, leading to aspiration with a very poor prognosis. This is avoided by feeding prescribed amounts and concentrations of liquid via either nasal/gastric or oral/gastric gavage.

We use 3.5-French catheters for nasal administration and 5 French for oral administration. Nasal/gastric administration is preferred because it avoids tube placement in the lungs, which is possible if the epiglottis is open during inhalation. However, administration is difficult and requires prolonged training. The technique also tends to cause nasal irritation.

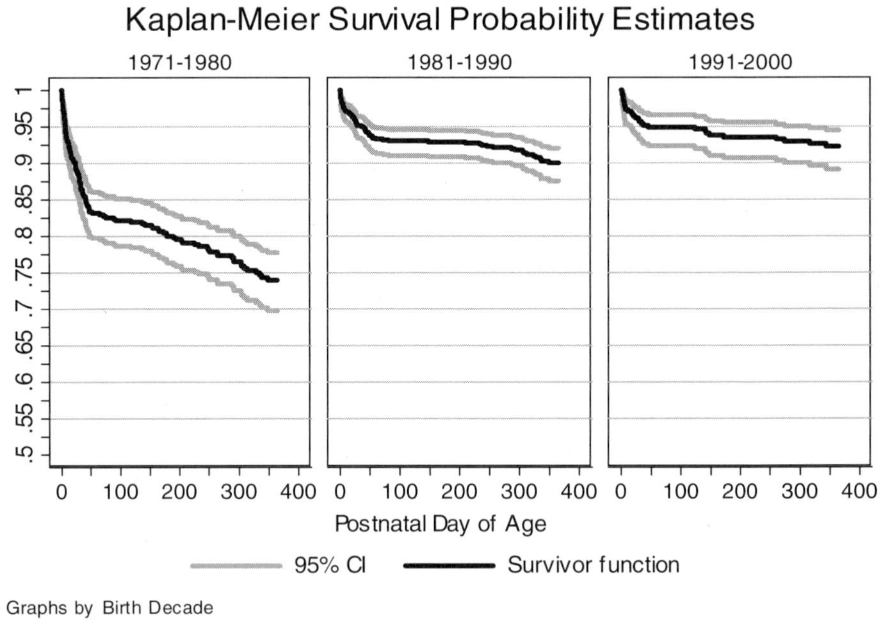

Figure 18–1. Survival functions for infants while in the IPRL over the three decades from 1971 through 2000.

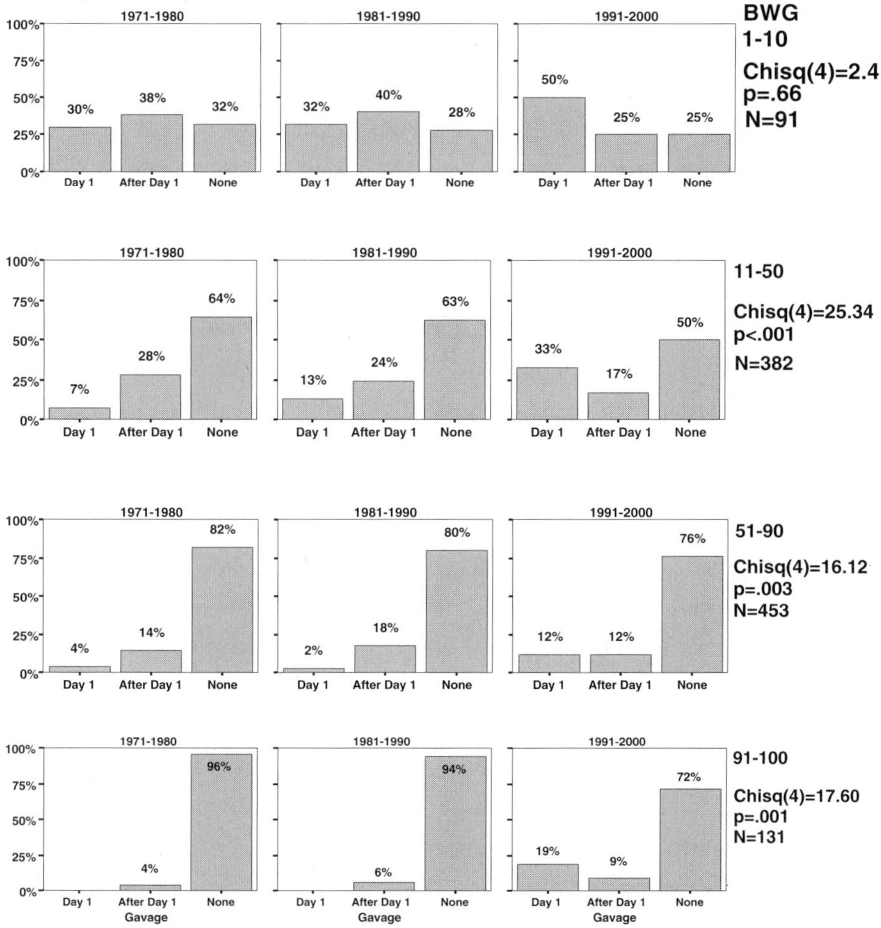

Figure 18–2. Percentage of infants that did or did not require gavage feeding by birth-weight percentile category.

Oral/gastric administration is easier to learn but the risk of placing the tube in the lung is greater. To reduce this possibility, after the oral/gastric feeding tube is inserted, 0.25 ml of saline is administered. If coughing occurs, the tube has been placed in a lung. In this case, the tube is withdrawn and reinserted and its position is tested again with saline. An alternative method is to rapidly push 1 ml of air through the tube while holding a stethoscope beneath the rib cage. If the tube is in the esophagus, air can be heard and the placement is correct.

After the tube has been correctly inserted by either method, residual stomach fluids are drawn and examined for quality and quantity; the amount is then recorded and the contents are reinserted into the stomach. Withdrawal of stomach contents to check for a clear or curdled residual provides information on how the infant is tolerating and digesting formula and tests for overfeeding. If the residual indicates overfeeding, indicated by undigested formula or 2–4 ml of fluid remaining, the concentration of formula and the amount fed at the next feeding are adjusted. The goal is to provide enough fluid so that a trace amount remains at the subsequent feeding. Regardless of the formula concentration, the initial feeding is 3–4 ml, then adjusted upward at the next feedings in 2- to 3-ml increments until a trace residual is present.

2.1.5b. Self-Feeding. A developmental milestone in the primate nursery is the attainment of self-feeding. When this milestone is reached, personnel time can be released for other purposes. For most monkeys, reaching this milestone indicates that the highest risk period has been passed, as the animal has attained the neural maturity underlying locomotor skills and their coordination with sensory and motivational systems.

We train the infants to self-feed using a specialized formula dispenser (Fig. 18–3) attached to the side of the cage (Sackett and Ruppenthal, 1992). When attached, the dispenser is laid on its side, which increases the surface area of the formula and decreases the effort required to suck the formula. The formula level is maintained below the hole in the nipple. The nipple is attached to a bent tube that is angled so that it reaches the bottom of the dispenser. When the infant stops sucking, the formula flows back into the dispenser rather than equilibrating pressure and leaking, as is the case with an inverted bottle. This allows for accurate measurement of the amount of formula consumed.

During training, we use a "surrogate mother" consisting of an inclined cylinder covered with soft diapers. The nipple passes through the cylinder. This setup allows the infant to grasp with its hands and maintain ventral contact for security as training to self-feed proceeds. Typically beginning on postnatal day 4, infants are held up to the nipple and allowed to suck. The amount consumed is recorded and infants are supplemented with a hand-held bottle if intake is less than desired. After the

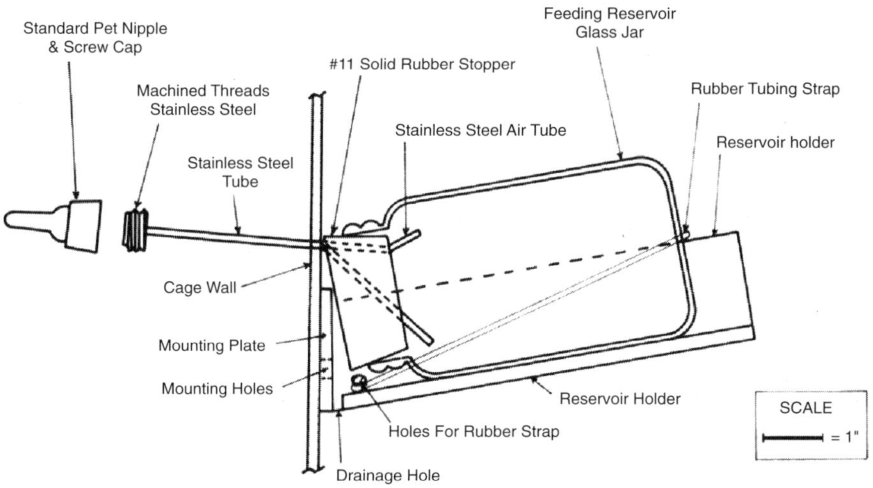

Figure 18–3. Feeding apparatus used for milk formula feeding in the IPRL. The design was developed to eliminate spillage and dripping so that intake could be measured accurately.

infant achieves consistent feeding, assistance is withheld and the infant is allowed to approach the feeder and suck on its own schedule. If a sufficient amount has not been obtained after a 1-hr period, the infant is held up to the feeder. Self-feeding has been achieved when the infant feeds without assistance during a full 24-hr period.

Our feeding schedule consists of 4 hr with fresh milk in the feeder, then 4 hr with no feeder. This schedule is superior to *ad libitum* feeding for a longer period or for a full 24 hr because it minimizes diarrhea from bacterial growth in the milk and maintains a high level of motivation to feed. We have found that the scheduled feeding supports growth as well as, and often better than, *ad libitum* feeding (Sackett and Ruppenthal, 1992).

Analysis of self-feeding data revealed an inverse relation of birth weight to self-feeding age, with all groups differing significantly from each other (Fig. 18–4). The two lower birth-weight groups took an especially long time to attain self-feeding relative to larger infants. Interestingly, although the difference was small, females required significantly less time than males to be able to feed themselves (Fig. 18–4).

There are at least three possible explanations for the lower-birth-weight infants' relative retardation in learning to self-feed. First, they required longer stays in incubators, and training to self-feed does not begin until the infant leaves the incubator. Second, they are less active early in life (Sackett *et al.*, 1979), and so are less likely to move to the feeding reservoir. And third, lower-birth-weight infants are more likely to be premature, needing more postnatal neural maturity to be able to learn self-feeding habits.

Age at self-feeding also is related to the infant's reason for being in the IPRL (Fig. 18–5). IPRL infants fall into one of five categories: (1) infants that are at risk due to inadequate mothering; (2) infants that have been injured; (3) "Infant Save" candidates owing to low birth weight/prematurity; illness, and/or failure to gain weight; (4) healthy stock infants whose mothers were reassigned or who were themselves assigned to fill out agemate playgroups; and (5) newborns assigned to experimental protocols including prenatal experiments. Stock infants attained self-feeding sooner than the other groups. This effect is not independent of birth-weight differences, as inadequately mothered and Infant Save monkeys are often born prematurely and/or at low birth

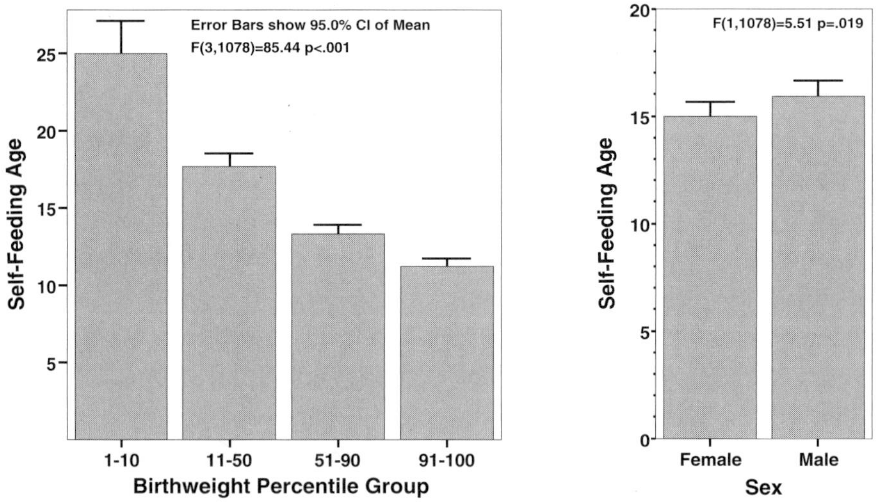

Figure 18–4. Self-feeding age by birth-weight percentile group and age.

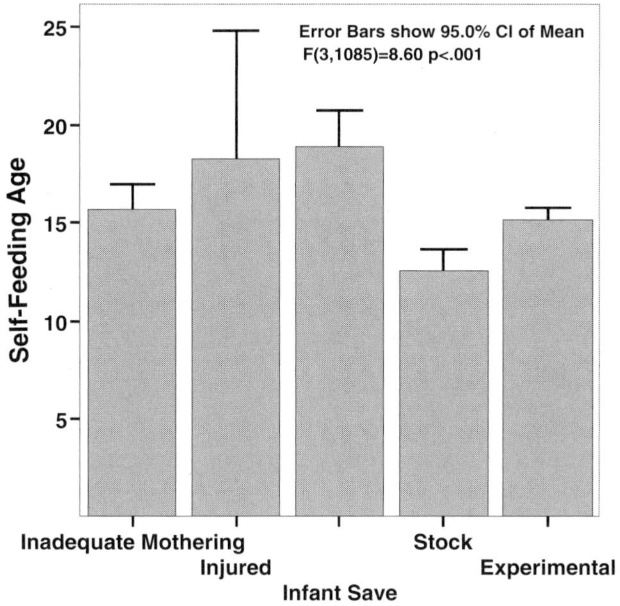

Figure 18–5. Self-feeding age for infants differing in reasons for requiring nursery rearing.

weight. The result does suggest, however, that during the neonatal period at-risk monkeys as a group are delayed in developing basic motor and learning skills.

2.1.6. Neonatal Weight Loss and Gain

Figure 18–6 shows patterns of weight loss and gain by the four birth-weight classes for males and females over the first 30 days. Weight loss is positively related to birth-weight group, with the heaviest infants of both sexes losing the most weight. Following the initial weight loss, heavier birth-weight infants gain more weight per day than those born at lighter weights. Figure 18–7 shows that low-birth-weight animals regain birth weight in significantly fewer days than animals in the other three birth-weight groups, which do not differ.

Weight loss during the first 3–4 postnatal days, with birth weight being regained by 6–8 days, is a well-known, apparently universal, primate

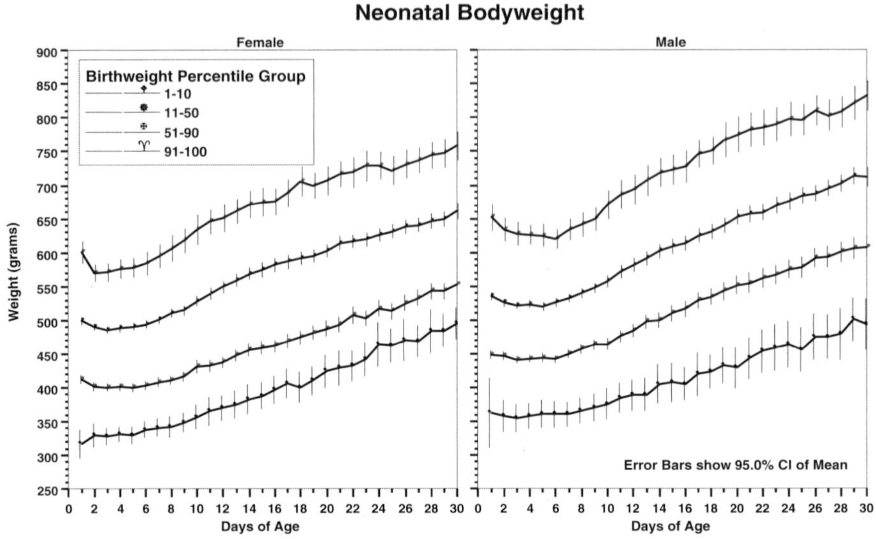

Figure 18–6. Ponderal growth by birth-weight percentile group during the neonatal period (days 1–30). Note the initial weight loss by all groups except low-birth-weight males.

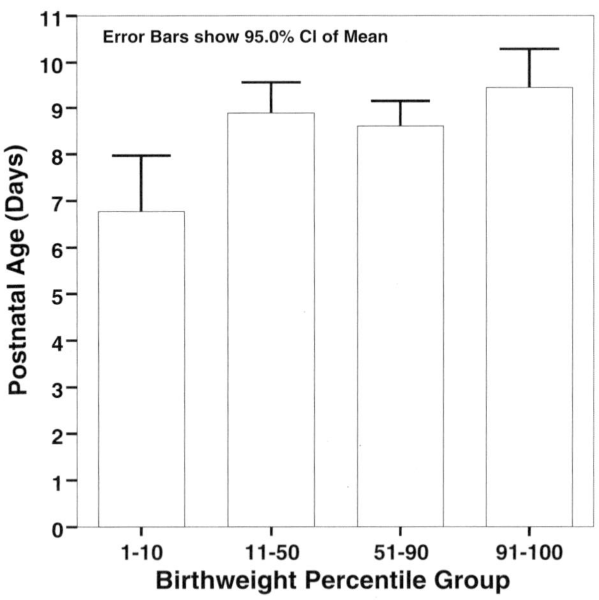

Figure 18–7. Days to regain birth weight by birth-weight percentile group.

phenomenon. This weight loss leads some facilities to emphasize early regaining of birth weight. We are not particularly concerned with caloric loading over the first week of life. Instead, our emphasis is placed on avoidance of hypoglycemia and hypothermia. While our infants may lose a greater percentage of body weight and take longer to regain lost weight than infants in facilities that emphasize weight gain, they equal or surpass the weights in those facilities by 3–4 weeks of age (Ruppenthal, 1985). Relative unconcern over weight loss is also reinforced in work by Guthrie *et al.* (1981), who studied lung maturation in premature infants delivered by cesarian section. Newborns were maintained for the first 24–72 hr on a restraint board and supplied with 5% dextrose via umbilical catheter. They experienced as much as a 15–20% weight loss, but upon completion of the protocol they regained the lost weight within 5–7 days, with no increase in morbidity. Interestingly, cesarian section infants lose a great deal of weight immediately after birth compared with vaginally delivered monkeys, and also take longer to regain birth weight than either infants with no postbirth intervention or those involved with invasive experiments (Fig. 18–8).

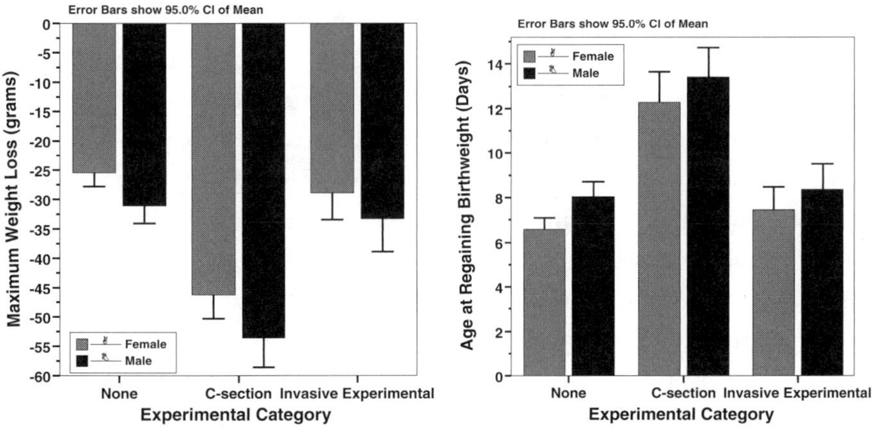

Figure 18–8. Maximum post-birth-weight loss and days to regain birth weight by infants receiving invasive experimental prenatal treatments, cesarian section delivery, or normal vaginal delivery.

3. POSTINCUBATOR HOUSING

After emergence from incubators, infants are placed in nursery cages measuring 0.78 m × 0.43 m × 0.51 m high. These cages are approximately a subclass 0.5 (USDA standards). The cages are intentionally low because macaque and baboon neonates have a very strong negative geotaxis (anti-gravity) reflex. They will climb to the highest point possible and do not have the voluntary motor control to climb down. Once at the cage ceiling, they will convulsively jerk, fear-grimace, and vocalize until exhausted. Having a lower cage increases the chance that the legs will be in contact with the substrate, allowing the infant to climb down.

When infants no longer require auxiliary temperature support, are healthy, and are self-feeding, usually at ~21 days of age, they graduate from the intensive care nursery and are moved to individual cages in a general infant colony room. In addition to a number of developmental assessments, they receive daily 30- to 60-min socialization sessions in groups of four to six animals in a large playroom. Temperatures are monitored once per day during week 1 in the new environment. If any weight loss or reduced food intake is observed, health examinations are performed.

4. CAUSES OF DEATH

Despite the best of care and intentions, some infants do die. The non-experimental causes of all deaths in the IPRL from birth through 12 months of age (or until infants left the laboratory due to experimental protocols or transfer to other facilities) are listed in Table 18–1. Empty cells indicate that the cause was not known. Dehydration, lung abnormalities and aspiration, and intestinal inflammation were the most frequent primary causes of death. Brain abnormalities and failure to thrive also had fairly high frequencies. As might be expected, in the lowest 10% of the birth-weight distribution death was caused most frequently by lung and aspiration pneumonia.

Table 18–2 shows the percentage of infants in each birth-weight group that died in each decade, as well as overall death percentages. It is clear that most deaths occurred in the low-birth-weight group; almost no deaths occurred among high-birth-weight infants. There was a

Table 18–1. Cause of Death by Birth Weight Percentile Category

	Birth weight percentile category				
	1–10	11–50	51–90	91–100	Total
Dehydration, diarrhea					
Count	10	12	10	1	33
% within cause-of-death	30.3%	36.4%	30.3%	3.0%	100.0%
% with birth weight percentile category	19.2%	15.4%	22.7%	20.0%	18.4%
Lun abnormality-NOS					
Count	13	10	9		32
% within cause-of-death	40.6%	31.3%	28.1%		100.0%
% with birth weight percentile category	25.0%	12.8%	20.5%		17.9%
Lung aspiration					
Count	10	7	4		21
% within cause-of-death	47.6%	33.3%	19.0%		100.0%
% with birth weight percentile category	19.2%	9.0%	9.1%		11.7%
Lung bacteria					
Count	4	5			9
% within cause-of-death	44.4%	55.6%			100.0%
% with birth weight percentile category	7.7%	6.4%			5.0%
Brain seizures, inflamation					
Count	3	5	7	1	16
% within cause-of-death	18.8%	31.3%	43.8%	6.3%	100.0%
% with birth weight percentile category	5.8%	6.4%	15.9%	20.0%	8.9%
Heart abnormality					
Count		2		1	3
% within cause-of-death		66.7%		33.3%	100.0%
% with birth weight percentile category		2.6%		20.0%	1.7%
Intestinal inflamation					
Count	4	20	6	1	31
% within cause-of-death	12.9%	64.5%	19.4%	3.2%	100.0%
% with birth weight percentile category	7.7%	25.6%	13.6%	20.0%	17.3%
Liver-kidney abnormality					
Count	1	1	3		5
% within cause-of-death	20.0%	20.0%	60.0%		100.0%
% with birth weight percentile category	1.9%	1.3%	6.8%		2.8%
Failure to thrive					
Count	6	8	1		15
% within cause-of-death	40.0%	53.3%	6.7%		100.0%
% with birth weight percentile category	11.5%	10.3%	2.3%		8.4%
Trachea abnormality					
Count	1	1	1		3
% within cause-of-death	33.3%	33.3%	33.3%		100.0%

Table 18–1. *Continued*

	Birth weight percentile category				
	1–10	11–50	51–90	91–100	Total
% with birth weight percentile category	1.9%	1.3%	2.3%		1.7%
Traumatic injury					
Count		4	1	1	6
% within cause-of-death		66.7%	16.7%	16.7%	100.0%
% with birth weight percentile category		5.1%	2.3%	20.0%	3.4%
Debilitating birth defect					
Count		3	2		5
% within cause-of-death		60.0%	40.0%		100.0%
% with birth weight percentile category		3.8%	4.5%		2.8%
Total					
Count	52	78	44	5	179
% within cause-of-death	29.1%	43.6%	24.6%	2.8%	100.0%
% with birth weight percentile category	100.0%	100.0%	100.0%	100.0%	100.0%

Table 18–2. Percent of IPRL Infants Dying during the Years 1971 through 2001 by Decade and Birth Weight Percentile Group

	Birth weight percentile category				
Decade	1–10	11–50	51–90	91–100	Overall
1971–80					
Number Dead	26	42	21	2	72
Total number	82	276	194	35	587
% dead	**31.7**	**15.2**	**10.8**	**5.7**	**12.3**
1981–90					
Number dead	17	27	20	0	64
Total number	67	349	350	81	847
% dead	**25.3**	7.7	**5.7**	**0.0**	7.6
1991–00					
Number dead	9	9	3	3	24
Total number	38	149	215	74	476
% dead	**23.7**	**6.0**	**1.4**	**4.1**	**5.0**
All years					
Number dead	52	78	44	5	179
Total number	187	774	759	190	1910
% dead	**27.8**	**10.1**	**5.8**	**2.6**	**9.4**

significant decline in deaths by decade in each birth-weight group, with the largest drop between the first and second decades. We believe that this strong trend is largely attributable to gavage feeding.

5. CONCLUSION

One important consideration in caring for infants with risk factors such as low birth weight, prematurity, injury, illness, or maternal rejection-abuse is that schedules are made to be altered. Close observation of the present status of infants should always supercede all else in dictating care. Even if infants are on a prescribed feeding schedule for formula constitution and type of feeding, immediate remedial attention is in order if weight loss occurs at inappropriate times, the infant has no intake when self-feeding, or the infant shows clinical symptoms. Being on a 4-hr feeding schedule does not mean that infants should not be inspected more frequently. Close attention to status will, in the long run, save animals that otherwise might die. This is why we maintain nursery personnel on a 24-hr basis, regardless of cost. Research monkeys do not come cheap and we do owe them the best life possible.

ACKNOWLEDGMENTS

The research and methods presented in this chapter were funded by NIH grants RR00166 from the NCRR to the Washington National Primate Research Center and HD02274 from the NICHHD to the University of Washington Center on Human Development and Disability. We are especially grateful to Dr. Orville Smith, Dr. Michael Guralnick, and all of the other directors of our two Centers who supported the work of the IPRL over its 33-year history.

REFERENCES

Guthrie, R.D., Standaert, R., Hodson, W.A., and Woodrum, D.E. (1981). Development of CO_2 sensitivity: Effects of gestational age, postnatal age, and sleep state. *J. Appl. Physiol.* 50:956–961.

Newell-Morris, L.L. (1979). Age determination in macaque fetuses and neonates. In: Ruppenthal, G.C., and Reese, D.J. (eds.), *Nursery Care of Nonhuman Primates.* Plenum Press, New York, pp. 93–115.

Pruitt, J.L., Palmer, S., Standaert, T.A., Luchtel, J.H., Murphy, J.H., and Hodson, W.L. (1979). Lung development in the fetal primate *Macaca nemestrina*. III. HMD. *Pediatr. Res.* 13:654–659.

Ruppenthal, G.C. (1985). Weight gain and intake requirements in nursery-reared macaques. In: Meehan, T.P., and Allen, M.E. (eds.), *Proceedings of the 4th Annual Dr. Scholl Conference on the Nutrition of Captive Wild Animals.* Lincoln Park Zoological Society, Chicago, pp. 20–35.

Ruppenthal, G.C., and Sackett, G.P. (1992). *Research Protocol and Technicians' Manual: A Guide to the Care, Feeding, and Evaluation of Infant Monkeys.* Infant Primate Research Laboratory, Seattle.

Ruppenthal, G.C., Goodlin, B.L., and Sackett, G.P. (1983). Perinatal hypothermia and maternal temperature declines during labor in pigtailed macaques. *Am. J. Primatol.* 4:81–92.

Sackett, G.P., and Ruppenthal, G.C. (1992). Growth of nursery-raised *Macaca nemestrina* infants: Effects of feeding schedules, sex, and birth weight. *Am. J. Primatol.* 25:189–204.

Sackett, G.P., Fahrenbruch, C.E., and Ruppenthal, G.C. (1979). Development of basic physiological parameters and sleep-wakefulness patterns in normal and at-risk neonatal pigtail macaques (*Macaca nemestrina*). In: Ruppenthal, G.C., and Reese, D.J. (eds.), *Nursery Care of Nonhuman Primates.* Plenum Press, New York, pp. 125–142.

Sackett, G.P., Ruppenthal, G.P., and Davis, A.E. (2002). Survival, growth, health, and reproduction following nursery rearing compared with mother rearing in pigtailed monkeys (*Macaca nemestrina*). *Am. J. Primatol.* 56:165–183.

Schrier, A.M., Povar, M.L., Susa, J.B., and Schwartz, R. (1983). Behavioral consequences of neonatal hypoglycemia in the rhesus monkey: A preliminary report. *Infant Behav. Dev.* 6:121–125.

Schrier, A.M., Wilhelm, P.B., Church, R.M., Povar, M.L., Schrier, J.E., Sehgal, P.K., Boylan, J.M., Schwartz, R., and Susa, J.B. (1990). Neonatal hypoglycemia in the rhesus monkey: Effect on development and behavior. *Infant Behav. Dev.* 13:189–207.

CHAPTER NINETEEN

A Quick and Effective Method for Establishing Self-Feeding in Stump-Tailed Macaques (*Macaca arctoides*)

Arnold S. Chamove

1. INTRODUCTION

In 1970 the Psychology Department of the University of Stirling, Scotland, inspired by the work of Harry F. Harlow and colleagues, established a colony of stump-tailed macaques (*Macaca arctoides*), a species that was believed to resemble rhesus macaques (*Macaca mulatta*) in social structure and behavior but was reported to be more tractable (Schrier, 1967). The objective was to breed infant stump-tailed macaques for long-term research on their behavioral development. The colony was founded with 31 adult macaques imported from Thailand. The animals were housed in harem groups of 10–15 females with one male in a 30-m^2 room with access to outdoor pens. The original plan was that infants would be raised by their mothers in these social groups. However, the colony soon proved to be affected by *Shigella* and *Salmonella*, which are a major problem in macaques, though reportedly less so in the stump-tailed species (Chamove *et al.*, 1979). Because these and other pathogens (e.g., B virus; Zwartouw *et al.*, 1984) are particularly virulent in young monkeys, it was decided to protect vulnerable infants by separating them from their mothers and raising them in a nursery environment.

2. NURSERY FACILITY

The infant nursery, kept at an ambient temperature of 27–32°C, contained the usual assortment of incubators, individual cages, and socialization equipment. The incubator, covered with Perspex, had a semicircular floor with a diameter of 0.58 m. Along the flat side was a mesh ramp with solid sides measuring 0.26 × 0.26 m, protruding 0.16 m outward, and set at an angle of 45°. A semicircular plastic disk prevented the infant from creeping behind the ramp and getting trapped. A Belcroy feeder was clipped to the top of the ramp, and a piece of toweling was placed under the feeder. Toweling also covered the floor.

Once the infants were consistently self-feeding (described in the following section), they were removed from the incubator and housed in plastic tub-cages that were made in-house from heavy-duty translucent white polyethylene tubs (WCB Containers, Stalybridge, UK). The tub, measuring 0.58 × 0.65 × 0.75 m, was turned on its side and a light wire-mesh door was fitted at the open side. A rigid clear-plastic sheet could be fitted to the inside of the mesh to prevent very young infants from climbing before they were competent to get down. The solid floor easily became soiled and wet, but this was not problematic as the young infants were supplied with soft cloths that absorbed the moisture. Ambient temperature was supplemented with heat lamps, which produced a warm spot on the plastic floor of the tub. The tub-cages had the advantage of being lightweight (15 kg), inexpensive, and draft-proof.

At 1 month of age the infants were moved to ready-made cages of galvanized steel (Associated Crates Ltd, Stockport, UK). These cages measured 0.65 × 0.60 × 0.60 m, adequate for monkeys up to 2 years of age. The cages were fitted with interconnecting, removable Perspex side panels, which could be transparent, translucent, opaque, or composed of a mirrored plastic. Panels between two adjoining cages were removed for about 2 hr per day so that the infants could play with each other, thereby encouraging the development of species-typical social behavior and contributing to the animals' psychological well-being (Chamove, 1973a,b).

3. SELF-FEEDING

Most techniques for hand-rearing infant primates are labor-intensive, involving many days and nights of hand-feeding. The literature suggests that hand-reared infants must be fed every 2–4 hr around the clock until they learn to self-feed. Thus, one of the first tasks of nursery personnel is to teach infants to feed themselves. According to the literature, this process can take as little as 3 days or as long as 8 weeks, depending on species and technique (Table 19–1). In our nursery, however, we developed a technique that involved neither late nights nor many days of hand-feeding. The method involved leaving a formula-filled bottle with the infant, so that the handler could feed the infant as often as desired while also allowing the infant the opportunity to feed itself (Chamove *et al.*, 1982).

Recall that a feeder was clipped to the top of a mesh ramp in the incubator. The feeder was filled with SMA formula supplemented with Abidec multivitamin (1 drop per day). Approximately once per hour between about 900 and 1700 hr, the infant was lifted onto the mesh ramp and positioned with its mouth on the feeder nipple. If the infant did not suck, the nipple was directed into its mouth and a small amount of formula (1 or 2 ml) was expressed. The infant was allowed to feed *ad libitum*, and the formula was renewed at least five times during the day. The last hourly feeding was at 1700. If by then the infant had not consumed a total of at least 20 ml of formula, it was persistently encouraged to consume the amount required to reach 20 ml. A dim light was left on during the night so that the infant could see the feeder. This procedure was continued until the infant had self-fed a minimum of 5 ml twice without assistance. Most infants achieved this milestone within just 30 hr.

At 16 weeks of age, the infant began receiving cow's milk in place of SMA formula. Supplements of dry chow and fresh fruit were offered as the amount of milk was gradually reduced.

Table 19–1. Time Required for Typical Nonhuman Primates to Become Self-Feeding

Species	Time	Reference
Macaques	1–2 weeks	Vickers (1968)
Rhesus macaques	3–5 days	Valerio *et al.* (1969)
	1–2 weeks	Kerr *et al.* (1969b)
	4 weeks	Blomquist and Harlow (1961)
Pigtailed macaques	1–3 weeks	Sackett *et al.* (1976)
Squirrel monkeys	3 weeks	Kinkle and Session (1972)
Guinea and yellow baboons	1–2 weeks	Voss *et al.* (1971)
	1–3 weeks	Vice *et al.* (1966)
	1–8 weeks	Buss and Voss (1971)
Baboons	8 weeks	Chorazyna (1972)

3.1. Effects of Preparation for Self-Feeding

Before infants were separated from their mothers, they often received supplementary feeding, called "prefeeding," in the form of about 34 ml SMA formula per day. For this procedure the mother was gently restrained in a crush-back cage and allowed to hold the infant while a technician fed the infant with a Belcroy Tube Feeder. To assess whether this supplementary feeding influenced the infants' adaptation to separation and their ability to self-feed as quickly as they did, we studied the records of 50 infant stump-tailed macaques and computed correlations to answer the following questions: Does the amount of prefeeding influence the time required for the infant to learn to self-feed? Does it influence either weight or intake? Does the time taken to reliably self-feed influence subsequent intake or weight?

To assess whether prefeeding the infants would substantially facilitate the development of self-feeding, we compared two groups of infants. In one, infants ($n = 12$) were prefed an average of 34 ml per day with their mothers and then moved to the nursery between days 3 and 10. In the other group ($n = 42$), the infants were separated from their mothers between days 1 and 10 without being prefed. Each infant's intake was recorded first at hourly and then at daily intervals. Body weight was recorded once daily at 1100 hr. For our analysis we used the total intake of formula during prefeeding and on the day of separation, an average of the daily intake on days 29–31, the weight of the infant on the day of separation, and the average of the infant's three daily weights on days 29–31. Finally, to measure the speed at which the infants learned to self-feed, we calculated the total number of hours that elapsed after separation from the mother until the infant fed itself twice, consuming at least 5 ml each time.

These measures and the age of the infant at separation were subjected to Pearson product-moment multiple correlation analysis. One analysis was performed on the data from the 12 monkeys that received supplementary prefeeding, one on the remaining 42 monkeys, and one on all subjects combined.

In general, the results showed that prefed infants, compared with non-prefed infants, consumed more formula on the day of separation (57 ml versus 47 ml or 104 ml/kg body weight versus 82 ml/kg) and mastered self-feeding more quickly (27 hr versus 41 hr, or 99 ml of formula versus 165 ml fed).

High intake on the day following separation led to rapid acquisition of self-feeding ($r = -0.27$, $p = 0.05$). Furthermore, higher intake on the day of separation was predictive of greater intake ($r = 0.48$, $p = 0.001$) and weight ($r = 0.33$, $p = 0.01$) at 1 month of age. These correlations were even higher in the prefed group ($r = 0.85$, $p = 0.001$, and $r = 0.68$, $p = 0.01$, respectively). However, in the prefed group, speed of achieving self-feeding was not related to the total amount ingested during prefeeding ($r = 0.03$, $p = 0.93$). The rapidity of learning to self-feed seemed unrelated to most other measures taken.

Most (60%) infants that were not prefed were separated from their mothers on day 8 or 9, and those 2 days showed the most dramatic benefit of prefeeding. Infants that were prefed while with their mothers were self-feeding by 6 hr on average, whereas the other infants took an average of 35 hr. As a comparison, two infants that were separated from their mothers at birth were self-feeding within 30 hr.

The weights of 62 stump-tailed macaques over the course of their first year and their total intake of SMA formula over 106 days are shown in Fig. 19–1. None of these infants had received any prefeeding and all were separated from their mothers on day 8 or 9. The computation of 3-day averages serves to reduce the variability of intake values.

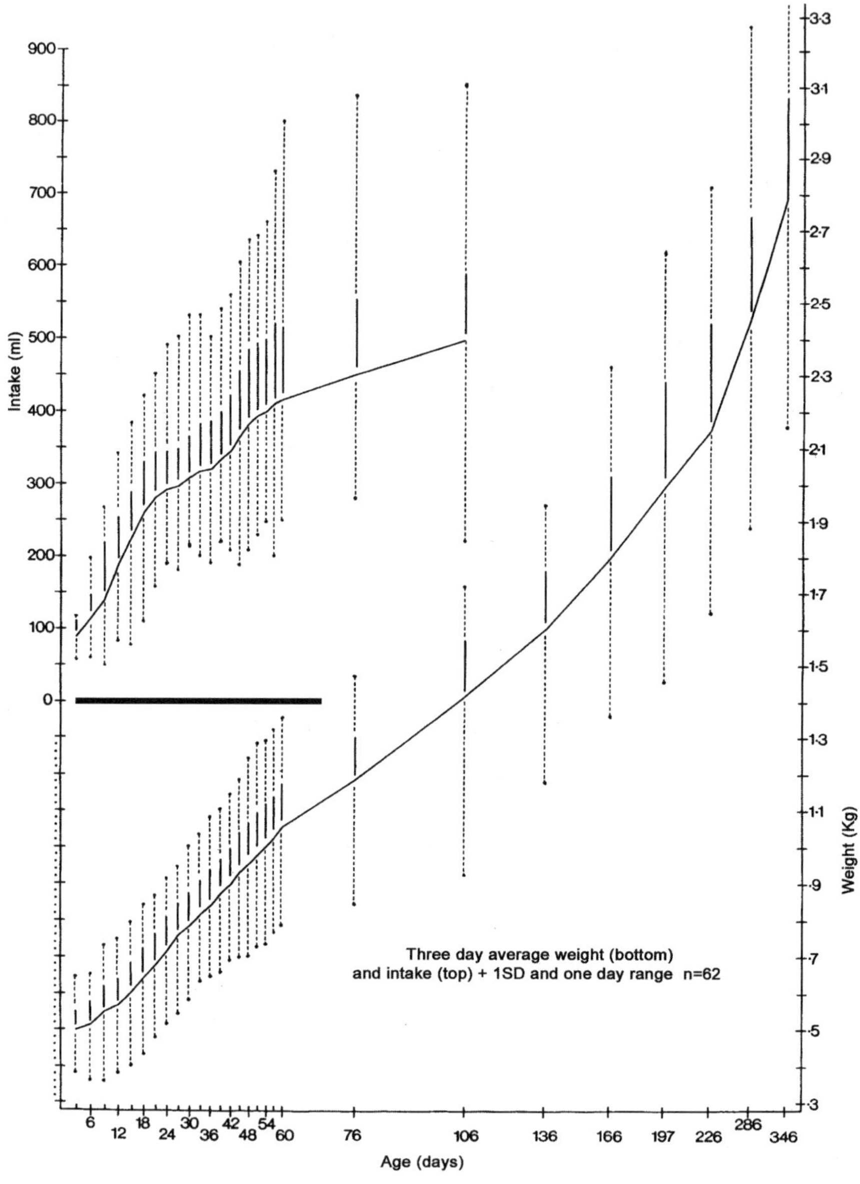

Figure 19–1. Intake of SMA formula (top) and body weights (bottom) of 62 stump-tailed macaques over 1 year reared on the basic self-feeding regimen.

3.2. Comparative Outcomes of Feeding Methods

The results of our feeding methods compared very favorably with those obtained in other species and with other methods. The growth patterns displayed by our infants were quite similar to those observed in 10 stump-tailed macaque infants that were hand-fed at 4-hr intervals round the clock but did not have food in front of them continuously (Scheffler and Kerr, 1975). Our animals were slightly heavier, by about 0.25 kg at day 60 and 0.3 kg at day 110. They also consumed more formula, reaching a maximum intake of 419 ml/kg/day versus 374 ml/kg/day, and increased their intake over a longer period, peaking at 48 versus 28 days of age.

Our results are also similar to those obtained with the same methods in rhesus macaques at the University of Birmingham, UK. Seven rhesus macaques that were separated from their mothers at 3–9 days of age were self-feeding within 33 hr (range 12–40 hr) and eight others that were separated from their mothers at 12–27 days of age were self-feeding within 21 hr (J. Marston, personal communication, 7/1979).

Compared with 20 rhesus macaques that were hand-fed every 4 hr round the clock and given a solid supplement beginning at 46 days of age (Blomquist and Harlow, 1961), the mean weight of our stump-tailed macaques was heavier by 2 SD at 1 month of age and by 3 SD at 1 year of age (see also Scheffler and Kerr, 1975). The growth pattern for our stump-tailed macaques was similar in some respects to that of 27 rhesus macaques that were hand-fed a Similac milk substitute every 4 hr round the clock (Kerr *et al.*, 1969b). Mean weights of our stump-tailed macaques exceeded those of the rhesus macaques by 2 SD at 1 month and by 1 SD at 1 year. The intake of our stump-tailed macaques during the first 100 days was also about 2 SD above that of the rhesus macaques. The relative intake reached a maximum value of 419 ml/kg/day for the stump-tailed macaques at 48 days of age, versus 416 ml/kg/day for the rhesus macaques at 42 days of age. In general, however, the relative intake was about 20 ml/kg/day lower in stump-tailed macaques than in rhesus macaques for the first 100 days.

We also compared the consumption of our stump-tailed macaques with that of larger baboons. Baboons that were hand-fed four times a day showed a peak of 350 ml/kg/day at 28 days of age, with most values similar to those from our macaques. Self-fed baboons showed a maximum value of 800 ml/kg/day at 56 days of age, about twice that of the stump-tailed macaques.

A comparison with values for mother-reared baboons (Buss and Voss, 1971) gives a quite different picture. The results of the radioassay method used by Buss and Voss suggest that mother-reared baboons consume less than rhesus or stump-tailed macaques. Mother-reared baboons showed no obvious peak, but reached a maximum intake of only 300 ml/kg/day, and this was not attained until 75 days of age. The average value was about 100 mg/kg/day below that for rhesus macaques.

Furthermore, it appears that self-fed or hand-fed infants consume more than mother-fed infants. When infant stump-tailed macaques were given the opportunity to obtain extra nutrition by leaving their mothers and entering a small tunnel to reach a bottle of formula, they consumed about 50% of the amount of formula that a self-fed infant takes each day. This suggests that mother-fed infants might obtain from their mothers only half of the amount that they would like to consume. Comparison of the weekly weights of individually housed (i.e., self-feeding) rhesus macaques (Kerr *et al.*, 1969a) with those of four mother-reared rhesus macaques (Griffin, 1966) also provides supporting evidence. The mother-reared infants steadily gained 25 g/week over the first year, whereas the individually housed subjects gained much more, >37 g/week. Our self-feeding stump-tailed macaques gained 46 g/week. Sackett *et al.* (1979) also reported that nursery-reared *M. nemestrina* are heavier than their colony-reared counterparts throughout the first year of life. The precise reasons for the differences remain unknown.

In a study that seemed to contradict other evidence, Faucheux *et al.* (1978) reported that mother-reared offspring were actually heavier than individually reared stump-tailed macaques after 2 years of age. We therefore examined the records of 13 mother-reared stump-tailed macaques that were weighed 16 times during the first year of life. Their weights were consistently 2 SD below the mean of self-feeding infants until >100 days of age. At about 200 days of age the means were equal, but by 3 years of age the mother-reared group was just significantly heavier than the self-feeding group (Mann–Whitney *U*, two-tailed, $p < 0.05$). Mean weights at 2 and 3 years of age were 4.7 and 6.9 kg for the mother-reared group versus 4.3 and 6.1 kg for the self-feeding group.

It was clear that the rate of growth was substantially lower during the first 100 days in the mother-reared group than in individually housed infants. The slope of the weight increase in the former group was 0.004 (angle = 23°), whereas for the self-feeding group it was 0.05 (angle = 86°). The slopes for the entire first year were very similar between groups, 0.008 (45°) for the mother-fed group and 0.006 (34°) for the self-feeding group. It may be that the period when the infant's diet is changed to chow alters the growth pattern. The rhesus macaques studied by Kerr *et al.* (1969b) were not given supplementary chow. Their slope for the first 100 days (0.006, angle = 34°) is similar to that over the first year (0.005, angle = 31°), suggesting that a change in diet does alter growth.

In conclusion, the early establishment of self-feeding as described here appears to have advantages over comparable regimens. Using these methods, we have been able to return infants to their mothers or to a group of adults within several weeks after separation. In these group situations, one can ensure that the infant has exclusive access to a bottle of formula by putting the bottle in a small enclosure or behind mesh with openings just large enough for the infant to enter. Infants trained to self-feed readily adapt to this arrangement.

Finally, when infant primates are hand-fed by humans, they often cough up the formula during the first 30 days of life (Blomquist and Harlow, 1961). With the self-feeding method described here, however, this problem is negligible (also see Valerio *et al.*, 1969). A further advantage of self-feeding is that automated techniques for recording intake (e.g., Walike *et al.*, 1971) can be used while infants are still very young.

4. CONCLUSION

The method used to establish self-feeding in infant macaques in our nursery led to substantial savings in time compared with more widely used methods. Our results suggest that if infants are to be separated from their mothers at the end of the first week or later, then prefeeding of the infant before separation leads to better adjustment to the nursery schedule. Among other factors, prefeeding means that the infant does not have to adjust to an unfamiliar-tasting diet. When infants are separated from their mothers earlier than 1 week of age, however, there appears to be no advantage to prefeeding.

Comparisons of our data with those obtained from other species indicate that there are species differences in response to feeding schedules. Although comparisons are hampered by differences in husbandry and species, it can be said that stump-tailed macaques are considerably heavier than rhesus macaques as adults, even though birth weights appear similar in the two species (Harvey *et al.*, 1979).

ACKNOWLEDGMENTS

I am grateful to Peter McEwen, Angus Annan, John Russell, Robert Brody, Robert Anstruther, Sheila Campbell, and Susan Lake for their help in maintaining the colony, and to Carol Chamove for help with the manuscript.

REFERENCES

Blomquist, A.J., and Harlow, H.F. (1961). The infant rhesus monkey program at the University of Wisconsin Primate Laboratory. *Proc. Anim. Care Panel* 11:57–64.

Buss, D.H., and Voss, W.R. (1971). Evaluation of four methods for estimating the milk yield of baboons. *J. Nutr.* 101:901–909.

Chamove, A.S. (1973a). Social behavior comparison in laboratory-reared stump-tailed and rhesus macaques. *Folia Primatol.* 19:35–40.

Chamove, A.S. (1973b). Varying infant rhesus housing. *J. Inst. Anim. Technol.* 24:5–15.

Chamove, A.S., Cameron, G., and Nash, V. (1979). Primate disease and breeding rates. *Lab. Anim.* 13:313–316.

Chamove, A.S., Anderson, J.R., Morgan-Jones, S.C., and Jones, S.P. (1982). Deep woodchip litter: Hygiene, feeding, and behavioural enhancement in eight primate species. *Int. J. Study Anim. Prob.* 3:308–318.

Chorazyna, H. (1972). Separation of baby monkeys from their mothers. In: Goldsmith, E.I., and Moor-Jankowski, J. (eds.), *Medical Primatology 1972*, Part 1. Karger, Basel, pp. 88–92.

Faucheux, B., Bertrand, M., and Bourliere, F. (1978). Some effects of living conditions upon the pattern of growth in the stumptailed macaque (*Macaca arctoides*). *Folia Primatol.* 30:220–236.

Griffin, G.A. (1966). The effects of multiple mothering on the infant-mother and infant-infant affectional systems. Ph.D. Dissertation, University of Wisconsin, Madison, WI.

Harvey, N.C., Rhine, R.J., and Bunyak, S.C. (1979). Weights and heights of stump-tailed macaques (*Macaca arctoides*) living in colony groups. *J. Med. Primatol.* 8:372–376.

Kerr, G.R., Chamove, A.S., and Harlow, H.F. (1969a). Environmental deprivation: Its effect on the growth of infant monkeys. *J. Pediatr.* 75: 833–837.

Kerr, G.R., Scheffler, G., and Waisman, H.A. (1969b). Growth and development of infant *M. mulatta* fed on a standardized diet. *Growth* 33:185–199.

Kinkle, D.K., and Session, H.L. (1972). A method for hand rearing of *Saimiri sciureus. Lab. Anim. Sci.* 22:207–209.

Sackett, G.P., Holm, R.A., Ruppenthal, G.C., and Fahrenbruch, C.E. (1976). The effects of total social isolation rearing on the behaviour of rhesus pigtail macaques. In: Walsh, R.N., and Greenough, W.T. (eds.), *Environments as Therapy for Brain Dysfunction*. Plenum Press, New York, pp. 115–131.

Sackett, G.P., Holm, R.A., and Fahrenbruch, C.E. (1979). Ponderal growth in colony- and nursery-reared pigtailed macaques. In: Ruppenthal, G.E., and Reese, D.C. (eds.), *Nursery Care of Nonhuman Primates*. Plenum Press, New York, pp. 187–201.

Scheffler, G., and Kerr, G.R. (1975). Growth and development of infant *M. arctoides* fed a standardized diet. *J. Med. Primatol.* 4:32–44.

Schrier, A.M. (1967). Learning-set formation by three species of macaque monkeys. In: Stark, S., Schneider, S., and Kuhn, A. (eds.), *Neve Ergebnisse der Primatologie*. Fisher, Stuttgart, pp. 60–75.

Valerio, D.A., Miller, R.L., Innes, J.R.M., Courtney, K.D., Pallota, A.J., and Guttmacher, R.M. (1969). *Macaca mulatta: Management of a Laboratory Breeding Colony*. Academic Press, New York.

Vice, T.E., Britton, H.A., Ratner, I.A., and Kalter, S.S. (1966). Care and raising of newborn baboons. *Lab. Anim. Care* 16:12–22.

Vickers, J.H. (1968). Primate pediatrics. *J. Small Anim. Pract.* 9:132–133.

Voss, W.R., Buss, D.H., and Cairol, L.W. (1971). A self-feeding device for infant baboon liquid diets. *Lab. Anim. Sci.* 21:901–903.

Walike, B.C., Campbell, D.J., and Hillman, R.A. (1971). A new liquid-diet feeder for primates. *J. Appl. Psychol.* 31:946–947.

Zwartouw, H.T., McArthur, J.A., Boulter, E.A., Seamer, J.H., Marston, J.H., and Chamove, A.S. (1984). The transmission of B virus infection between monkeys especially in relation to breeding colonies. *Lab. Anim.* 18:125–130.

CHAPTER TWENTY

Saliva as a Medium for Assessing Cortisol and Other Compounds in Nonhuman Primates: Collection, Assay, and Examples

Mark L. Laudenslager, Tamara Bettinger, and Gene P. Sackett

My mind sent a message to my hypothalamus, told it to release the hormone CRF into the short blood vessels connecting my hypothalamus and my pituitary gland. The CRF inspired my pituitary gland to dump the hormone ACTH into my bloodstream. My pituitary had been making and storing ACTH for just such an occasion. . . . And some of the ACTH in my bloodstream reached the outer shell of my adrenal gland, which had been making and storing glucocorticoids for emergencies. My adrenal gland added the glucocorticoids to my bloodstream. They went all over my body. (Kurt Vonnegut, *Breakfast of Champions*, 1973)

1. BACKGROUND

The body registers challenging, stressful, or traumatic events in a number of physiological systems. Most notable of these is the hypothalamic–pituitary–adrenal (HPA) axis (Miller and O'Callaghan, 2002; Seeman and McEwen, 1996), activation of which can profoundly influence brain development (Vazquez, 1998). True to Vonnegut's description three decades ago, glucocorticoids are found everywhere in bodily fluids, including saliva. Salivary cortisol is becoming increasingly popular as a means of assessing HPA activity, and meaningful relations exist between other steroids in serum/plasma and saliva (Vining *et al.*, 1983; Vining and McGinley, 1987; Kirschbaum *et al.*, 1992). A recent search of Medline revealed an almost exponential increase (Fig. 20–1) in the number of papers in which salivary cortisol appears as an index term for human studies in the 14 years since Kirschbaum and Hellhammer (1989) reviewed it as a biomarker of HPA activity. However, the use of saliva sampling for assessing steroid levels in nonhuman primates (NHP) has not reached the same level of enthusiasm.

Traditionally, investigators have relied on samples of blood, feces, or urine to obtain information on the levels of cortisol and other steroids in NHPs. Blood sampling is fraught with difficulties, however, whether humans, who are only marginally compliant, or NHPs, which are often not compliant with phlebotomy procedures, are sampled (Boccia *et al.*, 1992; Capitanio *et al.*, 1996; Blank *et al.*, 1983). While it is true that unrestrained NHPs can be trained, with the use of positive reinforcement, to participate in phlebotomy techniques without being restrained (Reinhardt, 1991), training is more complicated when the focus is on very young subjects. In developmental studies, the luxury of an extended training period is not typically available, and, thus, other methods need to be developed. One ingenious approach involves remote-control activation of an automated blood-collection device worn in a backpack attached to an indwelling catheter in freely moving baboons (Bentson *et al.*, 1999).

The primary means of assessing cortisol and sex steroids in free-ranging NHPs when capture is not an option relies on the collection of fecal and urine samples (Cavigelli, 1999; Pryce *et al.*, 1995; Robbins and

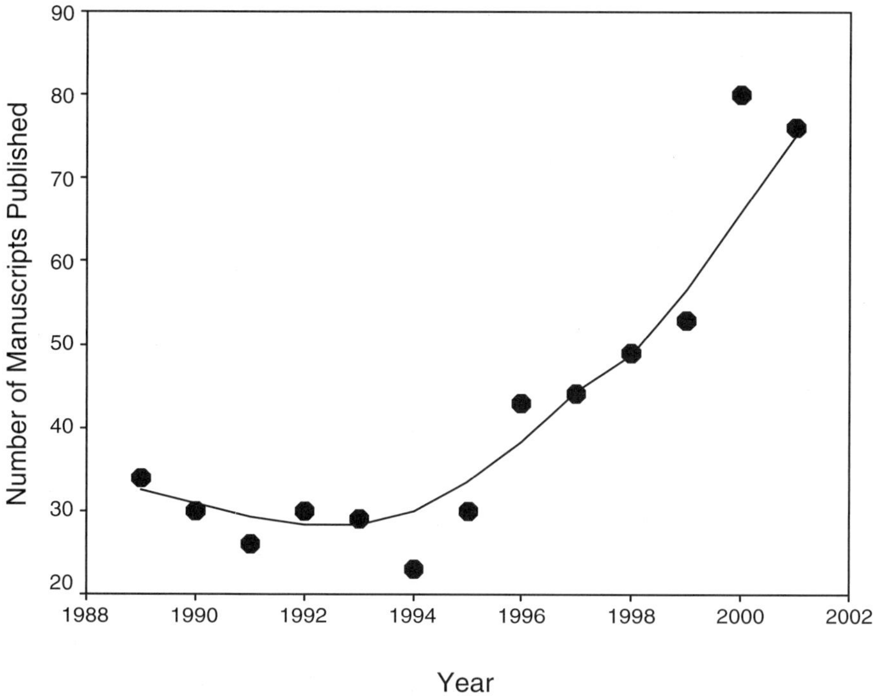

Figure 20–1. Graph of the nearly exponential increase in the number of human studies listed in Medline using saliva collection as a means of assessing free cortisol levels. A similar trend does not exist for nonhuman primates (data not shown).

Czekala, 1997; Smith and French, 1997; Stolinski *et al.*, 2002; Ziegler *et al.*, 1996). Fecal sampling provides an integrated perspective on steroid levels in the animals but it is generally not useful either for assessment of rapid changes or for tracking diurnal patterns. Transit time varies across species and diet. In addition, an extensive extraction procedure adds to the time required to process these samples.

Clearly, investigators working with NHPs need a simpler, less invasive method of obtaining information about steroid levels in their subjects, and saliva presents the solution. In this chapter we will describe methods of obtaining saliva samples from NHPs and address a number of factors that must be considered when using the samples to assess cortisol and other steroids and compounds.

1.1. Free Cortisol in Saliva

Cortisol in serum/plasma exists in two forms, bound and unbound. Cortisol may be bound to corticosteroid-binding globulin (CBG), plasma proteins and albumin, and platelets (Kirschbaum and Hellhammer, 1989; Vining *et al.*, 1987). Free or unbound cortisol is generally considered the biologically active component, although recent work suggests that cortisol bound to CBG also has localized effects on cellular activity (for review see Breuner and Orchinik, 2002). The unbound fraction represents 5–15% of the total cortisol concentration measured in plasma of most primates, though there is considerable species variation in CBG structure (Breuner and Orchinik, 2002).Variation in CBG's binding affinity for cortisol directly impacts levels of free hormone.

A unique aspect of salivary cortisol and sex steroids is that only the free component is measured because bound cortisol cannot enter the saliva compartment. Unconjugated cortisol moves via passive intracellular diffusion into the acinar cells of the salivary gland as a result of cortisol's high lipid solubility and then into the saliva following a very short delay [within 1 min after a bolus injection of cortisol (Vining *et al.*, 1983)]. However, the latency until the peak level is reached depends on the rate of change in serum level. Other unbound steroids move by similar processes into saliva.

Saliva is formed not by passive filtration of the serum but by an active process requiring energy. Small organic compounds found in saliva include urea, glucose, and steroids. Larger proteins such as enzymes and other glycoproteins are found in trace concentrations. In addition, a number of ions such as sodium are typically found in saliva in concentrations nearly isotonic with those in plasma. Saliva also contains bacteria, white blood cells, cells from the oral cavity, and mucins.

Saliva glands do not secrete saliva in the absence of stimulation, although even in sleep there is adequate stimulation to produce flow rates of ~0.05 ml/min. Secretion rate can increase to 5–10 ml/min in the presence of a taste stimulant (Vining and McGinley, 1987). Importantly, unconjugated steroid compounds such as cortisol and sex steroids are in equilibrium between saliva and plasma even at high flow rates. However, flow rate influences concentration of other compounds in saliva, such as secretory immunoglobulin A (sIgA), owing to the process by which they enter the saliva.

Compounds enter the saliva by a wide variety of processes, and these processes affect the extent to which samples accurately reflect serum/plasma levels and thus provide biologically meaningful information. Three physiological processes contribute to the presence of compounds in the saliva: (1) passive intracellular diffusion of lipid-soluble compounds such as steroids, which diffuse through acinar cells of the salivary glands; (2) ultrafiltration of small polar substances with a molecular weight less than 100, which pass between plasma and saliva; and (3) active transport or secretion of large proteins such as sIgA. Another mechanism by which compounds can appear in saliva is blood contamination such as that associated with oral injury, shedding of deciduous teeth, and gum disease. Such contamination results in elevated steroid concentrations and the presence of compounds that might not otherwise be found in saliva. There are commercial kits (Salimetrics, Inc.) available that can detect trace amounts of transferin, a signal of potential blood contamination, but only rarely do investigators actually check for blood contamination in saliva samples used for steroid determinations (Lac *et al.*, 1993).

Hormone concentrations in saliva can be affected by recent dietary intake. For example, consumption of a protein-rich diet elevates salivary cortisol in humans via HPA activation (Gibson *et al.*, 1999). Fluid consumption may dilute the saliva and lower cortisol concentrations, whereas milk and some formulas can raise cortisol levels. Residual milk or formula in the mouths of young monkeys can result in saliva samples that contain measurable cortisol and/or other compounds that cross-react with assay antisera (Magnano *et al.*, 1989). Food may also contain contaminants such as phytoestrogens, which can cross-react with highly sensitive steroid assays.

1.2. Other Compounds in Saliva

Transmitter metabolites have been measured in saliva, most notably methoxyhydroxyphenolglycol (MHPG), the primary metabolite of norepinephrine, and homovanillic acid (HVA), the primary metabolite of dopamine (Drebing *et al.*, 1989; Yang *et al.*, 1997). Salivary concentrations of metabolites show high correlations with plasma, and some investigators have suggested that they may also reflect central nervous system (CNS) activity of the noradrenergic and dopaminergic systems. Without a way of discriminating peripheral from central sources of these metabolites in saliva, such an interpretation for saliva samples may be stretching the point; in fact, even cerebrospinal fluid (CSF) has been challenged as a marker for CNS noradrenergic activity (Kopin *et al.*, 1983). These metabolites in saliva should be dismissed as a viable means of assessing peripheral sympathetic activation (Yang *et al.*, 1997).

Melatonin is another substance that is detectable in saliva. Although assays have been developed to measure salivary levels of melatonin (Leibenluft *et al.*, 1996; Voultsios *et al.*, 1997) and are available commercially, these techniques have not yet been applied to NHPs.

Peptide hormones have also been measured in saliva. In one study, growth hormone was present in measurable levels in samples collected from humans, albeit at levels 1/1000 of those present in serum (Rantonen *et al.*, 2000). Serum and saliva levels were correlated, $r = 0.59$, but serum/saliva cortisol level correlations were also quite low for the same samples, $r = 0.47$. The strength of these observations is questionable, however, because normally these correlations fall above 0.75, often approaching 0.95. In addition, growth hormone levels for many samples were below the detection limits for the assay. In a study of rhesus macaques (*Macaca mulatta*), salivary prolactin-like immunoreactivity was correlated ($r = 0.75$) with 5-hydroxyindoleacetic acid (5-HIAA) in the CSF (Lindell *et al.*, 1999), but saliva and plasma prolactin were not correlated. Because the magnitude of increase in plasma prolactin in response to fenfluramine challenge is considered a means of assessing CNS serotonergic tone (Coccaro *et al.*, 1997), the relation of salivary prolactin to CSF serotonergic functioning is of theoretical interest. Salivary prolactin therefore could prove a very useful noninvasive means of assessing CNS serotonergic activity in NHPs. Interestingly, an assay directed specifically toward human prolactin failed to detect the substance in saliva samples (Thijssen *et al.*, 2000, 2001). In a follow-up study with rhesus macaques, a more sensitive assay using a different antibody revealed that prolactin-like immunoreactivity was above the detection limit in 3 of 18 placebo samples and only 5 of 18 fenfluramine challenge samples (Dvoskin *et al.*, 2001). Since peptide hormones vary structurally between species, the antibodies recognizing human prolactin could very well have cross-reacted with an unknown peptide related to CNS serotonergic activity in the macaques. The samples that were positive for prolactin-like immunoreactivity were collected from ketamine-anesthetized monkeys via pipettes, which may have caused inadvertent blood contamination that could have confounded assessment.

On the basis of these observations, it is unlikely that these techniques will prove useful for assessing prolactin and other peptide hormone levels in the immediate future. Since there is no known active transport mechanism for prolactin (C. Kirschbaum, personal communication) or, probably, for growth hormone, saliva levels bear little relation to plasma/serum levels (Vining *et al.*, 1987). Thus, current assays may provide sufficient sensitivity to detect trace substances in saliva, but there are too many uncontrolled and unknown factors that contribute to the presence of large compounds in saliva. Currently, measurement of free steroids in saliva is a more informative and reliable use of NHP saliva samples.

2. METHODS FOR COLLECTING SALIVA FROM NONHUMAN PRIMATES

Even though the use of saliva sampling has been state-of-the-art in human studies for well over a decade, only recently has this approach been applied to NHPs. Aside from presentations at meetings, only four papers have described the use of saliva sampling for hormone assessment in NHPs (Arslan *et al.*, 1984; Boyce *et al.*, 1995; Fuchs *et al.*, 1997; Lutz *et al.*, 2000). Thus, for most NHP species normative data on salivary steroid hormones are seriously lacking.

What methods are available for collecting saliva samples from NHPs? Although human subjects can simply spit into a tube, the most popular method for obtaining samples from both humans and NHPs involves the use of a simple device called a Salivette (Sarstedt, Inc.). The Salivette consists of a cotton roll and a two-piece tube. The subject chews on the cotton roll (or, alternatively, on a strip of cotton dental rope), which absorbs saliva, and the roll (or rope) is put into the upper chamber of the tube and centrifuged, which expresses the saliva into the lower chamber.

The cotton is a great medium for collecting saliva but it poses a problem for measuring hormones other than cortisol (Granger *et al.*, 1999a,b,c). Cross-reacting compounds, either in the glue that binds the material together or in phytoestrogens present in the cotton, interfere with sex steroid and dehydroepiandrosterone (DHEA) assays and lead to inflated hormone levels. This is a problem for both the Salivette devices and cotton dental rope (Boyce *et al.*, 1995; Lutz *et al.*, 2000) frequently used in collection of saliva samples. We have found that we can either denature or extract this substance by soaking the cotton dental rope or Salivette rolls in 50% ethanol (1.3 liters for 25 standard 6-inch dental ropes) for 24 hr using three successive fresh alcohol cycles. After the third soak cycle, the ropes (or rolls) are dried completely in a 37°C dry incubator before use in saliva collection. After these soak cycles, the background levels of DHEA, progesterone, and testosterone are below the detection limits of the Salimetrics assay.

A major consideration in collecting saliva samples from NHPs is that these animals may harbor serious biohazards. Macaques, for instance, may carry the herpes B virus in their saliva, so samples must be collected with care. In the case of monkeys housed in standard cages, saliva samples can be obtained with the use of a pole device (Lutz *et al.*, 2000). In brief, a cotton dental rope is treated with Kool-Aid and attached to a pole. The monkey is permitted to chew on the rope until the rope is saturated with saliva, whereupon the pole is withdrawn and the saliva is expressed and stored for analysis. Alternatively, surgical gauze can be coated with Kool-Aid crystals and held in place with a screen arrangement while the monkey licks the gauze. The saliva-soaked gauze is then processed in much the same manner as the cotton dental rope. Of these two methods, the rope on a pole offers the more effective means of collecting saliva.

Because the pole device may seem threatening to some monkeys, we accustom them to the pole by placing similar-sized PVC pipes in their cages for 2 weeks before beginning saliva collection. Unfortunately, some monkeys cannot adapt to the pole and continue to react to it as a threatening object. We simply give these monkeys a flavored rope and allow them to chew on it until they drop it in a clean pan under the cage, whereupon the rope is quickly retrieved and centrifuged in the Salivette device.

3. FACTORS THAT AFFECT SALIVARY HORMONES

A number of factors must be considered with regard to handling, storage, and assessment of saliva samples. First, the saliva should be expressed as soon as practical after collection and transferred to O-ring-sealed tubes for long-term storage. Even with centrifugation, long-term storage in the Salivette tube is not recommended, nor should the cotton plug be frozen in this collection device without centrifugation or stored for longer than 2 weeks. Samples do not need to be stored at ultralow temperatures; storage at –15 to –20°C is adequate. In point of fact, cortisol (but not the sex steroids) is tolerant of a wide variety of temperatures and shipping conditions (Chen *et al.*, 1992; Clements and Parker, 1998). Samples for cortisol determinations may even be kept at room temperature for several days (Kirschbaum and Hellhammer, 1989), although longer storage without refrigeration promotes the growth of mold, making the specimen unpleasant to process. Transfer of the samples to O-ring-sealed tubes is particularly important if they are to be frozen for >3 months, as long-term storage at –20°C in the Salivette can freeze-dry the sample and inflate the levels of hormone measured in the saliva. Separate aliquots are recommended when possible, as the sex steroids and DHEA are not resistant to repeated freeze–thaw cycles (E. Schwartz, personal communication, 2002).

Second, the mucins in saliva make pipetting quite difficult. These substances can be denatured by freezing and thawing, with the added advantage that this process removes froth and the denatured mucins will absorb other particulate matter in the saliva. The mucins do not affect steroid concentrations in the saliva. Repeated freeze–thaw cycles do affect sex steroid concentrations, but cortisol seems to be resistant to freeze–thaw cycles.

Third, the flavors applied to the dental rope (e.g., Kool-Aid and a variety of food flavorings such as vanilla, butterscotch, almond, cherry, orange, banana, and rum) entice monkeys to chew on the ropes and stimulate saliva flow, but they can also affect assay results. Although these flavors do not interfere with the Salimetrics cortisol assay, it is possible that they could affect the performance of other assay kits. Indeed, a series of studies showed that not only do some flavors affect assay performance more than others, some assays are also more labile (Schwartz *et al.*, 1998). The addition of increasing concentrations (≥0.05 g/ml) of Crystal Light lemon-lime flavor with saliva was associated with the greatest interference in Pantex kits and the least in DPC and Corning cortisol kits. Using grape-flavored Kool-Aid to stimulate saliva flow in children was associated with minimal effects over the Kool-Aid concentration range (0.005–0.2 g/ml) tested for the Corning and DPC kits but the Pantex kit was affected by increasing concentration. These results, presented with parametric details by Schwartz *et al.* (1998), suggest that the potential impact of saliva flow stimulants on the proposed assay kit should always be tested.

Fourth, assay results can be affected by the pH of the sample (Schwartz *et al.*, 1998). Specifically, lowering the pH tends to inflate the apparent concentration of salivary cortisol. To track this effect, the Salimetrics kit includes internal pH indicators that alert the investigator to pH shifts that might affect assay results for that sample. Highly acidic flavorings (citrus flavors) have the potential to affect pH and should be carefully tested in a pilot study using the particular assay to be applied.

Finally, a few comments should be made with regard to the variety of assay techniques available for determining steroids in saliva. Early approaches relied predominantly on standard radioimmunoassays for serum/plasma hormones modified to accommodate the lower concentrations of these hormones in saliva. In our early studies in humans (Wamboldt *et al.*, 2003) we used a radioimmunoassay kit (Diagnostic Products Company, Cortisol Coat-A-Count RIA) modified to permit the determination of lower levels of cortisol found in saliva. A 100-μl sample (rather than the 25 μl recommended by the manufacturer) was transferred in duplicate to assay tubes coated with antibody. The standard curve was modified by serial dilutions of the standards supplied with the kit to include additional standards of 0.01, 0.05, 0.1, 0.25, 0.5, 1.0, 2.0, and 5.0 μg/dl. The standards were added in volumes of 100 μl. Incubation time was increased from 30 min to 3 hr. A calibrated low standard (range 4–5 μg/dl, Diagnostic Products Company) was run in every assay undiluted and diluted by a factor of 10 or 100 in Dulbecco's phosphate buffer to an approximate concentration of 0.4 and 0.04 μg/dl. Finally, saliva standards whose concentrations were confirmed by mass spectrography (kindly supplied by Dr. Clemens Kirschbaum) were included in initial assays to confirm assay reliability. The minimal detectable level for this modified procedure is ~0.05 μg/dl. Intra- and interassay coefficients of variation for this assay are <7%.

The newer assay kits are based on enzyme immunoassay (EIA) techniques that eliminate the use of radioactive materials. The detection limits of these newer assays are quite good, as low as 0.007 μg/dl for cortisol (Salimetrics) to an upper limit of 1.8 μg/dl. No serial dilution of the standards is required, as these kits were manufactured to cover the ranges seen in human saliva. However, the saliva of some NHPs, notably New World monkeys, requires dilution in assay buffer because the free levels are considerably higher than in humans. The use of assay buffer preserves the correct matrix for the performance of the assay. In our laboratory, the inter- and intraassay CVs obtained with the Salimetrics EIA kit tend to be ≤5%.

The protocol for the Salimetrics kit is simple and straightforward. Briefly, 50 μl of saliva is added in duplicate to the wells of a microtiter plate coated with rabbit anticortisol antibody. The unknowns compete with horseradish peroxidase-conjugated cortisol for the binding sites. The substrate for this enzyme, tetramethylbenzidine, is added and the reaction is stopped with H_2SO_4 and read at 450 nm in a microplate reader. Assay standard curves are fitted by a weighted regression analysis provided by commercial software for the EIA plate reader (Dynex MRX) using Revelation 3.2 software. From these curves, unknown values are computed. This kit shows minimal cross-reactivity (≤4%) with other steroids present in the saliva. As many samples as practical can be run in the same assay and subjects are never split across different assay plates. Salimetrics now has EIA kits available for DHEA, testosterone, and progesterone. We have used these kits for 4 years and are highly satisfied with their performance.

4. APPLICATION OF SAMPLING TECHNIQUES AND INTERPRETATION OF RESULTS

To reiterate, it is surprising that there are so few publications on the use of saliva as a biological medium for assessing steroids in NHPs. To date it appears that only macaques and squirrel monkeys have been studied in this way. In the earliest paper, Arslan *et al.* (1984) assessed salivary testosterone in anesthetized male pigtailed macaques (*M. nemestrina*) and rhesus macaques following a single intramuscular injection of testosterone proprionate. Testosterone levels in serum and saliva rose and fell in parallel over an 8-day period in response to a single 50-mg injection and were highly correlated with each other ($r = 0.92$, $n = 20$). Salivary levels ranged from 0.1 to 1.6% of serum total testosterone levels. The authors concluded that saliva was a valid and convenient means for determining testosterone levels in monkeys in much the same way as in humans.

Because of the necessity for caution in the handling of macaques, blood frequently is drawn with the animals under ketamine anesthesia. However, the use of ketamine elevates salivary cortisol compared with levels in saliva collected from the same monkeys on different days without ketamine (Fig. 20–2). A diurnal decline in salivary cortisol from morning to afternoon was noted in singly housed pigtailed macaques in the absence of ketamine. Others have also reported potential long-term impacts on cardiovascular (Bennett, 2001) and immune (Lockwood *et al.*, 1993; Worlein *et al.*, 1995) function associated with ketamine. Thus, collection of biological samples without the use of ketamine has distinct advantages.

Techniques for collecting saliva from unanasthetized NHPs have been described only recently (see also Chapter 21 by Roma, this volume). Saliva samples have been successfully collected on a cotton swab from adult male and female squirrel monkeys while they were seated in a restraining chair and performing a cognitive task (Fuchs *et al.*, 1997). Even after the animals were habituated to the task and sampling conditions, salivary cortisol levels were elevated at the end of each cognitive testing period, indicating that a rise in salivary cortisol was associated with testing and thereby validating the usefulness of this technique for

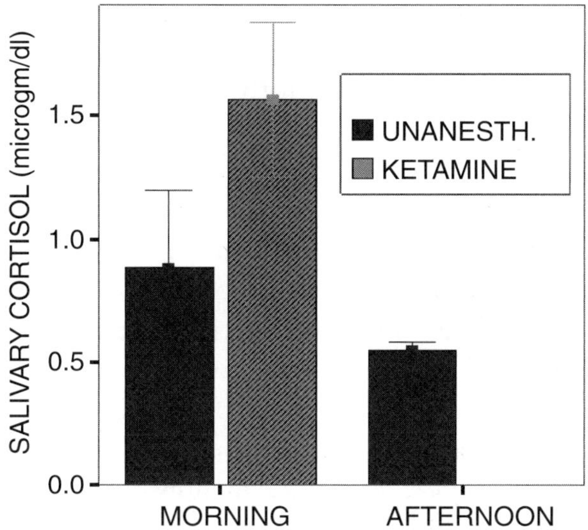

Figure 20–2. Influence of ketamine anesthesia on salivary cortisol in pigtailed macaques (*M. nemestrina*). Saliva samples were collected from four unanesthetized monkeys and from the same monkeys when they were anesthetized with 10 mg/kg ketamine. Saliva was collected as soon as the animal was unresponsive to handling. Vertical line indicates 1 SEM.

detecting HPA activation. Most studies involving New World monkeys have shown that levels of serum/plasma cortisol are quite high relative to those in other species of NHP. The well-adapted monkeys in the study by Fuchs *et al.* (1997) had salivary cortisol levels of <1 μg/dl when they were first placed in the restraining chair between 0730 and 1130 hr each morning and slightly >1 μg/dl 1 hr later. Indeed, salivary cortisol levels rose as much as 20-fold in the presence of an unfamiliar handler. Plasma total cortisol levels are typically reported to be ≥100 μg/dl for squirrel monkeys (Coe *et al.*, 1992), with 29–42% appearing in the unbound form (Klosterman *et al.*, 1986). Thus, the level observed in saliva of unhabituated squirrel monkeys should be in the range of 25–50 μg/dl. On the basis of the results reported by Fuchs *et al.* (1997), plasma levels in the literature probably do not reflect true resting levels. Using dental roll collection techniques with five aged (17-year-old) squirrel monkeys, we obtained salivary cortisol levels that were higher than those reported by Fuchs *et al.* (Fig. 20–3). Samples were collected between 0400 and 1600 hr. Salivary cortisol showed an expected diurnal pattern, albeit at a much higher level than in the study by Fuchs *et al.* The aged monkeys had not received extensive habituation to the collection methods, and this may have affected their salivary cortisol levels. Use of consistent salivary collection techniques with well-habituated animals is very likely to shed light on true resting levels of free cortisol in NHPs.

Indeed, monkeys may require a period of adaptation to these sampling techniques before basal levels are observed. The habituation effect in young pigtailed macaques, aged 2–12 months, is shown in Fig. 20–4. Cortisol levels in the initial samples were low but rose significantly in the second samples and remained elevated for the next sample or two before returning to the level noted for the first sample.

A single study has reported salivary cortisol in infant rhesus macaques (Boyce *et al.*, 1995). Saliva was collected on cotton dental rope that was flavored with sweetened Kool-Aid. After the rope was soaked with the monkey's saliva, it was retrieved and placed in a syringe. Then the plunger

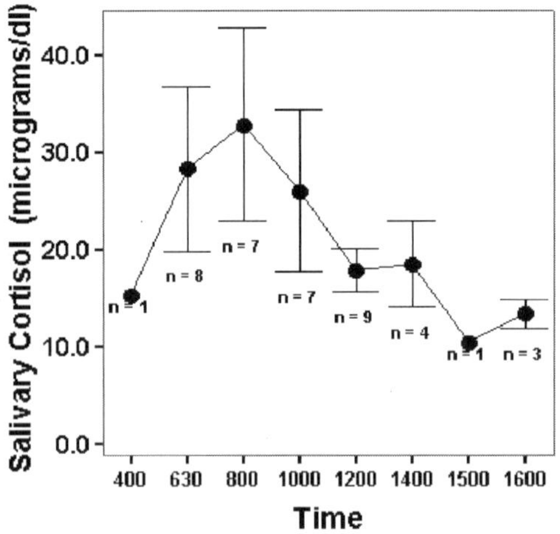

Figure 20–3. Mean (±SEM) diurnal variation of salivary cortisol in aged squirrel moneys (*Saimiri sciureus*) housed individually.

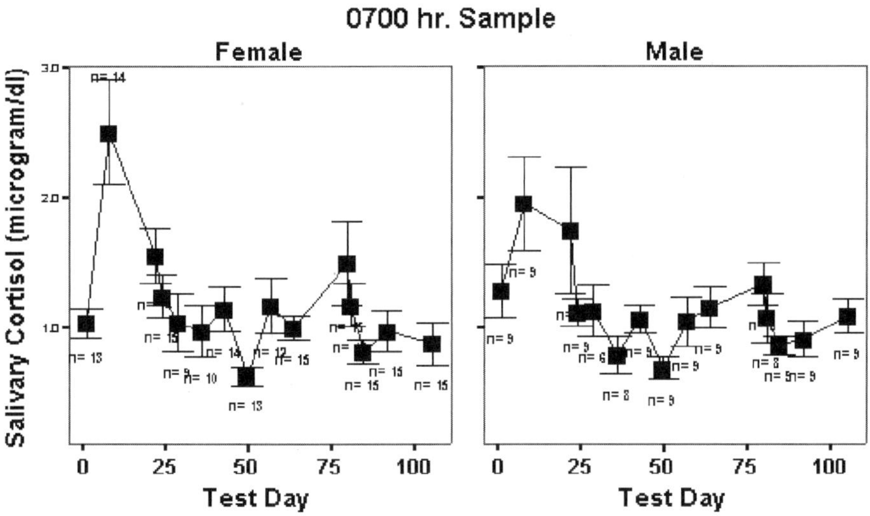

Figure 20–4. Habituation effect shown by young pigtailed macaques (*M. nemestrina*), aged 2–12 months. Note the higher salivary cortisol levels (mean ± SEM) on the two collection days following the first collection. Samples were collected at weekly intervals to begin and then more frequently and at multiple times during the day. Cortisol levels were within the range noted on the fist day by the fourth or fifth collection time.

of the syringe was depressed and the saliva thereby expressed, yielding a useful volume (~100 μl) of saliva in about 89% of attempts. We would indicate that the Salimetrics cortisol EIA kits require the least volume (50 μl) for duplicates, whereas 100 μl is required for DHEA. Saliva samples were analyzed with the Pantex kit, which, as noted earlier, is sensitive to acidic flavors. Cortisol levels in saliva collected after peer socialization were typically higher than those on nonsocialization days. Furthermore, salivary cortisol rose between 0800 and 1100 hr and declined between 1100 and 1600 hr. Thus, the diurnal rhythm of cortisol in these nursery-reared macaques did not follow the patterns seen in humans (Wamboldt *et al.*, 2003) and other NHPs. It may be that colony and nursery management practices interfere with a more typical diurnal decline after the initial morning sample. Despite the unexpected results, however, these observations suggest that saliva collection using flavored cotton dental rolls is a viable means of assessing free cortisol in nursery-housed macaques.

It should be noted that levels of plasma/serum total cortisol in non-human primates show a wide range of variation and some of this variance could be related to stress associated with sample collection. Total cortisol appears to be inversely related to body size; that is, smaller New World monkeys tend to have much higher levels than the considerably larger great apes (Coe *et al.*, 1992). Using banked serum from two zoos (Denver Zoological Garden and Cleveland Metroparks Zoo), we assessed total plasma/serum cortisol from NHPs representing a weight range of 0.5–150 kg (M.L. Laudenslager, T. Bettinger, C. Kuhar, D. England, and M. Goldstein, unpublished observations). We recognized that even under the best of conditions, there would be considerable variance in these samples due to each zoo's handling protocol and phlebotomy technique. Like Coe *et al.* (1992), we observed a nonlinear relation between body size and cortisol level (Fig. 20–5). Smaller New World species had higher total cortisol levels than species whose weight exceeds 1 kg. The percentage of free cortisol varies across species as well; in the smaller New World species it can be as high as 25–50% due to reduced binding affin-

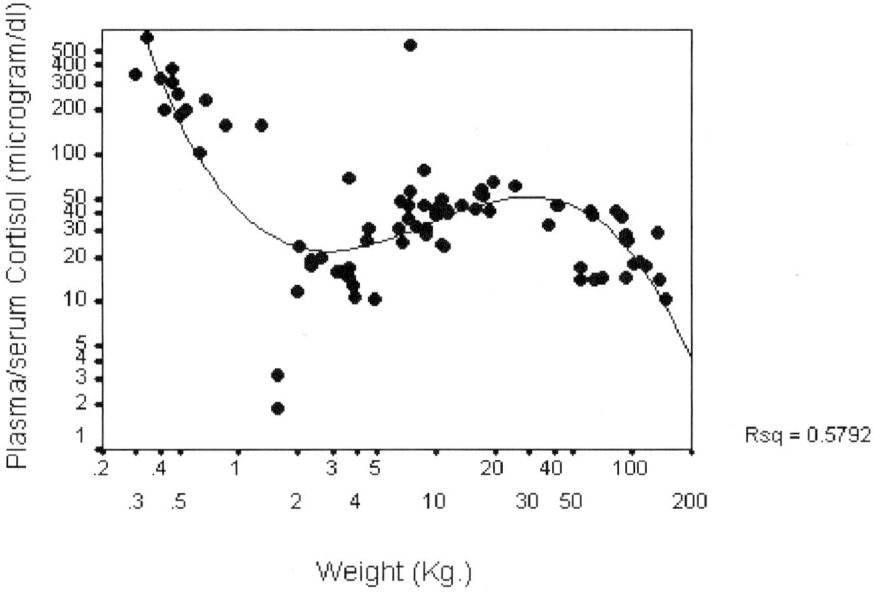

Figure 20–5. Data showing nonlinear relation between body weight and plasma/serum total cortisol. Cortisol levels tend to decline as body size increases. The relation is best fit by a cubic equation.

ity of CBG (Klosterman *et al.*, 1986). When establishing assay parameters for hormones in saliva samples, it is necessary to take into consideration the species, levels of total cortisol, and percentage that is unbound in order to keep sample concentrations within the linear range of the assay in use. If a sample is expected to run high, dilution with assay buffer often brings it into the linear range of the assay. When performing RIA and EIA procedures, it is preferable for the unknowns to fall within the midrange of the standard curve, as error is greater at the extremes of the curve.

The utility of saliva as a medium for assessing cortisol permitted us to assess diurnal declines in salivary cortisol levels in three captive young male western lowland gorillas (Bettinger *et al.*, 1999). Indeed, it would have been impossible to acquire this information using routine phlebotomy techniques. The apes were trained in a "trading" paradigm wherein the gorilla chewed on an unflavored cotton roll and returned it to the keeper in trade for fresh fruit treats. This approach is quite effective when working with the great apes. Over a 2-year period we obtained 133 samples per gorilla. Analysis of the data revealed that all three of these captive gorillas experienced a smooth decline in cortisol levels from morning to evening (Fig. 20–6). These findings do not differ significantly from values reported for healthy human populations (Wamboldt *et al.*, 2003). The nature of this decline and its overall level has been used as an important marker for normal HPA functioning in humans (Gunnar and Vazques, 2001) and is likely to be equally important for nonhuman primates.

As a final example of the utility of saliva collection in captive NHPs, we will describe repeated sampling from 1- to 11-month-old male and female pigtailed macaques that were individually housed under identical conditions according to the husbandry and rearing standards of the IPRL Nursery Manual (Ruppenthal and Sackett, 1992). Saliva samples were collected on Kool-Aid-flavored dental ropes at 0400, 0700, 1400, and 2100 hr and processed with the Salimetrics EIA kit. Cortisol levels were highest at 0700 hr and then declined until 2100 hr (Fig. 20–7). On the basis of these sampling intervals, it appears that salivary cortisol may begin rising as early as 0100 hr in these nursery-reared monkeys. Since sampling times were not identical to those reported by Boyce *et al.*

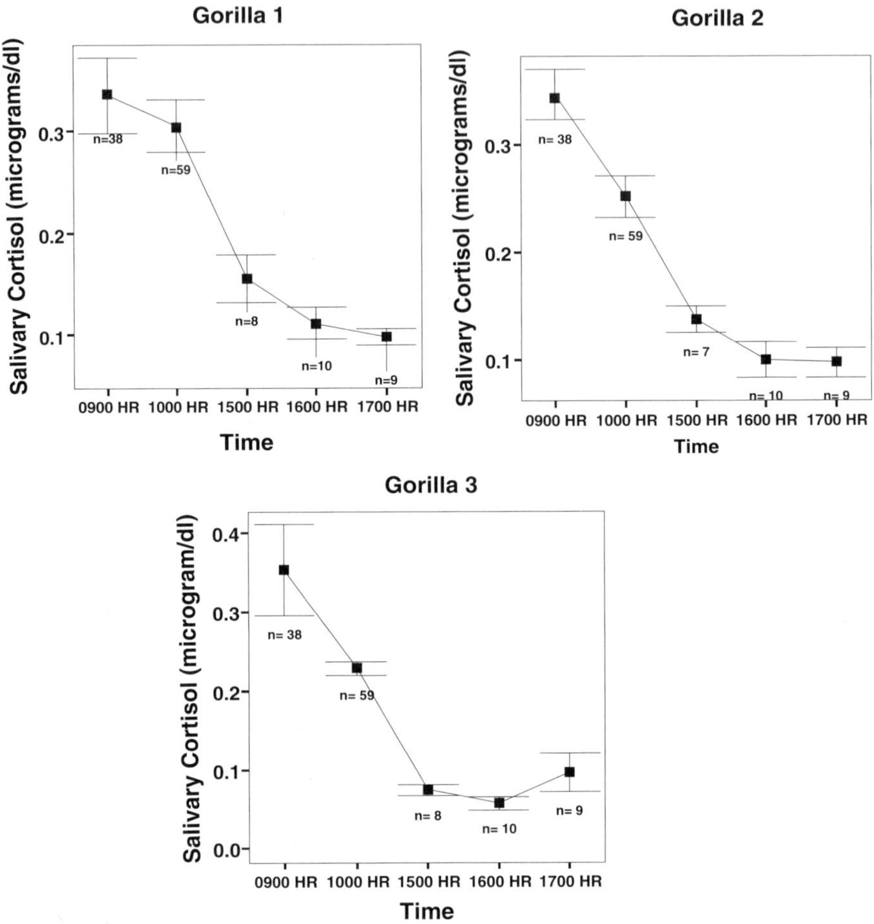

Figure 20–6. Mean (±SEM) diurnal declines in salivary cortisol in three male gorillas (*Gorilla gorilla*) maintained in the same social group. Cortisol concentrations noted in these gorillas are similar to those for human salivary cortisol (Kirschbaum and Hellhammer, 1994).

(1995), the presence of a peak in salivary cortisol at 1100 hr could not be verified. However, it is important to note that these young macaques did in fact demonstrate a circadian pattern in their salivary cortisol levels. Finally, a developmental decline in the 0700 hr sample level was noted between 1 month and 11 months of age (Fig. 20–8). This was not the habituation effect shown in Fig. 20–4, since the habituation pattern appears to have stabilized by the third or fourth collection interval.

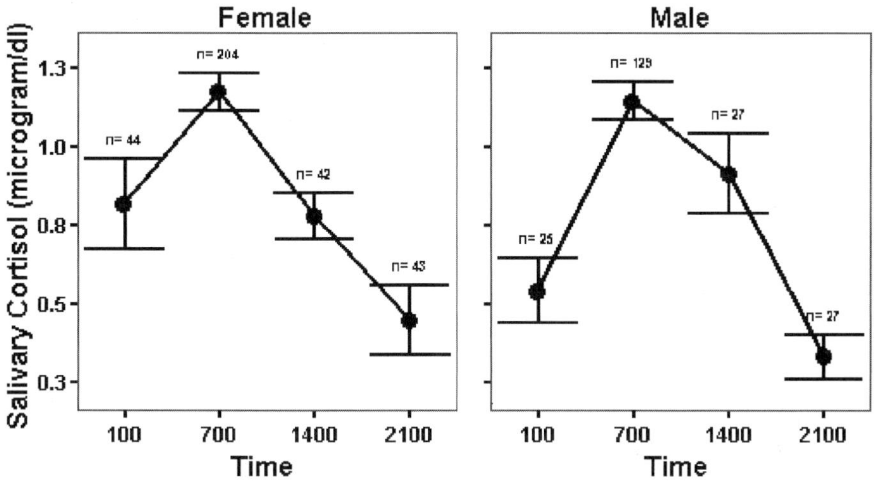

Figure 20–7. Mean (±SEM) diurnal decline in salivary cortisol in male and female nursery-reared pigtailed macaques (*M. nemestrina*) collapsed across 1 and 12 months of age. There were no differences based on gender for these monkeys.

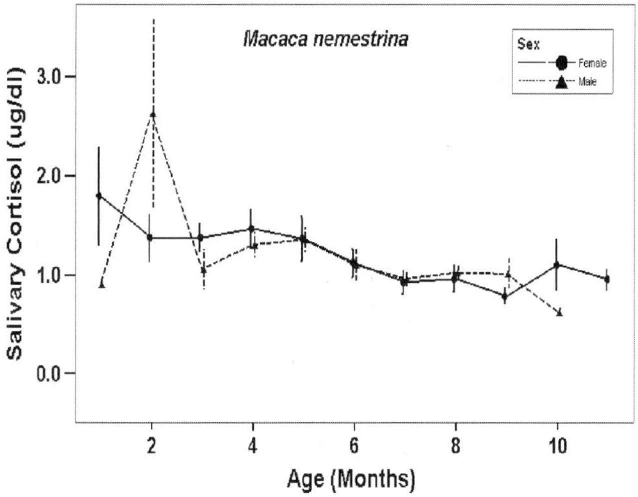

Figure 20–8. Mean (±SEM) 7 am salivary cortisol in nursery-reared pigtailed macaques (*M. nemestrina*) by age (1–12 months) and sex.

5. SUMMARY AND CONCLUSIONS

The ability to collect saliva using techniques described in this chapter opens the way for improved assessment of many steroid hormones in NHPs in the absence of stress typically encountered when collecting blood from either unrestrained or lightly anesthetized monkeys. There are, however, a number of factors that can contribute to variation in the hormone concentrations measured. A number of caveats have been described including the impact of blood contamination, an inability to measure many peptide compounds, adaptation to the collection procedures, flavoring used to either stimulate saliva flow or to increase the monkey's interest in chewing on the collection cotton, and so on. It is clear that any species, large or small, will cooperate with the researcher. However, the researcher must know how to ask the question and be patient enough to identify the conditions that are most suitable for a particular species—and often a particular individual. The ease with which saliva can be collected should be adequate motivation to encourage more primatologists to take advantage of this approach to less invasive sampling. The commercial availability of highly sensitive, nonradioactive EIA kits for assessing steroid hormones, such as the kits supplied by Salimetrics (WWW.Salimetrics.com), will permit laboratories to establish these procedures within a fairly short time frame. However, there are a number of factors that must be carefully considered in the interpretation of the data.

ACKNOWLEDGMENTS

This research was supported in part by NIH grant MH37373, an award from the UCHSC Developmental Psychobiology Endowment Fund, NCRR grant RR00166 to the Washington National Primate Research Center, and NICHHD grant HD02274 to the University of Washington Center on Human Development and Disability. The squirrel monkey data presented in Fig. 20–3 were collected at the primate colony of the National Institute of Aging, Laboratory of Experimental Gerontology (Donald Ingram, Director), Poolesville, MD, and supported by intramural resources. We gratefully acknowledge David England and Chris

Kuhar for their dedication in collecting and organizing the serum and saliva samples for data presented in Figs. 20–5 and 20–6, to Terri Rhyner, Cleveland Metroparks Zoo, for her diligence in training the gorillas and collecting the many saliva samples from them, and Mark Goldstein for his expertise in assaying saliva samples for the Behavioral Immunology and Endocrinology Laboratory, UCHSC.

REFERENCES

Arslan, M., Akhtar, F.B., and Nieschlag, E. (1984). Salivary testosterone levels in normal and testosterone treated monkeys. *J. Steroid Biochem.* 21:461–464.

Bennett, A. (2001). Cardiac signal analysis techniques in pharmacological and neurochemical studies with rhesus monkeys. *Am. J. Primatol.* 54:106.

Bentson, K.L., Miles, F.P., and Smith, O.A. (1999). A remote-controlled device for long-term blood collection from freely moving, socially housed animals. *Behav. Res. Methods Instrum. Comput.* 31:455–463.

Bettinger, T., Kuhar, C., Sironen, A., Goldstein, M., and Laudenslager, M.L. (2001). Behavior and salivary cortisol in an all-male gorilla group. *Am. J. Primatol.* 49:35.

Blank, M.S., Gordon, T.P., and Wilson, M.E. (1983). Effects of capture and venipuncture on serum levels of prolactin, growth hormone and cortisol in outdoor compound-housed female rhesus monkeys (*Macaca mulatta*). *Acta Endocrinol.* 102:190–195.

Boccia, M., Broussard, C., Scanlan, J., and Laudenslager, M. (1992). Practice makes predictable: The differential effect of repeated sampling on behavioral and physiological response in monkeys. In: Davis, H., and Balfour, D. (eds.), *The Inevitable Bond: Examining Scientist-Animal Interaction.* Cambridge University Press, New York, pp. 153–170.

Boyce, W.T., Champoux, M., Suomi, S.J., and Gunnar, M.R. (1995). Salivary cortisol in nursery-reared rhesus monkeys: Reactivity to peer interactions and altered circadian activity. *Dev. Psychobiol.* 28:257–267.

Breuner, C.W., and Orchinik, M. (2002). Plasma binding proteins as mediators of corticosteroid action in vertebrates. *J. Endocrinol.* 175:99–112.

Capitanio, J.P., Mendoza, S.P., and McChesney, M. (1996). Influences of blood sampling procedures on basal hypothalamic-pituitary-adrenal hormone levels and leukocyte values in rhesus macaques (Macaca mulatta). *J. Med. Primatol.* 25:26–33.

Cavigelli, S.A. (1999). Behavioural patterns associated with faecal cortisol levels in free-ranging female ring-tailed lemurs, *Lemur catta*. *Anim. Behav.* 57: 935–944.

Chen, Y., Cintron, N.M., and Whitson, P. (1992). Long-term storage of salivary cortisol at room temperature. *Clin. Chem.* 38:190.

Clements, A.D., and Parker, C.R. (1998). The relationship between salivary cortisol concentrations in frozen versus mailed samples. *Psychoneuroendocrinology* 23:613–616.

Coccaro, E.F., Kavoussi, R.J., Trestman, S.M., Gabriel, S.M., Cooper, T.B., and Siever, L.J. (1997). Serotonin function in human subjects: Intercorrelations among central 5-HT indices and aggressiveness. *Psychiatry Res.* 73:1–14.

Coe, C.L., Savage, A., and Bromley, L.J. (1992). Phylogenetic influences on hormone levels across the primate order. *Am. J. Primatol.* 28:81–100.

Drebing, C.J., Freedman, R., Waldo, M., and Gerhardt, G.A. (1989). Unconjugated methoxylated catecholamine metabolites in human saliva. Quantitation methodology and comparison with plasma levels. *Biomed. Chromatogr.* 3:217–220.

Dvoskin, R., Lindell, S.G., Higley, J.D., Suomi, S.J., and Shoaf, S. (2001). No detectable prolactin in human saliva: Reply. *Soc. Biol. Psychiatry* 50:9–10.

Fuchs, E., Kirschbaum, C., Benisch, D., and Bieser, A. (1997). Salivary cortisol: A non-invasive measure of hypothalamo-pituitary-adrenocortical activity in the squirrel monkey, *Saimiri sciureus*. *Lab. Anim.* 31:306–311.

Gibson, E.L., Checkley, S., Papadopoulos, A., Poon, L., Daley, S., and Wardle, J. (1999). Increased salivary cortisol reliably induced by a protein-rich midday meal. *Psychosom. Med.* 61:214–224.

Granger, D.A., Schwartz, E.B., Booth, A., and Arentz, M. (1999a). Salivary testosterone determination in studies of child health and development. *Horm. Behav.* 35:18–27.

Granger, D.A., Schwartz, E.B., Booth, A., Curran, M., and Zakaria, D. (1999b). Assessing dehydroepiandrosterone in saliva: A simple radioimmunoassay for use in studies of children, adolescents, and adults. *Psychoneuroendocrinology* 24:567–579.

Granger, D.A, Schwartz, E.B, Booth, A., Curran, M., and Zakaria, D. (1999c). Assessing dehydroepiandrosterone in saliva: A simple radioimmunoassay for use in studies of children, adolescents and adults. *Psychoneuroendocrinology* 24:567–579.

Gunnar, M.R., and Vazquez, D.M. (2001). Low cortisol and a flattening of expected daytime rhythm: Potential indices of risk in human development. *Dev. Psychopathol.* 13:515–538.

Kirschbaum, C., and Hellhammer, D.H. (1989). Salivary cortisol in psychobiological research: An overview. *Neuropsychobiology* 22:150–169.

Kirschbaum, C., and Hellhammer, D.H. (1994). Salivary cortisol in psychoneuroendocrine research: Recent developments and applications. *Psychoneuroendocrinology* 19:313–333.

Kirschbaum, C., Wust, S., and Hellhammer, D. (1992). Consistent sex differences in cortisol responses to psychological stress. *Psychosom. Med.* 54: 648–657.

Klosterman, L.L., Murai, J.T., and Siiteri, P.K. (1986). Cortisol levels, binding, and properties of corticosteroid-binding globulin in the serum of primates. *Endocrinology* 118:424–434.

Kopin, I.J., Gordon, E.K., Jimerson, D.C., and Polinsky, R.J. (1983). Relation between plasma and cerebrospinal fluid levels of 3-methoxy-4-hydroxyphenylglycol. *Science* 219:73–75.

Lac, G., Lac, N., and Robert, A. (1993). Steroid assays in saliva: A method to detect plasmatic contaminations. *Arch. Int. Physiol. Biochem. Biophys.* 101:257–262.

Leibenluft, E., Feldman-Naim, S., Turner, E.H., Schwartz, P.J., and Wehr, T.A. (1996). Salivary and plasma measures of dim light melatonin onset (DLMO) in patients with rapid cycling bipolar disorder. *Soc. Biol. Psychiatry* 40:731–735.

Lindell, S.G., Suomi, S.J., Shoaf, S., Linnoila, M., and Higley, J.D. (1999). Salivary prolactin as a marker for central serotonin turnover. *Soc. Biol. Psychiatry* 46:568–572.

Lockwood, L.L., Siebert, L.H., Laudenslager, M.L., Watkins, L.R., and Maier, S.F. (1993). Anesthesia-induced modulation of *in vivo* antibody levels. *Anesth. Analg.* 77:769–774.

Lutz, C.K., Tiefenbacher, S., Jorgensen, M.J., Meyer, J.S., and Novak, M.A. (2000). Techniques for collecting saliva from awake, unrestrained, adult monkeys for cortisol assay. *Am. J. Primatol.* 52:93–99.

Magnano, C.L., Diamond, E.J., and Gardner, J.M. (1989). Use of salivary cortisol measurements in young infants: A note of caution. *Child Dev.* 60:1099–1101.

Miller, D.B., and O'Callaghan, J.P. (2002). Neuroendocrine aspects of the response to stress. *Metabolism* 51:5–10.

Pryce, C.R., Schwarzenberger, F., Dobeli, M., and Etter, K. (1995). Comparative study of oestrogen excretion in female new world monkeys: An overview of non-invasive ovarian monitoring and a new application in evolutionary biology. *Folia Primatol.* 64:107–123.

Rantonen, P.J.F., Penttila, I., Meurman, J.H., Savolainen, K., Narvanen, S., and Helenius, T. (2000). Growth hormone and cortisol in serum and saliva. *Acta Odontol. Scand.* 58:259–303.

Reinhardt, V. (1991). Training adult male rhesus monkeys to actively cooperate during in-homecage venipuncture. *Anim. Technol.* 42:11–17.

Robbins, M.M., and Czekala, N.M. (1997). A preliminary investigation of urinary testosterone and cortisol levels in wild male mountain gorillas. *Am. J. Primatol.* 43:51–64.

Ruppenthal, G.C., and Sackett, G.P. (1992). *IPRL Research Protocol and Technicians Manual*, 2nd ed. University of Washington, Seattle. http://www.rprc.washington.edu/iprl/contents.htm.

Schwartz, E.B., Granger, D.A., Susman, E.J., Gunnar, M.R., and Laird, B. (1998). Assessing salivary cortisol in studies of child development. *Child Dev.* 69:1503–1513.

Seeman, T.E., and McEwan, B.S. (1996). Impact of social environment characteristics on neuroendocrine regulation. *Psychosomat. Med.* 58:459–471.

Smith, T.E., and French, J.A. (1997). Social and reproductive conditions modulate urinary cortisol excretion in black tufted-ear marmosets (*Callithrix kuhli*). *Am. J. Primatol.* 42:253–267.

Stolinski, T.S., Czekala, N., Lukas, K.E., and Maple, T.L. (2002). Urinary androgen and corticoid levels in captive, male, western lowland gorillas (*Gorilla g. gorilla*): Age- and social group-related differences. *Am. J. Primatol.* 56:73–87.

Thijssen, J.H.H., Van Goozen, S.H.M., Van Engeland, H., Matute, L.M.P., and Blankenstein, M.A. (2000). None of four commercially avaliable assays detects prolactin in human saliva. *Clin. Chem.* 46:1409–1410.

Thijssen, J.H.H., Matute, L.M.P., Van Goozen, S.H.M., Van Engeland, H., and Blankenstein, M.A. (2001). No detectable prolactin in human saliva. *Soc. Biol. Psychiatr.* 50:9.

Vazquez, D.M. (1998). Stress and the developing limbic-hypothalamic-pituitary-adrenal axis. *Psychoneuroendocrinology* 23:663–700.

Vining, R.F., and McGinley, R.A. (1987). The measurement of hormones in saliva: Possibilities and pitfalls. *J. Steroid Biochem.* 27:81–94.

Vining, R.F., McGinley, R.A., Maksvytis, J.J., and Ho, K.Y. (1983). Salivary cortisol: A better measure of adrenal cortical function than serum cortisol. *Ann. Clin. Biochem.* 20:329–335.

Voultsios, A., Kennaway, D.J., and Dawson, D. (1997). Salivary melatonin as a circadian phase marker: Validation and comparison to plasma melatonin. *J. Biol. Rhythms* 12:457–466.

Wamboldt, M., Laudenslager, M.L., Wambodlt, F.S., Kelsay, K., and Hewitt, J. (2003). Adolescents with atopic disorders have an attenuated response to laboratory stress. *J. Allerg. Clin. Immunol.* 111:509–514.

Worlein, J.M., Berger, C.L., and Laudenslager, M.L. (1995). Effects of ketamine hydrochloride anesthesia on endocrine and immune parameters in pigtail macaques (*Macaca nemestrina*). *Am. J. Primatol.* 36:166.

Yang, R., Yehuda, R., Holland, D.D., and Knott, P.J. (1997). Relationship between 3-methoxy-4-hydroxyphenylglycol and homovanillic acid in saliva and plasma of healthy volunteers. *Soc. Biol. Psychiatry* 42:821–826.

Ziegler, T.E., Scheffler, G., Wittwer, D.J., Schultz-Darken, N., Snowdon, C.T., and Abbott, D.H. (1996). Metabolism of reproductive steroids during the ovarian cycle in two species of Callitrichids, *Saguinus oedipus* and *Callithrix jacchus*, and estimation of the ovulatory period from fecal steroids. *Biol. Reprod.* 54:91–99.

CHAPTER TWENTY-ONE

The SPIT Method for Simultaneous and Unobtrusive Collection of Salivary Cortisol from Individually Housed Infant Monkeys

Peter G. Roma

1. INTRODUCTION

In recent years, an increasing number of researchers have turned to saliva sampling as a relatively noninvasive and acceptably accurate alternative to blood sampling. While a substantial amount of research in the biomedical sciences focuses on the composition and functions of saliva per se (e.g., Czegeny *et al.*, 2001; Won *et al.*, 2001), many scientists use saliva to assess endocrine, immunological, and even neurochemical activity (Riad-Fahmy *et al.*, 1982; Vining *et al.*, 1983; Stone *et al.*, 1987; Nishanian *et al.*, 1998; Lindell *et al.*, 1999). Saliva sampling allows researchers to collect multiple samples from the same subjects while limiting the expenditure of time and human resources required for venipuncture, or the potential confound of stressful sample acquisition (Riad-Fahmy *et al.*, 1982; Boyce *et al.*, 1995). These points are important for researchers studying dynamic short-term effects and pharmacokinetics, and they are especially salient for those engaged in baseline, circadian, or longitudinal investigations of hypothalamic–pituitary–adrenal (HPA) axis activity and biochemicals sensitive to physical and psychological stress.

While sample collection from humans is usually straightforward, obtaining saliva from laboratory animals can be considerably more challenging. Nonetheless, researchers have successfully collected saliva from a multitude of animals across taxa, including rodents such as mice, rats, and hamsters (*Mus musculus*, *Rattus norvegicus*, and *Mesocricetus auratus*, respectively; see Lin *et al.*, 2001; Guhad and Hau, 1996; McClure, 1999), tree shrews (*Tupaia belangeri*, Ohl *et al.*, 1999), dogs (*Canis familiaris*, e.g., Beerda *et al.*, 1999), pigs (*Sus scrofa*, e.g., de Groot *et al.*, 2000), horses (*Equus caballus*, e.g., Elsaesser *et al.*, 2001), and nonhuman primates, including squirrel monkeys (*Saimiri sciureus*; Carver and Hau, 2000) and rhesus macaques (*Macaca mulatta*; Boyce *et al.*, 1995). Of course, each species presents its own challenges for saliva sampling, even under anesthesia and subjected to invasive collection methods. Finding effective, noninvasive, and minimally stressful alternatives heightens the challenge, and this certainly holds true for nursery-reared nonhuman primates.

1.1. Ethics, Logistics, and Stress

Ethical concerns regarding psychological well-being and environmental enrichment of captive nonhuman primates are not exclusively modern-day issues, but are certainly a higher priority now than ever before (Novak, 1991; National Research Council, 1998; USDA, 1999). While most citizens, researchers, and lawmakers recognize the benefits of *in vivo* research models, the demand for animal welfare from inside and outside the research community is clear (Blum, 1994). This sentiment is amplified when the research involves infants, and primate researchers have turned to saliva sampling with perhaps more urgency than those who work with any other animal model. For both ethical and logistical reasons, saliva sampling is an especially desirable alternative to blood sampling in investigations of HPA axis activity, a fruitfully studied topic in primates of all ages.

The HPA axis, whose products have long been considered biochemical indices of stress, is sensitive to a wide variety of experiential and environmental influences. For example, primate researchers have produced transient effects on cortisol, the biochemical conclusion of HPA axis activity, with variables such as social introductions (Kimura *et al.*, 2000) and food availability (Lyons *et al.*, 2000; Champoux *et al.*, 2001). More profound experiences such as rearing condition (Shannon *et al.*, 1998) and social separations (Higley *et al.*, 1992) may produce a more pervasive effect, and it has been argued that dysregulation of the HPA axis may serve as a platform for psychopathology, namely depression and anxiety disorders (Anisman *et al.*, 1998; Arborelius *et al.*, 1999; Tiefenbacher *et al.*, 2000). Primate researchers have also found relations between cortisol and aging (Gust *et al.*, 2000; Goncharova and Lapin, 2002), temperament/personality (Byrne and Suomi, 2002), and even cerebral laterality (Westergaard *et al.*, 2001). While its relevance is broad and its utility clear, engaging the HPA axis in research with nonhuman primates, especially infants, presents some challenges that demand attention.

As Boyce *et al.* (1995) pointed out, typical blood-sampling procedures include a trio of aversive experiences: capture, handling, and venipuncture. With a lag time of as little as 1–2 min before cortisol changes in blood are reflected in saliva (cited in Kirschbaum and Hellhammer, 1989), it is imperative that investigators have less stressful alternatives at their disposal, especially when assessing baseline/pretreatment cortisol levels (Suzuki *et al.*, 2002). Individually housed adult macaques have been trained to present limbs voluntarily for blood sampling (see Reinhardt, 1997, for a review), but little is known of the plausibility of applying such techniques to infants. The ketamine anesthesia commonly used in primate biobehavioral research facilitates the blood-sampling process for the research staff, but essentially eliminates the possibility of taking multiple samples within the same day, as many investigators and veterinarians prefer to minimize infant exposure to anesthesia for both ethical and health reasons, while the experience itself still exacts a toll on cortisol values at the initial sampling and beyond (Elvidge *et al.*, 1976;

Lehmann *et al.*, 1997; Crockett *et al.*, 2000). Additionally, nursery-reared infants must be hand caught and handled for anesthesia administration. Simultaneous testing of many infants requires handling by additional nonnursery staff even if only one sample per animal is being taken. The presence of a bustling group of people, the novelty of unfamiliar handlers, and the experience of anesthetization and venipuncture all exert an influence on HPA axis data and compromise attempts to assess baseline/resting blood HPA activity. Clearly, the typical blood sampling procedure employed for research with nonhuman primates raises several ethical and logistical issues, especially for repeated daily and longitudinal sampling of infants, and thus warrants the development of alternative methods.

1.2. The Current State of the Art

The strong relation between plasma cortisol and salivary cortisol levels is the key to saliva sampling's appeal as an alternative to blood sampling. In their 1989 review, Kirschbaum and Hellhammer cited a litany of studies linking blood cortisol to salivary cortisol, with linear correlations as low as $r = 0.54$ and as high as $r = 0.97$, noting that most coefficients reported were ≥ 0.90. Since only non-protein-bound cortisol can filter from blood into saliva, correlating salivary cortisol with plasma free cortisol instead of total cortisol would likely yield higher coefficients, and is thus advisable for future studies of the relation between salivary and blood cortisol. Interestingly, the unbound cortisol in blood is a minority of total cortisol; however, only the unbound fraction can actually exert any effects upon the body's organs and subsequent behavior. Thus, measuring unbound cortisol—like that found in saliva—would provide researchers with a more functional assessment of HPA activity. Given these facts establishing the validity of salivary cortisol, over the past 15 years researchers have been developing and applying methods for collecting saliva.

Saliva sampling with cotton rolls was first developed for studies of cortisol in human infants (Gunnar *et al.*, 1989) as a less aversive alternative to the catheter, syringe, and direct pipette techniques for saliva collection that preceded it (e.g., Francis *et al.*, 1987; Riad-Fahmy *et al.*, 1982;

Hiramatsu, 1981). By effectively inhabiting the middle ground between the controlled invasiveness of blood sampling (e.g., Malone *et al.*, 1985) and the noninvasive but opportunistic nature of urine sampling (e.g., Tennes *et al.*, 1977), cotton roll-based sampling showed promise to researchers working with infants across primate species. As highly sensitive assaying techniques were developed, saliva sampling became a much more feasible alternative for researchers studying nonhuman primate models of development (Kirschbaum and Hellhammer, 1989, 1994). In 1995, Boyce *et al.* demonstrated the potential of cotton roll-based saliva sampling with 4- to 5.5-month-old infant rhesus macaques, and their method has become the standard for nursery-reared rhesus macaques. However, the cotton roll approach in general has not achieved widespread use in biobehavioral investigations of cortisol in nonhuman primates, and without broad interest the development of relevant techniques has been quite limited.

1.2.1. Saliva Collection Apparatuses

In their investigation of individual differences in circadian rhythm and responses to socialization, Boyce *et al.* (1995) manually offered the cotton rolls to their five infant monkeys rather than using any type of apparatus. This approach was effective for the pilot project, especially in minimizing experimenter presence during sampling, but it is vulnerable in several respects. Chiefly, permitting the infants unfettered access to the cotton rolls makes it possible for the rolls to be dropped onto the floor grating, or worse yet, fall through the grating and into the pan below. Both scenarios are risky, especially the latter, as cortisol concentrations systematically differ between saliva and other bodily excretions (Bahr *et al.*, 2000; Neary *et al.*, 2002), and contamination of saliva samples with urine or feces could adversely affect the data. Free access to the cotton rolls also allows cotton roll exchange or theft among animals housed in adjacent cages: if a cotton roll is mouthed by more than one animal, the original animal's sample is invalidated and a datum is lost.

Manually collecting the cotton rolls also poses logistical challenges. Boyce *et al.* (1995) offered their animals a food reward as they removed the cotton rolls, which worked well for the 2 weeks of data collection;

however, engaging in this practice over many samplings or longitudinally can lead to animals dropping their cotton rolls in anticipation of the reward—a desirable response strictly in terms of cotton roll retrieval, but still subject to the aforementioned contamination risks. Particularly anxious infants, impending reward or not, tend to simply drop their cotton rolls and retreat to their surrogates when approached.

A particularly strong impetus for development of collection apparatus is the 10-min time frame in which to acquire adequate sample volume and collect the rolls, especially when working with larger groups of animals. As Kirschbaum and Hellhammer (1989) report in humans, the latency between chemical intervention or psychosocial stress and peak cortisol response varies depending on the dose or stimulus intensity, respectively, ranging from 5 min for a low-dose injection of corticotrophin-releasing hormone to 20–30 min following the onset of mild laboratory stress. Details of these relationships in salivary cortisol are uncertain in humans and untested in infant macaques, so a conservative strategy is advisable, especially for baseline/resting cortisol samples. Since 10 min is usually sufficient to collect adequate sample volume, it has become the standard sampling time. Due to the uncertainty involved, this standard time limit makes the issue of cotton-roll collection one of data quality in addition to staff convenience and human resources management. From personal experience I can say that being solely responsible for quality longitudinal sampling of nine animals within a 10-min time frame was nearly impossible, given the individual variability in how infants handle their cotton rolls, and was a potent learning experience regarding the limits of the current methodology. I recall, more fondly now than at the time, on many occasions having to chase a particularly feisty monkey around the nursery in a vain attempt to retrieve the cotton rolls so firmly lodged in her cheek pouches. This disruptive set of activities often exceeded 10 min in length, on top of the time allowed for undisturbed mouthing of the cotton roll required for adequate sample volume, and certainly introduced unwanted arousal to our study of resting cortisol. The additional costs of hiring extra staff just for saliva collection and the scheduling conflicts and possible confound of recruiting other laboratory members for ad hoc assistance are not always practical, especially for longitudinal projects. The need for some sort of apparatus is clear, but the options available are scarce and inappropriate for groups of nursery-reared infant macaques.

No apparatus exists for simultaneous noninvasive saliva sampling of multiple laboratory animals. The only apparatus specifically designed for cotton roll-based saliva sampling of individually housed macaques comes from Lutz *et al.* (2000). In their "pole" design, a braided-cotton roll is secured to a long PVC pipe and offered to the animal; their alternative "screen" method is essentially the same, but instead offers the animal a mesh-covered gauze to lick. These apparatuses, actually designed for and tested on adult male rhesus macaques, are creative and effective in their own right, but they were designed for sampling one animal at a time. Sampling multiple animals simultaneously under this method would require as many technicians as subjects, or at best, half as many if each person could manage one apparatus per hand. Aside from the potential burden on human resources, the "pole" method's key methodological shortcoming for infant research is the inherent need for experimenter presence for the duration of the sampling, which may compromise subject participation and cortisol data, especially if administered by multiple unfamiliar research personnel.

1.2.2. Cotton Roll Preparation

In their study with rhesus macaque infants, Boyce *et al.* (1995) used Richmond braided-cotton dental rolls because of their strength and absorbency. To promote cooperation and salivation from the infants, the cotton rolls were soaked in a flavoring solution comprised of roughly 2 parts water to 1 part flavored drink mix crystals; the rolls were air dried and then refrigerated until use.

It is worth noting that the use of cotton rolls for saliva sampling has been shown to interfere with assays of several hormones, but not cortisol (Shirtcliff *et al.*, 2001). As far as cotton roll preparation, the greatest threat to the integrity of salivary cortisol assay results seems to be high sample acidity facilitated by the use of commercially available drink mixes. Lutz *et al.* (2000) assessed salivary cortisol using radioimmunoassay (RIA) techniques, but found no significant differences between plain cotton and rolls pretreated with Kool-Aid in the style of Boyce *et al.*

(1995). However, Schwartz *et al.* (1998), also using RIA, found dose-dependent artificial inflation effects of drink mix crystals on salivary cortisol levels, artificial variance in samples not controlled for flavor crystal concentration, and attenuated correlations between serum cortisol and flavor crystal-influenced salivary cortisol. I am unaware of any published reports specifically testing the effects of acidic drink mix flavorings on enzyme immunoassays (EIA) versus RIA for salivary cortisol, though Salimetrics, maker of a popular EIA kit and collaborator with several investigators cited in this chapter, prefers citric acid-free cotton rolls for saliva samples destined for cortisol assay (Salimetrics, 2000). The idea of using pretreated/flavored cotton rolls for saliva sampling is a good one, but the potential for error may offset the practicality of the current approach.

To summarize thus far, ethical and logistical concerns make cotton roll-based saliva sampling a promising alternative to blood sampling for researchers interested in single, repeated, and longitudinal assessments of HPA axis activity in infant nonhuman primates. However, investigators currently have few methodological options, both in terms of apparatuses and cotton roll preparation. When put into practice, the limits of the currently available methods emerge, as does the need for an innovative solution.

2. THE SPIT METHOD

The method described in this chapter was developed in response to some of the imperfections and potential confounds of the available cotton-roll saliva-collection methods designed for individually housed macaques. It has been affectionately dubbed the Saliva Procurement Integrated Tool, or SPIT. As its light-hearted name implies, the SPIT is a tool for procurement of saliva samples, and its design integrates solutions to several logistical and research methodological issues. The SPIT method is comprised of the SPIT Apparatus and SPIT Sticks; each will be discussed in turn.

2.1. The SPIT Apparatus

The purpose of the SPIT Apparatus is to offer an easy-to-use tool so a single researcher can collect saliva samples from multiple individually

Table 21–1. Materials for the SPIT Apparatus

Item	Size	Source	Product number	Contact information
CPVC rectangular Bar	4 feet long, 1 1/2 inches wide, 1/4 inch thick		8788K19	
Partially threaded wire eyebolt	1/4 inch thread, 1 inch long, 1/2 inch inner diameter	McMaster-Carr Supply Company	9489T117	www.mcmaster.com
Double-end bolt snap	3/8 inch opening, 4 inches long		3901T19	
ILSCO split-bolt connector	1/0 Strand–2 strand		McMaster-Carr # 6921K84 ILSCO Catalog # AK-1/0	
Optional: nylon-insert lock nuts and flat washers	Fit for 1/4 inch threading		Widely available	

housed animals simultaneously and unobtrusively. The raw materials for the SPIT Apparatus, listed with their sources in Table 21–1, are relatively simple. A single apparatus capable of simultaneous sample collection from two animals requires a CPVC rectangular bar, two partially threaded eyebolts, two double-end bolt snaps, and four split-bolt connectors; optional materials include nylon-insert lock nuts and flat washers.

Note that two ILSCO split-bolt connectors are required per animal. In actuality, only one split-bolt and two spacer bars are required; however, spacer bars are not sold separately and can be purchased only as part of the entire split-bolt connector unit. While this markedly increases the cost per SPIT Apparatus, the gripping power of two of the ridged spacer bars is far superior to that of one spacer bar against the smooth inner surface of the split bolt's head or against the rectangular bar itself. This feature becomes increasingly pertinent as infant monkeys age and grow stronger.

To assemble the SPIT Apparatus, start by drilling one 1/4-inch hole on each end of the rectangular bar for the eyebolt/bolt-snap grouping. Insert the eyebolts and install the provided nuts. For a more secure fit, discard the provided nuts and insert one flat washer above the rectangular bar and one below, then tighten with nylon-insert lock nuts (two washers and one lock nut per eyebolt). Before tightening, make sure the long top rim of the eyebolt runs parallel to the rectangular bar, so that the open "eye" of each eyebolt faces forward and not toward the other eyebolt. Once the eyebolts are secure, hook the bolt-snaps in place. The bolt-snaps should be installed with the snap end facing away from the animal, i.e., the smooth, mechanism-free side facing the animal; this makes cage mounting easier and reduces the likelihood of animals unhooking the apparatus from the cage. Next, using a router or several small drill holes, make two slots per split-bolt through the rectangular bar; the slots should be appropriately spaced and sufficient in size to allow the split-bolt effortless vertical movement. Finally, install the split-bolt itself through the rectangular bar, with the provided nut below the rectangular bar, and the spacer bars above it. The placement of the split-bolts on the rectangular bar depends on the size and placement of the cages with which the SPIT will be used.

2.2. The SPIT Sticks

The purpose of the SPIT Sticks is to provide the animals a flavorful cotton roll to promote adequate sample volume, while circumventing the potential influence of high sample acidity on assay results. The raw materials for the SPIT Sticks, detailed in Table 21–2, include braided-cotton dental rolls, flavoring oil, food coloring, granulated sugar, and water.

The method for preparing the SPIT Sticks is essentially the same as that employed by Boyce *et al.* (1995), except the flavoring solution for the SPIT Sticks is acid free and its components are measured precisely to ensure consistency across batches. First, make the flavoring solution by thoroughly combining 500 ml warm water, 200 ml granulated sugar, 1 ml flavoring, and 2 ml food coloring. Next, add 80 cotton rolls and thoroughly toss until all liquid has been absorbed. Finally, place the wet rolls individually onto a bed of absorbent material atop a flat surface, and let them air dry uncovered for 3–5 days or until completely dry and rigid. Once dry, the SPIT Sticks should be refrigerated until used.

The relative proportions of the solution's components are a rather awkward 71.12% water, 28.45% sugar, 0.14% flavoring, and 0.28% food

Table 21–2. Materials for the SPIT Sticks

Item	Size	Source	Product number	Contact information
Richmond dental braided-cotton rolls	4 inches long, 3/8 inch diameter	Sullivan-Schein Dental	200226	1-800-277-0377
LorAnn tropical punch flavoring oil	1 ounce bottle	LorAnn Oils	0500-0500	www.lorannoils.com
McCormick red food coloring	Any	Widely available		
Granulated Sugar	N/A			
Water	N/A			

coloring; however, SPIT Stick yield can be easily manipulated without compromising the formula's proportions by increasing or decreasing the absolute quantities of the original formula in 25% increments. For example, to make 60 SPIT Sticks instead of 80, combine 375 ml water, 150 ml sugar, 0.75 ml flavoring, and 1.5 ml food coloring (original quantities × 0.75), or for 160 SPIT Sticks, simply double the original quantities.

Also note that commercially available flavored drink mixes are highly acidic to both enhance flavor *and* preserve the product. The SPIT Sticks are low acid by design and the only preservative/antifungal agent present is Propylparaben (propyl 4-hydroxybenzoate; www.chemfinder.com, 2002), which is part of the McCormick food coloring. Given the minute contribution of food coloring to the total volume of the flavoring solution, the amount of Propylparaben present and its ability to preserve are negligible. Therefore, investigators and research staff should be particularly vigilant about how long and in what conditions the sticks are left out to air dry, as the high-sugar/no-acid SPIT Sticks provide a more hospitable environment for fungi than the high-sugar/high-acid alternative.

Once the apparatus is assembled and the cotton rolls are prepared, the SPIT is ready for use. Simply place each SPIT Stick between a split-bolt's spacer bars and tighten the associated nut either manually or with an electric drill or ratchet wrench fitted with a 1-inch hex head socket. A diagram

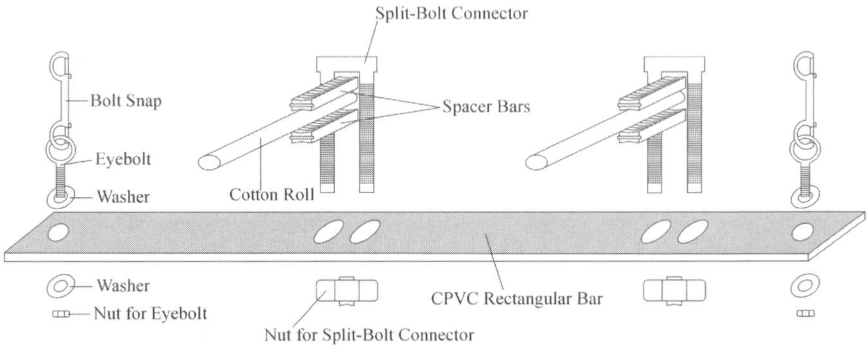

Figure 21–1. Diagram of the SPIT Apparatus.

of the SPIT Apparatus, drawn roughly to scale and showing the relative placement of all the SPIT's parts, is presented in Fig. 21–1. Once the SPIT Sticks are secure, mount the apparatus to the cage(s) for sample collection as depicted in Fig. 21–2.

Figure 21–2. Fully assembled SPIT, mounted and ready for sampling. Inset: closeup of the split-bolt assembly with SPIT Stick.

3. APPLICATION OF THE SPIT METHOD

The following section offers empirical details about the use of the SPIT method in studies involving salivary cortisol in individually housed infant rhesus macaques at the Laboratory of Comparative Ethology (LCE). Included is the standard operating procedure (SOP) employed for sampling via the SPIT method, the overall success of the SPIT method versus its predecessor, and a demonstration of the SPIT method's value for circadian and longitudinal investigations of HPA axis activity.

3.1. Standard Operating Procedure

The SOP for collecting baseline/resting saliva samples via the SPIT method is as follows. First, remove all milk bottles (even apparently empty bottles) 30 min before sampling time; water bottles may remain. Milk should be removed because of postprandial changes in cortisol (Spangler, 1991) as well as potential inflation effects on salivary cortisol assays produced by milk and baby formulas (Magnano *et al.*, 1989). Although the time course is uncertain for these effects, Kirschbaum and Hellhammer (1994) suggest collecting saliva samples "a considerable time following feeding," and 30 milk-free minutes between possible feeding and saliva sampling seems to be a reasonable, if conservative, way to shield the data. At this point, the SPIT Sticks should be secured to the SPIT Apparatus (if they are not already attached); this should be done in another room out of the infants' visual and auditory range, especially if it is done with tools.

By 10 min before sampling, the nursery area should be cleared of potential stressors, particularly people with whom the animals are not familiar. The infants should then be left completely undisturbed until sampling time.

For the actual saliva sampling period, outlined in Fig. 21–3, use a stopwatch/timer to ensure that the session begins, occurs, and ends within a 10-min window. Wait in an anteroom or somewhere out of the infants' visual and auditory range until the designated sampling time. At that time, start the watch and mount the SPIT Apparatuses to the cages. Once the apparatuses are secured and each animal has easy access to its SPIT

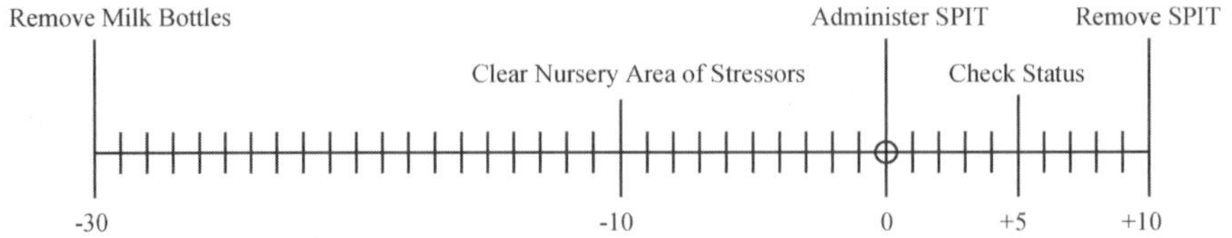

Figure 21–3. SPIT method SOP for baseline sampling. Each tick mark represents 1 min. The "Clear Nursery Area of Stressors" and "Check Status" steps are recommended, but may not be necessary.

Stick, exit the nursery area. At 5 min into the sampling, check to ensure that the procedure is progressing well. Some monkeys will have produced adequate sample volume by this point, while others may need the remaining 5 min. If a monkey has not yet approached the apparatus, this "status-check" provides the researcher an opportunity to salvage the data point by manually offering the infant a SPIT Stick. At the 10-min mark, remove the apparatuses from the cages.

To express the saliva from the cotton rolls, centrifuge in Salivette tubes (www.sarstedt.com). Each Salivette comes with a sterilized cotton roll; discard this roll and place the used SPIT Stick in its stead. If necessary, the SPIT Sticks may be placed wet-end-down into the Salivette's top chamber and snapped in half with heavy scissors for a better fit. Spin the samples for 20 min at 3000 rpm, then transfer the saliva into 2-ml CryoTubes (www.nuncbrand.com) using fine-tip pipettes. Freeze the samples immediately in liquid nitrogen and then refrigerate them at −70°C until ready for assay or shipment.

As far as specifically training the animals to use the SPIT, I wish I could take credit for designing and executing a finely crafted operant conditioning paradigm, but frankly, most infant rhesus macaques require very little encouragement to participate in this SOP. Despite individual differences in initial reactivity to the apparatus, their natural curiosity is eventually and quickly reinforced by the flavorful SPIT Sticks. Of the 12 monkeys I have trained under the SPIT method, all but 1 reliably cooperated within 10 practice sessions over 2 weeks.

3.2. Viability of the SPIT Method

Despite the sound theoretical impetus, careful design, and detailed SOP, the SPIT method can be only as successful as the quality of the datasets it produces. If a saliva-collection method cannot yield assay-worthy samples reflecting real individual differences, then it is of little use to researchers exploring the HPA axis in nursery-reared nonhuman primates. The data presented here are part of a project on the physiological and behavioral effects of early experience with contingent versus noncontingent reward. Though they were not collected specifically to test the viability of the SPIT method or to compare the SPIT method with its predecessor, the opportunity to look empirically at the "big picture" is a valuable one as it can provide insight on the strengths and weaknesses of the two saliva-collection methods while offering some basic information on HPA axis activity in nursery-reared macaques. Thus in this section I will present an archival look at the viability of the SPIT method in terms of subject participation, sample quality, and ability to discern differences in salivary cortisol activity.

All animals described henceforth were born and reared under standard nursery-rearing conditions (Ruppenthal, 1979; also see Boyce *et al.*, 1995) at the National Institutes of Health Animal Center. All rearing and experimental procedures were in compliance with NIH and NICHD guidelines.

All SPIT method data were collected from three groups of four surrogate-peer-reared rhesus macaques ($n = 12$; 8 males, 4 females) every Tuesday and Wednesday at 0645, 1000, and 1400 hr from week 10 of life through week 21 of life (mean age at first sampling = 69.83 days, range = 56–88 days). None of these groups was sampled concurrently at any time during their respective 12-week study periods.

The "predecessor" method was employed as part of a pilot study, with cotton roll preparation and saliva collection in the style of Boyce *et al.* (1995). One group of four and another group of five macaques that were reared with surrogates or peers were studied concurrently over a 15-week period ($n = 9$; 5 males, 4 females; mean age at first sampling = 99 days, range = 40–151 days). Samples were taken on average 3 days a week on various weekdays and at various times, with most sampling sessions initiated within 15 minutes of 0800, 1000, and 1400 hr.

All saliva samples were tested in duplicate via EIA. The interassay coefficient of variation was 23.32%, the intraassay coefficient of variation was 13.15%, and the correlation between duplications was extremely high (Pearson $r = 0.996$, $p \leq 0.0009$). For each sample, the mean value of the duplicate assays served as the single data point for analyses.

3.2.1. Subject Participation

The SPIT method is appealing for studies of HPA axis activity because its low-stress approach promotes truer baseline data than blood sampling procedures; however, there is a price to pay for this advantage. The SPIT method, by definition noninvasive and unobtrusive, cannot be forced on any animal. While the experience itself is nonaversive, and animals can be conditioned to further increase the likelihood of participation, researchers are ultimately at the mercy of their infant subjects. Fortunately, the SPIT method is generally successful in this regard. Of 723 sample attempts, only 22 (3.04%) were failures due to total nonparticipation or no expressible saliva. A 96.96% success rate, though, may be considered generous depending on one's interpretation of "success" and "failure." Only 22 of 723 attempts resulted in no sample, making them unqualified failures; however, the "successful" attempts include the samples of one particular infant (ZA02) collected via manual administration of the SPIT Stick. This infant's engagement with the SPIT Apparatus was sporadic and hesitant at best, and after 2 weeks of training and 5 weeks of testing in earnest, I finally resorted to manual administration as the default approach. If a sampling cannot be considered a "success" without engagement of the SPIT Apparatus, and "failure" includes samplings via manual cotton roll administration (despite adequate sample volume for assay), then 70 of 723 sample attempts were failures (9.68%).

In lieu of a focused conditioning effort, I offer as a rule of thumb to expect on average one "problem" infant in four—not an animal that cannot learn the procedure, but one that may need more encouragement with the SPIT Apparatus, whose full cooperation is less predictable, one that often provides low sample volume, or will provide a sample only with a manually presented cotton roll. Despite this note of caution, the SPIT method is still a highly effective approach to saliva sampling of nursery-reared nonhuman primates, boasting a successful participation rate of over 90% even under the strictest criteria.

The predecessor method is also generally effective, with only 42 of 660 sample attempts (6.36%) deemed failures due to total nonparticipation or lack of expressible saliva. Keep in mind that the predecessor method is based on manual administration of the cotton rolls, so no alternative presentation method exists if an animal does not participate in a sampling session. If, over many samplings of many animals, there is no inherent difference between the SPIT method and its predecessor in terms of subject participation, I submit that the SPIT method is superior because of its ease of use.

3.2.2. Sample Quality

In the preceding section I focused strictly on subject participation, without regard to the quality of the samples submitted for assay. Integrating issues of sample quality would then facilitate a more thorough, and ultimately more relevant, assessment of the SPIT method's viability. "Sample quality" as reported here refers to adequate sample volume and pH level for EIA and is based on the raw data logs and notes provided by those who performed the assays.

The SPIT method's performance is stellar in terms of adequate sample volume and pH level for EIA. Of 704 samples sent for assay, 701 (99.57%) were of sufficient volume for duplicate assay, and no samples were rejected because of pH interference.

The predecessor method is also effective in producing assay-worthy samples for EIA: 615 of 618 samples sent for assay (99.52%) were of sufficient quantity for duplicate testing. Interestingly, pH problems were minimal, as only 1 of 618 samples (0.16%) was not assayed due to pH interference. Of course, the risk posed by flavored drink mixes on assay results is more about artificial variance in the data than outright sample rejection due to high acidity; the vulnerability of RIA to this threat can be determined only experimentally.

Echoing an earlier point, if no inherent difference exists between the SPIT method and its predecessor in terms of sample quantity, the SPIT method is still superior because of its low acidity. Using neutral tap water (pH 7) and litmus paper (Hydrion Insta-Chek, Micro Essentials Laboratory, Brooklyn, NY), I tested the acidity of several flavoring solutions to support this assertion. The SPIT flavoring solution had a pH of 6, versus 3 for the predecessor (low pH = high acidity), even in equal proportion of dry to liquid as the SPIT formula. The potential risk posed by acidic samples on salivary cortisol EIA results can (and perhaps should) be tested empirically; however, the SPIT method manages this risk by essentially eliminating it, making such a pursuit more academic than practical. Overall, low sample acidity, along with successful subject participation and minimal use of human resources, make the SPIT method preferable to its predecessor. However, the predecessor method using SPIT Sticks is an acceptable substitute if needed, especially if used for only one animal at a time. For a more detailed quantitative comparison of the two methods in terms of logistical viability, see Table 21–3.

3.2.3. Circadian and Longitudinal Dynamics of Salivary Cortisol

With its logistic utility established, let us now examine the SPIT method's value in terms of actual data, specifically, the study of baseline cortisol activity within circadian and longitudinal frameworks. Indeed, the SPIT method proved successful in this regard, allowing differences in salivary cortisol to emerge on multiple levels.

As far as circadian rhythm, this group of 12 nursery-reared infant rhesus macaques exhibited both typical and atypical characteristics in their salivary cortisol (Fig. 21–4). A repeated-measures ANOVA revealed a significant Time-of-Day effect ($F = 24.13$, $p \leq 0.0009$), while Bonferroni corrected t-tests ($\alpha = 0.017$ for each of three comparisons) showed that salivary cortisol levels were significantly higher at 0645 hr than at 1000 hr or 1400 hr ($p \leq 0.0009$ for both). These results per se conform to the expected circadian rhythm (Boyce *et al.*, 1995); however, the expected rhythm was contradicted in one way. Instead of a continuous decrease in cortisol throughout the day after the early morning peak, these infants showed a significant *increase* in cortisol from 1000 hr to

Table 21–3. Viability of the SPIT Method and Its Predecessor for Longitudinal Sampling

	Method			
	SPIT		Predecessor[a]	
Number of animals tested	12		9	
Mean age in weeks at first sample	9.98		14.14	
Median age in weeks at first sample	9.64		11.00	
Age range in weeks at first sample	8.00–12.57		5.86–21.57	
	n	%	*n*	%
Sample attempts	723	—	660	—
Failed attempts[b]	22	3.04	42	6.36
Data-yielding[c] attempts	701	96.96	615	93.18
Samples sent for assay	704	—	618	—
Samples too acidic	0	0	1	0.16
Samples quantity not sufficient for assay	3	0.43	0	0
Data-yielding samples	701	99.57	615	99.52

[a] Cotton roll preparation and sample collection in the style of Boyce *et al.* (1995).
[b] Due to total nonparticipation or no expressible saliva; see text for caveats and variations based on one particular infant's performance.
[c] Sufficient sample quantity and pH level for duplicate assay results.

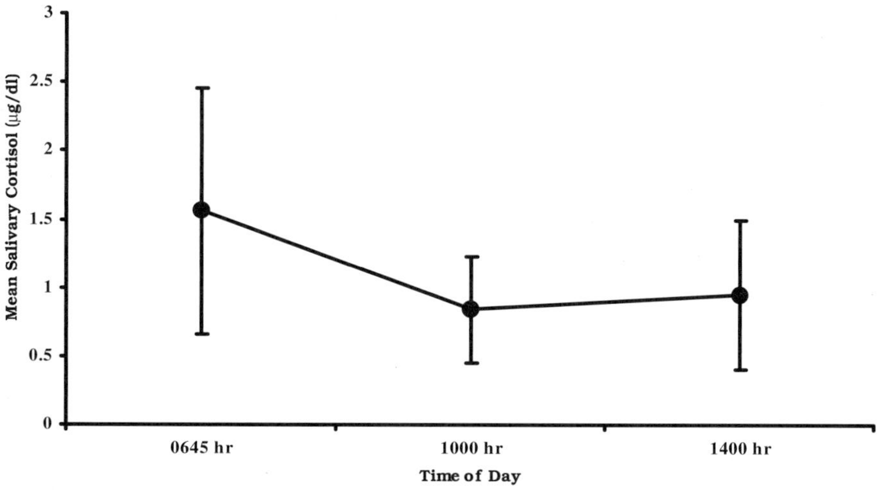

Figure 21–4. Circadian rhythm of salivary cortisol.

1400 hr ($p = 0.004$). This difference in means (0.84 versus 0.95 μg/dl), though not nearly of the same magnitude as the 0645 hr levels compared to the other two time points (1.56 versus 0.84 or 0.95 μg/dl), was consistent for each of the three groups of four infants, as well as longitudinally, as salivary cortisol levels were higher at 1400 hr than at 1000 hr for 9 of the 12 weeks of sampling. While the implications of these findings are beyond the scope of this particular chapter, their consistency is a testament to the SPIT method's ability to allow naturally occurring patterns and differences in HPA axis activity to be expressed and detected.

Longitudinally, the circadian rhythm just discussed did remain essentially intact throughout weeks 10–21, and salivary cortisol levels at each of the three time points showed no clear developmental trajectory during the 12 weeks of sampling (Fig. 21–5). The three time points, however, did differ significantly in terms of variability. A repeated-measures ANOVA comparing the subjects' standard deviations for each time point produced a significant main effect for Time of Day ($F = 13.35$, $p = 0.002$). Bonferroni corrected t-tests revealed significant differences between all three time points, with the most variability at 0645 hr ($p \leq 0.0009$ versus 1000 hr and $p = 0.008$ versus 1400 hr), followed by

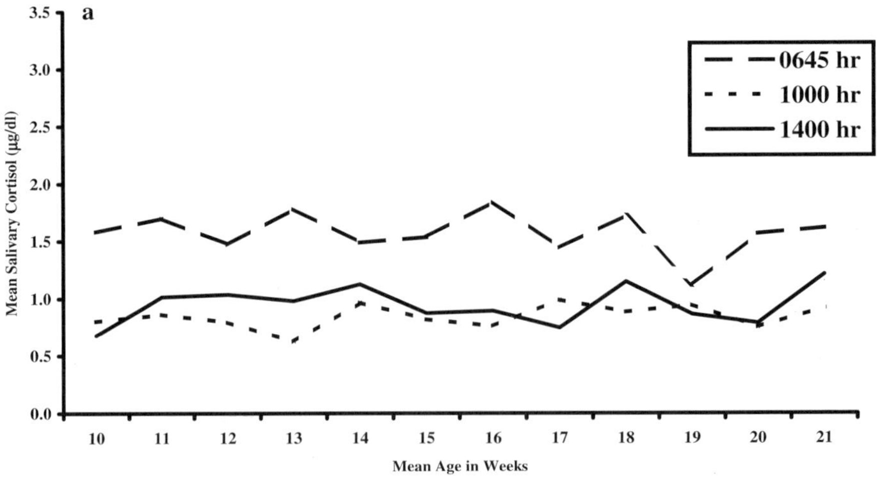

Figure 21–5. (**a**) Salivary cortisol by age and time of day. (**b**) Salivary cortisol at 0645 hr. (**c**) Salivary cortisol at 1000 hr. (**d**) Salivary cortisol at 1400 hr. Error bars represent standard deviation for that week.

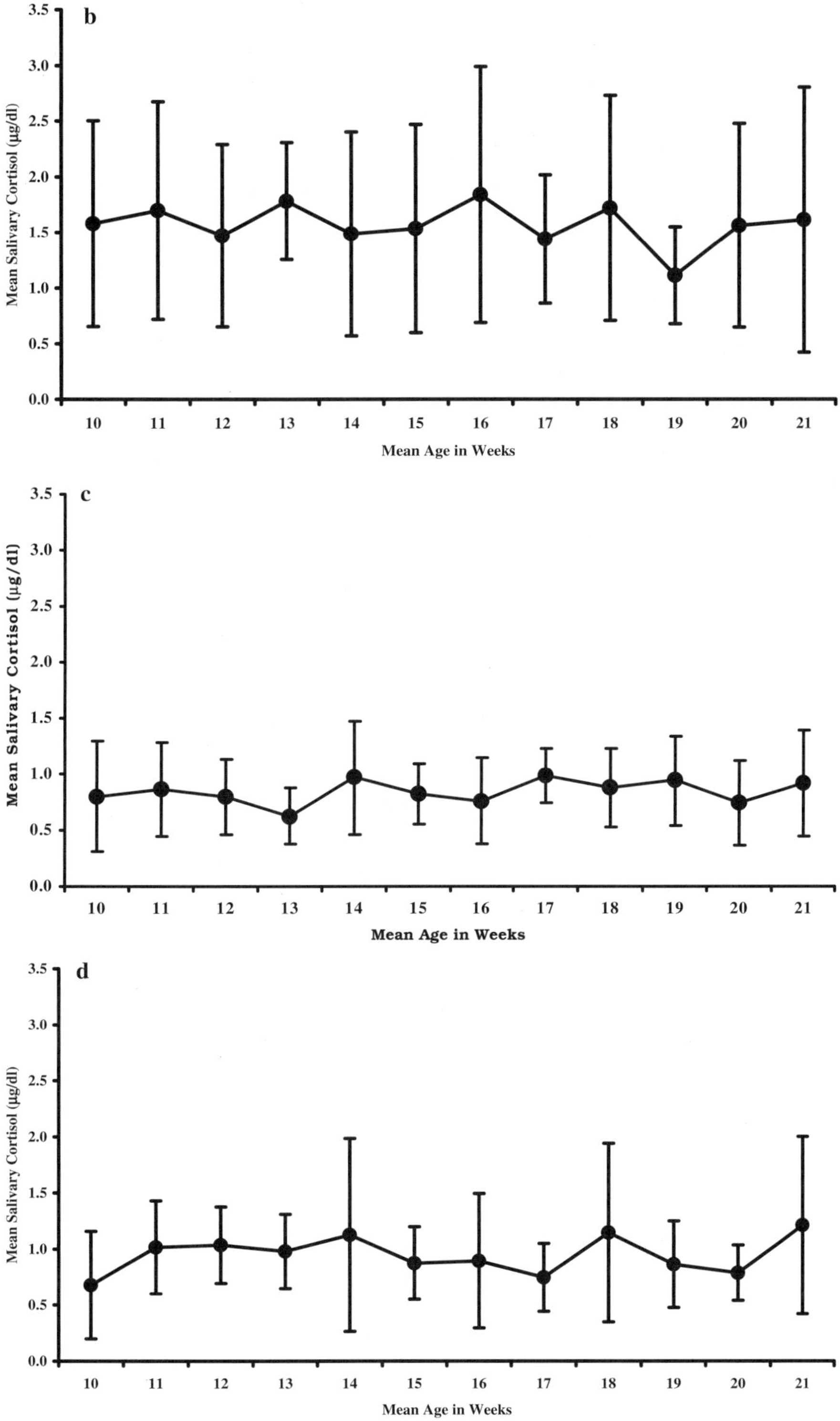

Figure 21–5. *Continued*

1400 hr ($p = 0.014$ versus 1000 hr), then 1000 hr. Taken together, these facts suggest that circadian rhythm of cortisol in nursery-reared rhesus macaques is set by no later than week 10 postnatal of life, and remains stable through at least week 21.

The SPIT method is also effective for the study of individual differences, even within a larger context such as circadian rhythm (Table 21–4). For example, while the 12 infants studied generally migrated in the same direction at the same time, thus exhibiting a rhythm, individual animals' cortisol levels relative to each other tended to remain stable within each circadian shift (Fig. 21–6); the stability of these individual differences was statistically significant between 1000 and 1400 hr (Pearson $r = 0.87$, $p \leq 0.0009$) and 0645 and 1400 hr ($r = 0.62$, $p = 0.032$), and nearly so between 0645 and 1000 hr ($r = 0.51$, $p = 0.09$).

Table 21–4. Individual Differences of Salivary Cortisol within Circadian Rhythm

Subject ID	Time of day: 0645 hr M	SD	1000 hr M	SD	1400 hr M	SD
Z06	1.41	(0.75)	0.75	(0.24)	0.93	(0.44)
Z09	1.92	(1.06)	1.13	(0.42)	1.20	(0.39)
Z13	2.24	(0.98)	1.08	(0.35)	1.33	(0.85)
Z16	1.70	(0.64)	1.12	(0.38)	1.26	(0.51)
Group 1	*1.82*	*(0.89)*	*1.02*	*(0.38)*	*1.18*	*(0.57)*
ZA01	1.88	(0.84)	1.03	(0.28)	1.04	(0.34)
ZA02	1.29	(0.41)	0.57	(0.17)	0.82	(0.61)
ZA03	1.88	(0.73)	0.72	(0.16)	0.70	(0.24)
ZA04	1.64	(0.96)	0.74	(0.59)	0.91	(0.66)
Group 2	*1.68*	*(0.78)*	*0.76*	*(0.37)*	*0.87*	*(0.49)*
ZA54	0.75	(0.30)	0.86	(0.53)	0.82	(0.34)
ZA56	1.13	(0.61)	0.70	(0.24)	0.80	(0.73)
ZA63	1.39	(1.20)	0.82	(0.35)	0.83	(0.46)
ZA65	1.53	(1.14)	0.57	(0.13)	0.78	(0.44)
Group 3	*1.20*	*(0.91)*	*0.74*	*(0.35)*	*0.81*	*(0.49)*
All	1.56	(0.90)	0.84	(0.39)	0.95	(0.54)

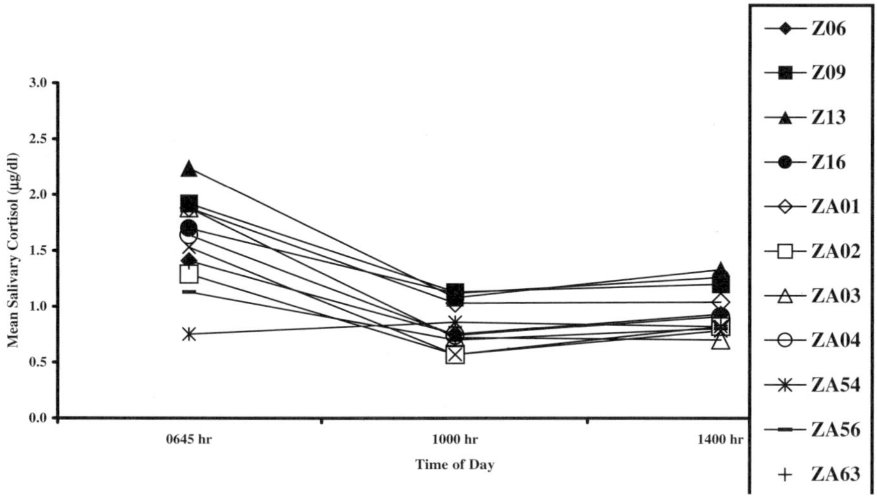

Figure 21–6. Stable individual differences within circadian rhythm.

To further support the SPIT method's utility, I performed a set of a posteriori comparisons to provide evidence of a sampling-method effect on salivary cortisol levels. The weeks of sampling when the mean ages of the SPIT and predecessor groups matched were compared (applicable only to the second, younger predecessor group of five infants, so SPIT $n = 12$ and predecessor $n = 5$). Using all data-yielding samples taken within 30 min of 1000 hr and 1400 hr during the age-matched periods (1000 hr = weeks 14–20, 1400 hr = weeks 11–15, 17, 19, and 20), I compared the SPIT and predecessor methods via 2 × 2 mixed ANOVA. Significant Time-of-Day and Time-of-Day × Sampling Method effects were unveiled ($F = 19.05$, $p = 0.001$ and $F = 10.13$, $p = 0.006$, respectively). Bonferroni corrected t-tests ($\alpha = 0.0125$ for each of four comparisons) showed that salivary cortisol values were significantly higher under the predecessor method than the SPIT method at 1400 hr, and nearly so at 1000 hr ($p \leq 0.0009$ and $p = 0.024$, respectively). Also, while cortisol levels under the SPIT method did not differ significantly from 1000 hr to 1400 hr during the age-matched periods ($p = 0.312$), cortisol did increase under the predecessor method ($p = 0.008$); see Fig. 21–7 for a summary. While these differences may be generally attributed to the sampling method

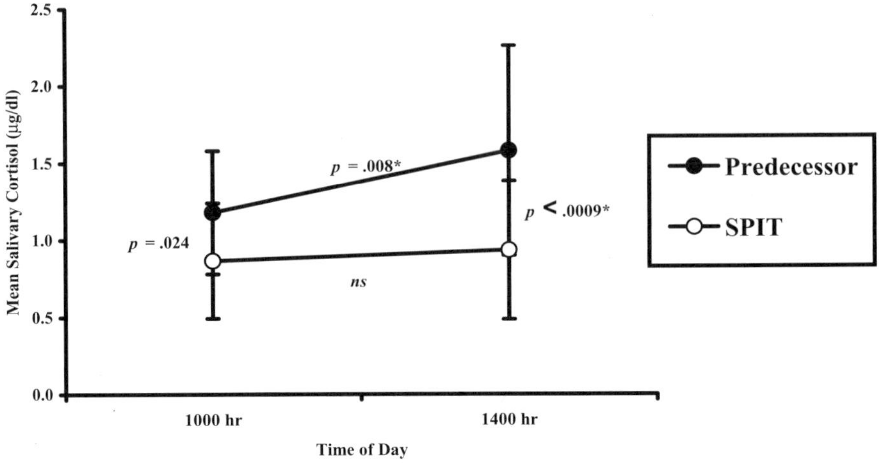

Figure 21–7. Sampling method effect.

employed, it is uncertain whether the effects were due to differences in cotton roll preparation (Kool-Aid versus SPIT flavoring solution) or sample collection (manual administration versus SPIT Apparatus); nonetheless, the SPIT method seems to better facilitate the expression of normal circadian rhythm of salivary cortisol in infant rhesus macaques, without exacerbating any possible preexisting abnormalities in the function of the HPA axis.

Overall, the SPIT method has proven itself an invaluable asset for the study of circadian and longitudinal dynamics of salivary cortisol in nursery-reared rhesus macaques. Group differences were consistently detectable based on time of day, as were individual differences in baseline cortisol output, all within a longitudinal framework. Using the SPIT method, investigators may now collect datasets of unprecedented detail and accuracy for studies of salivary cortisol activity in nursery-reared nonhuman primates.

4. DISCUSSION

Issues of animal welfare and data quality make saliva sampling an appealing alternative to blood sampling for investigations of HPA axis activity in nursery-reared nonhuman primates. However, the methodological options available for noninvasive saliva sampling of nonhuman primates are limited and fraught with complications. Fortunately, emerging technologies and techniques are now taking saliva sampling's appeal from the theoretical to the practical. Cotton roll-based saliva sampling is a most promising approach for studies of cortisol in infant primates because it is much less stressful than standard blood sampling procedures and, in conjunction with modern assaying techniques, has laid the foundation for a genuinely viable formalized saliva-sampling methodology.

4.1. Variations on a Theme

The SPIT Apparatus allows researchers to collect saliva samples from multiple animals simultaneously and unobtrusively while the SPIT Sticks help ensure high sample quality. The low-stress approach of the SPIT method is particularly well suited to investigations of baseline cortisol activity in nursery-reared nonhuman primates. Indeed, sample collection is not just nonaversive, it is reinforcing. All of the items for the SPIT method are relatively inexpensive, and most of the materials required for both the SPIT Apparatus and SPIT Sticks can be found at any hardware or grocery store, respectively. The SPIT is easy to assemble, use, and clean, with little to no maintenance required.

Note that the items listed in this chapter are those used for application of the SPIT method to individually housed infant rhesus macaques at the LCE, and that many variations of these items exist and may be utilized to create an infinite variety of SPITs as different species, housing environments, and research objectives warrant. For example, McMaster-Carr offers bars, sheets, and rods in dozens of sizes and various materials such as aluminum and stainless steel (sturdier, cage-washable alternatives to CPVC), as well as ILSCO split-bolt connectors in eight different sizes. Richmond braided-cotton rolls range in size from 3/4 inch long by 5/16 inch in diameter to 6 inches long by 1/2 inch in diameter. LorAnn oils offers over 75 different fat- and sugar-free flavorings. For animals on calorie-controlled diets, SPIT Sticks could be prepared without sugar or with noncaloric sweetener (pH of flavoring solution prepared with Equal instead of sugar is 5). To adapt the SPIT Apparatus for use with a single subject, simply use a shorter length of bar and one split-bolt assembly. Such an adaptation could be useful for studies conducted on a subject-by-subject basis, or even in opportunistic sample collection in field, zoo, or other group-housing settings.

4.2. Limitations and Future Directions

In all, the SPIT method is an effective and practical tool, but like any method, it is not perfect. For example, securing the cotton rolls to an apparatus greatly reduces the likelihood of sample contamination, but as long as the animals have any kind of access to the rolls, contamination is possible. The SPIT method's imperfections, though, lie more in its limitations than its current applications. For collecting resting/baseline saliva samples from multiple infants, the SPIT method is unparalleled; however, the voluntary nature of sample collection limits its usefulness for point assessments of poststress cortisol. This limitation became apparent during a series of tests assessing physiological and behavioral responses to 15- to 30-min sessions in a novel environment. The first group of four infant monkeys (from the aforementioned 12 trained under the SPIT method) were less willing to engage the apparatus upon return to their home cages from the novel environment than they had been in the preceding weeks of home-cage baseline sampling. Of 48 sample attempts, only 28 yielded data (58.33%, versus 96.96% of the longitudinal samplings). Given this rather dismal success rate, the remaining eight monkeys trained under the SPIT SOP were manually administered SPIT Sticks poststress on an as-needed basis, an approach that met with considerably more success, with 46 out of 48 attempts (95.83%) yielding data.

Another limitation, and perhaps the greatest one, is the vulnerability of braided-cotton dental rolls for sampling. Of course, a balance must be struck between absorbency and strength, and the braided rolls perform admirably when used with young rhesus macaques (as with humans of all ages), but even infants are capable of nipping and chewing off the wetted part of the rolls within 10 min, leaving little if any left for centrifugation and assay. A significant improvement of the SPIT method would be the discovery or development of a stronger but still absorbent material, or perhaps some type of sheath or outer layer to protect the cotton rolls without obscuring the flavor or damaging the animals' teeth.

This, along with a split-bolt or similar mechanism with a strong enough grip, will be key if researchers intend to use the SPIT method in studies with older animals and/or larger species.

5. CONCLUSION

The SPIT method, comprised of the SPIT Apparatus and SPIT Sticks, is a highly effective approach for saliva sampling of individually housed non-human primate infants. It has proven itself both methodologically valuable and logistically practical for the longitudinal study of circadian rhythm of cortisol. However, its practical appeal notwithstanding, the SPIT method is ultimately as much a concept as it is a concrete tool. The concept is simple—a noninvasive method to collect saliva samples from multiple animals simultaneously and unobtrusively. Researchers, clinicians, and animal care professionals interested in the concept are encouraged to explore the multitude of options available and manipulate the SPIT method presented here to suit their needs.

This presentation of the SPIT method is not a wake-up call, as many in the research community are keenly aware of the possibilities of saliva sampling. If anything, it is more of a rallying cry. At least for studies of baseline/resting cortisol, saliva sampling may be an alternative, but it is not a compromise. Given the tight relation between salivary and blood cortisol, the biobehavioral relevance of the unbound cortisol found in saliva, and of course HPA axis reactivity to conventional blood sampling procedures, saliva sampling may indeed be *preferable* to blood sampling. It is my hope that this presentation of the SPIT method will further stimulate interest in saliva sampling, compel other investigators to develop and expand the method, and allow saliva sampling to take its place as a common method for biobehavioral research in the twenty-first century.

ACKNOWLEDGMENTS

I wish to thank Steve Suomi and Maribeth Champoux for their support throughout the development of the SPIT method, as well as Jim Sackett and Gerry Ruppenthal for providing an opportunity to present it. Special thanks to Terry Lowery for prototype design suggestions, prototype fabrication, and help in fabrication of the final SPIT Apparatus; thanks also to Ernie Davis for prototype design suggestions and help in fabrication of the final SPIT Apparatus, Peter J. Pierre for prototype design suggestions and introducing me to McMaster-Carr, and Greg Lazor for help in prototype fabrication. Additional thanks to the LCE research, veterinary, animal care, and facilities teams for their cooperation and help during sample collection and manuscript preparation. The projects described herein were supported by intramural funds from the National Institute of Child Health and Human Development, National Institutes of Health, Public Health Service, United States Department of Health and Human Services.

REFERENCES

Anisman, H., Zaharia, M.D., Meaney, M.J., and Merali, Z. (1998). Do early-life events permanently alter behavioral and hormonal responses to stressors? *Int. J. Dev. Neurosci.* 16:149–164.

Arborelius, L., Owens, M.J., Plotsky, P.M., and Nemeroff, C.B. (1999). The role of corticotropin-releasing factor in depression and anxiety disorders. *J. Endocrinol.* 160:1–12.

Bahr, N.I., Palme, R., Möhle, U., Hodges, J.K., and Heistermann, M. (2000). Comparative aspects of the metabolism and excretion of cortisol in three individual nonhuman primates. *Gen. Comp. Endocrinol.* 117:427–438.

Beerda, B., Schilder, M.B.H., Bernadina, W., Van Hooff, J.A.R.A.M., De Vries, H.W., and Mol, J.A. (1999). Chronic stress in dogs subjected to social and spatial restriction. II: Hormonal and immunological responses. *Physiol. Behav.* 66:243–254.

Blum, D. (1994). *The Monkey Wars.* Oxford University Press, New York.

Boyce, W.T., Champoux, M., Suomi, S.J., and Gunnar, M.R. (1995). Salivary cortisol in nursery-reared rhesus monkeys: Reactivity to peer interactions and altered circadian activity. *Dev. Psychobiol.* 28:257–267.

Byrne, G., and Suomi, S.J. (2002). Cortisol reactivity and its relation to homecage behavior and personality rating in tufted capuchin (*Cebus apella*) juveniles from birth to six years of age. *Psychoneuroendocrinology* 27:139–154.

Carver, J.F., and Hau, J. (2000). Development of a saliva sampling method and a rocket immunoelectrophoretic assay for quantification of salivary IgA in the squirrel monkey (*Saimiri sciureus*). *In Vivo* 14:735–739.

Champoux, M., Hwang, L., Lang, O., and Levine, S. (2001). Feeding demand and plasma cortisol in socially-housed squirrel monkey mother-infant dyads. *Psychoneuroendocrinology* 26:461–477.

Chemfinder.com (2002). Cambridge, MA; www.chemfinder.com.

Crockett, C.M., Shimoji, M., and Bowden, D.M. (2000). Behavior, appetite, and urinary cortisol responses by adult female pigtailed macaques to cage size, cage level, room change, and ketamine sedation. *Am. J. Primatol.* 52:63–80.

Czegeny, Z.S., Chicharro, J.L., Fernandez, P., Gutierrez, A., and Camara, C. (2001). Homogeneity and stability studies on sodium, calcium, magnesium, and manganese in human saliva. *Biol. Trace Element Res.* 79:131–137.

de Groot, J., de Jong, I., Prelle, I.T., and Koolhaas, J.M. (2000). Immunity in barren and enriched housed pigs differing in baseline cortisol concentration. *Physiol. Behav.* 71:217–223.

Elsaesser, F., Klobasa, F., and Ellendorff, F. (2001). ACTH stimulationstest und Bestimmung von Cortisol im Blut und Speichel zur Bewertung des Trainingszustands/der Kondition beim Warmblutpferd. [Evaluation of salivary cortisol determination and of cortisol responses to ACTH as markers of the training status/fitness of warmblood sports horses.] *Deutsche Tierärztliche Wochenschrift* 108:31–36.

Elvidge, H., Challis, J.R.G., Robinson, J.S., Roper, C., and Thorburn, G.D. (1976). Influence of handling and sedation on plasma cortisol in rhesus monkeys (*Macaca mulatta*). *J. Endocrinol.* 70:325–326.

Francis, S.J., Walker, R.F., Riad-Fahmy, D., Hughes, D., Murphy, J.F., and Gray, O.P. (1987). Assessment of adrenocortical activity in term newborn infants using salivary cortisol determinations. *J. Pediatr.* 111:129–133.

Goncharova, N.D., and Lapin, B.A. (2002). Effects of aging on hypothalamic-pituitary-adrenal system function in non-human primates. *Mech. Aging Dev.* 123:1191–1201.

Guhad, F.A., and Hau, J. (1996). Salivary IgA as a marker of social stress in rats. *Neurosci. Lett.* 216:137–140.

Gunnar, M.R., Connors, J., and Isensee, J. (1989). Lack of stability in neonatal adrenocorticol reactivity because of rapid habituation of the adrenocortical response. *Dev. Psychobiol.* 22:221–233.

Gust, D.A., Wilson, M.E., Stocker, T., Conrad, S., Plotsky, P.M., and Gordon, T.P. (2000). Activity of the hypothalamic-pituitary-adrenal axis is altered by aging and exposure to social stress in female rhesus monkeys. *J. Clin. Endocrinol. Metab.* 85:2556–2563.

Higley, J.D., Suomi, S.J., and Linnoila, M. (1992). A longitudinal assessment of CSF monoamine metabolite and plasma cortisol concentrations in young rhesus monkeys. *Biol. Psychiatr.* 32:127–135.

Hiramatsu, R. (1981). Direct assay of cortisol in human saliva by solid phase radioummunoassay and its clinical applications. *Clin. Chim. Acta* 117:239–249.

Kimura, K., Shimizu, K., Hayashi, M., Ishikawa, T., and Ago, Y. (2000). Pituitary-adrenocortical responses to the first dyadic encounters in male rhesus monkeys: Effect of dominance relationship. *Am. J. Primatol.* 50:247–256.

Kirschbaum, C., and Hellhammer, D.H. (1989). Salivary cortisol in psychobiological research: An overview. *Neuropsychobiology* 22:150–169.

Kirschbaum, C., and Hellhammer, D.H. (1994). Salivary cortisol in psychoneuroendocrine research: Recent developments and applications. *Psychoneuroendocrinology* 19:313–333.

Lehmann, R., Wagner, J.L., Fernandez, L.A., Bourgoignie, J.J., Ricordi, C., Alejandro, R., and Kenyon, N.S. (1997). Effects of ketamine sedation on glucose clearance, insulin secretion, and counterregulatory hormone production in baboons (*Papio hamadryas*). *J. Med. Primatol.* 26:312–321.

Lin, A.L., Johnson, D.A., Wu, Y., Wong, G., Ebersole, J.L., and Yeh, C.-K. (2001). Measuring short-term γ-irradiation effects on mouse salivary gland function using a new saliva collection device. *Arch. Oral Biol.* 46:1085–1089.

Lindell, S.G., Suomi, S.J., Shoaf, S., Linnoila, M., and Higley, J.D. (1999). Salivary prolactin as a marker for central serotonin turnover. *Biol. Psychiatr.* 46:568–572.

Lutz, C.K., Tieffenbacher, S., Jorgensen, M.J., Meyer, J.S., and Novak, M.A. (2000). Techniques for collecting saliva from awake, unrestrained, adult monkeys for cortisol assay. *Am. J. Primatol.* 52:93–99.

Lyons, D.M., Fong, K.D., Schrieken, N., and Levine, S. (2000). Frustrative nonreward and pituitary-adrenal activity in squirrel monkeys. *Physiol. Behav.* 71:559–563.

Magnano, C.L., Diamond, E.J., and Gardner, J.M. (1989). Use of salivary cortisol measurements in young infants: A note of caution. *Child Dev.* 60:1099–1101.

Malone, S.M., Gunnar, M.R., and Fisch, R.O. (1985). Adrenocortical and behavioral responses to limb restraint in human neonates. *Dev. Psychobiol.* 18:435–446.

McClure, D.E. (1999). Clinical pathology and sample collection in the laboratory rodent. *Vet. Clin. N. Am.: Exotic Anim. Pract.* 2:565–590.

National Research Council. (1998). *The Psychological Well-Being of Nonhuman Primates.* National Academy Press, Washington, D.C.

Neary, J.P., Malbon, L., and McKenzie, D.C. (2002). Relationship between serum, saliva, and urinary cortisol and its implication during recovery from training. *J. Sci. Med. Sport* 5:108–114.

Nishanian, P., Aziz, N., Chung, J., Detels, R., and Fahey, J.L. (1998). Oral fluids as an alternative to serum for measurement of markers of immune activation. *Clin. Diag. Lab. Immunol.* 5:507–512.

Novak, M.A., and Petto, A.J. (eds.) (1991). *Through the Looking Glass: Issues of Psychological Well-Being in Captive Nonhuman Primates.* American Psychological Association, Washington, D.C.

Ohl, F., Kirschbaum, C., and Fuchs, E. (1999). Evaluation of hypothalamo-pituitary-adrenal activity in the tree shrew (*Tupaia belangeri*) via salivary cortisol measurement. *Lab. Anim.* 33:269–274.

Reinhardt, V. (1997). Training nonhuman primates to cooperate during blood collection: A review. *Lab. Primate Newsl.* 36(4).

Riad-Fahmy, D., Read, G.F., Walker, R.F., and Griffiths, K. (1982). Steroids in saliva for assessing endocrine function. *Endocrine Rev.* 3:367–395.

Ruppenthal, G.C. (1979). Survey of protocols for nursery rearing infant macaques. In: Ruppenthal, G.C. (ed.), *Nursery Care of Nonhuman Primates.* Plenum Press, New York, pp. 165–185.

Salimetrics, College Park, PA. (2002). http://www.salimetrics.com/pdf/CortisolKitInsert90601.pdf.

Schwartz, E.B., Granger, D.A., Susman, E.J., Gunnar, M.R., and Laird, B. (1998). Assessing salivary cortisol in studies of child development. *Child Dev.* 69:1503–1513.

Shannon, C., Champoux, M., and Suomi, S.J. (1998). Rearing condition and plasma cortisol in rhesus monkey infants. *Am. J. Primatol.* 46:311–321.

Shirtcliff, E.A., Granger, D.A., Schwartz, E., and Curran M.J. (2001). Use of salivary biomarkers in biobehavioral research: Cotton-based sample collection methods can interfere with salivary immunoassay results. *Psychoneuroendocrinology* 26:165–173.

Spangler, G. (1991). The emergence of adrenocortical circadian function in newborns and infants and its relationship to sleep, feeding and maternal adrenocortical activity. *Early Hum. Dev.* 25:197–208.

Stone, A.A., Cox, D.S., Valdimardottir, H., and Neale, J.M. (1987). Secretory IgA as a measure of immunocompetence. *J. Hum. Stress* (Fall 1987):136–140.

Suzuki, J., Ohkura, S., and Terao, K. (2002). Baseline and stress levels of cortisol in conscious and urestrained Japanese macaques (*Macaca fuscata*). *J. Med. Primatol.* 31:340–344.

Tennes, K., Downey, K., and Vernadakis, A. (1977). Urinary cortisol excretion rates and anxiety in normal 1-year-old infants. *Psychosom. Med.* 39:178–187.

Tiefenbacher, S., Novak, M.A., Jorgensen, M.J., and Meyer, J.S. (2000). Physiological correlates of self-injurious behavior in captive, socially-reared rhesus monkeys. *Psychoneuroendocrinology* 25:799–817.

United States Department of Agriculture (USDA). (1999). *Environmental Enrichment for Nonhuman Primates Resource Guide: January 1992–February 1999*. Beltsville, MD.

Vining, R.F., McGinley, R.A., and Symons, R.G. (1983). Hormones in saliva: Mode of entry and consequent implications for clinical interpretation. *Clin. Chem.* 10:1752–1756.

Westergaard, G.C., Champoux, M., and Suomi, S.J. (2001). Plasma cortisol is associated with handedness in infant rhesus monkeys. *Dev. Psychobiol.* 38:116–122.

Won, S.-H., Kho, H.-S., Kim, Y.-K., Chung, S.-C., and Lee, S.-W. (2001). Analysis of residual saliva and minor salivary gland secretions. *Arch. Oral Biol.* 46:619–624.

CHAPTER TWENTY-TWO

Actimetry Measurement of Behavioral Regulation and Sleep Cycles in Infant Rhesus Macaques (*Macaca mulatta*)

Peter J. Pierre, Allyson J. Bennett, and Stephen J. Suomi

1. INTRODUCTION

Activity patterning and circadian cycles have proven to be important measures for clinical and translation research. For example, circadian disruption and activity levels that deviate significantly from either baseline or a comparison group may serve as diagnostic criteria for a range of disorders including depression (Jean-Louis *et al.*, 2000), attention deficit disorder (Miller and Kraft, 1994), posttraumatic stress disorder (Glod *et al.*, 1997a,b), and substance abuse (Johnson and Breslau, 2001). At the same time, individual differences in activity patterns may serve as an early risk marker for those disorders.

A useful tool for the continuous measurement of activity and circadian cycles is automated actigraphy. Because it is sensitive, noninvasive, and relatively inexpensive, it can be used to obtain data for a wide variety of studies, including longitudinal assessments and discrimination of group or individual differences in activity levels and circadian cycles. It may also be beneficial for investigation of both acute and long-term effects of drugs, pharmacological interventions, stress, and environmental manipulation.

Although wristwatch actigraphy has been used to investigate activity patterning in humans for over a decade (Brown *et al.*, 1990), only recently has it been applied in nonhuman primate research (Zhdanova *et al.*, 2002). In this chapter, we describe a modified actigraphic technique for continuous measurement of behavioral activity in nursery-reared infant primates. Our approach is threefold. First, we will describe a viable method for using wristwatch actigraphy that ensures the safety of the animal and the integrity of the equipment. Second, we will present data from two experiments that demonstrate the reliability of this measurement technique and its relation to activity data collected by human observers. Finally, we will describe the activity cycles of infant rhesus macaques that were measured continuously over 120 hr, as well as group differences in the activity patterns of monkeys reared in two nursery conditions.

Although measurement of activity is widely used in behavioral studies of nonhuman primates, the ability to obtain continuous data over extended periods of time has been limited by practical issues. Options for recording activity have consisted primarily of behavioral observations obtained either directly or by subsequent coding of videotaped sessions. In both cases, the requisite resources present a significant constraint for nonautomated collection of activity and point to the need for an automated technique. On the other hand, the automated devices used to record activity in humans typically are not durable enough for use with nonhuman primates, especially in social groups. Our first goal, therefore, was to design a system that would allow us to place actimeters on group-housed monkeys without jeopardizing either the monkeys' safety or the integrity of the equipment. Advances in miniaturization and durability to meet the need of recording activity in human infants have resulted in equipment that is also feasible for use with nonhuman primates. The method we outline is relatively nonintrusive, permits extended sampling durations (potentially for months), and will provide the opportunity for repeated measurement in individuals across the lifespan.

Our second goal was to determine the reliability of the activity data collected with this system. Several studies have demonstrated that the technique provides accurate and reliable continuous measurement of activity in children and human infants (Thoman *et al.*, 1978; Sadeh *et al.*, 1994; Sadeh, 1996). To verify the system's accuracy in monkeys, we first tested for correspondence between data collected from two actimeters and then examined the relation between activity data recorded by human observers and by the actimeter.

Our third goal in this work was to examine continuous activity patterns in two groups of infant monkeys reared in a nursery environment. Previous studies have demonstrated that nursery-reared monkeys differ from their mother-reared counterparts in activity levels. Champoux *et al.* (1991), using a focal-animal sampling technique, found that nursery-reared monkeys were more active than mother-reared monkeys and explored more early in development (days 2–7) and from 1.5 to 5.0 months of age, though this group difference was not maintained past the early developmental period. When the monkeys were observed in social groups at 10–21 months of age, the same scoring protocol revealed no significant group differences. It should be noted that housing conditions for the two groups were different during the first 6 months of life, with mother-reared monkeys living in large indoor–outdoor cages and nursery-reared animals in standard quadrant cages indoors. During the latter observation period, the groups were housed together in large indoor–outdoor pens.

Further work is needed to determine whether housing conditions play a role in the expression of differences between mother- and nursery-reared animals in activity levels and, if so, to identify critical periods for the effect and for possible recovery. Nonetheless, other data suggest that nursery rearing may have long-lasting effects on activity levels and effects that persist when housing is held constant. For example, Suomi *et al.* (1971) reported that monkeys reared in partial isolation engaged in less locomotion and environmental exploration in adulthood than did their wild-caught (presumably mother-reared) counterparts. Along with the data of Champoux *et al.* (1991), these results provide evidence of the effects of early environmental experiences on activity levels and underscore the need for longitudinal studies to document these changes across the lifespan.

Among nursery-reared animals, differences in activity levels have also been detected between groups reared in varying social or environmental conditions. In one of the earliest studies of nursery-reared rhesus macaques tested in adulthood, Sackett (1967) found that isolation-reared animals were less active during a social interaction test than animals whose early experience had included visual contact, daily peer contact, or both. In this context, because activity was used primarily as a measure of response to challenge, the researchers did not address the question of rearing-group differences in general activity or activity expressed in a nonsocial context. In another study, differences were detected in the activity levels of two groups of nursery-reared monkeys whose early experiences were varied by the presence of either a moving or a stationary surrogate (Mason and Berkson, 1975). Across 200 consecutive daily observations, monkeys with a moving surrogate showed more play behavior than those with a stationary surrogate. In these two studies, by Sackett and by Mason and Berkson, the measurement of activity was secondary to evaluation of group differences in social behavior. Thus, while group differences in play behavior may have resulted in differences in activity, it is also possible that the groups did not differ in other forms of activity. Similarly, in the study by Sackett (1967) the observation of decreased activity in a social test may not have extended to activity in other contexts. At this point, therefore, it is not clear from existing data whether variation in nursery-rearing conditions leads to differences in activity levels between groups.

In the following study, we examined continuous activity patterns in two groups of nursery-reared monkeys. All monkeys were reared in identical conditions through 30 days of age and then assigned to either a peer-rearing (PR) or surrogate-peer-rearing (SPR) group. PR animals were housed in groups of four agemates through 6 months of age. SPR animals were housed individually except for a 2-hr socialization period with three other animals 5 days/week. Previous studies have demonstrated some differences between PR and SPR monkeys, including lower social dominance status in adolescence in SPR monkeys (Bastian *et al.*,

2003), but relatively little is known about other aspects of behavior. Our goal was to determine whether the two rearing conditions produced differences in overall activity levels and the pattern of activity exhibited across the day : night cycle.

The technique used here for continuous and extended measurement of activity can also be used to determine sleep : wake patterns. For this purpose we used actigraphy data to infer sleep onset on the basis of activity patterns. Previous work provides evidence of the high correspondence between sleep onset measured via activity analysis and sleep : wakefulness measured by polysomnographic techniques (Webster *et al.*, 1982; Cole *et al.*, 1992; Sadeh *et al.*, 1994; Jean-Louis *et al.*, 2001). Of particular interest was the possibility that nursery-reared monkeys differ in circadian rhythms and sleep : wake patterns. As discussed above, disruption of these processes provides both a potential marker and a window into developmental pathology. Thus, their early detection may prove useful in nonhuman primate research.

2. MATERIALS AND METHODS

2.1. Subjects

Subjects were 16 infant rhesus macaques (*Macaca mulatta*) born and housed in the Animal Center at the National Institutes of Health (Poolesville, MD). All animals were separated from their mothers soon after birth and raised in the nursery. For the first 15 days of life, each subject was housed singly in a temperature-controlled plastic cage (51 × 38 × 43 cm) that contained an inanimate surrogate. The surrogate was a fleece-covered plastic cylinder (20 cm long × 4 cm in diameter) mounted at a 110° angle on a spring attached to a heavy base that prevented the apparatus from tipping over or being moved by the monkey. The spring mount provided gentle movement contingent upon the infant's activity. For the infant's first 15 days of life, the surrogate was warmed with a heating pad.

At 15 days of age, subjects were moved into individual wire-mesh cages (64 × 61 × 76 cm) with the fleece-covered surrogates. At about 37 days of age they were placed into one of two rearing conditions. In the

PR condition, two groups of four age-matched monkeys (three males, five females) were housed together continuously with access to all four quadrants of a cage (63.5 cm deep × 147 cm wide × 163 cm high). In the SPR condition, eight infants (five males, three females) were housed singly in individual quadrants (63.5 × 74 × 76 cm) of a four-quadrant cage and were socialized in a large play cage (61 × 66 × 135 cm) for 2 hr each weekday. The nursery was maintained on an artificial light : dark cycle (9.5 hr light, 14.5 hr dark).

Data for the initial comparison of activity recorded by human observers and by the actimeter device were obtained when the subjects were 120 days old. Data for group comparisons and determination of circadian activity patterns were collected when the animals were 150 days of age.

2.2. Apparatus

2.2.1. *Actiwatch®*

The Actiwatch® activity monitor (AW-64, Minimitter Inc., Bend, OR) is a small accelerometer housed in a wristwatch-sized case (Fig. 22–1B). The accelerometer is connected to a weighted movement arm, which is attached to a piezoelectric film connected to a microprocessor that converts voltage changes to activity counts. Movement produces changes of variable magnitude in electrical current in the film and the microprocessor integrates the degree and speed of motion (Activity Count). The resultant data provide a measure of "gross locomotor activity" across time. The unit has a programmable range of temporal sampling epoch for data collection from 15 sec to 1 hr.

2.2.2. *Harness*

We modified a commercially available small nylon pet harness (Top Paw®, Pacific Coast Dist. Inc, Phoenix AZ; Fig. 22–1A) to accommodate the actimeter and provide a fit that would minimize potential for injury or removal by either the infant or its social group. The harness was altered to provide a snug fit and tailored to each individual to ensure its comfort for extended wear. The standard configuration of the harness consists of

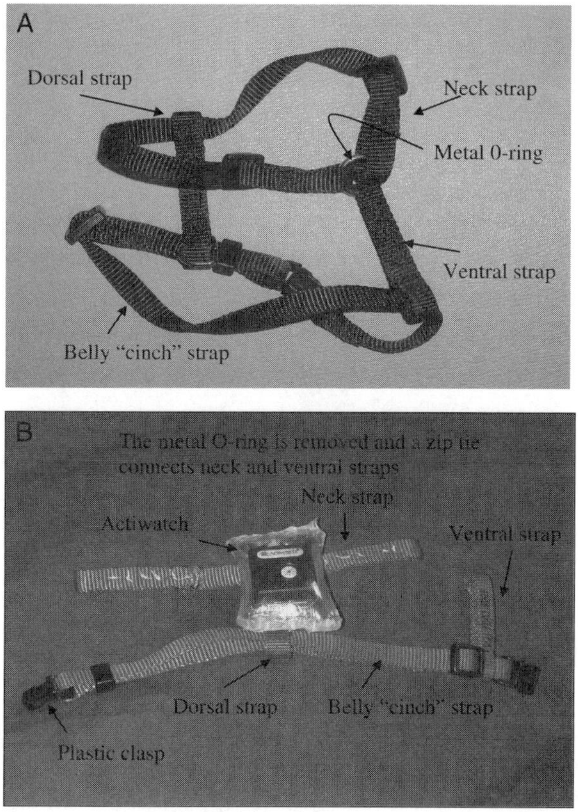

Figure 22–1. (**A**) Standard pet harness before alterations. Note location of the O-ring, the adjustable cinch strap, and the ventral and dorsal lengths of nylon strapping. (**B**) Completed harness ready for attachment to the infant monkey.

three pieces of woven polyester strapping (1.3 cm) attached to a metal O-ring (Fig. 22–1A). One length of strapping falls along the animal's midline and the other two lengths form a loop around the neck. An adjustable cinch strap circumscribes the belly girth of the animal and is attached with a detachable fastener. Two additional lengths of strapping extend from the center O-ring: one along the animal's ventrum and a second, parallel strap along the animal's dorsal length.

To modify the harness, we removed the center O-ring from the neck loop, shortened the remaining strapping material to approximate the circumference of the infant monkey's neck, and sewed each end of the strapping material into a loop (Fig. 22–1B). Monofilament fishing line (6–8 lb test, 0.011 diameter, Stren® Division of Remington Arms, Wilmington, DE) was used in all sewing and mending of the harnesses.

2.2.3. *Actiwatch/Harness Interface*

The Actiwatch was placed into a protective casing, namely, a short length of flexible vinyl tubing (3/4 inch diameter) trimmed to fit the external proportions of the Actiwatch case. A small hole 1 cm in length was made in the vinyl tubing on the bottom side of the Actiwatch aligning with each watch stay. To attach the Actiwatch to the harness, a zip tie (Boxer®, 4-inch cable tie, Marlon P. Jones Inc. FL, Item # TOA-5CTR) was inserted through the stay and the vinyl tubing, and then into the harness. Finally, each end of the tubing was sewn shut and any gaps were filled with clear silicone sealant (DAP Inc, Baltimore, MD).

2.2.4. *Application of the Interface*

The monkey was outfitted with the interface while under ketamine anesthesia (15 mg/kg IM). First, the cinch strap was connected around the belly girth of the infant and adjusted so that a single finger could be placed between the harness and the infant's body (Fig. 22–2A). Next, the two neck straps and the midline strap were joined under the animal's chin. These three converging straps were fastened and adjusted with a zip tie that had been covered with heat-shrink tubing (Radio Shack, Inc.) to guard against skin abrasion. The excess zip tie length was removed and the end was sewn into one of the three adjoining loops to prevent its movement. After all adjustments were made, the Actiwatch was positioned between the infant's shoulder blades and the base of the back strap was sewn shut. At this time any excess strapping also was sewn closed (Fig. 22–2B).

2.3. Data Analysis

2.3.1. *Selection of Recording Interval*

For the data collection discussed here, we used the smallest (15-sec) time bin supported by the Actiwatch, as well as a 1-min interval for extended

A

B

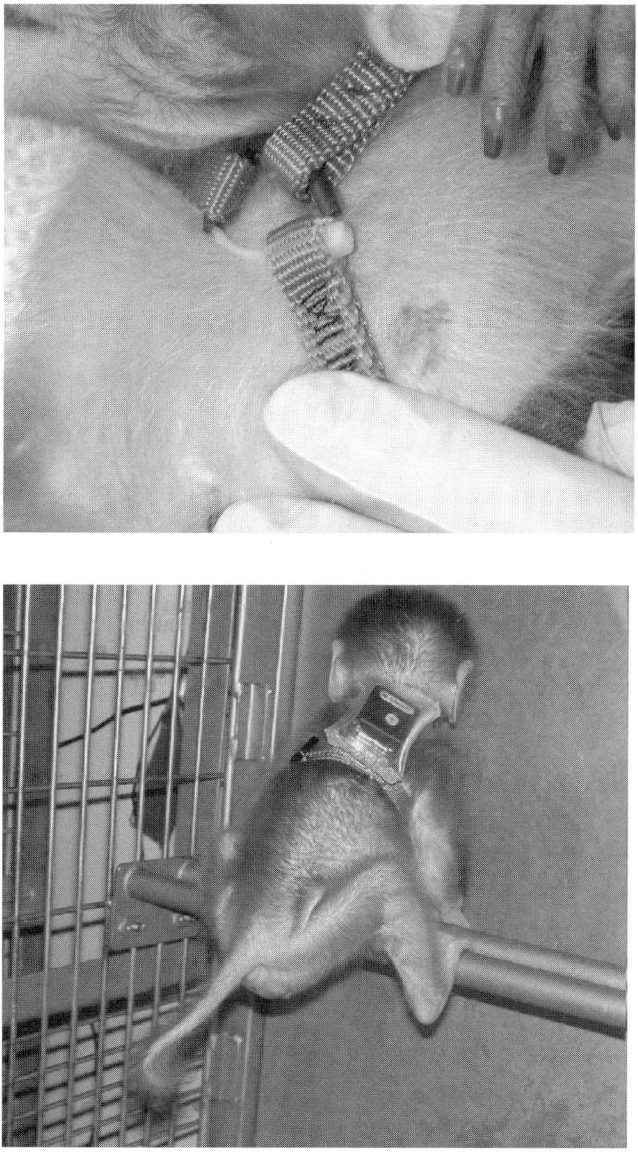

Figure 22–2. Critical aspects of attaching the Actiwatch harness to the infant macaque. (**A**) The harness is fastened below the neck with a zip tie. A small piece of electrical shrink-wrap tubing protects against skin abrasions from the zip tie. The knob on the ratchet of the zip tie is sewn into the nylon harness for further protection. (**B**) A freely moving infant with the harness interface attached.

recording. A limitation of the system is that the sample interval for data capture affects recording duration because the memory capacity of the Actiwatch is finite (64K memory with our units), so that recording duration can be increased only by increasing the sampling interval. With the Actiwatch set at a sampling interval of 15 sec one can record continuously for approximately 11 days. By contrast, using a sampling interval of 1 min permits recording for approximately 45 days.

In choosing the sampling interval, it is necessary to consider the most sensitive measurement required, while balancing the length of recording duration. For example, extended recording duration on the order of days or weeks may be needed to capture circadian cycles. By contrast, acute measurements for a shorter duration or concurrent measurement of activity and other physiological variables (i.e., heart rate, stress hormones) may require correspondence between timelines and accurate determination of activity within smaller time segments. The Actiware-Sleep™ software allows the recalculation of the sampling epoch to larger but not smaller bin sizes than the data-sampling interval.

2.3.2. Export of Data from Actimeter to Software

After the Actiwatch was removed from the animal, data were downloaded with the Actiwatch reader, a wireless telemetric device that transfers data through a standard serial communication port to the Actiware-Sleep™ software. The Actiware-Sleep software captures the actimeter data in a format that can be analyzed to provide primary dependent variables, or can be exported into another statistics package for further analysis. For the analysis presented here the primary dependent measures calculated from the Actiware-Sleep software included both frequency-based measures (minutes of mobility, minutes of immobility, total activity counts) and duration-based measures (latency to sleep onset).

2.3.3. Data Analysis for Nursery-Reared Group Comparisons

The total sampling duration for the Actiwatch measurements presented here was 4.5 days (108 hr). Analysis of group differences involved both patterns of the behavioral activity across the entire sampling period and specific analysis of the organization of activity in the light and dark components of the sampling period. First we assessed group differences in activity over 4.5 consecutive days from the first dark phase until noon prior to harness removal at a sampling resolution of 15 min. Next, we analyzed group differences in average activity counts (1-min resolution) across consecutive 24-hr periods, excluding the first dark phase following harness application (22:00–6:30) and the 5.5 hr of the light phase just prior to harness removal (6:30–12:00). We excluded these two periods because we wanted to begin the analysis at the first complete 24-hr cycle beginning at lights on (6:30) and include only complete 24-hr cycles. We used a time resolution of 1 min to correspond with 1-min intervals used in actigraphy studies of human infants (Korte *et al.*, 2001). The final analyses of nursery group differences assessed mobility in the dark phases of the light : dark cycle and latency to sleep onset. This analysis included the dark phase of day 1 following placement of the actimeter, thus producing five data points in the graphic representation of these data. Sleep onset was defined as the first 10-min period in which no more than one 1-min epoch was scored as mobile by the Actiware-Sleep program. The 10-min period reflected the moderate sensitivity setting for sleep detection by the Actiware-Sleep analysis software.

3. RESULTS

3.1. Preliminary Assessment of Automated Recording Method

To assess the continuity of measurement across two Actiwatch units, we placed the two units in the same vertical orientation, but opposed at 180°, beneath the fleece cover of a spring-mounted surrogate in a monkey's home cage. The Actiwatch units recorded the movement as it was animated by the monkey's activity. Data recorded over the course of a 48-hr period were consistent across the individual units ($r = 0.884$, $p < 0.05$; 15-sec sampling interval). Next we undertook a direct comparison of data recorded by the Actiwatch and data recorded by human observers across the same 5-min intervals. For this analysis, eight infants (four PR and four SPR) were fitted with the Actiwatch harness and observed in their home cages. Actiwatch sampling sensitivity was set for a 15-sec resolution. Experienced human observers used focal-animal sampling with a laptop-based cumulative recorder to code activity in real time with a five-point ethogram that ranked behavioral activity on a continuum from stationary to locomotor behavior for a sampling duration of 5 min.

For purposes of comparison with data from the actimeter, data recorded by the human observers were transformed into binomial categorization of "Active" and "Inactive" states. Each successive 15-sec time interval during which locomotor activity was scored was designated an Active State interval. The 15-sec intervals in which no locomotor activity occurred were designated Inactive State intervals. Raw activity counts from the actimeter recordings were similarly transformed such that Active State intervals were defined as the 15-sec interval values that exceeded the upper quartile range for a given group mean (i.e., PR versus SPR). The Inactive State response was defined as any value that was below the upper quartile value.

Two-tailed Pearson correlation showed modest concordance between data collected by human observers and Actiwatch sampling [$r(168) = 0.299$, $p < 0.01$]. Significant correlations between activity measures collected by the automated device and human observers suggest that the technique produces a reliable measure of activity in infant monkeys and serves as proof of concept. Nonetheless, the procedure we used to transform the human observer and Actiwatch data was designed to be a conservative estimate of the Active State and was less precise in describing the Inactive State, which we defined as activity counts less than the upper quartile range within a 15-sec interval. Further parametric analyses of this type are necessary as investigators adapt Actiwatch measurement to unique experimental designs and examine larger populations.

3.2. Comparison of SPR and PR Monkeys

Our preliminary use of the actimeter-harness method indicated that we could successfully increase the sampling duration of continuous activity monitoring to >4–5 days (approximately 120 hr) with a sampling epoch of 1 min. We were interested in the viability of the Actiwatch-harness data collection procedure to discriminate group differences in activity patterning of nursery-reared monkeys tested at 150 days of age. Of particular interest were potential group differences in activity during the dark portion of the circadian cycle that might reflect sleep disruption. We performed mixed-model analysis of variance with rearing condition (PR, SPR) and gender serving as between-groups comparisons and days as a repeated measure for three dependent measures: total activity, dark : light cycle comparisons, and latency to sleep onset.

For the purpose of demonstration, Fig. 22–3 shows the mean total activity counts across the four transitions of light : dark cycles for all monkeys. The circadian pattern of activity did not differentiate the rearing groups in this preliminary analysis, suggesting that all the animals developed a normal pattern of activity, consistently showing more activity during the light cycle than during the dark cycle. Repeated-measures analyses of variance over the entire sampling period showed a significant main effect of rearing group [$F(1,12) = 10.03$, $p < 0.01$]. No main effect of day or gender was detected, nor were there interactive effects between these variables (i.e., Day × Group or Gender × Day) in these or subsequent analyses.

Further data reduction in which mean activity levels were calculated in 1-min intervals showed that the group difference reflected increased behavioral activity during the dark portion of the circadian cycle, with the PR monkeys more active than the SPR monkeys [$F(1,12) = 5.16$, $p < 0.05$; see Fig. 22–4B]. However, the expression of behavioral activity (Fig. 22–4, upper panel) likely was affected by the presence of experimenters during testing (see Discussion).

Finally, we conducted a preliminary analysis from which to draw inferences concerning potential differences in sleep–wake patterning between rearing groups using the Actiwatch. In this analysis, we compared the number of 1-min intervals in which mobility was scored during the light : dark phases of our testing period, as well as the latency of sleep onset (defined as a 10-min period with no activity recorded). During the dark phase of the cycle, PR monkeys showed a decreased frequency of 1-min intervals in which they were immobile (mean = 306, SD ± 41) compared with SPR monkeys (mean = 380.2, SD ± 38) [Fig. 22–5, $F(1,12) = 18.01$, $p < 0.01$]. The number of 1-min epochs in which movement occurred was higher for PR (mean = 155.1, SD ± 27.1) than for SPR monkeys (mean = 108.5, SD ± 20.2) [Fig. 22–5; $F(1,12) = 30.48$, $p < 0.01$]. Finally,

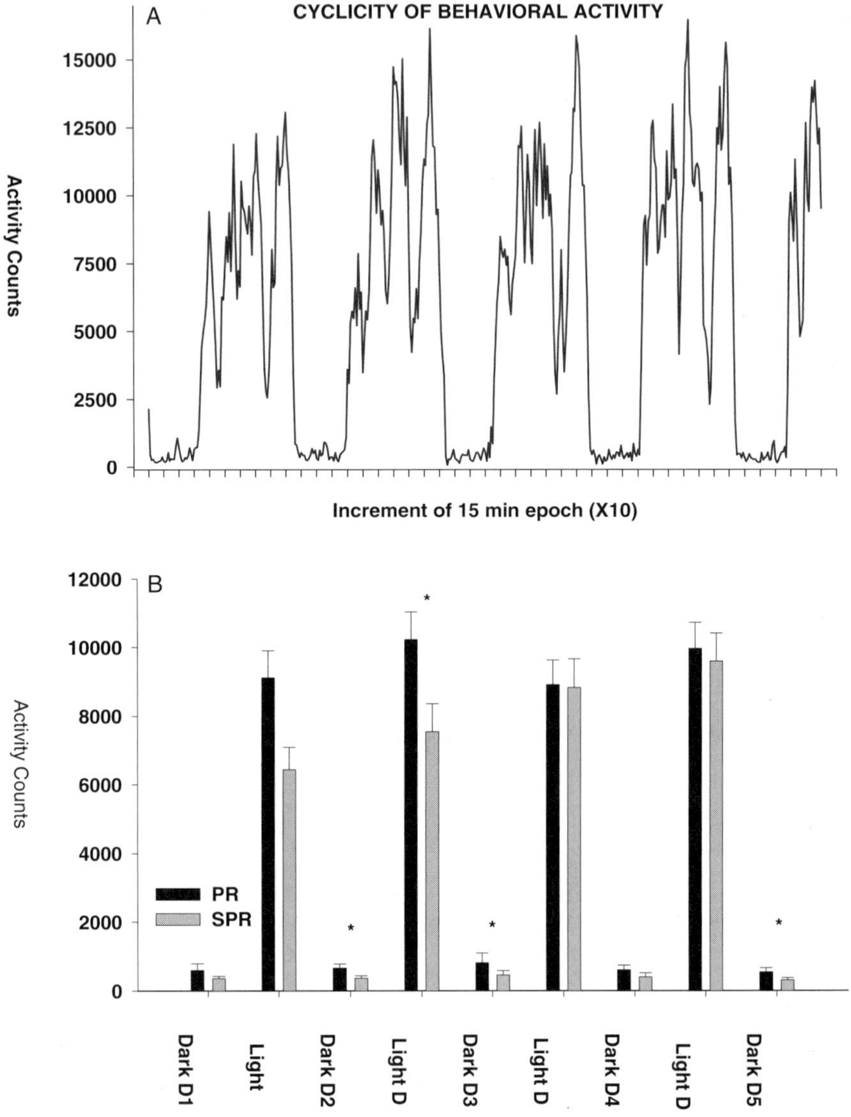

Figure 22–3. Total activity counts plotted over the duration of the experimental period for all subjects. (**A**) Each tick on the *x*-axis represents 10 15-min sampling epochs of mean activity counts from lights out on the first sampling day to noon on the final sampling day. (**B**) Average total activity counts per 15-min epoch plotted by group across each successive light : dark period. Overall, PR monkeys were more active than SPR counterparts and this effect was more pronounced during the dark portion of the circadian cycle. There were no interactions between gender, group, or day. Filled bars, PR monkeys; gray bars, SPR monkeys. Error bars = SEM. Asterisk denotes group comparisons at $p < 0.05$.

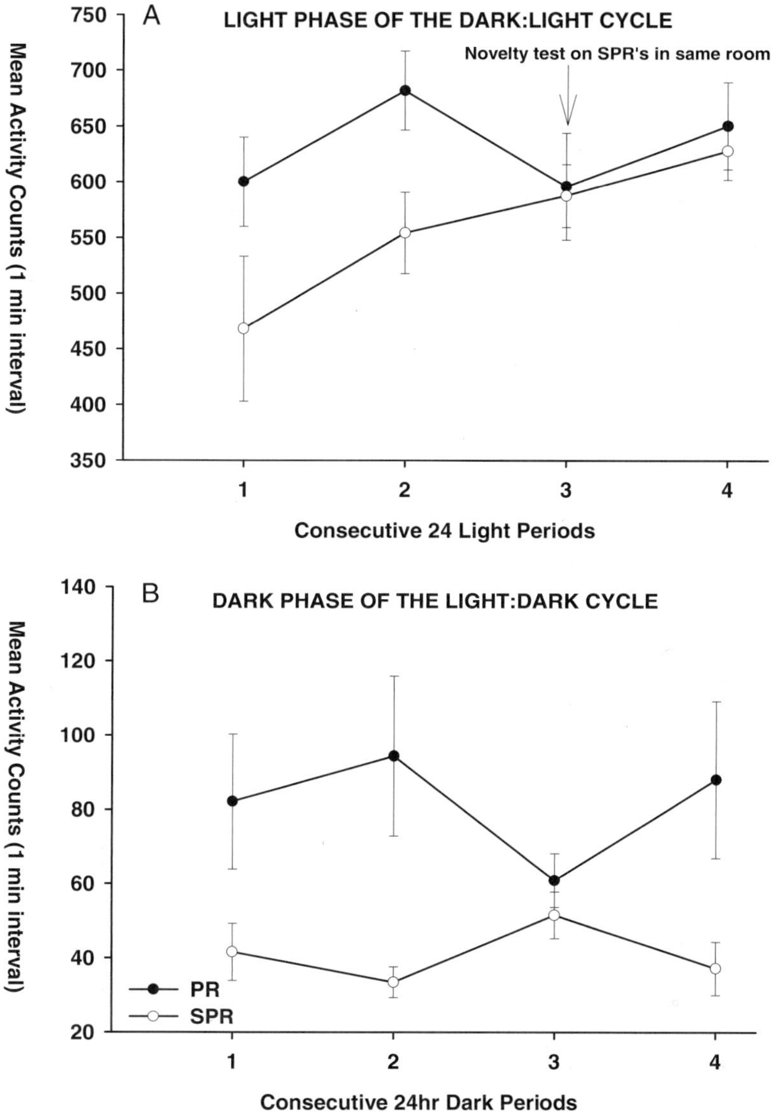

Figure 22–4. Mean activity count per 1-min interval for each nursery-rearing condition plotted across consecutive complete 24-hr periods separated by the light (**A**) and dark (**B**) periods of the circadian cycle. Activity counts for the light period did not differentiate nursery-rearing group, although the first two 24-hr periods suggest early differences (**A**). However, during the dark period, PR monkeys exhibited more activity counts per interval than their SPR counterparts (**B**). Note that the corresponding disruption of behavior denoted with the inverted arrow on the upper panel of day 3 may carry over to disrupt activity in the dark period for the PR monkeys (see Discussion). Filled circles, PR monkeys; open circles, SPR monkeys. Error bars = SEM. Group differences are significant at $p < 0.05$.

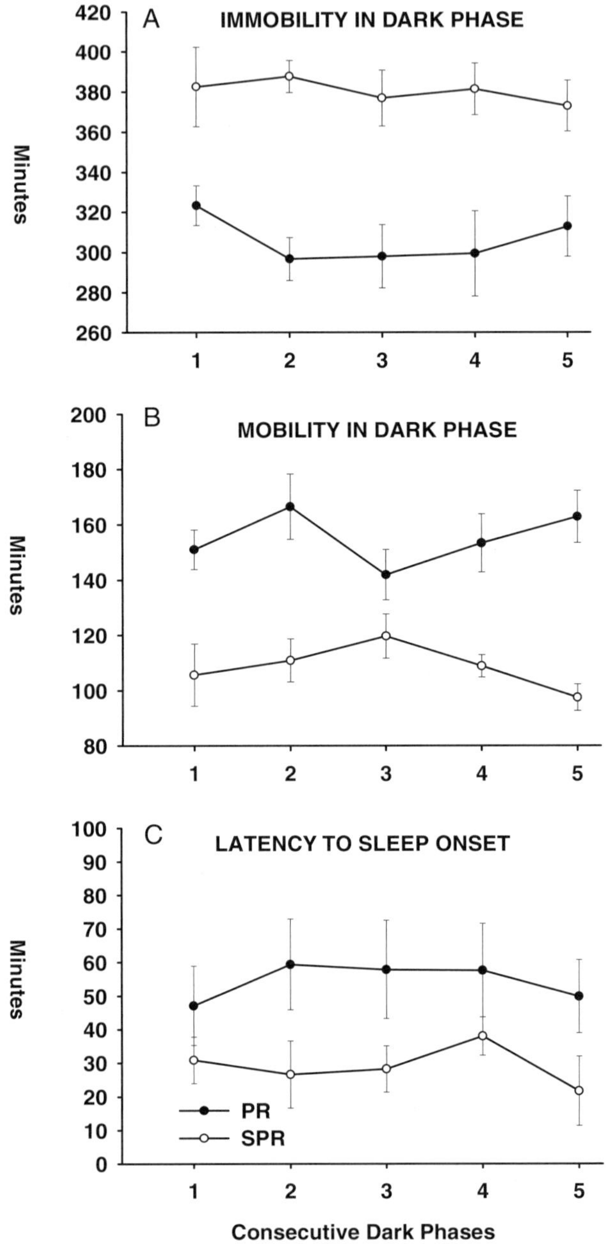

Figure 22–5. Components of actigraphic analysis corresponding to inferred aspects of sleep quality. (**A**, **B**) Number of minutes of mobility plotted by consecutive dark phases of the circadian cycle. PR monkeys were more active than SPR monkeys during the dark phase of the circadian cycle, which may suggest disruption of sleep. (**C**) Increased latencies to inferred sleep in PR versus SPR monkeys. Filled circles, PR monkeys; open circles, SPR monkeys. Error bars = SEM. Group differences are significant at $p < 0.05$.

latency to sleep onset was increased in PR monkeys (mean = 54.4, SD ± 36.6) compared with their SPR counterparts (mean = 29.1, SD ± 22.5), although this effect was not statistically significant [$F(1,12) = 3.62$, $p = 0.08$]. This effect did become statistically significant, however, when the data from the first dark cycle were excluded ($p < 0.05$). Preliminary data suggest that residual effects of ketamine anesthesia may have contributed to disruptions in activity during this period. Overall, our results demonstrate rearing-group differences in multiple measures of activity in PR and SPR monkeys tested at 150 days of age. PR monkeys showed more behavioral activity in both the light and dark portions of the circadian cycle than did SPR monkeys. The pattern of activity counts was consistent across days, lending further evidence for the reliability of the actimeter recording technique.

4. DISCUSSION

The data reported here illustrate the usefulness of automated actigraphy for characterizing 24-hr activity patterns in nursery-reared infant primates. Our preliminary evidence shows that there are rearing-group differences in sleep onset and activity during the dark portion of the circadian cycle. Together, these data suggest potential group differences in circadian activity cycles early in life.

In this initial application we demonstrated that a modified harness system encasing the Actiwatch could be worn safely over an extended period and withstand daily use in group-housed infant monkeys. Our preliminary comparisons showed interunit reliability and close correspondence between activity coded by human observers and activity measured by the automated device. Our data not only provide complete 24-hr records of activity with clear evidence of circadian cycles recorded over several days, but also indicate that multiple aspects of the activity cycle can differentiate monkeys reared in two nursery conditions.

Activity analyses for this extended period indicate that SPR monkeys are characterized by their relative inactivity in the home-cage environment. Previous studies have reported that nursery-reared infant monkeys exhibit increased clinging behaviors relative to mother-reared monkeys (Harlow and Zimmerman, 1959), and this type of behavior likely competes with more active components of behavioral expression. Similarly, dyad-housed monkeys have been reported to show deficits in tests of social interaction, likely resulting from attachment to the cagemate (Chamove *et al.*, 1973). Without continuous observation it is difficult to determine whether the observed clinging behavior may be due in part to an acute behavioral reaction to being observed, rather than an accurate reflection of decreased daily activity.

Our data suggest that behavioral activity differentiates groups primarily during dark phases of the circadian cycle, with PR infants more active than SPR infants at night. Previous behavioral analyses have generally not provided continuous sampling of behavior during this period. The measure of latency to sleep onset was greater in PR monkeys than in SPR monkeys, suggesting that PR monkeys had increased disruption in activity during the dark portion of the cycle. Such disruption likely reflects poorer sleep quality, which in humans is a factor comorbid with a range of disorders including attention-deficit hyperactivity disorder (ADHD) (Miller and Kraft, 1994), autism (Takase *et al.*, 1998; Diomedi *et al.*, 1999; Elia *et al.*, 2000), drug abuse (Johnson and Breslau, 2001), and depression (Jean-Louis *et al.*, 2000). However, the duration of the data set presented here is not long enough to permit this determination from actimetric sleep analyses (see Acebo *et al.*, 1999, for discussion) and we are currently in the process of cross-validating Actiwatch-Sleep measures with human observations in infant monkeys.

In general, actigraphic records of at least 7 consecutive nights should be used to identify patterns of circadian activity, whereas the data reported here were collected over 4 days and nights. In future analyses, we will investigate the circadian patterning of activity over a longer duration, but first we will verify the sleep : wake characterization in nursery-reared infant macaques with videotape analyses synchronized with the Actiwatch internal clock. Previous work suggests that sleep : wake scoring with visual media allows quantification of sleep : wake patterning (Carroll *et al.*, 1993; Zajicek *et al.*, 1997).

Longitudinal data of this type have significant potential for expanding our understanding of how the response of an animal in its early environment relates to later behavior. The automated system for recording activity and sleep cycles over relatively long periods should result in increased sensitivity of measurement. In turn, these data may be useful in uncovering individual differences in vulnerability or resilience to environmental manipulation.

Over the past 50 years a large body of literature has accumulated to show biobehavioral alterations in nursery-raised primates (Ruppenthal *et al.*, 1976; Suomi and Ripp, 1983). Recent evidence suggests, however, that some individuals are more sensitive than others to adverse environmental effects (e.g., see Bennett *et al.*, 2002; for review, Sackett *et al.*, 1999). Early predictive factors, or markers, for sensitivity have not been identified. Similarly, the pathways that convey risk for later biobehavioral pathology in nursery-reared animals are relatively unexplored.

Analysis of activity patterns in infant monkeys may provide useful information about the role that nursery rearing plays in later behavioral or physiological alterations. For example, the finding that SPR monkeys are less active than their PR counterparts suggests the possibility that lower activity might interact with other risk factors in some individuals and, in turn, serve to predispose or prime the onset of negative outcomes. The relative inactivity by SPR monkeys may represent the less optimal outcome. Certainly, lower activity levels may lead to decreased overall fitness and increased health risks, such as abnormalities in the regulation of eating and drinking behavior (Miller *et al.*, 1969). With the exception of one report that has documented earlier death in nursery-reared monkeys (Lewis *et al.*, 2000), relatively little is known about the long-term health consequences of early nursery rearing.

A recent article suggests that such long-term consequences do exist. In that study, nursery-reared rhesus macaques weighed less than their mother-reared counterparts in adolescence (Bastian *et al.*, 2003). Furthermore, preliminary analysis of body mass for the nursery cohort of some of the animals used in this experiment suggests that nursery-reared infants (both PR and SPR) were heavier than mother-reared counterparts at 150 days [$t(47) = 49.58$, $p < 0.01$]. Together these differences suggest relatively long-lasting differences in fitness, although further longitudinal study with multiple measures of "fitness" are needed to address this question.

One difficulty in our preliminary analysis provided useful information about the sensitivity of continuous activity recording for detecting response to environmental change. In this case, a large decrease in activity counts on day 3 (Fig. 22–4) corresponded to a time at which experimenters who were observing another group of animals noted that the Actiwatch-outfitted monkeys were disturbed by the experimenters' presence in the room. This serendipitous observation suggests that Actiwatch recording over a period that contains stimulus change could be used to evaluate acute behavioral change, and may provide a sensitive measure of the longevity of response to challenge conditions. This point is particularly salient because most previous analyses of measurements of behavioral activity in infant nursery-reared primates were based on direct human observation. With direct observation it is difficult to remove the effect of the experimenter's presence or other observational biases. This consideration may be especially relevant in a primate nursery because behavioral reactions among animals housed in the same room can affect behavioral responses of target animals under observation. The advantage of remote monitoring with the Actiwatch is that the researcher can poll specific time points to assess an individual activity state while removing any bias associated with the presence of an observer.

One final consideration in the application of our technique was to include analyses of activity in mother-reared infants. Previous work has suggested that mother-reared infants are less active than nursery-reared monkeys when measured for relatively short durations by direct human observation (Champoux *et al.*, 1991). While the addition of data from a mother-reared group will complete a range of effects of early rearing on Actiwatch measures of behavioral activity, there is a high likelihood that the mother would damage the Actiwatch. Therefore, in future analysis the inclusion of the maternal-rearing condition will begin when monkeys reach the juvenile period and are no longer with their mothers.

5. SUMMARY

The Actiwatch-harness interface we designed allowed us to place the actimeter on group-housed monkeys without jeopardizing either the monkeys' safety or the integrity of the equipment. The multimethod approach we used to investigate the viability of the interface for application in a neonatal nursery suggests that actigraphic analyses of behavioral activity will provide meaningful data that correspond to observable behavioral states. The resulting behavioral analyses differentiated two groups of infant monkeys reared in a nursery environment. Finally, we provide preliminary data suggesting that the Actiwatch is sensitive enough to quantify response to stimulus change. Thus, actigraphy is applicable to a variety of primate models and provides analysis techniques to investigate the effects of early experience on the circadian patterning of activity. Furthermore, analysis of activity profiles and sleep : wake patterns across development will provide translation of phenotypes for human disorders that often show comorbid dysregulation of behavioral activity or sleep parameters (e.g., ADHD, autism, depression, and drug abuse).

ACKNOWLEDGMENTS

We gratefully acknowledge the efforts of Diana Eldreth, Tristian Nicholson, and Clarissa Parker in methods development and data collection. We also thank the animal care and veterinary staff at the National Institutes of Health Animal Center and the neonatal nursery staff, especially Dr. Maribeth Champoux, Courtney Shannon, Pete Roma, and Rebecca Hommer. Finally, we acknowledge the National Institutes of Health Summer Internship Program for supporting the effort of Ms. Eldreth, Nicholson, and Parker.

REFERENCES

Acebo, C., Sadeh, A., Seifer, R., Tzischinsky, O., Wolfson, A.R., Hafer, A., and Carskadon, M.A. (1999). Estimating sleep patterns with activity monitoring in children and adolescents: How many nights are necessary for reliable measures? *Sleep* 22:95–103.

Bastian, M.L., Sponberg, A.C., Suomi, S.J., and Higley, J.D. (2003). Long-term effects of infant rearing condition on the acquisition of dominance rank in juvenile and adult rhesus macaques (*Macaca mulatta*). *Dev. Psychobiol.* 42:44–51.

Bennett, A.J., Lesch, K.P., Heils, A., Long, J.C., Lorenz, J.G., Shoaf, S.E., Champoux, M., Suomi, S.J., Linnoila, M.V., and Higley, J.D. (2002). Early experience and serotonin transporter gene variation interact to influence primate CNS function. *Mol. Psychiatry* 7:118–122.

Brown, A.C., Smolensky, M.H., D'Alonzo, G.E., and Redman, D.P. (1990). Actigraphy: A means of assessing circadian patterns in human activity. *Chronobiol. Int.* 7:125–133.

Carroll, D.A., Denenberg, V.H., and Thoman, E.B. (1993). Reliability and validity of computer scoring of behavioral sleep-wake states in rats and rabbits. *Physiol. Behav.* 54:269–273.

Chamove, A.S., Rosenblum, L.A., and Harlow, H.F. (1973). Monkeys (*Macaca mulatta*) raised only with peers. A pilot study. *Anim. Behav.* 21:316–325.

Champoux, M., Metz, B., and Suomi, S.J. (1991). Behavior of nursery/peer reared and mother-reared rhesus monkeys from birth through 2 years of age. *Primates* 32:509–514.

Cole, R.J., Kripke, D.F., Gruen, W., Mullaney, D.J., and Gillin, J.C. (1992). Automatic sleep/wake identification from wrist activity. *Sleep* 15:461–469.

Diomedi, M., Curatolo, P., Scalise, A., Placidi, F., Caretto, F., and Gigli, G.L. (1999). Sleep abnormalities in mentally retarded autistic subjects: Down's syndrome with mental retardation and normal subjects. *Brain Dev.* 21:548–553.

Elia, M., Ferri, R., Musumeci, S.A., Del Gracco, S., Bottitta, M., Scuderi, C., Miano, G., Panerai, S., Bertrand, T., and Grubar, J.C. (2000). Sleep in subjects with autistic disorder: A neurophysiological and psychological study. *Brain Dev.* 22:88–92.

Glod, C.A., Teicher, M.H., Hartman, C.R., and Harakal, T. (1997a). Increased nocturnal activity and impaired sleep maintenance in abused children. *J. Am. Acad. Child Adolesc. Psychiatry* 36:1236–1243.

Glod, C.A., Teicher, M.H., Hartman, C.R., Harakal, T., and McGreenery, C.E. (1997b). Enduring effects of early abuse on locomotor activity, sleep, and circadian rhythms. *Ann. N.Y. Acad. Sci.* 821:465–467.

Harlow, H.F., and Zimmerman, R.R. (1959). Affectional responses in the infant monkey. *Science* 130:421–432.

Jean-Louis, G., Mendlowicz, M.V., Gillin, J.C., Rapaport, M.H., Kelsoe, J.R., Zizi, F., Landolt, H., and von Gizycki, H. (2000). Sleep estimation from wrist activity in patients with major depression. *Physiol. Behav.* 70:49–53.

Jean-Louis, G., Kripke, D.F., Cole, R.J., Assmus, J.D., and Langer, R.D. (2001). Sleep detection with an accelerometer actigraph: Comparisons with polysomnography. *Physiol. Behav.* 72:21–28.

Johnson, E.O., and Breslau, N. (2001). Sleep problems and substance use in adolescence. *Drug Alcohol Depend.* 64:1–7.

Korte, J., Wulff, K., Oppe, C., and Siegmund, R. (2001). Ultradian and circadian activity-rest rhythms of preterm neonates compared to full-term neonates using actigraphic monitoring. *Chronobiol. Int.* 18:697–708.

Lewis, M.H., Gluck, J.P., Petitto, J.M., Hensley, L.L., and Ozer, H. (2000). Early social deprivation in nonhuman primates: Long-term effects on survival and cell-mediated immunity. *Biol. Psychiatry* 47:119–126.

Mason, W.A., and Berkson, G. (1975). Effects of maternal mobility on the development of rocking and other behaviors in rhesus monkeys: A study with artificial mothers. *Dev. Psychobiol.* 8:197–211.

Miller, L.G., and Kraft, I.A. (1994). Application of actigraphy in the clinical setting: Use in children with attention-deficit hyperactivity disorder. *Pharmacotherapy* 14:219–223.

Miller, R.E., Mirsky, I.A., Caul, W.F., and Sakata, T. (1969). Hyperphagia and polydipsia in socially isolated rhesus monkeys. *Science* 165:1027–1028.

Ruppenthal, G.C., Arling, G.L., Harlow, H.F., Sackett, G.P., and Suomi, S.J. (1976). A 10-year perspective of motherless-mother monkey behavior. *J. Abnorm. Psychol.* 85:341–349.

Sackett, G.P. (1967). Some persistent effects of different rearing conditions on preadult social behavior of monkeys. *J. Comp. Physiol. Psychol.* 64:363–365.

Sackett, G.P., Novak, M.F.S.X., and Kroeker, R. (1999). Early experience effects on adaptive behavior: Theory revisited. *Ment. Retard. Dev. Disabil. Res. Rev.* 5:30–40.

Sadeh, A. (1996). Evaluating night wakings in sleep-disturbed infants: A methodological study of parental reports and actigraphy. *Sleep* 19:757–762.

Sadeh, A., Sharkey, K.M., and Carskadon, M.A. (1994). Activity-based sleep-wake identification: An empirical test of methodological issues. *Sleep* 17:201–207.

Suomi, S.J., and Ripp, C. (1983). A history of motherless mother monkey mothering at the University of Wisconsin primate laboratory. In: *Child Abuse: The Nonhuman Primate Data*. Alan R. Liss, New York, pp. 49–78.

Suomi, S.J., Harlow, H.F., and Kimball, S.D. (1971). Behavioral effects of prolonged partial social isolation in the rhesus monkey. *Psychol. Rep.* 29:1171–1177.

Takase, M., Taira, M., and Sasaki, H. (1998). Sleep-wake rhythm of autistic children. *Psychiatry Clin. Neurosci.* 52:181–182.

Thoman, E.B., Freese, M.P., Becker, P.T., Acebo, C., Morin, V.N., and Tynan, W.D. (1978). Sex differences in the ontogeny of sleep apnea during the first year of life. *Physiol. Behav.* 20:699–707.

Webster, J.B., Kripke, D.F., Messin, S., Mullaney, D.J., and Wyborney, G. (1982). An activity-based sleep monitor system for ambulatory use. *Sleep* 5:389–399.

Zajicek, K.B., Higley, J.D., Suomi, S.J., and Linnoila, M. (1997). Rhesus macaques with high CSF 5-HIAA concentrations exhibit early sleep onset. *Psychiatry Res.* 73:15–25.

Zhdanova, I.V., Geiger, D.A., Schwagerl, A.L., Leclair, O.U., Killiany, R., Taylor, J.A., Rosene, D.L., Moss, M.B., and Madras, B.K. (2002). Melatonin promotes sleep in three species of diurnal nonhuman primates. *Physiol. Behav.* 75:523–529.

CHAPTER TWENTY-THREE

Noninvasive Neuroimaging Techniques for the Study of Primate Brain Development

James K. Rilling

1. INTRODUCTION

We now know that early environmental influences in childhood can affect adult mental health (Heim and Nemefoff, 2001; Johnson *et al.*, 2002; Repetti *et al.*, 2002). Presumably, this association is mediated by neurodevelopmental processes that alter function in the adult brain. Using nonhuman primate models, we can evaluate this hypothesis systematically by isolating and manipulating aspects of the early social environment and describing the neurobiological sequelae of the manipulations. For example, we can rear infants with their mothers or in a primate nursery and then examine the neurodevelopmental consequences of the two rearing conditions (Sanchez *et al.*, 1998).

Most of what we currently know about the development of the primate brain is based on postmortem studies (LaMantia and Rakic, 1990; Gibson, 1991; Rakic *et al.*, 1994; Gould *et al.*, 1999). These cross-sectional developmental studies have yielded valuable insights with respect to how aspects of brain anatomy—e.g., synaptic density, myelination, and neurogenesis—change over the course of development. However, one major disadvantage of this approach is that it precludes longitudinal studies of brain development, which are both more informative (because subjects are used as their own controls) and more efficient with respect to generating data. Fairly recently, a number of noninvasive *in vivo* neuroimaging methods have become available that permit longitudinal investigations of brain development, albeit with less detail than the postmortem histological methods. In this chapter, I will describe these *in vivo* neuroimaging methods and discuss how they can be used to study neurodevelopment in nonhuman primates. Where appropriate, I will also describe examples of studies that have successfully applied these methods to investigate the influence of the social environment on the development of the brain in nonhuman primates.

First, a word about some terminology may be useful. Brain images are composed of voxels. A voxel is the smallest-volume element in an image. It is similar to a pixel, except that it has depth in addition to length and width. The resolution of brain images is determined by the voxel size. Smaller voxels provide better spatial resolution and, hence, more detail. Voxels also have intensity values that vary based on the type of tissue or its functional activity, depending on the method being used. It is the contrast in voxel intensity between different tissue types that produces anatomical images.

2. ANATOMICAL IMAGING

2.1. Magnetic Resonance Imaging (MRI)

MRI is currently the best tool available for noninvasive imaging of brain anatomy *in vivo*. Subjects or patients are placed within the bore of a large magnet that exposes them to a strong magnetic field (Fig. 23–1). Different tissue types within the brain respond differently to perturbations of this magnetic field, which is reflected as differences in signal intensity in the resulting images. MRI is particularly effective in discriminating between gray and white matter in the brain because of differences in their

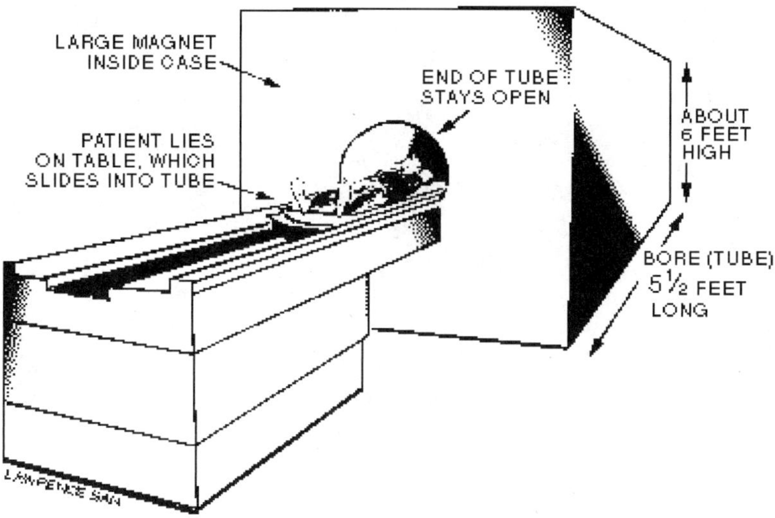

Figure 23–1. Drawing of MRI scanner.

water and fat composition (Westbrook and Kaut, 1993) (Fig. 23–2). One disadvantage of MRI compared with postmortem histological techniques is that it reveals only gross anatomical features of the brain, without access to microscopic, cytoarchitectonic (i.e., cellular architecture) details.

Several investigators have used MRI to examine the impact of stress early in life on the neuroanatomy of nonhuman primates. For example, to test the hypothesis that excessive glucocorticoid exposure can produce lasting damage to the hippocampus, Uno *et al.* (1994) measured hippocampal volume in MRIs from eight infant monkeys (20 months of age) that had been exposed to dexamethasone *in utero*. In support of the hypothesis, they found a 30% reduction in hippocampal volume in dexamethasone-treated infants compared with controls. Lyons (2002) measured prefrontal cortical volume in MRI scans from young adult squirrel monkeys that had been intermittently separated from their social group during infancy, and found that the volume of the right ventral medial prefrontal cortex was larger in the separated infants than in the controls. Whether these volumetric changes reflect alterations in the number and size of neurons or in the neuropil space between them is a question that MRI currently cannot answer (Sapolsky, 2000).

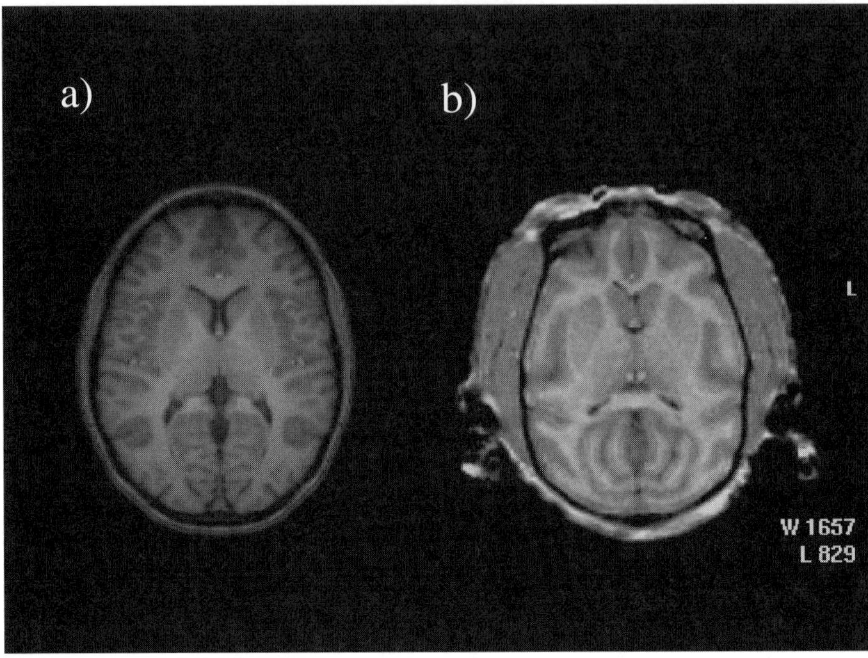

Figure 23–2. T1-weighted anatomical MRIs of axial (or horizontal) slice through the striatum. (**a**) Adult human male. The image was collected in 10 min on a 3-T magnet with a voxel size of 1 mm in all dimensions. A single volume was collected. (**b**) Adult female *Cebus* monkey. The image was collected in 40 min on a 1.5-T magnet with a voxel size of 0.7 mm × 0.7 mm × 1.2 mm. The image is the average of 8 volumes or acquisitions. Images are shown at the same size (not to scale) so that anatomical detail can be compared.

MRI can also be used to describe developmental changes in myelination, the process by which axons accumulate a fatty insulating substance that increases conduction velocity and improves functionality (Fig. 23–3). This is a particularly intriguing application of MRI because the timing of myelination in various white matter tracts correlates with the emergence of various cognitive, motor, and emotional abilities (Konner, 1991). Increases in myelination are reflected in MR images as increases in signal intensity and/or volume of white matter (Paus *et al.*, 1999). MRI has been used to describe developmental sequences of myelination in both humans (reviewed by Paus *et al.*, 2001) and baboons (Miot-

Noirault *et al.*, 1997). MRI has also been used to describe alterations in the area of the mid-sagittal corpus callosum in rhesus macaque infants exposed to early-life stressors (Sanchez *et al.*, 1998; Coe *et al.*, 2002) (Fig. 23–4). The corpus callosum is the major interhemispheric commissure in the primate brain, consisting of fibers that link the two cerebral hemispheres, and changes in its area could reflect changes in either myelination or the extent of axonal pruning. The inability of MRI to discriminate between these two possibilities is another illustration of its limitation in the realm of microscopic neuronal details.

Because the largest application of MRI is clinical, most of the related technology is designed to optimize image quality in human patients. Thus, imaging of nonhuman primates can pose challenges. First, head movement produces image artifacts, so it is important that subjects remain motionless throughout the scan. Human subjects can be asked to lie still, but nonhuman primates must be either anesthetized or habituated to a restraining apparatus. For structural MRI, where we are interested only in brain anatomy, anesthesia is the easier option.

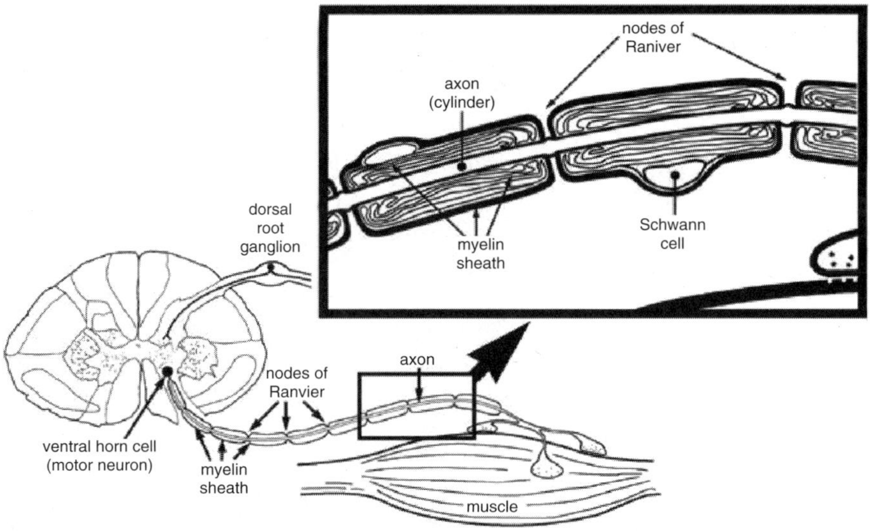

Figure 23–3. Schematic illustration of the myelin sheath that forms around axons during neurodevelopment.

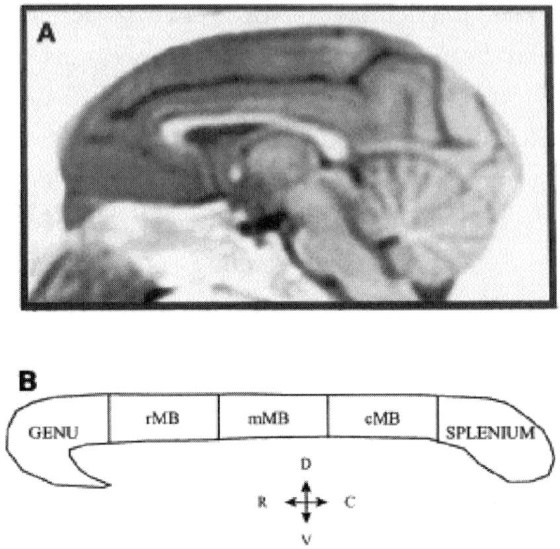

Figure 23–4. (**A**, **B**) Midsagittal T2-weighted MR image through the corpus callosum of an 18-month-old infant rhesus macaque. (From Sanchez *et al.*, 1998.)

Nonhuman primate brains are several times smaller than human brains. For example, a typical rhesus macaque brain is about 80 cm^3, whereas a typical human brain is about 1300 cm^3 (Rilling and Insel, 1999). This means that if the same *spatial* resolution is used for the two species (i.e., the same voxel size), the monkey brain images will show much less detail, that is, they will have poorer *anatomical* resolution (as seen in Fig. 23–1). In some cases, a single voxel may span two different tissue types (e.g., gray matter and white matter) so that its signal intensity will be the average of the two tissues. This is known as the partial voluming. One solution is to decrease the voxel size, which is equivalent to increasing the spatial resolution. However, MRI is all about trade-offs. Smaller voxels produce less signal, which actually degrades a different aspect of image quality, namely, image contrast. Nevertheless, a number of techniques are available to increase the signal : noise (or contrast : noise) ratio of images. One approach is to increase the number of signals that is averaged to generate the image, but this has its costs in terms of scanning time (Fig. 23–5). Doubling the scanning time will increase the signal :

noise ratio by the square root of 2 (Westbrook and Kaut, 1993). Another approach to increasing the signal:noise ratio is to use a magnet with higher field strength—and this costs only money! Magnetic field strength is measured in units called Tesla (T). Many 1.5-T scanners that were state-of-the-art a few years ago are now being replaced by 3-T scanners. Some facilities even have research scanners as high as 9.4 T that have been used for nonhuman primate imaging (Ferris *et al.*, 2001).

Yet another factor that can affect signal:noise ratios is the type of coil used to receive the MR signal (Westbrook and Kaut, 1993). Volumetric coils surround the subject's head (Fig. 23–6). Since the signal falls off with the square of the distance from the head, it is best if the head fits

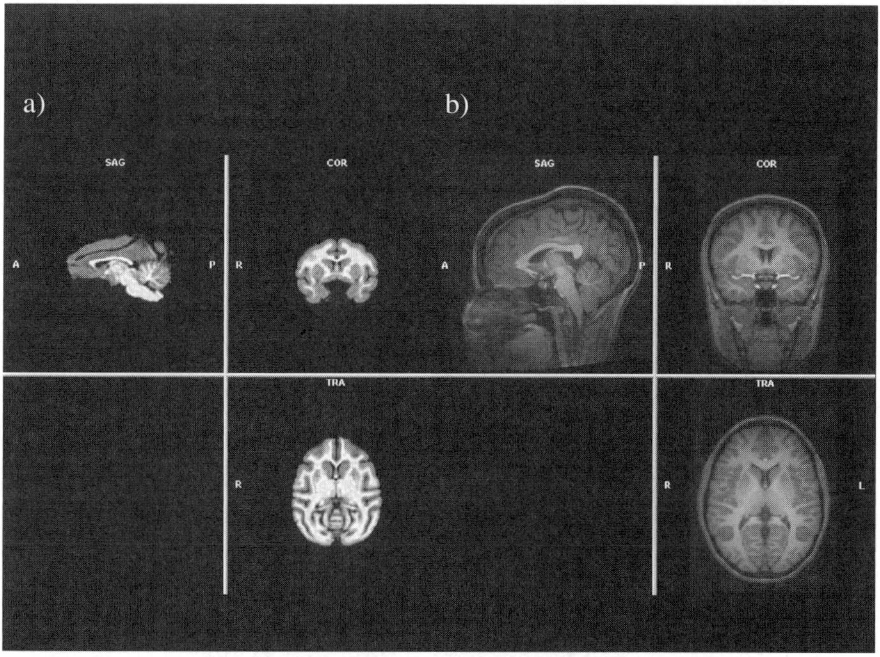

Figure 23–5. T1-weighted MRI scans. (**a**) Adult rhesus macaque (courtesy of Mark Pinsk, Princeton University). Images were acquired on a 3-T magnet with a voxel size of 0.5 mm × 0.5 mm × 1.0 mm. Twenty volumes were acquired over 3.5 hr to counter the low signal:noise ratio. (**b**) Adult human. Images were acquired on a 3-T magnet with a voxel size of 1.0 mm in all directions. A single volume was acquired in 10 min. Images are not to scale.

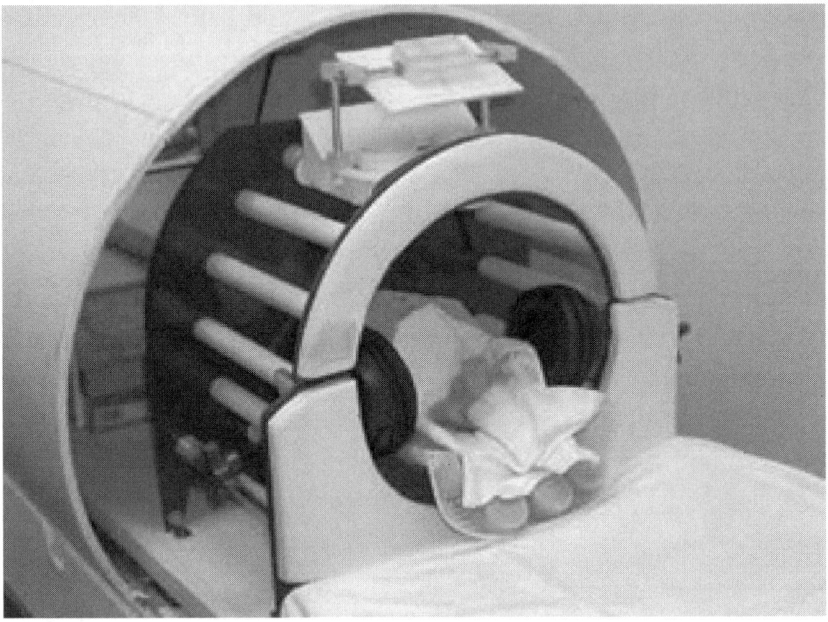

Figure 23–6. Photograph of volumetric head coil. The subject's head is positioned inside the coil.

snuggly within the coil. Monkey heads do not fit snuggly within standard coils designed for the human head. Volumetric coils designed for the human knee often provide a better fit, thereby improving the signal:noise ratio. Another type of coil is a surface coil. These are placed directly on the subject's head and offer an even greater improvement in signal:noise ratio in the vicinity of the coil (Vanduffel *et al.*, 2001; Andersen *et al.*, 2002). However, the signal fades as it moves away from the coil, creating an inhomogeneous image.

2.2. Diffusion Tensor Imaging (DTI)

Although standard MRI protocols can image white matter, they do not reveal the trajectory of specific fiber tracts within the white matter. Recently, an MRI procedure known as diffusion tensor imaging (DTI) has been used for *in vivo* tracking of fiber pathways (Basser and Jones, 2002; Mori and Van Zijl, 2002). DTI can measure the direction of water diffusion in brain tissue. For each brain voxel, a diffusion vector is calculated. In white matter, diffusion of water is restricted to the direction of axonal projection, yielding a vector with a large magnitude. These voxels have high signal intensity on DTI images so that white matter appears bright. In contrast, voxels in which diffusion of water is random, or at least less restricted (e.g., gray matter), have small vectors and low signal intensity. Tractography methods have been developed that use the direction and magnitude of these diffusion vectors to map fiber tract pathways in the brain. These methods have been employed to track fiber pathways *in vivo* in both human and nonhuman primates (Parker *et al.*, 2002) (Fig. 23–7). However, the nonhuman primate work has suffered from low anatomical resolution (as discussed above), which has led to some inaccuracies in the results. Nevertheless, there is no reason why

a)

b)

Figure 23–7. Comparison of DTI tractography of the corticospinal tract in human (**a**) and macaque (**b**). Projections leading from motor cortex to the cerebral peduncles below are shown in both (**a**) and (**b**). (From Parker *et al.*, 2002.)

these limitations cannot be overcome by enhancing signal:noise ratio using approaches described for MRI. It should be noted that this method may never achieve the level of anatomical specificity achieved by classic tract-tracing methods that involve injection of tracers followed by postmortem histology. Their advantage is that they, like other neuroimaging methods, allow for longitudinal studies and preclude the need for terminal studies.

DTI has the potential, or soon will, to reveal how early environmental influences might alter patterns of connectivity in nonhuman primate brains. It should be noted that most of these connections are probably laid down prenatally in monkeys (LaMantia and Rakic, 1990), so that environmental influences that affect this aspect of neurodevelopment would likely be operating *in utero*.

2.3. Manganese Imaging

Very recently, a new *in vivo* tract-tracing technique has been developed that is closely analogous to methods used in standard postmortem, histological tract-tracing techniques. Manganese (Mn^{2+}) has a high-intensity signal on T1-weighted MRI scans, making it easy to detect. It also has the fortunate property of being transported in an anterograde manner from cell bodies down their axons and across the synapse into the next neuron of the pathway. Manganese imaging involves injecting Mn^{2+} into a specific brain region, followed by collection of an MRI several hours later to track the tracer's movement. Like DTI, manganese imaging reveals connections between brain areas. However, unlike DTI, it also reveals the direction of fiber projection between the two areas. Saleem *et al.* (2002) injected $MnCl_2$ into the caudate and putamen of adult rhesus macaques and used MRI to follow its movement into the globus pallidus (Fig. 23–8, see color plate) substantia nigra, and thalamus. The results were in good agreement with histological results later collected from the same animals (Fig. 23–8, see color plate). Thus, manganese imaging can be used to visualize neuronal connections in developing monkeys and to study how these may be altered by environmental influences.

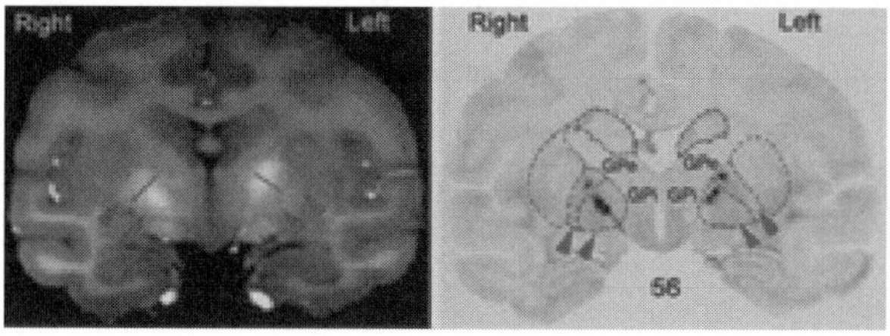

Figure 23–8. MRI and histological visualization of connections in the globus pallidus following $MnCl_2$ injection of caudate and putamen 45 hr earlier. (From Saleem *et al.*, 2002.) (See color plate)

3. FUNCTIONAL IMAGING

In addition to investigating developmental changes in brain anatomy, neuroimaging can be used to describe developmental changes in brain function. The two most common functional brain imaging techniques are positron emission tomography and functional magnetic resonance imaging.

3.1. Positron Emission Tomography (PET)

3.1.1. [^{18}F]FDG PET

PET involves the injection of radioactive compounds, followed by detection of the distribution and concentration of radioactivity in the brain (Votaw, 1995). One of PET's most common applications is the measurement of local cerebral glucose metabolism (lCMRglu), which is known to be a correlate of synaptic activity in the brain (Sokoloff, 1999). PET has been used to measure lCMRglu across development in both humans (Chugani and Phelps, 1986) and monkeys (Jacobs *et al.*, 1995; Moore *et al.*, 2000). These studies have generated several interesting findings. For example, Chugani and colleagues showed that levels of lCMRglu in humans are very low at birth, increase to a maximum between 3 and 10 years of age, and then slowly decline to adult levels. A similar pattern was described in monkeys (Jacobs *et al.*, 1995; Moore *et al.*, 2000), but with an earlier maximum between 2 and 8 months, followed by a decline starting as early as 10 months of age (Fig. 23–9). This

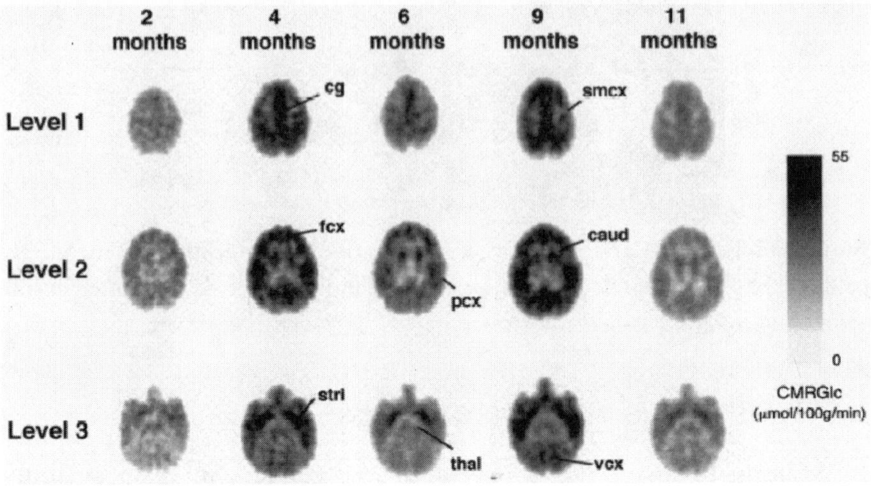

Figure 23–9. Representative [^{18}F]FDGPET images across development from five different animals displayed at three transverse levels of the brain. The gray scale represents absolute cerebral metabolic rates for glucose (lCMRglu). Note the increase of lCMRglu at 4 and 9 months of age in all structures and the decrease in lCMRglu at 11 months. (From Moore *et al.*, 2000.)

pattern may well reflect changes in synaptic density since most of the energy produced by glucose metabolism is used to restore resting membrane potentials at synapses (Sokoloff, 1999). Furthermore, histological studies in humans describe similar developmental trajectories for synaptic density, with substantial synaptic pruning in the second decade of life (Huttenlocher, 1979). In monkeys, however, synaptic density peaks between 2 and 3 months of age (Rakic *et al.*, 1986, 1994), whereas lCMRglu continues to increase until about 8 months of age. These data prompted Moore *et al.* (2000) to speculate that lCMRglu may track synaptic remodeling in addition to synaptic density, given evidence that remodeling is ongoing during this period.

Beyond developmental studies, PET's greatest application has been to describe cerebral metabolic correlates of various neurological and psychiatric disorders (Newberg *et al.*, 2002). Though less common, PET can also be used for activation studies at a fixed point in development. Activation studies involve comparing lCMRglu in response to two different conditions or manipulations of interest. For example, to examine changes in lCMRglu in response to psychological stress, a subject would be scanned once in a stressful condition and once in a nonstressful condition, and the two images would be subtracted to identify regions uniquely responsive to stress.

It is instructive to consider the method in some detail in order to understand its limitations for imaging nonhuman primates. The chemical used to measure lCMRglu is a modified glucose molecule that has been labeled with the positron-emitting ^{18}F isotope. Its official name is [^{18}F]fluorodeoxyglucose (or [^{18}F]FDG). There are two types of [^{18}F]FDG studies, quantitative and qualitative. A quantitative [^{18}F]FDG experiment involves intravenous injection of [^{18}F]FDG, followed by collection of blood samples at approximately 5-min intervals throughout the experiment. After the injection, [^{18}F]FDG accumulates in cells proportional to their rate of glucose metabolism, and all of the [^{18}F]FDG is trapped within cells by about 45 min postinjection. Only after completion of this brain uptake phase is the scan started. The scan normally takes between 20 and 30 min. The image produced by the scan shows the distribution of radioactivity in the brain, a reflection of regional brain metabolism during the previous brain uptake phase—a neural footprint, if you will (CD. Kilts, personal communication). Tracer kinetic models use these radioactivity data, along with plasma glucose and [^{18}F]FDG concentrations throughout the brain uptake phase, to estimate lCMRglu in amount of glucose per gram of tissue per unit time (Phelps *et al.*, 1979). Thus, quantitative [^{18}F]FDG studies require collection of blood samples throughout the brain uptake phase.

This process can be challenging when the subjects are awake, mobile nonhuman primates, which may be fearful and aggressive and can remove the catheters. One solution is to collect data from anesthetized animals (Jacobs *et al.*, 1995), but anesthetics obviously affect brain glucose metabolism and we are more interested in brain activity in conscious animals. Moore *et al.* (2000) used a very creative methodology to obtain lCMRglu from conscious vervet infants. After injecting [^{18}F]FDG, they collected blood samples from the infants as they clung to their anesthetized mothers. This was effective in limiting the infant's movement. Nevertheless, it is perhaps heartening that the average lCMRglu in the study by Moore *et al.* was similar to that found for anesthetized vervets (Jacobs *et al.*, 1995), suggesting that anesthesia may not dramatically alter findings in these types of developmental studies.

Another approach to circumventing the difficulty of blood collection in awake monkeys is to conduct nonquantitative [^{18}F]FDG studies that do not require collection of blood samples. In contrast to quantitative studies, these studies cannot provide data in units of glucose consumption per unit tissue per unit time. Instead, they provide data in radioactivity counts per unit tissue per unit time. Although absolute rates of glucose metabolism cannot be quantified, regional concentrations of radioactivity can be compared with the global average for the entire brain to provide a measure of local metabolism relative to the global mean. It is assumed that global cerebral metabolism does not change much in response to moment-to-moment changes in psychological states. Nonquantitative methods can therefore be useful for activation studies that examine how one condition or state changes regional cerebral glucose metabolism relative to another.

For example, to examine the neural correlates of maternal separation in juvenile rhesus macaques, my colleagues and I (Rilling *et al.*, 2001) briefly separated juveniles from their mothers and then injected the juveniles with [^{18}F]FDG intramuscularly and studied the neural responses in three conditions. In the first condition, juveniles were reunited with their mothers for the duration of the 40-min brain uptake phase; in the second condition, they were still separated from their mothers but allowed visual contact; and in the third condition, they were separated from their mother and not allowed visual or auditory contact. Following the 40-min brain uptake phase, the juveniles were sedated and scanned while anesthetized. Importantly, the anesthetic had no effect on the images since the pattern of radioactivity detected in the image reflects glucose uptake from the preceding phase when the animals were awake. Compared with the condition in which mother and juvenile were together, the separation conditions involved greater activation in right dorsolateral prefrontal and right ventral temporal regions. Activation in these regions was also positively correlated with plasma cortisol concentrations, suggesting that these activations may relate to the stress of maternal separa-

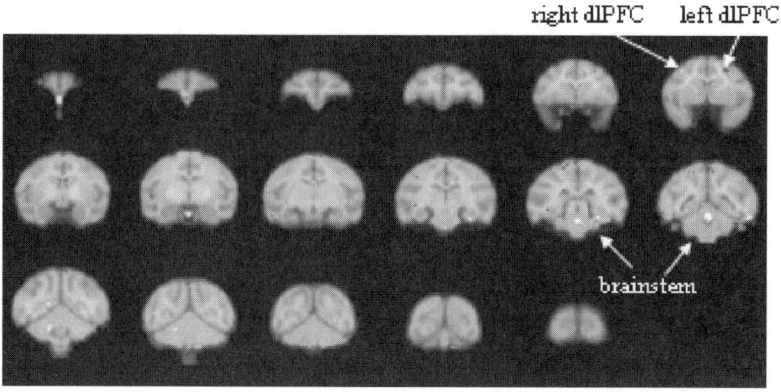

Figure 23–10. Voxel-by-voxel correlations between brain activity and plasma cortisol concentrations in juvenile monkeys following separation from their mothers. Voxels that are positively correlated with cortisol are colored orange ($0.01 < p < 0.05$) or yellow ($p < 0.01$) and those that are negatively correlated with cortisol are colored light blue ($0.01 < p < 0.05$) or dark blue ($p < 0.01$). (From Rilling *et al.*, 2001.) (See color plate)

tion. Positive correlations with plasma cortisol concentrations were also observed in brainstem targets of corticotropin-releasing factor (CRF) projections from the amygdala and hypothalamus (Fig. 23–10, see color plate).

As noted for the methods already discussed, one of the difficulties of applying PET imaging to nonhuman primates is the relatively small size of the monkey brain and the consequent reduction in anatomical resolution. In postmortem studies with 2-deoxy-D-glucose (2-DG), animals are injected with 2-DG and sacrificed after the brain uptake phase, and then radioactivity is quantified using autoradiographic procedures. Like the histological methods described above, this approach provides greater anatomical detail than [^{18}F]FDG imaging. But again, there are strategies for improving anatomical resolution. In PET, spatial resolution is limited by three factors: the energy of the emitted positron, the scanner diameter, and the size of the scanner detectors. Chatziioannou *et al.* (1999) developed a dedicated PET scanner that uses a reduced ring diameter and smaller detectors to markedly improve spatial resolution for use with nonhuman primates. This scanner, known as microPET, has the additional advantage of costing much less than conventional PET scanners.

3.1.2. ^{15}O Water PET

In addition to measuring regional brain glucose metabolism, PET can be used to measure cerebral blood flow, another correlate of neuronal activity (Kety, 1965). Blood flow increases in response to increases in neural activity. The tracer used to measure blood flow is a radioactively labeled water molecule, ^{15}O water. One advantage of ^{15}O water PET over [^{18}F]FDG PET is improved temporal resolution. Whereas a single [^{18}F]FDG scan integrates neural activity over 30–45 min, a single ^{15}O water scan integrates neural activity over 60–90 sec. Moreover, ^{15}O has a much shorter half-life than ^{18}F (2 min versus 110 min), which means it takes only about 10 min for ^{15}O to decay to negligible levels before another scan can be started. Thus, as many as 12 ^{15}O water scans can be collected in a single scanning session, whereas [^{18}F]FDG sessions typically involve acquisition of only a single scan (but see Strome *et al.*, 2002). With these advantages, ^{15}O water PET is obviously a much more efficient means of data collection. On the other hand, ^{15}O water PET has poorer spatial resolution than [^{18}F]FDG PET because ^{15}O positrons have more energy than ^{18}F positrons and therefore travel further before being detected by the tomograph. In addition, [^{18}F]FDG PET may be better than ^{15}O water PET for imaging more stable psychological states (e.g., depression) because it integrates over a larger temporal window.

In contrast to [^{18}F]FDG, there is no protracted brain uptake phase following injection of ^{15}O water, and no opportunity to scan an anesthetized animal and image a neural footprint left during the preceding conscious state. Neural activity is measured in real time. If the goal is to image brain activity in conscious animals, subjects must be awake for the scan. This poses the challenge of restraining the animal in a way that minimizes stress and discomfort, and preventing head movement that would cause image artifacts. A number of investigators have succeeded in habituating monkeys to restraining chairs that immobilize their heads in the PET scanner (Fig. 23–11) (Takechi *et al.*, 1994; Howell *et al.*, 2001; Noda

Figure 23–11. Photograph of head-restraint device mounted atop cradle used to support the body during PET imaging of awake monkeys. The head-restraint device consists of foam packed around the subject's head. The apparatus is placed on its side so that the monkey lies supine during the PET scan. (From Howell *et al.*, 2001; see that publication for additional details.)

et al., 2002), proving that blood flow activation studies are feasible in awake monkeys.

3.1.3. Receptor and Transporter Imaging with PET

There is yet a third application of PET in neuroimaging. PET can be used to image the density and distribution of neurotransmitter receptors and transporters in the brain (Kegeles and Mann, 1997). Much like classic autoradiography, this involves the injection of a radioactively labeled ligand for a specific receptor followed by detection of the location and intensity of the radioactivity. A great deal of research is devoted to synthesizing radiolabeled ligands that have the appropriate binding characteristics, in terms of their specificity and affinity. Ligands are currently available for several subtypes of dopamine and serotonin receptors, as well as their transporters. Ligands for other receptors are also available and the number of available ligands will continue to grow in the future. These radioligands are appropriate for nonhuman primate use, particularly since evaluation of potential ligands is often conducted in nonhuman primates. An interesting study by Morgan *et al.* (2002) used this method to detect changes in receptor density consequent to alterations of the social environment. These researchers used the radioligand [^{18}F]FCP to image D2 dopamine receptors in the basal ganglia of cynomolgus macaques. When individually housed monkeys were transitioned from solitary to social housing, dominant animals exhibited an increase in D2 receptor density or availability 3 months later, whereas subordinate monkeys showed no change. Ligands for other receptors could be even more informative with respect to neurobiological consequences of early-life stressors. Given the importance of the CRF system in stress and anxiety (Arborelius *et al.*, 1999), a ligand for the CRF receptor would be particularly desirable.

Some ligands, such as [^{11}C]raclopride, are displaceable by their endogenous ligands, thereby providing the intriguing ability to image endogenous neurotransmitter release. For example, [^{11}C]raclopride, a displaceable D2 receptor ligand, has been used to demonstrate striatal dopamine release in the brains of human volunteers playing a video game (Koepp *et al.*, 1998). The number of displaceable neurotransmitter receptor ligands is currently limited, but the eventual prospect of imaging endogenous transmission of a range of brain chemicals is an exciting one with clear applications for developmental studies of nonhuman primate behavioral biology.

I end this section by calling attention to one of the main drawbacks of PET imaging (i.e., [^{18}F]FDG, ^{15}O water, receptor, and transporter imaging), namely, that it requires a cyclotron staffed with trained personnel for the production of radioactive tracers. For most tracers, the half-life is such that they need to be synthesized on site. However, because of its long half-life, [^{18}F]FDG is sometimes transported from cyclotron facilities to PET scanning facilities that lack a cyclotron.

3.2. Functional Magnetic Resonance Imaging (fMRI)

Another method for noninvasively monitoring brain function *in vivo* is fMRI. fMRI measures local changes in blood flow secondary to changes in neural activity (Kwong *et al.*, 1992; Ogawa *et al.*, 1992; Logothetis, 2002). In response to increases in synaptic activity, there is an increased flow of oxygen-rich blood and a corresponding increase in the relative concentration of oxygenated hemoglobin. This hemodynamic response is rather sluggish, peaking about 5 sec after the onset of neural activity (Kwong *et al.*, 1992). Since oxygenated and deoxygenated hemoglobin have different magnetic properties, changes in their relative concentrations can be detected by the scanner and reflected in the signal intensity of the resulting images.

Compared with PET, fMRI has both advantages and disadvantages for functional neuroimaging of nonhuman primates. One advantage of fMRI over PET is that it relies on the brain's own endogenous contrast agent, hemoglobin (Ogawa *et al.*, 1990). Radioactive tracers need not be injected, and IV catheters need not be placed. Thus fMRI is a less invasive method than PET. Another advantage of fMRI over PET is superior temporal resolution. fMRI can generate functional images of the entire brain in as little as 2 sec. Since no tracer is injected, there is no need to allow for decay of radioactivity before starting the next scan. Therefore, scans can be collected continuously at 2-sec intervals (or even shorter, depending on the spatial resolution of the scans).

The main disadvantage of fMRI for imaging awake animals is the scanning environment, which is very loud and confining. The sound is loud enough that human subjects must wear earplugs to protect their ears. The space where subjects must lie is inside a long, narrow cylindrical tube (i.e., the magnet bore; see Fig. 23–1). Animals must be habituated to this environment to minimize stress and movement throughout the scan. Investigators commonly use operant conditioning and mock-up scanning environments to train subjects to remain motionless in the scanner (Stefanacci *et al.*, 1998; Dubowitz *et al.*, 2001; Vanduffel *et al.*, 2001; Andersen *et al.*, 2002). Minimizing head movement is the biggest challenge for nonhuman primate fMRI studies since it can cause severe problems for image analysis. Some type of head holder that restrains head

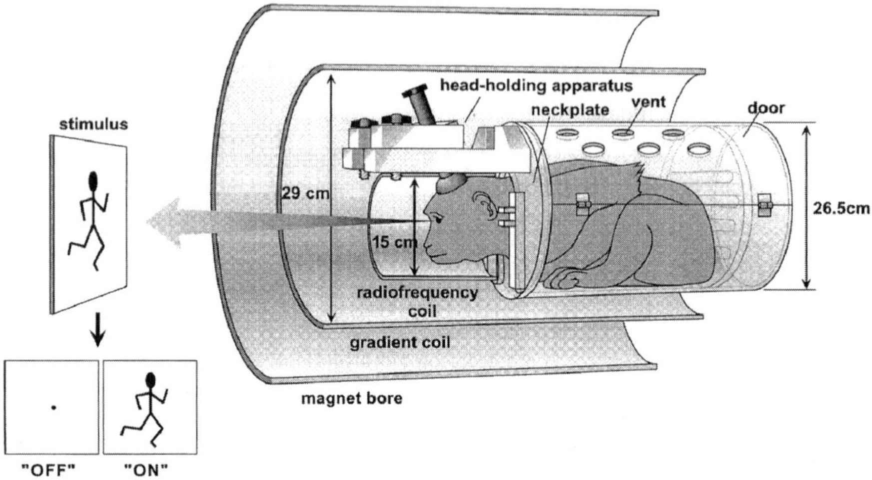

Figure 23–12. Set-up for monkey fMRI. The monkey sits in the sphinx position inside the volumetric receiving coil, with its head immobilized by a head-holding apparatus. (From Stefanacci *et al.*, 1998.)

movement is needed (Fig. 23–12). Obviously, these head holders must be made with nonferrous material since ferrous materials would be susceptible to the strong magnetic field force and would disrupt that field, adversely affecting image quality. Several groups have succeeded in using fMRI to image brain function in awake monkeys in response to visual stimuli (Dubowitz *et al.*, 1998; Stefanacci *et al.*, 1998; Logothetis *et al.*, 1999; Vanduffel *et al.*, 2001). Most of these teams have implanted headsets in the monkey's skull that then attach to a head-holder in the magnet bore (Fig. 23–12). This, of course, involves surgery and carries with it the risk of infection.

Most MRI scanners are horizontally oriented so that human patients can lie on their backs during the scan. As this is an unnatural position for monkeys, most investigators have designed restraining apparatuses that allow the monkeys to sit on their haunches in a "sphinx" position (Fig. 23–12), viewing stimuli that are presented in front of them. However, Logothetis *et al.* (1999) used a vertical bore magnet that allowed the subject to sit upright, and suggested that this position reduces unwanted body movements in awake animals.

As with other imaging modalities, anatomical resolution suffers and partial volume effects can be problematic in monkey brains because of their small size. As mentioned earlier, magnets with higher field strength permit higher spatial resolution because of their enhanced signal : noise ratio. For example, Logothetis *et al.* (1999) were able to detect activation in response to visual stimuli with superior anatomical resolution using a 4.7-T magnet (Fig. 23–13, see color plate). And, in an intriguing experiment, Ferris *et al.* (2001) scanned male common marmosets in a 9.4-T scanner while exposing them to odors from ovariectomized or ovulating female marmosets. They found that the odors from ovulating females were associated with activation in the preoptic area and the anterior hypothalamus, areas known to be involved in male sexual behavior.

Another approach to dealing with the small size of monkey brains is to increase sensitivity by injecting an exogenous contrast agent. Vanduffel *et al.* (2001) found that intravenous injection of an iron oxide contrast agent resulted in an approximately 10-fold increase in percentage of signal change in response to moving visual stimuli. Finally, the signal : noise ratio can be improved by using surface coils that abut against the head surface rather than standard volume coils that surround the

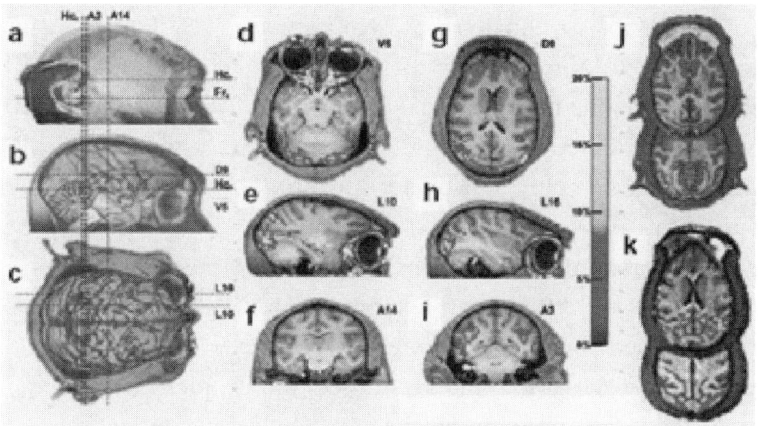

Figure 23–13. (**a–k**) fMRI activation of lateral geniculate nucleus and visual cortex in rhesus macaques in response to visual stimulation. (From Logothetis *et al.*, 1999.) (See color plate)

whole head but do not make direct physical contact with the head (Vanduffel *et al.*, 2001; Logothetis, 2002).

To summarize, impressive efforts by a handful of researchers have demonstrated the feasibility of using fMRI to measure brain function in awake monkeys. Nevertheless, this is clearly a highly demanding, labor-intensive endeavor that requires substantial resources for successful implementation.

4. CONCLUSIONS

This chapter has reviewed methods that are available to researchers interested in investigating the impact of early environmental influences on primate neurodevelopment. All of the methods discussed in this chapter are noninvasive *in vivo* techniques that allow investigators to track developmental changes within subjects, something that is not possible with standard cross-sectional approaches involving postmortem histology. Although imaging methods do not currently provide the level of anatomical detail provided by postmortem studies, we have not yet reached the limits of anatomical resolution with these methods. The level of anatomical detail will likely improve in the coming years. Techniques available for anatomical imaging include MRI, DTI, and manganese imaging. MRI is useful for quantifying changes in the size of brain structures and in the degree of myelination of white matter. DTI and manganese imaging can reveal the trajectory of white matter projections in the brain. Techniques available for functional imaging include PET and fMRI. PET can be used to describe changes in regional brain glucose metabolism and the density of neurotransmitter receptors over the course of development. It can also be used for activation studies that examine the impact of a particular manipulation on either glucose metabolism or blood flow. Finally, fMRI, which provides much better temporal resolution than PET, can also be used for activation studies. However, formidable technical challenges must be overcome to successfully implement this method. Collectively, these methods will help provide a clearer picture of how early environmental influences can alter neurodevelopment and, in turn, how alterations in brain development affect adult cognition and social behavior.

ACKNOWLEDGMENT

I would like to thank Mark Pinsk for assistance with this manuscript.

5. REFERENCES

Andersen, A.H., Zhang, Z., Barber, T., Rayens, W.S., Zhang, J., Grondin, R., Hardy, P., Gerhardt, G.A., and Gash, D.M. (2002). Functional MRI studies in awake rhesus monkeys: Methodological and analytical strategies. *J. Neurosci. Methods* 118:141–152.

Arborelius, L., Owens, M.J., Plotsky, P.M., and Nemeroff, C.B. (1999). The role of corticotropin-releasing factor in depression and anxiety disorders. *J. Endocrinol.* 160:1–12.

Basser, P.J., and Jones, D.K. (2002). Diffusion-tensor MRI: Theory, experimental design and data analysis—a technical review. *NMR Biomed.* 15:456–467.

Chatziioannou, A.F., Cherry, S.R., Shao, Y., Silverman, R.W., Meadors, K., Farquhar, T.H., Pedarsani, M., and Phelps, M.E. (1999). Performance evaluation of microPET: A high-resolution lutetium oxyorthosilicate PET scanner for animal imaging. *J. Nucl. Med.* 40:1164–1175.

Chugani, H.T., and Phelps, M.E. (1986). Maturational changes in cerebral function in infants determined by 18FDG positron emission tomography. *Science* 231:840–843.

Coe, C.L., Lulbach, G.R., and Schneider, M.L. (2002). Prenatal disturbance alters the size of the corpus callosum in young monkeys. *Dev. Psychobiol.* 41(2):178–185.

Dubowitz, D.J., Chen, D.Y., Atkinson, D.J., Grieve, K.L., Gillikin, B., Bradley, W.G., Jr., and Andersen, R.A. (1998). Functional magnetic resonance imaging in macaque cortex. *Neuroreport* 9:2213–2218.

Dubowitz, D.J., Chen, D.Y., Atkinson, D.J., Scadeng, M., Martinez, A., Andersen, M.B., Andersen, R.A., and Bradley, W.J. (2001). Direct comparison of visual cortex activation in human and non-human primates using functional magnetic resonance imaging. *J. Neurosci. Methods* 107:71–80.

Ferris, C.F., Snowdon, C.T., King, J.A., Duong, T.Q., Ziegler, T.E., Ugurbil, K., Ludwig, R., Schultz-Darken, N.J., Wu, Z., Olson, D.P., Sullivan, J.M., Jr., Tannenbaum, P.L., and Vaughan, J.T. (2001). Functional imaging of brain activity in conscious monkeys responding to sexually arousing cues. *Neuroreport* 12:2231–2236.

Gibson, K.R. (1991). Myelination and behavioral development: A comparative perspective on questions of neoteny, altriciality and intelligence. In: Gibson, K.R., and Petersen, A.C. (eds.), *Brain Maturation and Cognitive Development*. Aldine de Gruyter, New York, pp. 29–63.

Gould, E., Reeves, A.J., Graziano, M.S., and Gross, C.G. (1999). Neurogenesis in the neocortex of adult primates. *Science* 286:548–552.

Heim, C., and Nemefoff, C. (2001). The role of childhood trauma in the neurobiology of mood and anxiety disorders: Preclinical and clinical studies. *Biol. Psychiatry* 49:1023–1039.

Howell, L.L., Hoffman, J.M., Votaw, J.R., Landrum, A.M., and Jordan, J.F. (2001). An apparatus and behavioral training protocol to conduct positron emission tomography (PET) neuroimaging in conscious rhesus monkeys. *J. Neurosci. Methods* 106:161–169.

Huttenlocher, P.R. (1979). Synaptic density in human frontal cortex—developmental changes and effects of aging. *Brain Res.* 163:195–205.

Jacobs, B., Chugani, H.T., Allada, V., Chen, S., Phelps, M.E., Pollack, D.B., and Raleigh, M.J. (1995). Developmental changes in brain metabolism in sedated rhesus macaques and vervet monkeys revealed by positron emission tomography. *Cereb. Cortex* 5:222–233.

Johnson, J.G., Cohen, P., Gould, M.S., Kasen, S., Brown, J., and Brook, J.S. (2002). Childhood adversities, interpersonal difficulties and risk for suicide during late adolescence and early adulthood. *Arch. Gen. Psychiatry* 59: 741–749.

Kegeles, L.S., and Mann, J.J. (1997). In vivo imaging of neurotransmitter systems using radiolabeled receptor ligands. *Neuropsychopharmacology* 17: 293–307.

Kety, S.S. (1965). Biochemistry and mental function. *Nature* 208:1252–1257.

Koepp, M.J., Gunn, R.N., Lawrence, A.D., Cunningham, V.J., Dagher, A., Jones, T., Brooks, D.J., Bench, C.J., and Grasby, P.M. (1998). Evidence for striatal dopamine release during a video game. *Nature* 393:266–268.

Konner, M. (1991). Universals of behavioral development in relation to brain myelination. In: Gibson, K.R., and Petersen, A.C. (eds.), *Brain Maturation and Cognitive Development*. Aldine de Gruyter, New York, pp. 29–63.

Kwong, K., Belliveau, J., Chesler, D., Goldberg, I., Weisskoff, R., Poncelet, B., Kennedy, D., Hoppel, B., Cohen, M., and Turner, R. (1992). Dynamic magnetic resonance imaging of human brain activity during primary sensory stimulation. *Proc. Natl. Acad. Sci. USA* 89:5675–5679.

LaMantia, A.S., and Rakic, P. (1990). Axon overproduction and elimination in the cerebral cortex of the developing rhesus monkey. *J. Neurosci.* 10:2156–2175.

Logothetis, N.K. (2002). The neural basis of the blood-oxygen-level-dependent functional magnetic resonance imaging signal. *Phil. Trans. R Soc. Lond. B Biol. Sci.* 357:1003–1037.

Logothetis, N.K., Guggenberger, H., Peled, S., and Pauls, J. (1999). Functional imaging of the monkey brain. *Nat. Neurosci.* 2:555–562.

Lyons, D.M. (2002). Stress, depression, and inherited variation in primate hippocampal and prefrontal brain development. *Psychopharmacol. Bull.* 36:27–43.

Miot-Noirault, E., Barantin, L., Akoka, S., and Le Pape, A. (1997). T2 relaxation time as a marker of brain myelination: Experimental MR study in two neonatal animal models. *J. Neurosci. Methods* 72:5–14.

Moore, A.H., Hovda, D.A., Cherry, S.R., Villablanca, J.P., Pollack, D.B., and Phelps, M.E. (2000). Dynamic changes in cerebral glucose metabolism in conscious infant monkeys during the first year of life as measured by positron emission tomography. *Brain Res. Dev. Brain Res.* 120:141–150.

Morgan, D., Grant, K.A., Gage, H.D., Mach, R.H., Kaplan, J.R., Prioleau, O., Nader, S.H., Buchheimer, N., Ehrenkaufer, R.L., and Nader, M.A. (2002). Social dominance in monkeys: Dopamine D2 receptors and cocaine self-administration. *Nat. Neurosci.* 5:169–174.

Mori, S., and Van Zijl, P.C. (2002). Fiber tracking: Principles and strategies—a technical review. *NMR Biomed.* 15:468–480.

Noda, A., Ohba, H., Kakiuchi, T., Futatsubashi, M., Tsukada, H., and Nishimura, S. (2002). Age-related changes in cerebral blood flow and glucose metabolism in conscious rhesus monkeys. *Brain Res.* 936:76–81.

Ogawa, S., Lee, T.M., Kay, A.R., and Tank, D.W. (1990). Brain magnetic resonance imaging with contrast dependent on blood oxygenation. *Proc. Natl. Acad. Sci. USA* 87:9868–9872.

Ogawa, S., Tank, D.W., Menon, R., Ellermann, J.M., Kim, S.G., Merkle, H., and Ugurbil, K. (1992). Intrinsic signal changes accompanying sensory stimulation: Functional brain mapping with magnetic resonance imaging. *Proc. Natl. Acad. Sci. USA* 89:5951–5955.

Parker, G.J., Stephan, K.E., Barker, G.J., Rowe, J.B., MacManus, D.G., Wheeler-Kingshott, C.A., Ciccarelli, O., Passingham, R.E., Spinks, R.L., Lemon, R.N., and Turner, R. (2002). Initial demonstration of in vivo tracing of axonal projections in the macaque brain and comparison with the human brain using diffusion tensor imaging and fast marching tractography. *Neuroimage* 15:797–809.

Paus, T., Zijdenbos, A., Worsely, H., Collins, D.L., Blumenthal, J., Giedd, J.N., Rapoport, J.L., and Evans, A.C. (1999). Structural maturation of neural pathways in children and adolescents: In vivo study. *Science* 283:1908–1911.

Paus, T., Collins, D.L., Evans, A.C., Leonard, G., Pike, B., and Zijdenbos, A. (2001). Maturation of white matter in the human brain: A review of magnetic resonance studies. *Brain Res. Bull.* 54:255–266.

Phelps, M.E., Huang, S.C., Hoffman, E.J., Selin, C., Sokoloff, L., and Kuhl, D.E. (1979). Tomographic measurement of local cerebral glucose metabolic rate in humans with (F-18)2-fluoro-2-deoxy-D-glucose: Validation of method. *Ann. Neurol.* 6:371–388.

Rakic, P., Bourgeois, J.P., Eckenhoff, M.F., Zecevic, N., and Goldman-Rakic, P.S. (1986). Concurrent overproduction of synapses in diverse regions of the primate cerebral cortex. *Science* 232:232–235.

Rakic, P., Bourgeois, J., and Goldman-Rakic, P. (1994). Synaptic development of the cerebral cortex: Implications for learning, memory and mental illness. *Progr. Brain Res.* 102:227–243.

Repetti, R.L., Taylor, S.E., and Seeman, T.E. (2002). Risky families: Family and social environments and the mental and physical health of offspring. *Psychol. Bull.* 128:330–366.

Rilling, J.K., and Insel, T.R. (1999). The primate neocortex in comparative perspective using magnetic resonance imaging. *J. Hum. Evol.* 37:191–223.

Rilling, J.K., Winslow, J.T., O'Brien, D., Gutman, D.A., Hoffman, J.M., and Kilts, C.D. (2001). Neural correlates of maternal separation in rhesus monkeys. *Biol. Psychiatry* 49:146–157.

Saleem, K.S., Pauls, J.M., Augath, M., Trinath, T., Prause, B.A., Hashikawa, T., and Logothetis, N.K. (2002). Magnetic resonance imaging of neuronal connections in the macaque monkey. *Neuron* 34:685–700.

Sanchez, M.M., Hearn, E.F., Do, D., Rilling, J.K., and Herndon, J.G. (1998). Differential rearing affects corpus callosum size and cognitive function of rhesus monkeys. *Brain Res.* 812:38–49.

Sapolsky, R.M. (2000). Glucocorticoids and hippocampal atrophy in neuropsychiatric disorders. *Arch. Gen. Psychiatry* 57:925–935.

Sokoloff, L. (1999). Energetics of functional activation in neural tissues. *Neurochem. Res.* 24:321–329.

Stefanacci, L., Reber, P., Costanza, J., Wong, E., Buxton, R., Zola, S., Squire, L., and Albright, T. (1998). fMRI of monkey visual cortex. *Neuron* 20:1051–1057.

Strome, E.M., Wheler, G.H., Higley, J.D., Loriaux, D.L., Suomi, S.J., and Doudet, D.J. (2002). Intracerebroventricular corticotropin-releasing factor increases limbic glucose metabolism and has social context-dependent behavioral effects in nonhuman primates. *Proc. Natl. Acad. Sci. USA* 99:15749–15754.

Takechi, H., Onoe, H., Imamura, K., Onoe, K., Kakiuchi, T., Nishiyama, S., Yoshikawa, E., Mori, S., Kosugi, T., Okada, H., *et al.* (1994). Brain activation study by use of positron emission tomography in unanesthetized monkeys. *Neurosci. Lett.* 182:279–282.

Uno, H., Eisele, S., Sakai, A., Shelton, S., Baker, E., DeJesus, O., and Holden, J. (1994). Neurotoxicity of glucocorticoids in the primate brain. *Horm. Behav.* 28:336–348.

Vanduffel, W., Fize, D., Mandeville, J.B., Nelissen, K., Van Hecke, P., Rosen, B.R., Tootell, R.B., and Orban, G.A. (2001). Visual motion processing investigated using contrast agent-enhanced fMRI in awake behaving monkeys. *Neuron* 32:565–577.

Votaw, J.R. (1995). The AAPM/RSNA physics tutorial for residents. Physics of PET. *Radiographics* 15:1179–1190.

Westbrook, C., and Kaut, C. (1993). *MRI in Practice.* Blackwell Science, Oxford.

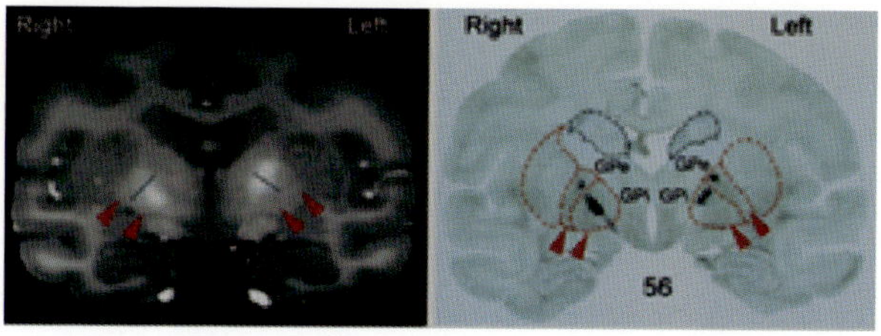

Figure 23–8. MRI and histological visualization of connections in the globus pallidus following $MnCl_2$ injection of caudate and putamen 45 hr earlier. (From Saleem *et al.*, 2002.)

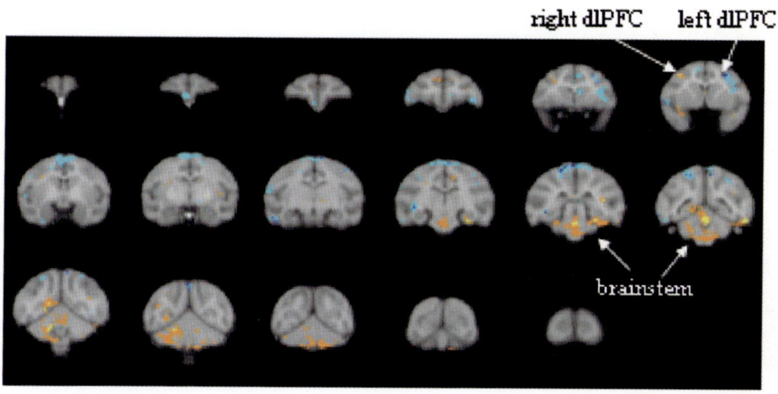

Figure 23–10. Voxel-by-voxel correlations between brain activity and plasma cortisol concentrations in juvenile monkeys following separation from their mothers. Voxels that are positively correlated with cortisol are colored orange ($0.01 < p < 0.05$) or yellow ($p < 0.01$) and those that are negatively correlated with cortisol are colored light blue ($0.01 < p < 0.05$) or dark blue ($p < 0.01$). (From Rilling *et al.*, 2001.)

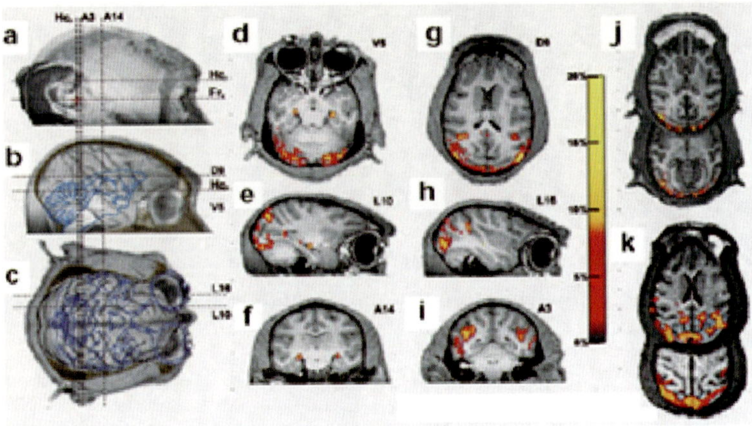

Figure 23–13. (**a–k**) fMRI activation of lateral geniculate nucleus and visual cortex in rhesus macaques in response to visual stimulation. (From Logothetis *et al.*, 1999.)

CHAPTER TWENTY-FOUR

Tethering with Maternal and Fetal Catheterization as a Model for Studying Pre- to Postnatal Continuities

Matthew Francis Stuart Xavier Novak

1. WHAT IS TETHERING?

Tethering with maternal and fetal catheterization is a preparation that gives researchers direct access to the physiological status of both mother and fetus without the need for anesthesia or restraint. The technique involves the use of a flexible hollow metal cable that attaches at one end to a nylon jacket worn by the female and at the other end to a freely rotating turntable at the center-top of the cage. The tether could also exit the cage through the back wall, but the top-of-cage configuration allows the instrumented mother to move about the cage as freely as a noninstrumented animal (Morton *et al.*, 1987). Catheters protected by the cable can be implanted in both mother and fetus as well as in the amniotic cavity. Using this preparation, a researcher can manipulate the fetus, deliver drugs, draw blood or amniotic fluid, and monitor intra-uterine pressure along with maternal and fetal blood pressure and heart rate. With the use of electrodes implanted to measure muscle function, fetal behavior can even be assessed. In addition, it may soon be possible to visualize the fetus *in utero* using a specialized catheter adapted for use with ultrasound technology.

The technology and procedures for tethering, developed mainly in sheep and adapted for baboons and rhesus macaques, are not new (McNamee *et al.*, 1984; Nathanielsz *et al.*, 1984; Stark *et al.*, 1989). The majority of tethering and catheterization studies in nonhuman primates have focused on prenatal physiological questions. Studies have addressed fetal cardiorespiratory measurements (Novy *et al.*, 1971), uterine blood flow in response to intravenous cocaine administration (Morgan *et al.*, 1991), fetal and maternal endocrine responses to reduced uteroplacental blood flow (Shepherd *et al.*, 1992), fetal and maternal origin of leukocytes in the amniotic fluid (Marcias *et al.*, 2000), the assembly of fetal fatty acids (Su *et al.*, 2001), timing, inhibition, and induction of myometrial activity and parturition (Baguma-Nibasheka *et al.*, 2000; Farber *et al.*, 1997; Giussani *et al.*, 1996a; Haluska *et al.*, 1997; Morgan *et al.*, 1994; Sadowsky *et al.*, 2000), the effect of cytokine inflammation and chorionic-decidual infection on the degradation of fetal membranes in relation to myometrial activity (Vadillo-Ortega *et al.*, 2002), maternal and fetal endocrine or immune responses to intraamniotic or choriodecidual infection (Gravett *et al.*, 1994, 1996; Witkin *et al.*, 1994), the maternal hypothalamo-pituitary-adreno-placental axis (Giussani *et al.*, 1996b), and androstenedione-induced forward shift in maternal estradiol relative to progesterone and increased contractions during late pregnancy (Giussani *et al.*, 2000). Few of these studies included any follow-up on the postnatal development of the infant.

2. EFFECTS OF TETHERING

Tethering with catheterization has the potential for combining direct fetal manipulation and measurement with postnatal developmental follow-up. Before this can happen, however, the impact on pregnancy and infant development needs to be determined. Although the effects of tethering and catheterization on the fetus have been assessed, the effects on postnatal development have rarely been of interest. In one study that also involved chair restraint of the pregnant female, tethering and catheterization seemingly caused no changes to the uterine environment (Murata *et al.*, 1984). However, 34 of 67 fetuses died within 48 hr after catheterization surgery. This study did not address the long-term development of the fetuses, nor did it examine postnatal developmental variables.

Postnatal data from infants whose mothers were exposed to azidothymidine (AZT; zidovudine) therapy via gastric catheter revealed effects on postnatal growth, neonatal reflexive behavior, simple learning, and cognitive tasks, but the interaction of tethering and AZT was not evaluated (Ha *et al.*, 1998). In studies of maternal–fetal transmission of simian immunodeficiency virus (SIV) (Ochs *et al.*, 1991) and human immunodeficiency virus (HIV) (Ho *et al.*, 2001) in pigtailed macaques, infants were monitored for virus levels and clinical symptoms of infection postnatally, but the interactions between catheterization procedures and drug or disease processes were not assessed.

2.1. Maternal Effects

As part of a research project to describe the acute effects of prenatal psychosocial stress on both the mother and the fetus (Novak, 2002), I collected several datasets to examine the invasiveness of tethering with maternal and fetal catheterization by comparing the pregnancies of instrumented and noninstrumented animals. Three datasets about maternal effects of chronic catheterization are reported here: cortisol, estradiol, and behavior.

2.1.1. Subjects

Five pregnant females, proven breeders, were catheterized at 120 (±1) days of gestation (DG). Catheters in mothers and fetuses were maintained in a tethering system until birth occurred via natural delivery. One pregnancy (monkey 94088) was spontaneously aborted during gestational week 18. A second pregnancy (monkey 99163) ended in premature delivery at 21 weeks of gestation. The infant was euthanized 3 weeks after birth due to complications with pulmonary edema secondary to oxygen therapy for prematurity. Two subjects (monkeys 391396 and 97077) gave birth at 22 weeks of gestation and one (monkey 94043) at 23 weeks of gestation.

These experimental pregnancies were compared with eight control pregnancies in proven breeders, which were similarly housed and tested but experienced no jacketing, tethering, or surgical procedures. Our first objective was to establish whether there were differences in pregnancy due to the use of these procedures, so we started by testing "all conditions" versus "no conditions." If differences occurred, we would run intermediate controls to evaluate which aspects of our procedures were causing what types of deficit. In the absence of differences, further breakdown of experimental conditions would not be necessary. One control female (monkey 92214) spontaneously aborted during week 16 and was replaced by monkey 93146, which gave birth at 24 weeks gestation. Control monkey 94069 also spontaneously aborted. Two control animals (monkeys 97067 and 998067) gave birth at 23 weeks of gestation, two (monkeys 99134 and 94054) at 24 weeks, and two (monkeys 92241 and 92212) after 25 weeks.

2.1.2. Procedures

At 50–60 DG, the subjects were moved to single cages in a multianimal room (6.1 m × 7.6 m × 2.1 m) with overhead fluorescent lighting turned on for 14 hr each day at 0600 hours. All animals in the room were adult or subadult female pigtailed macaques. Animal cages lined the walls facing the center of the room. Including the subject, there were usually five animals in the room (range 3–8). The subjects' cages (0.64 m × 0.64 m × 0.81 m) were fronted with Plexiglas. The tops were modified to accommodate a freely rotating lazy-susan tethering apparatus.

From GD 50 through the end of pregnancy (~172 GD), the females were observed directly several times a day and by video throughout the night. At 80 GD (±2), they were fitted with a nylon jacket to begin adaptation. The jacket was tailored to fit each animal and adjusted as pregnancy progressed. When a jacket became worn, it was replaced. At 95 GD (±2), a flexible stainless-steel tether was attached to the nylon jacket and anchored to a rotating turntable at the center top of the cage.

Habituation to the jacket and tether continued until surgery for catheterization at 120 GD (±2).

With the female under general anesthesia, six catheters were implanted: two in the mother (femoral artery and femoral vein), two in the fetus (internal carotid artery and jugular vein), and two in the amniotic cavity. The six catheters exited the uterus and were passed via subcutaneous tunneling to the upper mid-thoracic back area, where a small puncture site allowed the catheters to be externalized. The catheters then traveled through the hollow tethering cable to the turntable at the top of the cage, where the four blood lines were attached to a pumping system (Gilson Minipuls 3, Middleton, WI) that infused heparinized saline (3 U/ml) into the catheters to prevent clotting. In addition to the pumping system, the two arterial lines and one of the amniotic lines were attached to pressure transducers (Argon Medical, Athens, TX). The animals remained catheterized until delivery. Maintenance of pregnancy was achieved, when necessary, with the use of tocolytic drugs, terbutaline and/or magnesium, administered intravenously via the pumping system.

Two methods were used for blood collection. First, females in both groups were anesthetized periodically and blood samples were drawn immediately upon immobilization. Second, once the experimental animals were catheterized, their blood was drawn via the catheters without anesthesia or restraint.

2.1.3. Analysis and Results

Plasma cortisol and estradiol values across pregnancy were studied in the instrumented versus noninstrumented groups. Between-group analyses focused on samples from tethered versus nontethered animals. Separate within-subject analyses focused first on change across pregnancy, and then on samples from anesthetized versus unanesthetized animals, testing the interaction between the potential invasiveness of tethering procedures and disturbances due to anesthetization. Because data were not available for all animals at all time points, to maximize the sample sizes I conducted repeated-measures analyses of variance on each individual pair of time points. Main effects consisted of gestational age and tethered versus control animals.

Figure 24–1 shows fluctuation but no trend in maternal cortisol values from 60 GD through the end of pregnancy. Although sample sizes were quite small, no differences were found in baseline cortisol values between catheterized and control macaques. In addition, the baseline values of tethered animals were the same whether the samples were drawn via femoral puncture while the animals were anesthetized or via the catheter without anesthesia.

Maternal estradiol values increased across gestation in most animals, although not all pairwise comparisons of time blocks were significant (Fig. 24–2). Visual inspection of the data suggests a transitory drop in estrogen values for tethered animals following surgery. However, similar

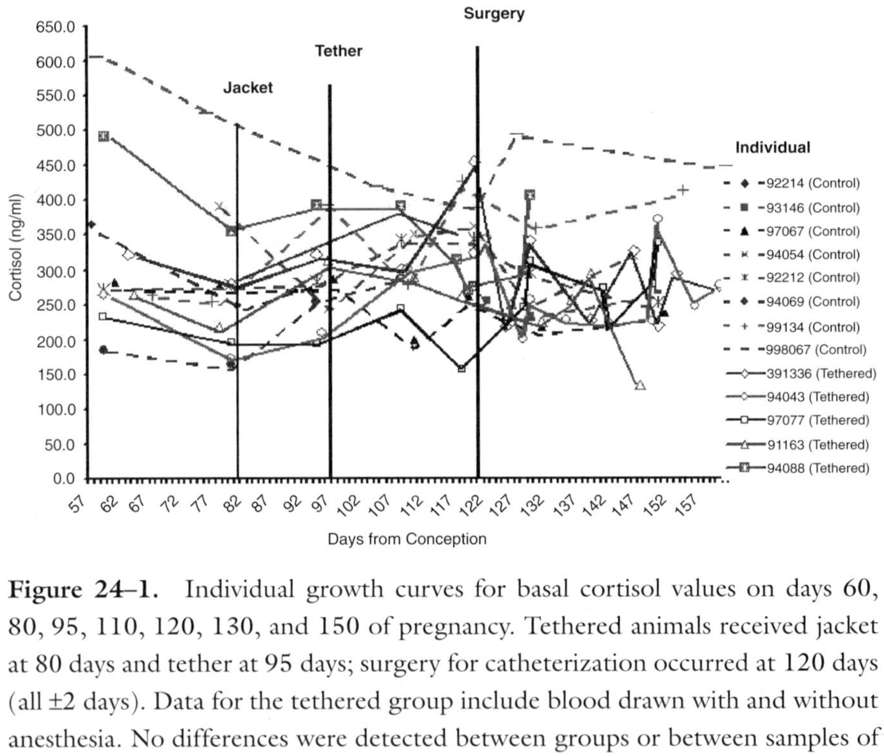

Figure 24–1. Individual growth curves for basal cortisol values on days 60, 80, 95, 110, 120, 130, and 150 of pregnancy. Tethered animals received jacket at 80 days and tether at 95 days; surgery for catheterization occurred at 120 days (all ±2 days). Data for the tethered group include blood drawn with and without anesthesia. No differences were detected between groups or between samples of tethered animals with and without anesthesia. Developmentally, cortisol values dropped between days 60 and 80 and increased between days 80 and 110. No other pairwise time comparisons were significant. No time-by-group interactions were significant ($\alpha = 0.05$).

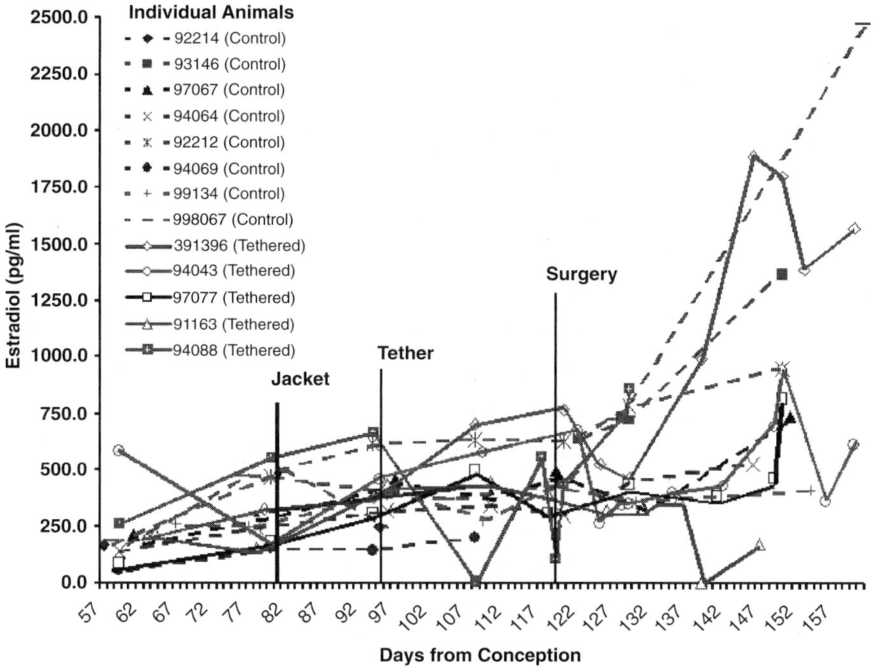

Figure 24–2. Individual growth curves for maternal estrogen values on days 60, 80, 95, 110, 120, 130, and 150 of pregnancy. Conditions are as described for Fig. 24–1.

to cortisol measurements, there were no main effects or interactions between samples drawn from tethered and control females, although within-group variability was quite large particularly late in gestation. As for cortisol, there were no differences within the tethered group, whether the samples were drawn from anesthetized or unanesthetized animals. However, the sample sizes for both hormonal data sets were quite small.

Levels of estrogen plummeted suddenly in two tethered females (Fig. 24–2). Low estrogen level during human pregnancy is often regarded as an indicator of fetal distress (Diczfalusy, 1974), and the low levels in these monkeys may reflect the invasiveness of the tethering procedures. Although the values recovered in both females, it is provocative that one (monkey 94088) aborted spontaneously within 2 weeks after catheterization surgery and the other (monkey 91163) delivered a premature infant that was later euthanized for pulmonary edema caused by oxygen therapy for prematurity. It is important to note, however, that in monkey 94088 the drop in estrogen values occurred at 118 days, i.e., before catheterization surgery, and that in monkey 91163 the drop did not occur until after 137 GD. Several factors in addition to fetal distress may account for fluctuations in estrogen level (Novy and Walsh, 1981).

In addition to these physiological parameters, the behavioral effects of jacketing, tethering, and catheterization procedures were observed. Data were collected during 5-min focal animal sampling of postural and behavior changes twice a day, in the morning and in the afternoon, 3 days per week. Observers collected data while seated 1–2 m from the front of the animals' Plexiglas-fronted homecages. Animals were habituated to observers for at least 10 days before data collection began. When multiple animals were observed, the order of observation was randomized. Before collecting data, the observers demonstrated interobserver reliability by achieving κ values above 0.65 on three consecutive sessions using real-time second-by-second coding.

Only minor differences in behavior were detected between experimental and control animals. Following recovery from surgery, but not jacketing and tethering, catheterized females exhibited more passive sitting than controls (Fig. 24–3). This effect appears, however, to be driven by a decrease in the amount of passive sitting on the part of control animals rather than an increase by the experimental animals. Toward the end of pregnancy, catheterized females exhibited more physical exploration and manipulation of their environment (Fig. 24–4). Although their exploration was directed primarily to the jacket and their own bodies following recovery from surgery, this increased interest in jacket and self did not occur in response to initial jacketing or tethering. Animals rarely attended to the tether other than to sit on it and occasionally use it as a swing. There were no differences in the intake of food or water between tethered and control groups at any time point during gestation (Fig. 24–5). Finally, it was thought that the experience of jacketing and tethering might affect the females' interactions with other monkeys in the room and with experimenters. In fact, the animals showed no social changes in response to either the jacket or tether adaptation, although they did show a slight increase in threat and dominance-related behavior following recovery from catheterization surgery (Fig. 24–6).

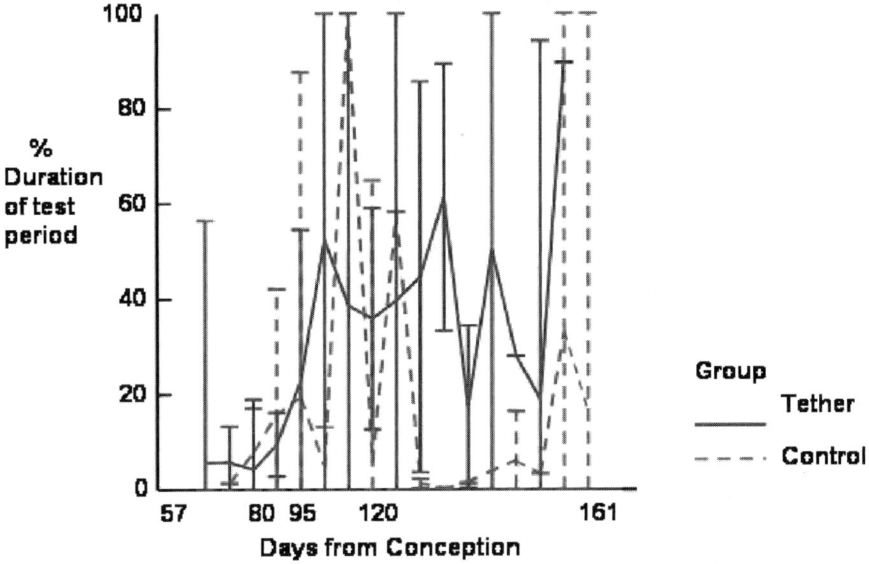

Figure 24–3. Postural behaviors in conditions described for Fig. 24–1. There was considerable individual variability within groups and tethered animals as a group exhibited more passive posture after surgery than controls. Error bars = 95% CIs.

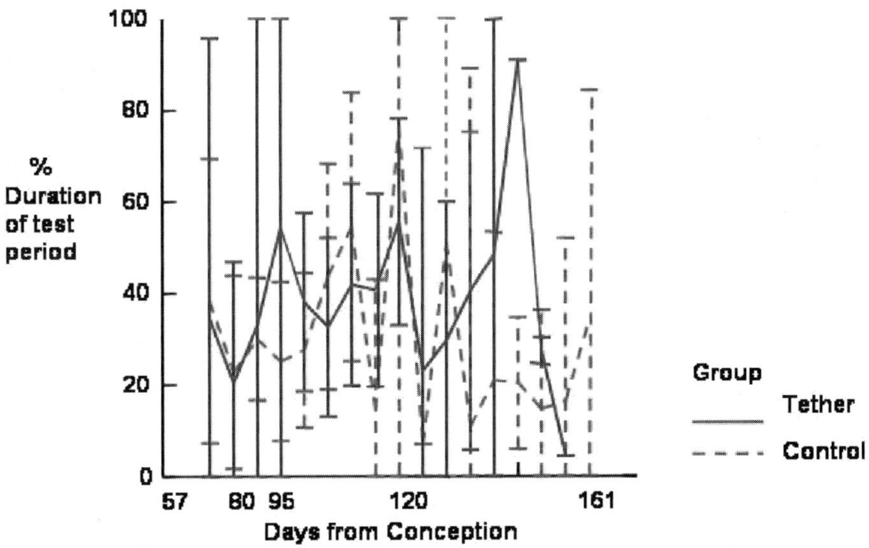

Figure 24–4. Exploratory behaviors in conditions as described for Fig. 24–1. There was considerable variability within groups and tethered females as a group exhibited more exploration late in pregnancy, after recovery from surgery (week 17), than controls. Error bars = 95% CIs.

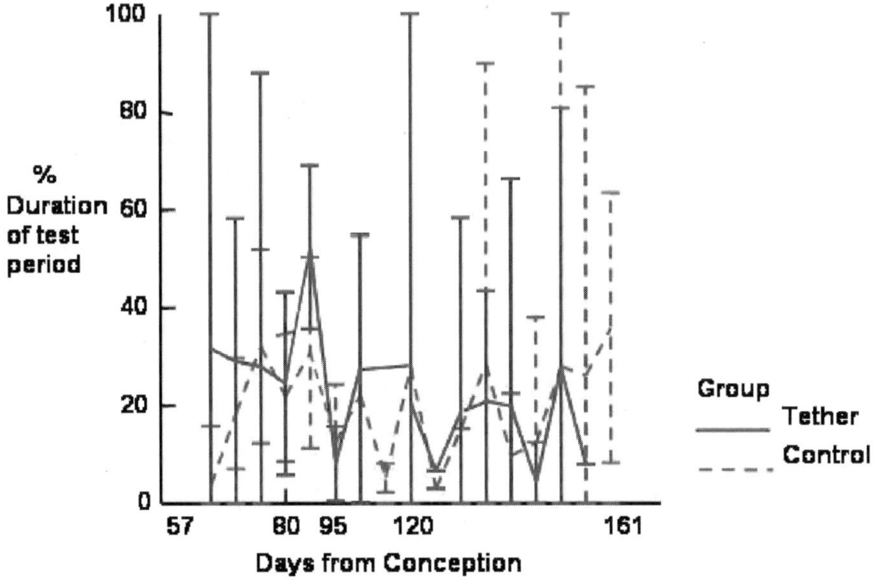

Figure 24–5. Food intake in conditions as described for Fig. 24–1. There were no differences between the two groups. Error bars = 95% CIs.

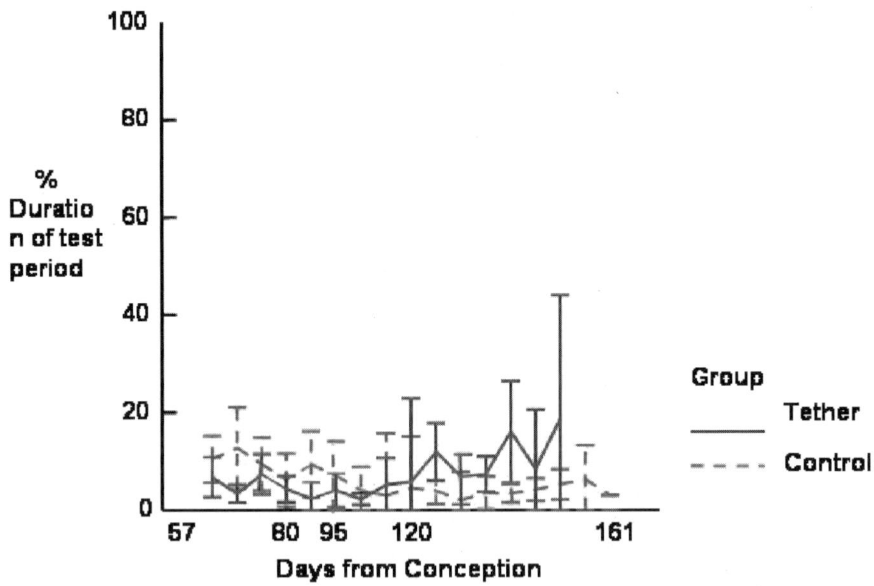

Figure 24–6. Dominance and threat behaviors. Following surgery tethered females spent more time in dominance and threat-related behavior than controls. Error bars = 95% CIs.

As for the procedure's direct effects on the fetus, catheterized infants were born earlier in gestation than controls (mean = 158 versus 173 days, respectively) and therefore had lower birth weights. There were no differences in the rates of live births between tethered and control groups. However, the offspring of tethered monkey 94043 was born microcephalic. Although it is tempting to assume that catheterization procedures induced the microcephaly, review of clinical and historical records revealed that this female had produced several anomalous pregnancies and should not have been classified as a proven breeder. Because of this confound, the cause of microcephaly remains unknown.

3. MATERNAL AND FETAL CARDIOVASCULAR FUNCTIONING

An example of cardiovascular data collected from tethered and catheterized mothers and fetuses is shown in Fig. 24–7. Chronic catheterization allows real-time continuous sampling from both the mother and the fetus via the pressure signal in the catheter. Because the blood pressure signal is muted by the pliability of the catheter material used, the systolic and diastolic values need to be interpreted with care. Mean arterial values are not affected. Only late-term pregnancy can be studied due to the limitations of catheter diameter used in fetal blood vessels. Heart rate and heart rate variability can be calculated from the blood pressure signal.

Data were collected at a rate of 50 samples per second, checked for errors, and collapsed into 5-min time blocks. With these procedures, overall data error rates were 1.3% of all signals. This included 2.5% of the fetal blood pressure and 3.1% of the fetal heart rate and variability signals. For maternal data, 0.04% of maternal blood pressure and 0.06% of maternal heart rate and variability data were classified as error. In addition to the 70-min samples illustrated in Fig. 24–7, complete diurnal 24-hr samples were collected. These datasets are currently being summarized.

The expected values of data, transformed to quintiles and plotted with the use of log-linear modeling, are shown in Fig. 24–8. The rationale for this transformation is twofold. First, there are a multitude of influences on these cardiovascular variables. The central nervous system inputs that accelerate heart rate differ from those that decelerate heart rate (Randall and Smith, 1984; Rowell, 1993). The reaction to a manipulation may be

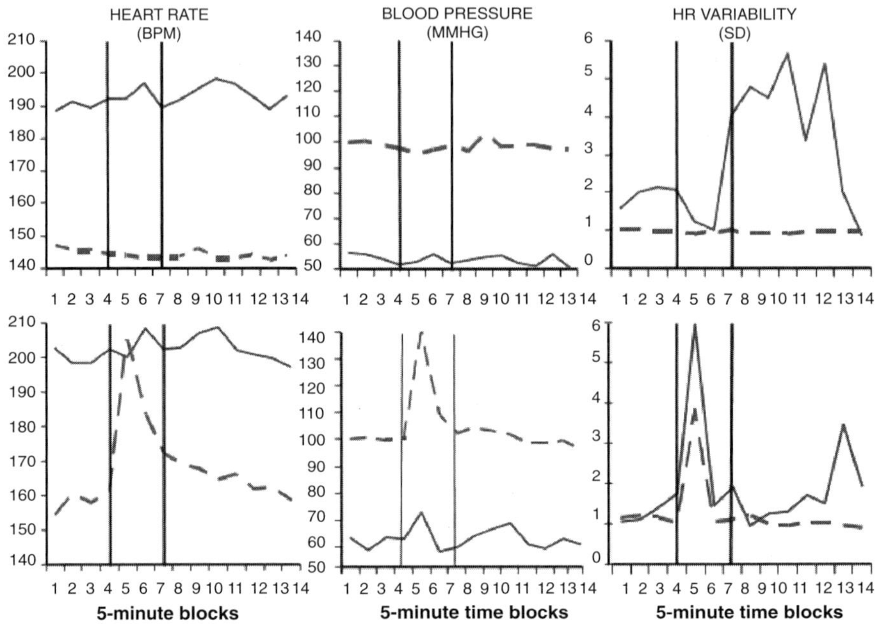

Figure 24–7. Examples of cardiovascular dimensions measured via pressure transducers using chronic tethering with maternal and fetal catheterization in one animal on a baseline day (top), and in response to anticipation of maternal capture (bottom). Data are collapsed into mean values per 5-min time block across 70 min of data collection. Manipulation occurred during time block 5. Dotted lines, maternal values; solid lines, fetal values.

bimodal, resulting in both low and high heart rates. This effect on two different aspects of heart rate may or may not be correlated, and the amount of correlation may or may not change within similar environments or across different environments (Katz, 2001). For blood pressure data, systolic and diastolic values are often analyzed separately and each may exhibit different associations with independent and dependent variables (Kagan *et al.*, 1999). The results of one such study showed that systolic but not diastolic levels were related to fetal variables (Monk *et al.*, 2000). Because separate analyses ignore interaction effects that may be present, it is likely that a single descriptor of central tendency does not capture the degree and types of variation occurring in cardiovascular data.

Log-linear analyses of the frequency data divided into fifths allows analysis of how multiple aspects of the cardiovascular variables are performing, maintaining information about the total distribution of values that is lost with single measures such as mean and standard deviation. This transformation also allows a functional comparison between maternal and fetal values on an ordinal scale, as there is often a large difference between their interval or ratio scale values and their ranges.

The change in data across time and a maternal-by-fetal interaction included in the model are illustrated in Fig. 24–8. The factors in the model include four time blocks, the frequency of each maternal quintile, and the frequency of each fetal quintile. These examples from a single animal illustrate the weakness of the mean as a representative descriptive

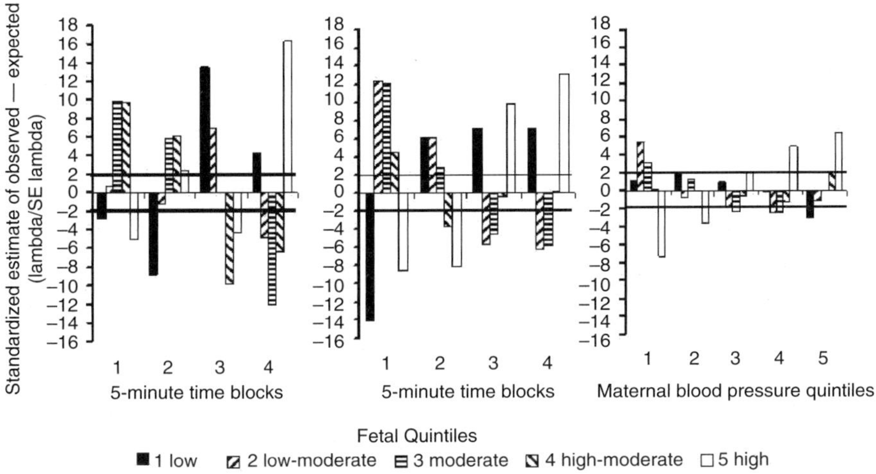

Figure 24–8. Pattern of significance for maternal and fetal cardiovascular data transformed into quintiles. Estimates greater than or less than ±1.96 are significant, indicating that quintile occurs more or less frequently than expected using a best fit log-linear model. On the left, fetal heart rate values are bimodal during time block 4 on a baseline day; center, fetal heart rate variability values are bimodal during time blocks 3 and 4 on a baseline day; right, across time blocks maternal and fetal blood pressure values interact on baseline day. Maternal and fetal blood pressure are in sync; when maternal blood pressure is low, fetal blood pressure is also low; as maternal blood pressure increases, so does fetal blood pressure.

statistic because of bimodality in the data. These data also show maternal–fetal interactions in blood pressure measurements, illustrating the weakness of analyzing systolic and diastolic blood pressure data separately.

In a single tethered and catheterized mother–fetus pair analyzed with these techniques, both bimodality within 5-min time blocks and maternal–fetal interactions were common. Bimodality never exceeded 22% of baseline time blocks for the mother and 19% of the baseline time blocks for the fetus. Inclusion of the maternal–fetal interaction, in order to improve the fit of the model beyond the maternal and fetal changes over time, was necessary for both heart rate data (30% of time blocks), and blood pressure data (60% of time blocks), but not for heart rate variability. For heart rate, interactions were more common in response to or following mild laboratory stresses. For blood pressure, interactions were more common during baseline time periods.

More animals need to be assessed to ensure that the patterns observed in this mother–fetus pair were not unique. It is possible that the bimodality observed in different quintiles was a function of quantitative shifts in cardiovascular values from low to high, or from high to low, within the 5-min time block. However, visual inspection of the data suggests this is not the case for these data. Furthermore, it is tempting to infer that the interactions between maternal and fetal blood pressures were a function of maternal movement. Although this is a possibility, at least some of the interactions were negative, with the maternal values high and the fetal values low. Others were positive with the two distributions matching each other. This makes the argument for a movement artifact less likely. From these data, it is clear that simple analyses of the mean, and/or analyses that assess only certain parts of cardiovascular distributions, are probably inadequate for many questions.

3.1. Prenatal Stress and the Origins of Infant Reactivity and Sensitivity to the Environment

One potentially useful application of tethering technology involves the assessment of one of the most slippery definitional issues for developmental psychologists interested in the origins of individual differences.

Methods for objectively assessing subjective psychological experience are rare, and the task is often ignored because it is so difficult. Partly due to this difficulty, the literature on the effects of prenatal psychosocial stress on infant development is somewhat ambiguous, with results running the gamut of negative and null effects (for a review see Lobel, 1994) and a subset of studies even reporting positive effects of prenatal manipulations and stress (Denenberg *et al.*, 1998; Francis *et al.*, 1999).

The specific mechanism by which prenatal stress alters the developmental trajectory in a fetus remains unknown. Proposed mechanisms include impairment of uteroplacental blood flow (Morishima *et al.*, 1978), anoxia to the fetus resulting in organ sparing, which then causes asymmetrical development relative to organs not spared (Thornburg, 1991), and alteration of hypothalamic–pituitary–adrenal (HPA) axis glucocorticoid function mediated by placental transfer of hormones (Weinstock, 1997; Zarrow *et al.*, 1970). In addition, alterations of the hypothalamic–pituitary–gonadal axis resulting in abnormally low testosterone levels in males (Ward, 1984), enhanced humoral immune responses (Klein and Rager, 1995), altered endocrine or behavioral profiles, and endocrine-induced alterations of immune function (Coe *et al.*, 1996) have been proposed. In nonhuman primates the mechanism most studied concerns alterations of the HPA axis. However, the studies designed thus far have involved indirect measurements, which cannot answer acute questions.

Research on stress and stress phenomena requires that both the environmental context of the stress (Selye, 1980) and the response of the individual (Coyne and Lazarus, 1980) be defined and quantified. For the fetus, the simplest model of psychological stress involving the environment is a mother that either does or does not respond to its environment, with the consequence that the fetus also does or does not respond. Although it is possible for the fetus to respond directly to the environment outside the uterus, independent of the mother, for the sake of this illustration let us limit our discussion to maternally mediated responses. Using this characterization, a 2×2 research design, we can cross the presence or absence of an acute stress response in the mother with the presence or absence of an acute stress response in the fetus. This generates four types of fetal response (Table 24–1).

Table 24–1. Assessing Maternal and Fetal Acute Stress Responses[a]

	Fetal environment (maternal response)	
Fetal response	Stress response	No stress response
Stress response	Stressed (developmentally capable but outside pacer range)	Chaotic (developmentally capable and within pacer range)
No stress response	Resistant (developmentally capable and within pacer range; or developmentally incapable)	Unstressed (developmentally capable but outside pacer range; developmentally incapable; developmentally capable and within pacer range)

[a] Stress responses may describe individual differences in maternal–fetal relationship or developmentally appropriate responses across pregnancy. Four fetal environments are described.

The first type is what is usually thought to be the traditional subject of prenatal stress studies: demonstrating that stress to the mother produces stress to the fetus. The underlying assumption of most prenatal stress studies is represented by the "Stress response" column in Table 24–1. In traditional studies, fetuses whose mothers respond to a stressor, but exhibit no stress response themselves, usually would be classified as an error variance. However, these fetuses also may be classifed as a "resistant" group, potentially different from the stressed fetuses. With this approach, these animals become part of the analysis and no longer part of the error term. Fetuses whose mothers do not experience the stressor and so cannot initiate a stress response themselves make up the proper control group for studies of stress. A final group (upper right cell in Table 24–1) is composed of fetuses that exhibit a stress-response profile even though their mothers did not experience the targeted stressor. This group may be said to be chaotic, perhaps even pathological. The stress responses in this group may vary in chronicity, or the organism may be responding to an unmeasured or latent stressor.

The four fetal response classifications shown in Table 24–1 may represent individual differences in the maternal–fetal relationship. However, it is also possible that there is a developmental progression. For example, until the response systems in a fetus are functioning, a maternal response without fetal response is developmentally appropriate. It is also possible that immature stress responses of the fetus are more likely to follow the mother's response. In this case, the maternal stress responsivity functions as scaffolding on which the fetal responsivity develops, and stressed and unstressed classifications are developmentally appropriate. Finally, as fetal stress systems become mature and capable of more independent functioning, fetal responsivity would be less likely to match maternal reactions across various situations. In this case, chaotic and resistant classifications are developmentally appropriate. The unstressed condition is developmentally appropriate in several situations.

We used tethering with maternal and fetal catheterization to assess the acute effects of potentially stressful maternal experiences. Heart rate, blood pressure, and heart rate variability in the mother and the fetus show that the fetal response to maternal stressors is rapid in response to some types of environmental challenges (Fig. 24–7). At other times, the mother shows a cardiovascular response but the fetus does not. And at still other times, the fetal cardiovascular parameters change although there is no apparent maternal response. We are currently investigating whether there are individual differences in the degree to which mothers and fetuses are correlated in response to environmental demands on the mother.

The rapidity of response observed in the fetus does not preclude changes in the HPA axis as the mechanism by which negative psychosocial experiences of the mother affect fetal development. However, it is evidence that there may be additional mechanisms. The acute fetal response to maternal psychosocial stress responses is too rapid to be mediated by hormones alone. Candidate mechanisms can be any of the fetal sensory systems that are functioning this late in pregnancy (Gottlieb, 1971); most likely, no one mechanism is soley responsible.

Within the ontogeny of a prenatal stress response may exist the origins of infant reactivity and sensitivity to environmental stimulation. Sackett *et al.* (1999; see also Dember and Earl, 1957) described the postnatal origins of a "pacer range," which constitutes the range of environmental complexity in which an organism willingly responds adaptively. The environmental complexity of rearing experiences, maturation level of the organism, and complexity of the current environment all mediate both the developmental ability of the organism and the range of variation in environments to which an organism responds adaptively (Table 24–1). This is particularly true when we characterize the response of the fetus relative to the response of the mother. How the fetus responds when the mother is in a particular cell of Table 24–1, and subsequently how the fetus recovers (Fig. 24–7), may shed light on the origins of mother–infant reciprocity (Tronick, 1989).

3.1.1. Fetal Conditioning

Using principles from classical conditioning, it is possible to hypothesize a testable developmental model about origins of fetal responsivity to maternal clues and bidirectional infant–environment interactions. As fetal response systems start to function, sounds and somatosensory inputs to the fetus are a random jumble of information. Experience and conditioning begin the process of pattern differentiation, laying the framework for many postnatal developmental processes. Hormonal responses of the mother, representing an unconditioned stimulus to the fetus, provoke unconditioned hormonal responses in the fetus, either by directly crossing the placental barrier or by mediating the placenta to produce hormones that affect the fetus. This type of maternal–fetal communication is, initially, rudimentary and functions from the moment fetal tissues develop receptors for the maternally mediated messages. Maturation of fetal sensory systems, however, brings about new capabilities. The fetus becomes able to detect and then differentiate environmental stimuli. Under normal developmental conditions, the onset of these stimuli predictably precedes the hormonal influences from the mother, becoming conditioned stimuli. These conditioned stimuli include changes in the sound of the heart rate, gustatory sounds, external environmental noise, changes in heat, somatosensory information, and even possible acute changes in immune status and visual input. These stimuli can precede, and perhaps overlap, the hormonal disturbance that is generated in the mother and transferred to the fetus via the placenta. Over time and repeated exposure, classical conditioning requires that a link be formed between these acute predictors, as long as the information improves the prediction of the hormonal disturbance to the fetus, such that these conditioned stimuli produce the response in the fetus that previously was produced only by the hormonal disturbance to the fetus, a conditioned response.

Maternal–fetal pairs that learn this type of anticipatory "sensitivity" to each other will have an advantage postnatally in developing maternal–infant reciprocity and laying the groundwork for desired attachment patterns. The necessary formation of this maternal–fetal link may help explain the diversity of results from prenatal stress research. Some stress is necessary and helps to optimize development, some seems to have no effect, and some seems to be detrimental. Using this model, I argue that predictable forms of stress that help to classically condition mother–infant reciprocal responsiveness will not be harmful to the fetus, even in intense and/or chronically stressful situations. Only unpredictable forms of stress, in which the classical conditioning of the acute aspects of the stress is prevented, would be damaging to the fetus, perhaps even in relatively acute and modest exposures.

For the infant to attach appropriately to its mother, classical conditioning must occur. The effect of this learning is increased sensitivity to the subtle cues in behavior and physiology by both the mother and the fetus. Consistent with the literature that finds null and/or positive effects of prenatal stress, some disturbance of the maternal–fetal unit is necessary for the fetus to learn maternal physiological and behavioral predictors of stress. Without predictable stress during pregnancy, fetal sensitivity to the immediate environment for both reactivity and soothing purposes does not develop appropriately. Acute aspects of maternal stress responses become classically conditioned to produce anticipatory reactivity in "normally developing" fetuses. Likewise, acute aspects of the maternal soothing response (e.g., the slowing of heart rate) may also play a role. Furthermore, disturbances created by the fetus may classically condition the mother to be sensitive to her offspring.

4. TETHERING WITH MATERNAL AND FETAL CATHETERIZATION FOR THE FUTURE

Tethering with maternal and fetal catheterization can be used to characterize the interaction or correlation between the physiology and behavior of the mother and fetus. However, current technological limitations of this methodology preclude use outside of late-term pregnancy. Advances in surgical techniques and postsurgical care will continue to reduce the risks to typical pregnancy. This includes the possibility of moving fetal catheters to the femoral blood vessels to further reduce the potential for impeding the blood supply to the rapidly developing fetal brain. In one attempt to perform this surgery, we were unable to catheterize fetal femoral blood vessels because they were so small and inflexible. Although the use of smaller catheters is one option, there appears to be a lower limit in inner diameter of the catheter that will yield a reasonable pressure signal. In addition, smaller diameter catheters increase the likelihood of blood clotting.

Technological advances in tethering, such as the ability to visualize the fetus, may also increase the power of tethering studies considerably. Visualization of the fetus in human mothers restricted to lying in a recumbent position is a common technique. The extent to which studies of fetal development and fetal–maternal interaction can replicate the variables of interest in human pregnancy will determine the relevance and value of these studies for human health and well-being.

Some mother–fetus pairs of interest may be vulnerable to complications in response to the invasiveness of fetal surgery and therefore untestable by methods requiring instrumentation. Such pairs can be studied by a variety of other methods, either in combination with tethering or separately. Direct fetal manipulation, magnetic resonance technology, magnetoencephalography, and perhaps even nanotechnology may help describe the developmental progression of maternal–fetal correlations as a precursor for individual differences in pacer range and infant reactivity or sensitivity to its environment.

REFERENCES

Baguma-Nibasheka, M., Mecenas, C.A., Giussani, D.A., and Nathanielsz, P.W. (2000). Inhibitory effect of iloprost on the contractility of lower uterine segment myometrium from rhesus monkeys in normal-term and androstenedione-induced preterm labor. *J. Soc. Gynecol. Invest.* 7:167–169.

Coe, C.L., Lubach, G.R., Karaszewski, J.W., and Ershler, W.B. (1996). Prenatal endocrine activation alters postnatal cellular immunity in infant monkeys. *Brain Behav. Immun.* 10:221–234.

Coyne, J.C., and Lazarus, R.L. (1980). Cognitive style, stress perception, and coping. In: *Handbook on Stress and Anxiety: Contemporary Knowledge, Theory, and Treatment.* Jossey-Bass Inc., San Francisco, pp. 144–158.

Dember, N.W., and Earl, R.W. (1957). Analysis of exploratory, manipulatory, and curiosity behaviors. *Psychol. Rev.* 64:91–96.

Denenberg, V.H., Hoplight, B.J., and Mobraaten, L.E. (1998). The uterine environment enhances cognitive competence. *NeuroReport* 9:619–623.

Diczfalusy, E. (1974). Endocrine function of the human fetus. *Am. J. Obstet. Gynecol.* 119:419–433.

Farber, D.M., Giussani, D.A., Jenkins, S.L., Mecenas, C.A., Winter, J.A., Wentworth, R.A., and Nathanielsz, P.W. (1997). Timing of the switch from myometrial contractures to contractions in late-gestation pregnant rhesus monkeys as recorded by myometrial electromyogram during spontaneous term and androstenedione-induced labor. *Biol. Reprod.* 56:557–562.

Francis, D., Diorio, J., Liu, D., and Meaney, M.J. (1999). Nongenomic transmission across generations of maternal behavior and stress responses in the rat. *Science* 286:1155–1158.

Giussani, D.A., Jenkins, S.L., Mecenas, C.A., Winter, J.A., Honnebier, M.B.O.M., Wu, W., and Nathanielsz, P.W. (1996a). Daily and hourly temporal association between Δ4-androstenedione-induced preterm myometrial contractions and maternal plasma estradiol and oxytocin concentrations in the 0.8 gestation rhesus monkey. *Am. J. Obstet. Gynecol.* 174:1050–1055.

Giussani, D.A., Jenkins, S.L., Mecenas, C.A., Winter, J.A., Pedro, J.G., Farber, D.M., Goland, R.S., and Nathanielsz, P.W. (1996b). Effect of androstenedione administration on the maternal hypothalamo-pituitary-adreno-placental axis in the pregnant rhesus monkey. *Endocrinology* 137:608–614.

Giussani, D.A., Jenkins, S.L., Winter, J.A., Tame, J.D., and Nathanielsz, P.W. (2000). Androstenedione treatment of pregnant baboons at 0.7–0.8 of gestation promotes a premature forward shift in the nocturnal maternal plasma estradiol surge relative to progesterone and increases myometrial contraction activity. *Endocrinology* 141:3296–3303.

Gottlieb, G. (1971). Ontogenesis of sensory function in birds and mammals. In: Tobach, E., Aronson, L.A., and Shaw, E. (eds.), *Biopsychology of Development*. Academic Press, New York, pp. 8–43.

Gravett, M.G., Witkin, S.S., Haluska, G.J., Edwards, J.L., Cook, M.J., and Novy, M.J. (1994). An experimental model for intraamniotic infection and preterm labor in rhesus monkeys. *Am. J. Obstet. Gynecol.* 171:1660–1667.

Gravett, M.G., Haluska, G.J., Cook, M.J., and Novy, M.J. (1996). Fetal and maternal responses to experimental intrauterine infection in rhesus monkeys. *Am. J. Obstet. Gynecol.* 174:1725–1731.

Ha, J.C., Nosbisch, C., Abkowitz, J.L., Conrad, S.H., Mottet, N.K., Ruppenthal, G.C., Robinette, R., Sackett, G.P., and Unadkat, J.D. (1998). Fetal, infant, and maternal toxicity of zidovudine (azidothymidine) administered throughout pregnancy in *Macaca nemestrina*. *J. Acquir. Immune Defic. Syndr. Hum. Retrovirol.* 18:27–38.

Haluska, G.J., Cook, M.J., and Novy, M.J. (1997). Inhibition and augmentation of progesterone production during pregnancy: Effects on parturition in rhesus monkeys. *Am. J. Obstet. Gynecol.* 176:682–691.

Ho, R.J.Y., Larsen, K., Kinman, I., Sherbert, C., Wang, X.Y., Finn, E., Nosbisch, C., Schmidt, A., Anderson, D., Hu, S.L., Agy, M., Ochs, H., Morton, W.R., and Unadkat, J.D. (2001). Characterization of a maternal-fetal HIV transmission model using pregnant macaques infected with HIV-2287. *J. Med. Primatol.* 30:131–140.

Kagan, J., Snidman, N., Zentner, M., and Peterson, E. (1999). Infant temperament and anxious symptoms in school age children. *Dev. Psychopathol.* 11:209–224.

Katz, S. (2001). Bidirectional experimental effects. *Psychol. Methods* 6:270–281.

Klein, S.L., and Rager, D.R. (1995). Prenatal stress alters immune function in the offspring of rats. *Dev. Psychobiol.* 28:321–336.

Lobel, M. (1994). Conceptualizations, measurement, and effects of prenatal maternal stress on birth outcomes. *J. Behav. Med.* 17:225–272.

Macias, A.E., Wong, S.W., Sadowsky, D.W., Luetjens, C.M., Axthelm, M.K., Gravett, M.G., Haluska, G.J., and Novy, M.J. (2001). Maternal or fetal origin of rhesus monkey (*Macaca mulatta*) amniotic fluid leukocytes can be identified by polymerase chain reaction using the zinc finger Y gene. *Am. J. Primatol.* 55:159–170.

McNamee, G.A., Jr., Wannemacher, R.W., Jr., Dinterman, R.E., Rozmiarek, H., and Montrey, R.D. (1984). A surgical procedure and tethering system for chronic blood sampling, infusion, and temperature monitoring in caged nonhuman primates. *Lab. Anim. Sci.* 34:303–307.

Monk, C., Fifer, W.P., Myers, M.M., Sloan, R.P., Trien, L., and Hurtado, A. (2000). Maternal stress responses and anxiety during pregnancy: Effects on fetal heart rate. *Dev. Psychobiol.* 36:67–77.

Morgan, M.A., Silavin, S.L., Randolph, M., Payne, G.G., Sheldon, R.E., Fishburne, J.I., Jr., Wentworth, R.A., and Nathanielsz, P.W. (1991). Effect of intravenous cocaine on uterine blood flow in the gravid baboon. *Am. J. Obstet. Gynecol.* 164:1021–1030.

Morgan, M.A., Wentworth, R.A., Silavin, S.L., Jenkins, S.L., Fishburne, J.I., Jr., and Nathanielsz, P.W. (1994). Intravenous administration of cocaine stimulates gravid baboon myometrium in the last third of gestation. *Am. J. Obstet. Gynecol.* 170:1416–1420.

Morishima, H.O., Pederson, H., and Finster, M. (1978). The influence of maternal psychological stress on the fetus. *Am. J. Obstet. Gynecol.* 131:286–290.

Morton, W.R., Knitter, G.H., Smith, P.M., Susor, T.G., and Schmitt, K. (1987). Alternatives to chronic restraint of nonhuman primates. *J. Am. Vet. Med.* 191:1282–1286.

Murata, Y., Martin, C.B. Jr., Ikenoue, T., Socol, M.L., and Druzin, M.L. (1984). Intrauterine growth of chronically instrumented rhesus monkey fetuses. *Am. J. Obstet. Gynecol.* 50:40–46.

Nathanielsz, P.W., Frank, D., Gleed, R., Dillingham, L., Figueroa, J.P., and Poore, E.R. (1984). Methods for investigation of the chronically instrumented pregnant rhesus monkey preparation maintained on a tether and swivel system. In: Nathanielsz, P.W. (ed.), *Animal Models in Fetal Medicine.* Perinatology Press, New York, pp. 109–159.

Novak, M.S. (2002). A model experimental system for studying prenatal stress in pigtailed macaque monkeys. Doctoral Dissertation. University of Washington, Seattle, WA.

Novy, M.J., and Walsh, S.W. (1981). Regulation of fetoplacental steroidogenesis in rhesus macaques. In: Novy, M.J., and Resko, J.A. (eds.), *Fetal Endocrinology.* Academic Press, New York, pp. 65–94.

Novy, M.J., Piasecki, G.J., Hill, J.D., and Jackson, B.T. (1971). Cardiorespiratory measurements in fetal monkeys obtained by chronic catheters. *J. Appl. Physiol.* 31:788–791.

Ochs, H.D., Morton, W.R., Tsai, C-C., Thouless, M.E., Zhu, Q., Kuller, L.D., and Wu, Y.P. (1991). Maternal-fetal transmission of SIV in macaques: Disseminated adenovirus infection in an offspring with congenital SIV infection. *J. Med. Primatol.* 20:193–200.

Randall, D.C., and Smith, O.A. (1984). Neural control of the heart in intact conscious animal with implications concerning the etiology of cardiovascular disease. In: Randall, W.C. (ed.), *Nervous Control of Cardiovascular Function.* Oxford University Press, New York, pp. 391–413.

Rowell, L.B. (ed.). (1993). *Human Cardiovascular Control.* Oxford University Press, New York.

Sackett, G.P., Novak, M.F.S.X., and Kroeker, R. (1999). Early experience effects on adaptive behavior: Theory revisited. *Ment. Retard. Dev. Disabil. Res. Rev.* 5:30–40.

Sadowsky, D.W., Haluska, G.J., Gravett, M.G., Witkin, S.S., and Novy, M.J. (2000). Indomethacin blocks interleukin 1β-induced myometrial contractions in pregnant rhesus monkeys. *Am. J. Obstet. Gynecol.* 183:173–180.

Selye, H. (ed.). (1980). *Selye's Guide to Stress Research.* Van Nostrand Reinhold, San Francisco.

Shepherd, R.W., Stanczyk, F.Z., Bethea, C.L., and Novy, M.J. (1992). Fetal and maternal endocrine responses to reduced uteroplacental blood flow. *J. Clin. Endocrinol. Metab.* 75:301–307.

Stark, R.I., Daniel, S.S., James, L.S., MacCarter, G., Morishima, H.O., Niemann, W.H., Rey, H., Tropper, P.J., and Yeh, M.N. (1989). Chronic instrumentation and long term investigation in the fetal and maternal baboon: Tether system, conditioning procedures and surgical techniques. *Lab. Anim. Sci.* 39(1):25–32.

Su, H.-M., Huang, M.-C., Saad, N.M.R., Nathanielsz, P.W., and Brenna, J.T. (2001). Fetal baboons convert 18:3n-3 to 22:6n-3 in vivo: A stable isotope tracer study. *J. Lipid Res.* 42:581–586.

Thornburg, K.L. (1991). Fetal response to intrauterine stress. In: *The Childhood Environment and Adult Disease.* Ciba Foundation Symposium 156. Wiley, Chichester, pp. 17–37.

Tronick, E.Z. (1989). Emotional and emotional communication in infants. *Am. Psychol.* 44:112–119.

Vadillo-Ortega, F., Sadowsky, D.W., Haluska, G.J., Hernandez-Guerrero, C., Guevara-Silva, R., Gravett, M.G., and Novy, M.J. (2002). Identification of matrix metalloproteinase-9 in amniotic fluid and amniochorion in spontaneous labor and after experimental intrauterine infection or interleukin-1β infusion in pregnant rhesus monkeys. *Am. J. Obstet. Gynecol.* 186:128–138.

Ward, I.L. (1984). The prenatal stress syndrome: Current status. *Psychoneuroendocrinology* 9:3–11.

Weinstock, M. (1997). Does prenatal stress impair coping and regulation of hypothalamic-pituitary-adrenal axis? *Neurosci. Biobehav. Rev.* 21(1):1–10.

Witkin, S.S., Gravett, M.G., Haluska, G.J., and Novy, M.J. (1994). Induction of interleukin-1 receptor antagonist in rhesus monkeys after intraamniotic infection with group B streptococci or interleukin-1 infusion. *Am. J. Obstet. Gynecol.* 171:1668–1672.

Zarrow, M.X., Philpott, J.E., and Denenberg, V.H. (1970). Passage of ^{14}C-corticosterone from the rat mother to the fetus and neonate. *Nature* 226:1058–1059.

SECTION FIVE

Introduction to Section 5: Hematology and Serum Chemistry Values

In this section, Chapters 25–28, we present four samples of representative hematology and serum chemistry values for several of the primate species most commonly reared under nursery conditions. Originally we intended to present growth and blood assay data for a wide range of nursery-reared primates, but we learned that growth data are readily available for many of these species, so they are not included here.

Other than behavioral and body temperature indices, hematology and serum chemistry parameters are the primary measures of health for any mammalian species. Although blood assay parameter values are readily available for adults and juveniles of many primate species, we could find almost no examples of such data in any reasonable sample size for primate neonates and young infants. Furthermore, with the exception of some immunological measures, we could find almost no published comparisons of blood values for nursery-reared versus mother-reared infants. Although we advertised widely in an attempt to gather such information, with the exception of these four samples we were unsuccessful. However, Howell and her colleagues not only provided detailed data for nursery-

and mother-reared chimpanzees, they also contributed an excellent summary of typical methods used to assay blood and provided a review of sex and rearing differences on these measures. We hope that these data will be useful to others, and perhaps serve to stimulate the systematic collection of this critical health information for more species in the future.

As to the results, with the exception of immunological measures, it appears that most blood parameters do not differ markedly between nursery- and mother–reared infants of these four primate species. Also, sex differences appear to be minimal during infancy.

CHAPTER TWENTY-FIVE

Hematology and Serum Chemistry in Young Captive Chimpanzees (*Pan troglodytes*)

Susan M. Howell, Kathleen Hoffman, Jo Fritz, and Melanie Schwandt

1. INTRODUCTION

Data on hematology and serum clinical chemistry (H/SCC) values in captive chimpanzees (*Pan troglodytes*) have been published for a number of institutions in the United States, including Holloman Air Force Base, now called the Alamagordo Primate Facility (Hodson *et al.*, 1967, 1968); Yerkes National Primate Research Center (Huser and Olberding, 1972; McClure *et al.*, 1973; Herndon and Tigges, 2001); Southwest Foundation for Biomedical Research, now called the Southwest National Primate Research Center (Hainsey *et al.*, 1993); New Iberia Research Center (Stone *et al.*, 2000); and the M.D. Anderson Cancer Center at the University of Texas (Ihrig *et al.*, 2001). Values at the Primate Foundation of Arizona (PFA) were submitted for publication this year (S. Howell *et al.*, unpublished observations).

All of the published data show that there are sex differences in H/SCC values of chimpanzees. Among hematology measures, research shows that males have higher levels of red blood cells, hemoglobin, and hematocrit than females (McClure *et al.*, 1973; Herndon and Tigges, 2001; Ihrig *et al.*, 2001; S. Howell *et al.*, unpublished observations). Males also have higher levels of MCHC (Herndon and Tigges, 2001) and neutrophils (Ihrig *et al.*, 2001). On the other hand, females have higher levels of white blood cells (McClure *et al.*, 1973; Herndon and Tigges, 2001) and of lymphocytes and eosinophils (Herndon and Tigges, 2001) than males. Among SCC measures, research suggests that males have higher levels of creatinine, bilirubin, alanine aminotransferase (ALT), and aspartate aminotransferase (AST) (Ihrig *et al.*, 2001) and of uric acid, sodium, and creatine kinase (S. Howell *et al.*, unpublished observations). Sedimentation rates are also higher in males than in females (McClure *et al.*, 1973).

The published data also document age-related changes in H/SCC values of chimpanzees. Among hematology values, red and white blood cell counts increase from infant to juvenile age periods (McClure *et al.*, 1973) and there is a general increase with age in hemoglobin (McClure *et al.*, 1973), hematocrit and monocytes (Herndon and Tigges, 2001), mean cell volume (MCV) and mean corpuscular hemoglobin (MCH) (Herndon and Tigges, 2001; S. Howell *et al.*, unpublished observations), and neutrophils (Ihrig *et al.*, 2001). Lymphocytes (Ihrig *et al.*, 2001) and MCH concentration (MCHC), platelets, and neutrophils (Herndon and Tigges, 2001) all decrease with age. Several SCC measures increase with age, including urea nitrogen and albumin (Ihrig *et al.*, 2001), total protein, globulin, and triglycerides (Ihrig *et al.*, 2001; S. Howell *et al.*, unpublished observations), and γ-glutamyl transferase (GGT), ALT, creatine kinase, bilirubin, and iron (S. Howell *et al.*, unpublished observations). On the other hand, several SCC measures decrease with age, including creatinine (Ihrig *et al.*, 2001), cholesterol and alkaline phosphatase (Ihrig *et al.*, 2001; S. Howell *et al.*, unpublished observations), and potassium, chloride, calcium, phosphorus, and anion gap (S. Howell *et al.*, unpublished observations).

In contrast to earlier reports (Hodson *et al.*, 1967, 1968; Huser and Olberding, 1972; McClure *et al.*, 1973), recent publications have provided H/SCC value tables by age and sex (Stone *et al.*, 2000; Herndon and Tigges, 2001; Ihrig *et al.*, 2001; S. Howell *et al.*, unpublished observations). However, no prior study has considered the effect of early rearing history on H/SCC values or provided normal reference interval tables by age, sex, and rearing. Such tables can be a valuable diagnostic tool for veterinarians to assess health status, rule out clinical problems, diagnose disease, and monitor therapies and prognoses. Nutritional intake can vary for nursery- and mother-reared captive chimpanzees owing to differences between formula and mother's milk, and these differential dietary regimens may be an important factor to consider because they can affect H/SCC values and subsequent diagnoses of nutritional and nonnutritional diseases and anemia. In this chapter we will compare H/SCC values for mother-reared and nursery-reared infant (0.5–3.9 years), juvenile (4–6.9 years), and adolescent (7–9.9 years) chimpanzees at the Primate Foundation of Arizona (PFA) and provide reference interval tables for H/SCC variables by sex, age, and rearing history.

2. MATERIALS AND METHODS

Data were collected at the PFA, a private, nonprofit biomedical research facility. The housing and care program exceeds current standards (National Research Council, 1996) and the institution is accredited by the Association for Assessment and Accreditation of Laboratory Animal Care International. For a complete description of the PFA environmental enrichment and psychological well-being program see Fritz and Howell (1993).

2.1. Mother-Reared Chimpanzees

Most infants are reared by their mothers in social groups of six to eight chimpanzees. They are separated from their mothers only in instances of illness, injury, or severe maternal neglect (Fritz and Howell, 1993). Mother-reared chimpanzees usually remain in their natal social groups until they are naturally weaned (3.3–5 years of age) and/or reach sexual maturity (5–7 years of age). When infants consistently move away from their mothers and approach the cage front (6–9 months of age), fresh fruits and vegetables are introduced. At this time, infants are gradually conditioned to drink a vitamin cocktail (sodium ascorbate in a powdered fruit-flavored drink mixed in water) from a soda straw and to take a children's chewable multivitamin. We begin with a dose of 500 mg sodium ascorbate in 10 ml fruit drink and increase the dose and mix at 60-day intervals to a final dose, at age 78 weeks, of 1 g sodium ascorbate in 20 ml water. By 78 weeks of age, the chimpanzees also receive a varied diet of seasonal fruits and vegetables along with pelleted monkey chow biscuits (Purina Lab Diet #5045) and forage treats.

2.2. Nursery-Reared Chimpanzees

Nursery-rearing methods at the PFA are responsive and intensive (for details see Fritz and Fritz, 1982, 1985). Young infants receive 24-hr attention by a primary caregiver. They are carried in a ventro-ventral position throughout much of the 24-hr day and are fed on demand. Beginning at 2 weeks of age, infants are carried throughout the colony to become accustomed to the presence and vocalizations of other chimpanzees. They may also be provided canine companions for comfort, play, and stimulation, especially infants without peer contact.

The PFA nutritional guidelines are described in detail by Fritz *et al.* (1985). For the first 16 weeks of life, infants receive a low-iron human infant formula on demand at 1- to 3-hr intervals. Beginning at 3 weeks, they receive 125 mg of dissolved sodium ascorbate crystals. At 17 weeks, an iron-fortified formula is initiated and sodium ascorbate is increased to 620 mg/day. Rice cereal, mixed with formula, is introduced. During this period, infants are conditioned to drink through a soda straw. At the same time, fresh vegetables and fruit are introduced. Formula intake is gradually decreased as solid food intake is increased. At approximately 37 weeks, one children's chewable multivitamin tablet is added and sodium ascorbate is increased to 750 mg/day. At 1 year of age, formula is gradually replaced with powdered, reconstituted cow's milk and pelleted monkey chow biscuits (Purina Lab Diet #5045) are introduced. By 78 weeks of age, nursery-reared chimpanzees, like their mother-reared peers, receive a varied diet of seasonal fruits and vegetables along with pelleted monkey chow biscuits (Purina Lab Diet #5045) and forage treats. They also receive a daily supplement of 1 g sodium ascorbate and a chewable multivitamin.

2.3. H/SCC Reference Intervals

H/SCC values were determined from blood and serum samples collected during routine biannual physical examinations. Meals and environmental enrichment foods were withheld from infants and young juveniles for 4–8 hr prior to sedation and from older juveniles and adolescents for 15 hr before sedation. Young mother-reared subjects were allowed to nurse prior to sedation, but were not provided additional food. Chimpanzees were sedated with an intramuscular injection of either ketamine HCl (Ketaset®, Fort Dodge, Iowa) at a dosage of 11 mg/kg body weight or tiletamine HCI/Zolazepam (Telazol®, Fort Dodge, Iowa) at a dosage of 3.0–4.0 mg/kg body weight. Injections were given by Telinject Vario I–V dart pistol or manually by syringe and needle. Immobilization and anesthetic may affect results presented here but those effects could not be ascertained in this study. Blood was collected from the femoral vein into Vacutainer® tubes (Becton Dickinson, Rutherford, NJ) with 21-gauge needles. Hematology samples were collected in EDTA anticoagulant tubes and SCC samples were collected in tubes with no additives. H/SCC tests were performed by a commercial laboratory (Sonora Quest Laboratory, Tempe, AZ). Automated complete blood counts with differential and platelet counts were run using flow cytometry and impedance counting. Differential cell counts were visually evaluated only if abnormalities were detected with the automated counter.

H/SCC values were determined from samples collected longitudinally from 1975 to 2000, although most samples were collected between 1989 and 2000. Subjects included 64 chimpanzees (32 males, 32 females) ranging in age from 6 months to 9.9 years. The data set included 717 samples (357 samples for males, 360 samples for females; 209 for infants, 244 for juveniles, 264 for adolescents). On average, subjects were sampled longitudinally over an 8-year period (range 1.41–9.9 years). Data for females during pregnancy were eliminated, as were data for all subjects with any wounding or other injury and/or any appearance of clinical disease. Thus the data reported here reflect H/SCC values in normal, healthy chimpanzees. The data include nine standard hematology variables and 25 SCC variables, including those related to hematology, leukocyte, and serum electrolyte values and to pancreatic, renal, and liver function (Table 25–1).

Table 25–1. Hematology and Serum Chemistry Variables and Measurement Units

Variable	Unit of measurement	Method
Hematology		
White blood cell	th/mm^3	Flow cytometry
Red blood cell	th/mm^3	Impedance counting
Hemoglobin	g/dl	Cyanmethemoglobin
Hematocrit	Percent	Calculation
MCV	μm^3	Calculation
MCH	pg	Calculation
MCHC	g/dl	Calculation
Platelets	th/mm^3	Impedance
Neutrophils	th/mm^3	Flow cytometry
Lymphocytes	th/mm^3	Flow cytometry
Monocytes	th/mm^3	Flow cytometry
Eosinophils	th/mm^3	Flow cytometry
Basophils	th/mm^3	Flow cytometry
Serum chemistry		
Glucose	mg/dl	Hexokinase
Urea nitrogen	mg/dl	Glutamate dehydrogenase
Creatinine	mg/dl	Enzymatic
Uric acid	mg/dl	Enzymatic
Sodium	mEq/liter	Ion selective electrode
Potassium	mEq/liter	Ion selective electrode
Chloride	mEq/liter	Ion selective electrode
Carbon dioxide	mEq/liter	Spectrophotometry
Anion gap	Gap	Calculation
Total protein	g/dl	Colorimetry
Albumin	g/dl	Colorimetry
Globulin	g/dl	Calculation
Albumin/globulin	Ratio	Calculation
Cholesterol (total)	mg/dl	Photometric
Triglyceride	mg/dl	Colorimetry
Calcium	mg/dl	Spectrophotometry
Ionized calcium, calculated	mg/dl	Spectrophotometry
Phosphorus	mg/dl	Photometry
γ-glutamyl transferase (GGT)	U/liter	Enzymatic
Alkaline phosphatase	U/liter	Enzymatic
Alanine aminotransferase (ALT, SGPT)	U/liter	Enzymatic
Asparate aminotransferase (AST, SGOT)	U/liter	Enzymatic
Lactic dehydrogenase (LDH)	U/liter	Enzymatic
Creatine kinase	U/liter	Enzymatic
Direct bilirubin	mg/dl	Diazo, spectrophotometry
Iron	μg/dl	Colorimetry
Sedimentation rate	mm/h	Auto. mod. Westergren

2.4. H/SCC Analysis

Mean H/SCC values and interval ranges were calculated and sorted by age, sex, and rearing history. Range values were estimated as the mean ± 2 SD. When the calculated lower range value was less than zero, the actual minimum value was used. A two-sample *t*-test was used to assess rearing effects by comparing values for standard nursery- and mother-reared subjects. *p*-values were evaluated for significance using the Bonferroni adjustment for multiple comparisons (overall $\alpha = 0.05$). The statistical analysis was conducted in SAS® version 8.

3. RESULTS

Reference interval tables of hematology variables are presented for chimpanzee infants (0.5–3.9 years of age) in Table 25–2, for juveniles (4–6.9 years of age) in Table 25–3, and for adolescents (7–9.9 years of age) in Table 25–4. Reference interval tables of SCC variables for chimpanzee infants are shown in Table 25–5, for juveniles in Table 25–6, and for adolescents in Table 25–7. The tables also provide the total number of subjects sampled for age, sex, and rearing categories, along with the total number of samples included in the calculation.

Two-sample *t*-test results are provided in Table 25–8. Results suggest that there were no significant differences in H/SCC values for nursery- and mother-reared subjects between 6 months and 9.9 years of age.

4. DISCUSSION

Results presented here represent the first publication of normal reference interval tables for H/SCC by age, sex, and rearing. However, this study is only the first step toward understanding the effect of the early rearing environment on H/SCC variables. While *t*-test results suggest that H/SCC values are generally similar for nursery- and mother-reared chimpanzees, further study is warranted. Owing to a limited subject sample size for some age/sex/rearing categories (e.g., infant male nursery-reared subjects), we were not always able to compare H/SCC values by age, sex, and rearing. In particular, larger subject samples at relatively

Table 25–2. Infant (0.5–3.9 Years) Hematology Values by Sex and Rearing Category

	Total subjects	Total samples	Mean	SD	Minimum	Maximum	Confidence intervals	
							Lower 99%	Upper 99%
White blood cells (th/mm^3)								
Female—nursery reared	5	33	10.22	3.28	4.50	21.10	3.66	16.78
Female—mother reared	15	63	10.94	6.01	4.10	52.60	4.10	22.97
Male—nursery reared	2	14	7.91	1.56	5.40	11.20	4.80	11.03
Male—mother reared	16	89	9.95	2.85	4.80	18.20	4.25	15.65
Red blood cells (th/mm^3)								
Female—nursery reared	5	33	5.01	0.38	4.35	5.83	4.25	5.77
Female—mother reared	15	63	5.17	0.36	4.60	6.47	4.44	5.89
Male—nursery reared	2	14	5.40	0.32	4.85	5.82	4.76	6.05
Male—mother reared	16	89	5.08	0.46	3.88	6.28	4.17	6.00
Hemoglobin (g/dl)								
Female—nursery reared	5	33	12.93	0.68	11.80	14.70	11.58	14.29
Female—mother reared	15	63	12.99	0.77	10.70	14.90	11.45	14.53
Male—nursery reared	2	14	13.71	0.84	11.80	14.60	12.03	15.40
Male—mother reared	16	89	12.30	1.11	5.38	14.10	10.07	14.52
Hematocrit (%)								
Female—nursery reared	5	33	39.62	2.29	36.10	44.90	35.04	44.20
Female—mother reared	15	63	39.84	2.23	33.90	46.20	35.38	44.31
Male—nursery reared	2	14	42.11	2.51	36.80	45.00	37.09	47.13
Male—mother reared	16	89	38.51	2.60	30.50	44.50	33.30	43.72

MCV (μm^3)								
Female—nursery reared	5	33	79.29	3.20	72.40	86.00	72.89	85.69
Female—mother reared	15	63	77.25	3.75	64.60	85.00	69.74	84.76
Male—nursery reared	2	14	78.00	3.41	73.90	84.20	71.19	84.81
Male—mother reared	16	89	76.12	6.07	56.00	93.30	63.98	88.26
MCH (pg)								
Female—nursery reared	5	33	25.91	1.19	23.70	28.60	23.52	28.29
Female—mother reared	15	63	25.35	1.94	21.70	36.40	21.47	29.22
Male—nursery reared	2	14	25.40	0.98	24.10	27.20	23.44	27.36
Male—mother reared	16	89	24.47	2.38	16.20	32.40	19.71	29.24
MCHC (g/dl)								
Female—nursery reared	5	33	32.69	1.01	30.50	34.50	30.67	34.71
Female—mother reared	15	63	32.60	0.81	30.80	34.20	30.98	34.23
Male—nursery reared	2	14	32.58	0.59	31.60	33.50	31.41	33.75
Male—mother reared	16	89	32.13	1.14	27.70	34.80	29.85	34.41
Platelets (th/mm^3)								
Female—nursery reared	5	33	290.58	56.11	154.00	384.00	178.36	402.79
Female—mother reared	15	60	341.43	94.51	175.00	747.00	152.41	530.46
Male—nursery reared	2	14	318.50	72.14	239.00	490.00	174.21	462.79
Male—mother reared	16	88	307.64	74.99	170.00	495.00	157.66	457.61
Neutrophils (th/mm^3)								
Female—nursery reared	4	17	4.10	3.63	0.95	14.37	0.95	11.36
Female—mother reared	6	26	3.53	1.41	1.81	7.42	0.70	6.36
Male—nursery reared	2	14	2.48	1.02	1.41	5.70	0.43	4.52
Male—mother reared	13	53	3.45	1.51	1.29	8.01	0.43	6.47

(Continued)

Table 25–2. *Continued*

	Total subjects	Total samples	Mean	SD	Minimum	Maximum	Confidence intervals	
							Lower 99%	Upper 99%
Lymphocytes (th/mm^3)								
Female—nursery reared	5	22	5.75	1.56	2.85	9.63	2.62	8.88
Female—mother reared	13	36	5.40	1.49	2.89	8.31	2.42	8.38
Male—nursery reared	2	14	4.53	0.99	2.32	6.16	2.54	6.52
Male—mother reared	16	59	5.32	1.85	2.35	11.06	1.62	9.02
Monocytes (th/mm^3)								
Female—nursery reared	4	18	0.43	0.31	0.18	1.39	0.18	1.05
Female—mother reared	6	26	0.29	0.14	0.13	0.64	0.02	0.57
Male—nursery reared	2	14	0.33	0.10	0.20	0.50	0.13	0.53
Male—mother reared	16	56	0.36	0.14	0.13	0.85	0.07	0.65
Eosinophils (th/mm^3)								
Female—nursery reared	4	17	0.23	0.11	0.08	0.43	0.01	0.46
Female—mother reared	6	26	0.33	0.49	0.02	2.40	0.02	1.30
Male—nursery reared	2	14	0.12	0.07	0.02	0.26	0.02	0.26
Male—mother reared	13	53	0.26	0.15	0.06	0.77	0.06	0.56
Basophils (th/mm^3)								
Female—nursery reared	4	17	0.06	0.03	0.00	0.15	0.00	0.13
Female—mother reared	6	26	0.06	0.03	0.00	0.13	0.00	0.12
Male—nursery reared	2	14	0.05	0.02	0.03	0.10	0.01	0.09
Male—mother reared	13	53	0.06	0.03	0.00	0.15	0.00	0.13

Table 25–3. Juvenile (4–6.9 Years) Hematology Values by Sex and Rearing Category

							Confidence intervals	
	Total subjects	Total samples	Mean	SD	Minimum	Maximum	Lower 99%	Upper 99%
White blood cells (th/mm^3)								
Female—nursery reared	10	46	9.24	2.40	4.90	18.00	4.44	14.04
Female—mother reared	17	85	10.31	3.69	4.80	26.70	2.94	17.68
Male—nursery reared	4	15	10.74	3.17	7.60	17.80	4.41	17.07
Male—mother reared	19	97	9.19	3.76	3.50	23.60	1.66	16.72
Red blood cells (th/mm^3)								
Female—nursery reared	10	46	5.03	0.39	4.43	6.00	4.26	5.81
Female—mother reared	17	85	4.90	0.54	0.76	5.73	3.81	5.98
Male—nursery reared	4	15	5.22	0.34	4.38	5.62	4.54	5.90
Male—mother reared	19	97	4.98	0.27	4.42	5.56	4.44	5.52
Hemoglobin (g/dl)								
Female—nursery reared	10	46	13.45	0.92	11.70	15.40	11.60	15.29
Female—mother reared	17	85	12.93	0.85	9.70	14.80	11.22	14.64
Male—nursery reared	4	15	13.80	1.09	11.30	15.50	11.63	15.97
Male—mother reared	19	97	13.14	0.54	12.00	14.90	12.05	14.23
Hematocrit (%)								
Female—nursery reared	10	46	41.25	2.69	36.70	47.70	35.87	46.64
Female—mother reared	17	85	39.78	2.63	32.20	45.90	34.52	45.04
Male—nursery reared	4	15	42.12	3.14	35.40	46.20	35.84	48.40
Male—mother reared	19	97	40.11	1.70	36.70	44.70	36.72	43.50

(*Continued*)

Table 25–3. *Continued*

	Total subjects	Total samples	Mean	SD	Minimum	Maximum	Confidence intervals	
							Lower 99%	Upper 99%
MCV (μm^3)								
Female—nursery reared	10	46	82.08	2.67	77.70	90.30	76.75	87.41
Female—mother reared	17	85	80.55	3.42	70.50	90.00	73.71	87.39
Male—nursery reared	4	15	80.69	3.38	74.40	85.20	73.93	87.46
Male—mother reared	19	97	80.65	2.81	72.60	87.00	75.02	86.28
MCH (pg)								
Female—nursery reared	10	46	26.74	1.01	24.60	29.00	24.71	28.76
Female—mother reared	17	85	26.18	1.28	21.20	28.80	23.61	28.74
Male—nursery reared	4	15	26.45	1.37	23.10	29.10	23.71	29.20
Male—mother reared	19	97	26.22	2.35	5.60	29.20	21.51	30.92
MCHC (g/dl)								
Female—nursery reared	10	46	32.58	1.03	29.50	34.80	30.52	34.65
Female—mother reared	17	85	32.52	0.96	29.40	34.80	30.61	34.43
Male—nursery reared	4	15	32.77	1.10	31.10	34.50	30.57	34.97
Male—mother reared	19	96	32.76	0.70	30.80	34.20	31.35	34.17
Platelets (th/mm^3)								
Female—nursery reared	10	46	302.80	66.33	192.00	477.00	170.14	435.47
Female—mother reared	17	83	309.24	68.47	155.00	493.00	172.31	446.17
Male—nursery reared	4	15	298.87	68.65	234.00	498.00	161.56	436.17
Male—mother reared	19	97	278.82	52.28	174.00	521.00	174.27	383.38

Neutrophils (th/mm^3)								
Female—nursery reared	7	23	3.81	2.16	1.48	11.70	1.48	8.13
Female—mother reared	9	41	5.36	3.92	1.42	22.56	1.42	13.21
Male—nursery reared	2	4	6.63	4.58	2.61	13.10	2.61	15.80
Male—mother reared	12	70	4.53	3.04	1.26	15.48	1.26	10.61
Lymphocytes (th/mm^3)								
Female—nursery reared	9	26	4.27	1.44	1.76	7.26	1.39	7.16
Female—mother reared	16	55	3.88	1.39	1.33	7.54	1.10	6.67
Male—nursery reared	4	6	4.38	1.53	2.21	6.26	1.32	7.44
Male—mother reared	17	79	3.91	1.93	1.42	14.86	0.04	7.77
Monocytes (th/mm^3)								
Female—nursery reared	7	23	0.35	0.17	0.14	0.90	0.01	0.69
Female—mother reared	10	44	0.35	0.17	0.09	0.96	0.01	0.70
Male—nursery reared	2	4	0.58	0.24	0.41	0.93	0.10	1.06
Male—mother reared	14	75	0.35	0.20	0.07	1.14	0.07	0.76
Eosinophils (th/mm^3)								
Female—nursery reared	7	23	0.19	0.11	0.04	0.38	0.04	0.40
Female—mother reared	9	41	0.23	0.16	0.03	0.78	0.03	0.55
Male—nursery reared	2	4	0.11	0.04	0.08	0.16	0.04	0.19
Male—mother reared	12	70	0.20	0.19	0.00	0.84	0.00	0.58
Basophils (th/mm^3)								
Female—nursery reared	7	23	0.04	0.03	0.00	0.09	0.00	0.11
Female—mother reared	9	41	0.06	0.03	0.00	0.15	0.00	0.12
Male—nursery reared	2	4	0.05	0.04	0.00	0.09	0.00	0.13
Male—mother reared	12	70	0.05	0.05	0.00	0.23	0.00	0.14

Table 25–4. Adolescent (7–9.9 Years) Hematology Values by Sex and Rearing Category

	Total subjects	Total samples	Mean	SD	Minimum	Maximum	Confidence intervals Lower 99%	Upper 99%
White blood cells (th/mm^3)								
Female—nursery reared	10	45	10.37	4.12	5.80	31.60	2.13	18.60
Female—mother reared	16	81	10.35	3.41	5.00	21.90	3.53	17.17
Male—nursery reared	9	40	8.88	2.61	5.20	15.60	3.66	14.10
Male—mother reared	19	96	8.85	2.78	3.20	21.60	3.28	14.41
Red blood cells (th/mm^3)								
Female—nursery reared	10	45	4.79	0.34	3.96	5.57	4.11	5.47
Female—mother reared	16	81	4.93	0.31	4.12	5.68	4.30	5.56
Male—nursery reared	9	40	5.34	0.44	4.60	6.18	4.46	6.22
Male—mother reared	19	96	5.27	0.41	4.45	6.38	4.45	6.09
Hemoglobin (g/dl)								
Female—nursery reared	10	45	13.01	0.79	11.60	15.00	11.42	14.60
Female—mother reared	16	81	13.21	0.79	11.60	15.10	11.63	14.80
Male—nursery reared	9	40	14.49	0.99	12.90	16.80	12.52	16.47
Male—mother reared	19	96	14.38	1.02	11.00	16.80	12.33	16.43
Hematocrit (%)								
Female—nursery reared	10	45	39.86	2.51	33.50	45.50	34.85	44.87
Female—mother reared	16	81	40.31	2.24	34.80	44.40	35.84	44.79
Male—nursery reared	9	40	43.94	2.92	38.90	49.60	38.10	49.79
Male—mother reared	19	96	43.74	2.98	34.00	51.40	37.77	49.70

MCV (μm³)								
Female—nursery reared	10	45	83.19	2.53	77.90	88.00	78.13	88.26
Female—mother reared	16	81	81.86	3.51	75.20	89.50	74.84	88.88
Male—nursery reared	9	40	82.39	2.89	77.70	90.60	76.62	88.16
Male—mother reared	19	96	83.10	3.57	70.00	89.70	75.95	90.24
MCH (pg)								
Female—nursery reared	10	45	27.18	1.36	24.80	29.80	24.46	29.89
Female—mother reared	16	81	26.84	1.48	23.60	29.70	23.88	29.80
Male—nursery reared	9	40	27.21	1.18	23.70	29.60	24.84	29.57
Male—mother reared	19	96	27.33	1.44	22.60	30.80	24.45	30.21
MCHC (g/dl)								
Female—nursery reared	10	45	32.66	1.12	29.80	35.20	30.42	34.91
Female—mother reared	16	81	32.74	0.92	30.60	34.90	30.91	34.57
Male—nursery reared	9	40	33.02	0.99	29.90	34.80	31.04	35.00
Male—mother reared	19	96	32.88	0.88	30.70	35.70	31.13	34.64
Platelets (th/mm³)								
Female—nursery reared	10	44	278.50	53.13	165.00	420.00	172.23	384.77
Female—mother reared	16	81	281.79	53.46	157.00	426.00	174.87	388.71
Male—nursery reared	9	40	297.73	89.31	179.00	544.00	119.11	476.34
Male—mother reared	19	96	250.02	48.70	138.00	376.00	152.62	347.42
Neutrophils (th/mm³)								
Female—nursery reared	8	30	4.54	1.72	2.28	8.74	1.10	7.98
Female—mother reared	13	60	5.72	3.48	1.99	18.40	1.99	12.69
Male—nursery reared	4	20	4.44	2.73	1.80	12.95	1.80	9.90
Male—mother reared	14	64	4.73	2.77	1.00	16.80	1.00	10.28

(*Continued*)

Table 25–4. *Continued*

	Total subjects	Total samples	Mean	SD	Minimum	Maximum	Confidence intervals	
							Lower 99%	Upper 99%
Lymphocytes (th/mm^3)								
Female—nursery reared	9	32	5.05	4.07	2.07	25.80	2.07	13.18
Female—mother reared	16	66	3.82	1.42	2.05	10.50	0.98	6.66
Male—nursery reared	8	26	3.61	0.98	2.03	6.13	1.64	5.57
Male—mother reared	17	69	3.24	1.08	0.54	8.37	1.07	5.40
Monocytes (th/mm^3)								
Female—nursery reared	8	30	0.37	0.19	0.12	0.91	0.12	0.75
Female—mother reared	13	60	0.36	0.23	0.09	1.60	0.09	0.83
Male—nursery reared	4	20	0.38	0.14	0.18	0.62	0.09	0.66
Male—mother reared	15	65	0.31	0.19	0.00	1.00	0.00	0.68
Eosinophils (th/mm^3)								
Female—nursery reared	8	30	0.25	0.17	0.03	0.82	0.03	0.59
Female—mother reared	13	60	0.26	0.30	0.00	2.18	0.00	0.87
Male—nursery reared	4	20	0.16	0.13	0.00	0.41	0.00	0.42
Male—mother reared	14	64	0.15	0.10	0.00	0.69	0.00	0.35
Basophils (th/mm^3)								
Female—nursery reared	8	30	0.06	0.05	0.00	0.27	0.00	0.17
Female—mother reared	13	60	0.04	0.04	0.00	0.21	0.00	0.12
Male—nursery reared	4	20	0.04	0.03	0.00	0.08	0.00	0.11
Male—mother reared	14	63	0.03	0.04	0.00	0.21	0.00	0.11

Table 25–5. Infant (0.5–3.9 Years) Serum Clinical Chemistry Values by Sex and Rearing Category

	Total subjects	Total samples	Mean	SD	Minimum	Maximum	Confidence intervals	
							Lower 99%	Upper 99%
Glucose (mg/dl)								
Female—nursery reared	5	32	78.16	13.61	55.00	106.00	50.95	105.37
Female—mother reared	15	62	88.42	17.06	54.00	151.00	54.29	122.55
Male—nursery reared	2	14	74.14	9.47	60.00	90.00	55.20	93.08
Male—mother reared	15	87	84.94	12.16	55.00	112.00	60.63	109.26
Urea nitrogen (mg/dl)								
Female—nursery reared	5	32	11.84	3.91	4.00	19.00	4.02	19.67
Female—mother reared	15	62	9.98	3.74	3.00	25.00	2.50	17.47
Male—nursery reared	2	14	17.36	9.32	7.00	36.00	7.00	36.00
Male—mother reared	15	87	10.91	4.09	4.00	21.00	2.74	19.08
Creatinine (mg/dl)								
Female—nursery reared	5	32	0.64	0.15	0.30	1.00	0.33	0.95
Female—mother reared	15	62	0.68	0.18	0.40	1.30	0.32	1.04
Male—nursery reared	2	14	0.72	0.11	0.60	0.90	0.50	0.95
Male—mother reared	15	87	0.55	0.15	0.10	0.90	0.25	0.84
Uric acid (mg/dl)								
Female—nursery reared	5	32	2.03	0.48	1.30	3.30	1.08	2.99
Female—mother reared	15	59	2.25	0.74	0.90	4.70	0.77	3.73
Male—nursery reared	2	14	2.53	0.93	1.80	5.50	0.67	4.38
Male—mother reared	15	86	2.12	0.41	1.10	3.20	1.29	2.94

(*Continued*)

Table 25–5. *Continued*

	Total subjects	Total samples	Mean	SD	Minimum	Maximum	Confidence intervals	
							Lower 99%	Upper 99%
Sodium (mEq/liter)								
Female—nursery reared	5	32	137.94	1.76	134.00	141.00	134.42	141.45
Female—mother reared	15	62	139.02	2.69	132.00	146.00	133.64	144.39
Male—nursery reared	2	14	137.86	0.95	136.00	139.00	135.96	139.76
Male—mother reared	15	87	137.94	3.32	114.00	146.00	131.30	144.59
Potassium (mEq/liter)								
Female—nursery reared	5	32	3.97	0.33	2.90	4.70	3.31	4.63
Female—mother reared	15	62	3.85	0.40	2.80	4.50	3.04	4.66
Male—nursery reared	2	14	3.93	0.30	3.30	4.40	3.33	4.53
Male—mother reared	15	87	4.05	0.35	3.30	5.10	3.35	4.76
Chloride (mEq/liter)								
Female—nursery reared	5	32	102.41	2.23	98.00	107.00	97.95	106.86
Female—mother reared	15	62	103.24	4.20	88.00	113.00	94.84	111.65
Male—nursery reared	2	14	102.50	1.61	101.00	107.00	99.29	105.71
Male—mother reared	15	87	104.38	3.50	88.00	113.00	97.37	111.39
Carbon dioxide (mEq/liter)								
Female—nursery reared	5	32	25.09	2.23	18.00	30.00	20.63	29.56
Female—mother reared	15	62	24.08	3.29	16.00	31.00	17.51	30.65
Male—nursery reared	2	14	24.79	3.31	19.00	30.00	18.17	31.40
Male—mother reared	15	87	23.06	2.89	15.00	32.00	17.28	28.84

Anion gap (gap)								
Female—nursery reared	5	32	10.44	2.30	4.00	15.00	5.84	15.04
Female—mother reared	15	62	11.89	4.12	7.00	28.00	3.65	20.12
Male—nursery reared	2	14	10.57	2.74	6.00	15.00	5.10	16.05
Male—mother reared	15	87	10.55	2.63	5.00	18.00	5.30	15.81
Total protein (g/dl)								
Female—nursery reared	5	32	6.70	0.52	5.70	8.10	5.66	7.74
Female—mother reared	15	62	6.81	0.59	5.80	8.50	5.63	7.99
Male—nursery reared	2	14	6.56	0.45	5.70	7.20	5.66	7.46
Male—mother reared	15	87	6.71	0.51	4.90	8.00	5.69	7.72
Albumin (g/dl)								
Female—nursery reared	5	32	3.79	0.29	2.80	4.30	3.21	4.38
Female—mother reared	15	62	3.65	0.39	2.70	4.40	2.88	4.42
Male—nursery reared	2	14	3.76	0.29	3.30	4.40	3.19	4.34
Male—mother reared	15	87	3.58	0.31	2.80	4.30	2.96	4.20
Globulin (g/dl)								
Female—nursery reared	5	32	2.91	0.66	1.70	5.30	1.58	4.24
Female—mother reared	15	62	3.16	0.58	2.40	5.50	1.99	4.32
Male—nursery reared	2	14	2.79	0.35	2.00	3.20	2.09	3.49
Male—mother reared	15	87	3.13	0.52	2.10	4.90	2.09	4.16
Albumin/globulin (ratio)								
Female—nursery reared	5	32	1.39	0.38	0.50	2.40	0.62	2.16
Female—mother reared	15	62	1.20	0.24	0.50	1.70	0.71	1.69
Male—nursery reared	2	14	1.38	0.24	1.10	1.90	0.91	1.85
Male—mother reared	15	87	1.18	0.22	0.60	1.80	0.74	1.62
Cholesterol (total) (mg/dl)								
Female—nursery reared	5	32	184.34	29.71	137.00	271.00	124.93	243.76
Female—mother reared	15	62	249.68	48.01	143.00	378.00	153.66	345.70
Male—nursery reared	2	14	190.86	16.12	162.00	209.00	158.61	223.10
Male—mother reared	15	87	243.48	40.90	167.00	342.00	161.68	325.29

(*Continued*)

Table 25–5. *Continued*

	Total subjects	Total samples	Mean	SD	Minimum	Maximum	Confidence intervals	
							Lower 99%	Upper 99%
Triglyceride (mg/dl)								
Female—nursery reared	5	32	63.31	24.36	27.00	151.00	14.60	112.03
Female—mother reared	15	59	71.58	47.11	27.00	243.00	27.00	165.80
Male—nursery reared	2	14	52.14	17.04	22.00	84.00	18.06	86.23
Male—mother reared	15	86	57.24	23.06	21.00	134.00	11.13	103.36
Calcium (mg/dl)								
Female—nursery reared	5	32	9.45	0.42	8.60	10.30	8.61	10.29
Female—mother reared	15	62	9.35	0.48	8.20	10.60	8.39	10.31
Male—nursery reared	2	14	9.56	0.27	9.10	10.00	9.02	10.10
Male—mother reared	15	87	9.32	0.44	7.70	10.90	8.44	10.20
Ionized calcium, calculated (mg/dl)								
Female—nursery reared	5	32	4.29	0.28	3.50	4.80	3.74	4.85
Female—mother reared	15	59	4.21	0.32	3.20	4.80	3.58	4.84
Male—nursery reared	2	14	4.39	0.20	4.00	4.70	3.99	4.80
Male—mother reared	15	86	4.23	0.25	3.70	5.10	3.73	4.73
Phosphorus (mg/dl)								
Female—nursery reared	5	32	4.96	0.80	2.70	6.50	3.35	6.56
Female—mother reared	15	62	4.60	0.88	2.20	6.60	2.83	6.36
Male—nursery reared	2	14	4.77	0.64	3.20	5.70	3.49	6.05
Male—mother reared	15	87	4.69	0.68	3.00	6.00	3.33	6.06

γ-Glutamyl transferase (GGT) (U/liter)								
Female—nursery reared	5	31	19.65	8.51	5.00	35.00	2.62	36.67
Female—mother reared	15	59	17.07	7.78	2.00	52.00	1.51	32.63
Male—nursery reared	2	14	31.79	12.55	14.00	52.00	6.68	56.89
Male—mother reared	15	86	17.62	7.32	6.00	52.00	2.97	32.26
Alkaline phosphatase (U/liter)								
Female—nursery reared	5	31	482.90	164.19	101.00	906.00	154.52	811.28
Female—mother reared	15	62	820.42	515.62	52.00	3014.00	52.00	1851.66
Male—nursery reared	2	14	399.93	84.48	237.00	535.00	230.97	568.88
Male—mother reared	15	86	846.28	598.63	341.00	3850.00	341.00	2043.55
Alanine aminotransferase (ALT, SGPT) (U/liter)								
Female—nursery reared	5	32	38.31	10.08	15.00	55.00	18.16	58.46
Female—mother reared	15	62	40.27	21.09	16.00	178.00	16.00	82.45
Male—nursery reared	2	14	53.64	16.14	27.00	86.00	21.36	85.93
Male—mother reared	15	87	37.90	15.15	12.00	82.00	7.61	68.19
Asparate aminotransferase (AST, SGOT) (U/liter)								
Female—nursery reared	5	32	20.97	4.78	13.00	33.00	11.40	30.53
Female—mother reared	15	62	21.87	10.90	8.00	95.00	0.07	43.67
Male—nursery reared	2	14	25.57	6.93	16.00	43.00	11.72	39.42
Male—mother reared	15	87	21.56	6.12	8.00	52.00	9.32	33.81
Lactic dehydrogenase (LDH) (U/liter)								
Female—nursery reared	5	32	399.66	113.41	232.00	695.00	172.83	626.49
Female—mother reared	15	62	513.82	152.92	204.00	1062.00	207.99	819.65
Male—nursery reared	2	14	508.36	81.11	388.00	615.00	346.14	670.57
Male—mother reared	15	87	486.09	82.87	339.00	699.00	320.35	651.84

(*Continued*)

Table 25–5. *Continued*

	Total subjects	Total samples	Mean	SD	Minimum	Maximum	Confidence intervals Lower 99%	Upper 99%
Creatine kinase (U/liter)								
Female—nursery reared	5	32	160.66	136.17	73.00	802.00	73.00	433.00
Female—mother reared	15	62	206.31	100.35	13.00	614.00	5.61	407.00
Male—nursery reared	2	14	238.79	181.71	98.00	720.00	98.00	602.21
Male—mother reared	15	87	241.63	150.04	59.00	739.00	59.00	541.72
Direct bilirubin (mg/dl)								
Female—nursery reared	5	32	0.01	0.03	0.00	0.10	0.00	0.08
Female—mother reared	15	61	0.01	0.03	0.00	0.10	0.00	0.08
Male—nursery reared	2	14	0.01	0.04	0.00	0.10	0.00	0.09
Male—mother reared	15	86	0.01	0.03	0.00	0.10	0.00	0.06
Iron (μg/dl)								
Female—nursery reared	5	32	87.84	30.57	17.00	152.00	26.71	148.98
Female—mother reared	15	59	74.17	31.41	15.00	150.00	11.35	136.99
Male—nursery reared	2	14	78.14	24.60	37.00	111.00	28.94	127.34
Male—mother reared	15	86	60.20	24.52	14.00	128.00	11.15	109.24
Sedimentation rate (mm/hr)								
Female—nursery reared	3	17	10.00	13.03	0.00	53.00	0.00	36.07
Female—mother reared	6	15	13.80	9.27	3.00	34.00	3.00	32.33
Male—nursery reared	2	11	4.00	3.46	0.00	10.00	0.00	10.93
Male—mother reared	9	50	22.26	19.23	1.00	84.00	1.00	60.72

Table 25–6. Juvenile (4–6.9 Years) Serum Clinical Chemistry Values by Sex and Rearing Category

	Total subjects	Total samples	Mean	SD	Minimum	Maximum	Confidence intervals	
							Lower 99%	Upper 99%
Glucose (mg/dl)								
Female—nursery reared	10	45	84.71	15.28	53.00	132.00	54.14	115.28
Female—mother reared	17	83	89.51	13.17	58.00	119.00	63.16	115.85
Male—nursery reared	4	15	85.40	13.53	59.00	110.00	58.35	112.45
Male—mother reared	19	97	86.48	12.39	53.00	130.00	61.70	111.27
Urea nitrogen (mg/dl)								
Female—nursery reared	10	45	12.20	4.57	6.00	26.00	3.07	21.33
Female—mother reared	17	83	12.30	3.40	4.00	24.00	5.50	19.11
Male—nursery reared	4	15	18.73	4.65	10.00	26.00	9.43	28.04
Male—mother reared	19	97	13.59	2.98	6.00	21.00	7.63	19.54
Creatinine (mg/dl)								
Female—nursery reared	10	45	0.73	0.18	0.40	1.30	0.38	1.09
Female—mother reared	17	83	0.72	0.13	0.40	1.10	0.46	0.97
Male—nursery reared	4	15	0.85	0.19	0.60	1.10	0.47	1.22
Male—mother reared	19	97	0.65	0.10	0.50	0.90	0.45	0.86
Uric acid (mg/dl)								
Female—nursery reared	10	45	1.94	0.80	1.00	5.50	0.33	3.54
Female—mother reared	17	81	2.37	0.49	1.50	4.40	1.39	3.36
Male—nursery reared	4	15	2.91	0.38	1.90	3.40	2.15	3.67
Male—mother reared	19	97	2.26	0.44	1.30	3.20	1.37	3.15
Sodium (mEq/liter)								
Female—nursery reared	10	45	139.18	2.82	130.00	145.00	133.53	144.82
Female—mother reared	17	83	138.89	1.93	134.00	145.00	135.04	142.74
Male—nursery reared	4	15	139.67	2.32	135.00	144.00	135.03	144.31
Male—mother reared	19	97	139.48	2.36	133.00	147.00	134.77	144.20

(*Continued*)

Table 25–6. *Continued*

	Total subjects	Total samples	Mean	SD	Minimum	Maximum	Confidence intervals	
							Lower 99%	Upper 99%
Potassium (mEq/liter)								
Female—nursery reared	10	45	4.05	0.33	3.30	4.80	3.39	4.71
Female—mother reared	17	83	3.89	0.33	3.20	4.80	3.23	4.56
Male—nursery reared	4	15	4.04	0.38	3.60	4.80	3.28	4.80
Male—mother reared	19	97	4.08	0.33	3.40	5.00	3.42	4.75
Chloride (mEq/liter)								
Female—nursery reared	10	45	101.96	2.94	96.00	111.00	96.08	107.83
Female—mother reared	17	82	102.68	3.02	92.00	108.00	96.64	108.73
Male—nursery reared	4	15	100.87	2.53	96.00	105.00	95.80	105.93
Male—mother reared	19	97	102.52	2.91	96.00	110.00	96.69	108.34
Carbon dioxide (mEq/liter)								
Female—nursery reared	10	45	25.93	3.22	11.00	31.00	19.49	32.38
Female—mother reared	17	81	25.81	2.65	18.00	31.00	20.51	31.12
Male—nursery reared	4	15	27.47	2.23	24.00	32.00	23.00	31.93
Male—mother reared	19	97	26.19	2.27	19.00	32.00	21.65	30.73
Anion gap (gap)								
Female—nursery reared	10	44	10.98	3.30	1.00	18.00	4.39	17.57
Female—mother reared	16	80	10.33	2.85	3.00	21.00	4.63	16.02
Male—nursery reared	4	15	11.33	2.53	8.00	18.00	6.28	16.39
Male—mother reared	19	97	10.78	2.85	4.00	19.00	5.07	16.49
Total protein (g/dl)								
Female—nursery reared	10	45	7.11	0.27	6.40	7.70	6.57	7.66
Female—mother reared	17	83	7.12	0.55	5.60	9.40	6.02	8.22
Male—nursery reared	4	15	7.04	0.61	6.10	8.30	5.81	8.27
Male—mother reared	19	97	7.05	0.41	6.20	8.30	6.23	7.87

Albumin (g/dl)								
Female—nursery reared	10	45	3.94	0.25	3.50	4.50	3.44	4.45
Female—mother reared	17	83	3.89	0.49	2.60	7.30	2.90	4.88
Male—nursery reared	4	15	3.83	0.32	3.10	4.30	3.19	4.46
Male—mother reared	19	97	3.93	0.27	3.20	4.60	3.39	4.47
Globulin (g/dl)								
Female—nursery reared	10	45	3.17	0.36	2.50	4.00	2.45	3.89
Female—mother reared	17	82	3.23	0.53	2.20	5.80	2.17	4.30
Male—nursery reared	4	15	3.21	0.77	2.50	5.20	1.66	4.76
Male—mother reared	19	97	3.12	0.47	2.30	5.10	2.19	4.05
Albumin/globulin (ratio)								
Female—nursery reared	10	45	1.28	0.22	0.90	1.80	0.84	1.71
Female—mother reared	17	83	1.22	0.24	0.40	2.00	0.74	1.69
Male—nursery reared	4	15	1.25	0.25	0.60	1.40	0.75	1.75
Male—mother reared	19	97	1.29	0.22	0.60	1.80	0.85	1.74
Cholesterol (total) (mg/dl)								
Female—nursery reared	10	45	198.62	40.43	138.00	304.00	117.77	279.47
Female—mother reared	17	83	202.46	38.89	133.00	337.00	124.68	280.23
Male—nursery reared	4	15	203.40	21.62	167.00	234.00	160.17	246.63
Male—mother reared	19	97	206.75	27.82	143.00	281.00	151.11	262.40
Triglyceride (mg/dl)								
Female—nursery reared	10	45	87.87	36.18	37.00	188.00	15.51	160.22
Female—mother reared	17	81	87.52	39.11	29.00	246.00	9.30	165.74
Male—nursery reared	4	15	80.20	37.73	24.00	149.00	4.74	155.66
Male—mother reared	19	97	87.46	28.07	29.00	160.00	31.33	143.60
Calcium (mg/dl)								
Female—nursery reared	10	45	9.47	0.40	8.60	11.00	8.66	10.27
Female—mother reared	17	83	9.37	0.36	8.30	10.40	8.66	10.08
Male—nursery reared	4	15	9.62	0.42	9.00	10.40	8.79	10.45
Male—mother reared	19	97	9.39	0.34	8.60	10.30	8.71	10.06

(*Continued*)

Table 25–6. *Continued*

	Total subjects	Total samples	Mean	SD	Minimum	Maximum	Confidence intervals	
							Lower 99%	Upper 99%
Ionized calcium, calculated (mg/dl)								
Female—nursery reared	10	45	4.17	0.29	3.80	5.60	3.58	4.75
Female—mother reared	17	81	4.12	0.25	3.50	5.30	3.61	4.63
Male—nursery reared	4	15	4.26	0.32	3.60	4.70	3.62	4.90
Male—mother reared	19	97	4.14	0.20	3.50	4.70	3.74	4.54
Phosphorus (mg/dl)								
Female—nursery reared	10	45	5.05	0.80	3.20	6.70	3.46	6.65
Female—mother reared	17	83	4.79	0.72	3.20	6.20	3.34	6.23
Male—nursery reared	4	15	4.48	0.68	3.40	5.80	3.12	5.84
Male—mother reared	19	97	4.91	0.68	3.00	6.20	3.56	6.26
γ-Glutamyl transferase (GGT) (U/liter)								
Female—nursery reared	10	45	23.78	6.77	8.00	39.00	10.24	37.31
Female—mother reared	17	81	20.79	7.44	10.00	42.00	5.92	35.66
Male—nursery reared	4	15	29.47	8.80	15.00	44.00	11.87	47.06
Male—mother reared	19	97	27.56	10.09	10.00	59.00	7.38	47.73
Alkaline phosphatase (U/liter)								
Female—nursery reared	10	45	607.78	205.83	211.00	1218.00	196.12	1019.44
Female—mother reared	17	83	639.06	230.39	104.00	1833.00	178.29	1099.83
Male—nursery reared	4	15	530.80	243.60	157.00	1050.00	43.61	1017.99
Male—mother reared	19	96	528.83	122.47	205.00	1026.00	283.90	773.77
Alanine aminotransferase (ALT, SGPT) (U/liter)								
Female—nursery reared	10	45	38.29	7.47	24.00	56.00	23.35	53.23
Female—mother reared	17	83	41.11	11.72	19.00	77.00	17.67	64.55
Male—nursery reared	4	15	47.73	16.79	24.00	81.00	14.16	81.31
Male—mother reared	19	97	45.54	17.04	19.00	135.00	11.46	79.61

Asparate aminotransferase (AST, SGOT) (U/liter)								
Female—nursery reared	10	45	19.78	6.43	11.00	51.00	6.92	32.64
Female—mother reared	17	83	20.96	6.17	9.00	41.00	8.62	33.31
Male—nursery reared	4	15	23.60	6.57	14.00	35.00	10.47	36.73
Male—mother reared	19	97	20.55	5.82	11.00	39.00	8.91	32.19
Lactic dehydrogenase (LDH) (U/liter)								
Female—nursery reared	10	45	406.02	179.45	194.00	928.00	47.11	764.93
Female—mother reared	17	82	437.76	120.10	236.00	994.00	197.55	677.96
Male—nursery reared	4	15	422.20	85.43	286.00	571.00	251.34	593.06
Male—mother reared	19	96	437.79	102.49	281.00	775.00	232.81	642.77
Creatine kinase (U/liter)								
Female—nursery reared	10	44	174.39	95.51	73.00	551.00	73.00	365.40
Female—mother reared	16	82	220.94	157.56	64.00	976.00	64.00	536.07
Male—nursery reared	4	15	183.93	83.31	57.00	332.00	17.31	350.56
Male—mother reared	19	97	210.46	141.64	78.00	1024.00	78.00	493.75
Direct bilirubin (mg/dl)								
Female—nursery reared	10	44	0.01	0.03	0.00	0.10	0.00	0.08
Female—mother reared	16	80	0.02	0.04	0.00	0.20	0.00	0.09
Male—nursery reared	4	15	0.03	0.05	0.00	0.10	0.00	0.13
Male—mother reared	19	92	0.01	0.03	0.00	0.10	0.00	0.08
Iron (μg/dl)								
Female—nursery reared	10	44	118.73	38.17	53.00	222.00	42.39	195.06
Female—mother reared	16	80	101.81	36.97	32.00	175.00	27.88	175.75
Male—nursery reared	4	15	91.80	38.47	23.00	181.00	14.87	168.73
Male—mother reared	19	96	93.26	27.83	27.00	176.00	37.59	148.93
Sedimentation rate (mm/hr)								
Female—nursery reared	5	22	6.36	7.36	0.00	28.00	0.00	21.08
Female—mother reared	7	39	14.38	15.13	3.00	88.00	3.00	44.64
Male—nursery reared	2	8	7.25	17.31	0.00	50.00	0.00	41.87
Male—mother reared	11	73	11.75	11.74	0.00	50.00	0.00	35.24

Table 25–7. Adolescent (7–9.9 Years) Serum Clinical Chemistry Values by Sex and Rearing Category

	Total subjects	Total samples	Mean	SD	Minimum	Maximum	Confidence intervals	
							Lower 99%	Upper 99%
Glucose (mg/dl)								
Female—nursery reared	10	43	84.16	12.84	56.00	123.00	58.48	109.84
Female—mother reared	16	76	89.12	15.11	62.00	158.00	58.90	119.34
Male—nursery reared	9	40	84.83	9.98	63.00	102.00	64.86	104.79
Male—mother reared	19	96	90.33	13.22	58.00	136.00	63.89	116.78
Urea nitrogen (mg/dl)								
Female—nursery reared	10	43	11.42	2.52	7.00	20.00	6.38	16.46
Female—mother reared	16	76	11.99	3.17	4.00	21.00	5.64	18.33
Male—nursery reared	9	40	12.80	4.05	7.00	25.00	4.70	20.90
Male—mother reared	19	96	11.69	2.21	5.00	17.00	7.27	16.10
Creatinine (mg/dl)								
Female—nursery reared	10	43	0.79	0.16	0.40	1.00	0.48	1.10
Female—mother reared	16	76	0.80	0.13	0.50	1.10	0.55	1.06
Male—nursery reared	9	40	0.94	0.20	0.60	1.40	0.54	1.33
Male—mother reared	19	96	0.86	0.13	0.50	1.10	0.59	1.12
Uric acid (mg/dl)								
Female—nursery reared	10	43	2.01	0.56	1.20	3.50	0.90	3.12
Female—mother reared	16	76	2.26	0.50	1.30	3.70	1.25	3.26
Male—nursery reared	9	40	3.06	0.50	1.80	4.20	2.05	4.06
Male—mother reared	19	96	2.65	0.47	1.40	4.00	1.72	3.59

Sodium (mEq/liter)								
Female—nursery reared	10	43	138.44	2.65	133.00	147.00	133.14	143.74
Female—mother reared	16	76	138.79	2.17	134.00	147.00	134.44	143.14
Male—nursery reared	9	40	140.20	1.54	136.00	143.00	137.12	143.28
Male—mother reared	19	96	140.57	2.05	135.00	146.00	136.48	144.66
Potassium (mEq/liter)								
Female—nursery reared	10	43	3.92	0.39	3.10	4.80	3.14	4.69
Female—mother reared	16	76	3.91	0.36	3.40	5.20	3.19	4.62
Male—nursery reared	9	40	4.05	0.34	3.20	4.70	3.37	4.73
Male—mother reared	19	96	4.00	0.30	3.10	4.80	3.39	4.61
Chloride (mEq/liter)								
Female—nursery reared	10	43	101.88	2.81	93.00	108.00	96.26	107.51
Female—mother reared	16	76	102.34	2.81	96.00	109.00	96.72	107.97
Male—nursery reared	9	40	101.33	2.36	97.00	108.00	96.61	106.04
Male—mother reared	19	96	102.01	2.28	95.00	107.00	97.45	106.57
Carbon dioxide (mEq/liter)								
Female—nursery reared	10	43	25.74	3.36	18.00	38.00	19.02	32.46
Female—mother reared	16	76	25.92	2.60	19.00	34.00	20.72	31.12
Male—nursery reared	9	40	27.85	1.96	24.00	32.00	23.94	31.76
Male—mother reared	19	96	27.69	2.14	22.00	33.00	23.41	31.97
Anion gap (gap)								
Female—nursery reared	10	43	10.81	3.48	4.00	21.00	3.85	17.77
Female—mother reared	16	76	10.53	2.81	5.00	20.00	4.90	16.15
Male—nursery reared	9	40	11.03	2.39	5.00	16.00	6.24	15.81
Male—mother reared	19	96	10.88	2.73	5.00	18.00	5.42	16.33

(Continued)

Table 25–7. *Continued*

	Total subjects	Total samples	Mean	SD	Minimum	Maximum	Confidence intervals	
							Lower 99%	Upper 99%
Total protein (g/dl)								
Female—nursery reared	10	43	7.23	0.34	6.60	8.00	6.54	7.92
Female—mother reared	16	76	7.29	0.43	6.40	8.40	6.43	8.15
Male—nursery reared	9	40	7.09	0.40	6.20	8.00	6.29	7.89
Male—mother reared	19	96	7.26	0.40	6.30	8.90	6.46	8.06
Albumin (g/dl)								
Female—nursery reared	10	43	3.95	0.27	3.10	4.40	3.41	4.50
Female—mother reared	16	76	3.96	0.30	3.10	4.60	3.36	4.55
Male—nursery reared	9	40	4.08	0.31	3.50	4.80	3.45	4.71
Male—mother reared	19	96	4.05	0.34	2.70	4.70	3.37	4.74
Globulin (g/dl)								
Female—nursery reared	10	43	3.27	0.35	2.50	4.10	2.58	3.97
Female—mother reared	16	76	3.33	0.44	2.50	4.90	2.45	4.21
Male—nursery reared	9	40	3.01	0.52	2.00	4.40	1.97	4.05
Male—mother reared	19	96	3.21	0.42	2.30	5.50	2.36	4.05
Albumin/globulin (ratio)								
Female—nursery reared	10	43	1.22	0.17	0.80	1.60	0.89	1.55
Female—mother reared	16	76	1.21	0.21	0.70	1.70	0.79	1.63
Male—nursery reared	9	40	1.40	0.30	0.80	2.10	0.79	2.01
Male—mother reared	19	96	1.29	0.22	0.60	1.90	0.85	1.73

Cholesterol (total) (mg/dl)								
Female—nursery reared	10	43	198.19	35.25	130.00	281.00	127.68	268.69
Female—mother reared	16	76	195.82	35.72	101.00	317.00	124.37	267.27
Male—nursery reared	9	40	197.10	21.75	156.00	247.00	153.60	240.60
Male—mother reared	19	96	191.34	32.73	114.00	281.00	125.88	256.81
Triglyceride (mg/dl)								
Female—nursery reared	10	43	87.91	34.54	28.00	159.00	18.83	156.99
Female—mother reared	16	76	102.66	35.70	37.00	180.00	31.26	174.05
Male—nursery reared	9	40	94.45	34.53	41.00	175.00	25.38	163.52
Male—mother reared	19	96	98.35	35.67	32.00	220.00	27.02	169.68
Calcium (mg/dl)								
Female—nursery reared	10	43	9.27	0.41	8.60	10.30	8.44	10.09
Female—mother reared	16	76	9.31	0.39	8.40	10.10	8.52	10.10
Male—nursery reared	9	40	9.58	0.38	8.80	10.80	8.82	10.34
Male—mother reared	19	96	9.44	0.36	8.60	10.20	8.71	10.16
Ionized calcium, calculated (mg/dl)								
Female—nursery reared	10	43	4.02	0.22	3.60	4.50	3.57	4.47
Female—mother reared	16	76	4.03	0.23	3.30	4.50	3.57	4.49
Male—nursery reared	9	40	4.21	0.22	3.80	4.80	3.77	4.64
Male—mother reared	19	96	4.09	0.20	3.60	4.60	3.70	4.48
Phosphorus (mg/dl)								
Female—nursery reared	10	43	4.57	0.76	2.90	5.90	3.05	6.09
Female—mother reared	16	76	4.32	0.90	2.20	6.60	2.53	6.12
Male—nursery reared	9	40	4.58	0.71	2.90	6.20	3.16	6.00
Male—mother reared	19	96	4.76	0.88	2.80	6.50	3.00	6.53

(Continued)

Table 25–7. *Continued*

	Total subjects	Total samples	Mean	SD	Minimum	Maximum	Confidence intervals	
							Lower 99%	Upper 99%
γ-Glutamyl transferase (GGT) (U/liter)								
Female—nursery reared	10	43	25.12	13.58	10.00	86.00	10.00	52.28
Female—mother reared	16	75	23.31	10.94	5.00	84.00	1.42	45.20
Male—nursery reared	9	40	24.15	8.55	10.00	52.00	7.04	41.26
Male—mother reared	19	96	26.00	20.58	5.00	205.00	5.00	67.17
Alkaline phosphatase (U/liter)								
Female—nursery reared	10	43	346.98	169.17	130.00	760.00	8.64	685.32
Female—mother reared	16	76	402.96	195.26	88.00	1146.00	12.43	793.49
Male—nursery reared	9	40	558.63	248.97	167.00	1212.00	60.69	1056.56
Male—mother reared	19	96	550.27	244.01	112.00	1410.00	62.25	1038.29
Alanine aminotransferase (ALT, SGPT) (U/liter)								
Female—nursery reared	10	43	34.77	10.82	23.00	84.00	13.13	56.40
Female—mother reared	16	76	35.13	12.46	20.00	113.00	10.21	60.06
Male—nursery reared	9	40	34.88	9.18	20.00	58.00	16.51	53.24
Male—mother reared	19	95	36.40	9.60	17.00	75.00	17.19	55.61
Asparate aminotransferase (AST, SGOT) (U/liter)								
Female—nursery reared	10	43	17.60	5.01	9.00	38.00	7.58	27.63
Female—mother reared	16	76	17.41	4.50	8.00	33.00	8.41	26.41
Male—nursery reared	9	40	20.95	5.39	12.00	35.00	10.17	31.73
Male—mother reared	19	96	22.06	7.21	12.00	51.00	7.64	36.48

Lactic dehydrogenase (LDH) (U/liter)								
Female—nursery reared	10	43	382.19	241.76	223.00	1225.00	223.00	865.72
Female—mother reared	16	76	385.13	148.45	196.00	929.00	88.24	682.02
Male—nursery reared	9	40	355.85	58.67	244.00	480.00	238.52	473.18
Male—mother reared	19	96	404.98	108.55	238.00	820.00	187.88	622.08
Creatine kinase (U/liter)								
Female—nursery reared	10	43	153.56	122.08	61.00	680.00	61.00	397.72
Female—mother reared	16	75	176.81	104.54	51.00	617.00	51.00	385.88
Male—nursery reared	9	40	216.75	99.17	72.00	517.00	18.40	415.10
Male—mother reared	19	96	230.48	147.89	59.00	1115.00	59.00	526.27
Direct bilirubin (mg/dl)								
Female—nursery reared	10	41	0.01	0.03	0.00	0.10	0.00	0.07
Female—mother reared	16	72	0.02	0.04	0.00	0.10	0.00	0.10
Male—nursery reared	9	39	0.02	0.04	0.00	0.10	0.00	0.10
Male—mother reared	18	90	0.02	0.04	0.00	0.10	0.00	0.11
Iron (μ/dl)								
Female—nursery reared	10	43	122.58	36.31	64.00	202.00	49.96	195.20
Female—mother reared	16	75	112.47	37.24	38.00	220.00	37.99	186.94
Male—nursery reared	9	40	105.70	39.66	50.00	224.00	26.38	185.02
Male—mother reared	19	96	111.77	33.02	31.00	215.00	45.73	177.81
Sedimentation rate (mm/hr)								
Female—nursery reared	5	29	12.79	8.10	1.00	34.00	1.00	29.00
Female—mother reared	12	52	13.13	15.44	0.00	78.00	0.00	44.02
Male—nursery reared	4	17	1.88	2.29	0.00	9.00	0.00	6.46
Male—mother reared	13	60	9.45	14.06	0.00	62.00	0.00	37.57

Table 25–8. Two Sample t-Test Results

	t	p[1]	Nursery Reared Mean	Nursery Reared SD	Mother Reared Mean	Mother Reared SD
Hematology						
White Blood Cells (th/mm^3)	0.53	0.6000	9.62	3.15	9.85	3.80
Red Blood Cells (th/mm^3)	0.18	0.8546	5.08	0.43	5.05	0.42
Hemoglobin (g/dl)	1.98	0.0518	13.52	1.04	13.19	1.09
Hematocrit (%)	1.87	0.0666	41.34	3.07	40.46	2.96
MCV (um^3)	2.87	0.0049	81.52	3.29	80.08	4.67
MCH (pg)	2.46	0.0156	26.68	1.31	26.11	2.09
MCHC (g/dl)	0.78	0.4377	32.72	1.02	32.61	0.94
Platelets (th/mm^3)	0.58	0.5609	294.91	68.21	291.28	70.48
Neutrophils (th/mm^3)	−0.32	0.7493	4.11	2.54	4.64	3.00
Lymphocytes (th/mm^3)	1.28	0.2069	4.62	2.42	4.14	1.73
Monocytes (th/mm^3)	1.59	0.1203	0.38	0.20	0.34	0.19
Eosinophils (th/mm^3)	−0.34	0.7360	0.19	0.13	0.23	0.24
Basophils (th/mm^3)	0.21	0.8366	0.05	0.04	0.05	0.04
Serum Clinical Chemistry						
Glucose (mg/dl)	−3.17	0.0021	82.77	13.21	88.09	13.79
Urea Nitrogen (mg/dl)	2.07	0.0439	12.99	4.98	11.86	3.44
Creatinine (mg/dl)	2.77	0.0074	0.78	0.19	0.71	0.17
Uric Acid (mg/dl)	0.45	0.6578	2.33	0.77	2.33	0.53
Sodium (meq/l)	−0.46	0.6482	138.96	2.37	139.16	2.58
Potassium (meq/l)	1.63	0.1074	4.00	0.35	3.98	0.35
Chloride (meq/l)	−2.48	0.0153	101.84	2.58	102.83	3.19
Carbon Dioxide (meq/l)	2.00	0.0488	26.19	2.97	25.57	3.02
Anion Gap (gap)	−0.88	0.3789	10.86	2.89	10.79	2.99
Total Protein (g/dl)	0.53	0.5955	7.02	0.46	7.05	0.52
Albumin (g/dl)	1.17	0.2449	3.93	0.30	3.86	0.39
Globulin (g/dl)	−0.21	0.8321	3.09	0.51	3.19	0.49
Albumin/Globulin (ratio)	0.89	0.3751	1.31	0.27	1.24	0.23
Cholesterol (total) (mg/dl)	−2.72	0.0076	195.59	31.45	213.12	43.01
Triglyceride (mg/dl)	0.03	0.9741	81.86	34.92	84.77	37.83
Calcium (mg/dl)	2.02	0.0483	9.46	0.41	9.37	0.39
Ionized Calcium, Calc. (mg/dl)	0.91	0.3682	4.19	0.28	4.13	0.25
Phosphorus (mg/dl)	1.55	0.1247	4.76	0.77	4.70	0.81
GGT	1.80	0.0768	24.53	10.29	22.52	12.67
Alkaline Phosphatase (u/l)	−2.23	0.0286	495.46	219.90	623.06	382.17
ALT, SPGT	−0.85	0.3951	38.66	11.93	39.50	15.03
AST, SGOT	−1.47	0.1463	20.47	6.03	20.77	6.99
Lactic Dehydrogenase (LDH) (u/l)	−1.60	0.1148	397.77	160.73	441.32	125.35
Creatine Kinase (u/l)	−2.50	0.0145	181.86	119.01	215.90	138.68
Direct Bilirubin (mg/dl)	0.08	0.9391	0.01	0.04	0.02	0.04
Iron (mcg/dl)	2.24	0.0282	106.41	38.71	93.12	37.16
Sedimentation Rate (mm/hr)	−1.32	0.1934	7.84	9.66	13.80	15.23

[1] *sign. p # 0.00125; **trend = p # 0.00250

young ages (0–72 weeks of age) will help us to understand whether differences can be explained by variations in dietary intake afforded by maternal milk versus baby formula. Concomitant consideration of differential growth rates may also help to explain H/SCC differences across nursery- and mother-reared chimpanzees.

ACKNOWLEDGMENTS

We gratefully acknowledge PFA staff members, Lyle Bartel, D.V.M., Veterinarian, James Murphy, Assistant Director/Colony Manager, Kelly Carbone, Assistant Colony Manager, and the dedicated carestaff. We also thank Scott Burnett, M.A., and Lucilla Spini, Ph.D., for their assistance in data entry and data management for this project. This project was supported in part by the University of Texas, M.D. Anderson Cancer Research Center, as a subcontract (1U42RR15090-02) within the National Institutes of Health Biomedical Research Program. PFA is accredited by the Association for Assessment and Accreditation of Laboratory Care International. The data collection protocol was reviewed and approved by the PFA Institutional Animal Care and Use Committee.

REFERENCES

Fritz, J., and Fritz, P. (1982). Great ape hand-rearing with a goal of normalcy and a reproductive continuum. *Proceedings of the Annual Meeting of the American Association of Zoo Veterinarians*, New Orleans, pp. 27–31.

Fritz, J., and Howell, S. (1993). Psychological wellness for captive chimpanzees: An evaluative program. *Humane Innovations Alternatives* 7:426–434.

Fritz, J., Ebert, J.W., and Carland, J.F. (1985). Nutritional management of infant apes. In: Graham, C.E., and Bowen, J.A. (eds.), *Clinical Management of Infant Great Apes.* Alan R. Liss, Inc., New York, pp. 141–156.

Hainsey, B.M., Hubbard, G.B., Leland, M.M., and Brasky, K.M. (1993). Clinical parameters of the normal baboons (*Papio* species) and chimpanzees (*Pan troglodytes*). *Lab. Anim. Sci.* 43:236–243.

Herndon, J.G., and Tigges, J. (2001). Hematologic and blood biochemical variables of captive chimpanzees: Cross-sectional and longitudinal analyses. *Comp. Med.* 51:60–69.

Hodson, H.H., Lee, B.D., Wisecup W.G., and Fineg, J. (1967). Baseline blood values of the chimpanzee. I. The relationship of age and sex and hematological values. *Folia Primatol.* 7:1–11.

Hodson, H.H., Wisecup, W.G., Faulkner, M.F., and Felts, P.E. (1968). Baseline blood values of the chimpanzees. II. The relationship of age and sex and serum chemistry values. *Folia Primatol.* 8:77–86.

Huser H.-J., and Olberding, B.A. (1972). The hematology of the chimpanzee. In: Bourne, G.H. (ed.), *The Chimpanzee*, Vol. 5: *Histology, Reproduction and Restraint.* Karger, Basel, pp. 193–225.

Ihrig, M., Tassinary, L.G., Bernacky, B., and Keeling, M.E. (2001). Hematologic and serum biochemical reference intervals for the chimpanzee (*Pan troglodytes*) categorized by age and sex. *Comp. Med.* 51:30–37.

McClure, H.M., Guilloud, N.B., and Keeling, M.E. (1973). Clinical pathology data for the chimpanzee and other anthropoid apes. In: Bourne, G.H. (ed.), *The Chimpanzee*, Vol. 6: *Anatomy and Pathology.* Karger, Basel, pp. 121–181.

National Research Council. (1996). *Guide for the Care and Use of Laboratory Animals.* National Academy Press, Washington, D.C.

Stone, G.A., Johnson, B.K., Druilhet, R., Garza, P.B., and Gibbs, C.J. (2000). Immunophenotyping of peripheral blood, ranges of serum chemistries and clinical hematology values of healthy chimpanzees (*Pan troglodytes*). *J. Med. Primatol.* 29:324–329.

CHAPTER TWENTY-SIX

Hematology and Serum Chemistry Reference Values for Rhesus Macaque (*Macaca mulatta*) Infants

John P. Capitanio

The rhesus macaque infants whose data are described in these tables were assessed for various hematology parameters at the mean age of 107.8 days (SD = 10.3, range = 89–130). Note that information presented by Capitanio *et al.* in Chapter 11 of this book reflects a single year's data; in the present set of tables (Tables 26–1, 26–2, and 26–3), data from a second year of assessments are included. Infants were hand captured from their field cage, corn crib, nursery, or indoor cage, separated from their companions, relocated to an unfamiliar indoor room, and placed in individual cages. Blood sampling occurred between 1030 and 1100 hr, approximately 2 hr after the separation and relocation. Each animal was hand captured from its individual cage and manually restrained while 1 ml of blood was drawn into an unheparinized syringe from a femoral vein. Blood was transferred to EDTA tubes, and hematology and flow cytometry were performed by California National Primate Research Center's (CNPRC) Clinical Laboratory. Complete blood counts were performed using a Serono Baker

Table 26–1. Rhesus Macaque Total White Cell and CD4 and CD8 T Cell Counts

Rearing group	Mean	95% CI Lower	95% CI Upper	Median	Number of subjects
Total white cells ($10^3/\mu l$)					
Female corncrib	16.50	14.50	18.50	15.15	30
Female field cage	17.95	17.15	18.75	16.60	238
Female nursery	14.71	13.47	15.94	14.20	46
Female indoor mother	15.90	13.84	17.95	14.65	32
Male corncrib	15.22	11.58	18.85	13.60	18
Male field cage	16.70	15.95	17.46	15.85	186
Male nursery	14.78	13.00	16.57	15.05	26
Male indoor mother	16.08	14.32	17.84	14.70	19
$CD4^+$ T cells (cells/μl)					
Female corncrib	1972.77	1423.00	2522.53	1766.00	30
Female field cage	1912.29	1813.58	2010.99	1802.00	237
Female nursery	3368.80	2964.49	3773.12	3345.50	46
Female indoor mother	3542.56	3045.64	4039.49	3456.50	32
Male corncrib	1848.44	1486.78	2210.11	1836.00	18
Male field cage	1885.32	1774.81	1995.84	1731.00	186
Male nursery	3302.23	2798.71	3805.75	3363.00	26
Male indoor mother	3741.68	3028.36	4455.00	3425.00	19
$CD8^+$ T cells (cells/μl)					
Female corncrib	666.50	539.93	793.07	616.00	30
Female field cage	776.31	721.91	830.71	717.00	236
Female nursery	1100.61	941.64	1259.58	976.00	46
Female indoor mother	1376.22	1106.48	1645.96	1236.00	32
Male corncrib	747.67	560.67	934.66	700.50	18
Male field cage	758.27	697.44	819.11	662.00	186
Male nursery	1178.69	954.34	1403.04	1114.00	26
Male indoor mother	1347.05	1030.03	1664.07	1163.00	19

Table 26–2. Rhesus Infant Red Cell Measures

Rearing group	Mean	95% CI Lower	95% CI Upper	Median	Number of subjects
Total red cells ($10^6/\mu l$)					
Female corncrib	5.61	5.43	5.78	5.62	30
Female field cage	5.77	5.69	5.84	5.71	238
Female nursery	5.81	5.62	6.00	5.79	46
Female indoor mother	6.08	5.81	6.35	6.04	32
Male corncrib	5.39	5.18	5.60	5.44	18
Male field cage	5.84	5.75	5.94	5.79	186
Male nursery	6.09	5.77	6.40	5.95	26
Male indoor mother	5.96	5.66	6.27	5.86	19

Table 26–2. *Continued*

Rearing group	Mean	95% CI Lower	95% CI Upper	Median	Number of subjects
Hemoglobin (g/dl)					
Female corncrib	13.48	12.96	14.00	13.45	30
Female field cage	13.64	13.46	13.81	13.60	238
Female nursery	14.35	13.93	14.78	14.25	46
Female indoor mother	14.58	13.84	15.32	14.30	32
Male corncrib	13.14	12.40	13.89	13.00	18
Male field cage	13.66	13.43	13.89	13.50	186
Male nursery	16.79	12.35	21.24	14.40	26
Male indoor mother	13.88	13.28	14.49	13.80	19
Hematocrit (%)					
Female corncrib	39.50	38.17	40.84	39.00	30
Female field cage	40.07	39.57	40.57	39.90	238
Female nursery	41.11	40.01	42.21	40.85	46
Female indoor mother	42.38	40.18	44.59	41.85	32
Male corncrib	38.02	36.60	39.43	38.50	18
Male field cage	39.97	39.31	40.62	39.50	186
Male nursery	43.06	40.52	45.60	41.45	26
Male indoor mother	40.69	39.04	42.34	41.40	19
Mean cell volume (fl)					
Female corncrib	70.73	69.61	71.86	71.00	30
Female field cage	69.59	69.19	69.99	70.00	238
Female nursery	71.17	70.31	72.04	71.00	46
Female indoor mother	69.75	68.16	71.34	70.00	32
Male corncrib	70.67	69.46	71.87	71.00	18
Male field cage	68.60	68.01	69.19	69.00	186
Male nursery	70.69	69.79	71.60	71.00	26
Male indoor mother	68.32	66.48	70.16	69.00	19
Corpuscular hemoglobin (pg)					
Female corncrib	23.81	23.41	24.21	23.60	30
Female field cage	23.65	23.49	23.81	23.70	237
Female nursery	24.81	24.51	25.10	24.85	46
Female indoor mother	23.99	23.32	24.66	24.55	32
Male corncrib	23.89	23.40	24.39	23.95	18
Male field cage	23.44	23.20	23.67	23.50	186
Male nursery	24.39	24.06	24.73	24.35	26
Male indoor mother	23.40	22.48	24.32	24.00	19
MCH concentration (pg/fl)					
Female corncrib	33.74	33.22	34.26	33.35	30
Female field cage	34.04	33.89	34.19	34.20	238
Female nursery	34.90	34.62	35.17	35.10	46
Female indoor mother	34.45	33.93	34.97	34.50	32
Male corncrib	33.86	33.19	34.52	33.70	18
Male field cage	34.15	33.98	34.33	34.30	186
Male nursery	34.55	34.15	34.96	34.65	26
Male indoor mother	34.15	33.37	34.93	34.40	19

Table 26–3. Other Hematology Measures for Rhesus Infants

Rearing group	Mean	95% CI Lower	95% CI Upper	Median	Number of subjects
Platelets ($10^5/\mu l$)					
Female corncrib	6.05	5.56	6.55	6.19	30
Female field cage	6.14	5.92	6.36	6.15	238
Female nursery	7.13	6.62	7.64	7.08	46
Female indoor mother	6.57	5.98	7.17	6.27	32
Male corncrib	6.38	5.66	7.10	6.55	18
Male field cage	6.17	5.91	6.43	6.05	186
Male nursery	6.47	5.80	7.14	6.52	26
Male indoor mother	5.67	5.00	6.35	5.81	19
Plasma protein (g/dl)					
Female corncrib	6.95	6.82	7.09	6.90	30
Female field cage	7.01	6.95	7.07	7.00	236
Female nursery	6.94	6.77	7.10	6.90	46
Female indoor mother	7.06	6.93	7.19	7.10	32
Male corncrib	6.86	6.67	7.05	6.80	18
Male field cage	6.97	6.90	7.04	6.90	186
Male nursery	6.85	6.71	6.98	6.90	26
Male indoor mother	7.00	6.85	7.14	7.00	19
Neutrophils (cells/μl)					
Female corncrib	10861.77	9445.81	12277.72	10044.00	30
Female field cage	11711.16	11021.53	12400.80	10723.00	238
Female nursery	6825.74	6004.93	7646.55	6381.00	46
Female indoor mother	7872.88	6420.30	9325.45	7100.50	32
Male corncrib	9574.11	6415.11	12733.11	7968.00	18
Male field cage	10661.48	9990.10	11332.87	9788.00	186
Male nursery	7158.08	5579.07	8737.08	7434.50	26
Male indoor mother	8220.16	6690.32	9750.00	7599.00	19
Lymphocytes (cells/μl)					
Female corncrib	5264.20	4281.93	6246.47	4988.50	30
Female field cage	5811.51	5503.80	6119.21	5478.00	237
Female nursery	7344.80	6518.81	8170.80	6784.00	46
Female indoor mother	7659.34	6512.03	8806.65	6820.00	32
Male corncrib	5345.44	4193.61	6497.28	4949.00	18
Male field cage	5489.20	5187.24	5791.16	5248.50	186
Male nursery	7251.42	6326.05	8176.80	7156.50	26
Male indoor mother	7610.21	6344.25	8876.17	7056.00	19
Monocytes (cells/μl)					
Female corncrib	236.77	122.84	350.69	143.50	30
Female field cage	277.57	222.07	333.06	143.50	238
Female nursery	343.89	207.46	480.32	215.50	46
Female indoor mother	241.13	74.60	407.65	93.00	32
Male corncrib	475.89	–234.85	1186.63	89.50	18
Male field cage	364.26	276.88	451.64	190.00	186
Male nursery	217.15	111.25	323.06	112.50	26
Male indoor mother	179.89	75.00	284.79	148.00	19

Table 26–3. *Continued*

Rearing group	Mean	95% CI Lower	95% CI Upper	Median	Number of subjects
Eosinophils (cells/μl)					
Female corncrib	129.97	41.73	218.20	0.00	30
Female field cage	107.69	83.60	131.78	0.00	238
Female nursery	191.04	115.85	266.23	97.00	46
Female indoor mother	123.53	68.51	178.56	93.00	32
Male corncrib	424.39	−288.76	1137.54	0.00	18
Male field cage	108.63	75.80	141.47	0.00	186
Male nursery	157.96	92.89	223.03	146.00	26
Male indoor mother	68.68	21.21	116.16	0.00	19

Diagnostic System (Allentown, PA), and all electronic counts were verified by a manual differential.

Aliquots of whole blood (50μl) were labeled directly with phycoerythrin (anti-CD4-M-T477; BD Pharmingen), peridinin chlorophyll-α protein (anti-CD8-SK1; BD Pharmingen), and fluorescein isothiocyanate (anti-CD3-SP34; BD Pharmingen). A Coulter Q-prep (Coulter Corp., Miami, FL) was used to lyse the red blood cells and fix the sample in paraformaldehyde. Lymphocytes were gated by forward and side light scatter. A FACS Calibur flow cytometer (BD Pharmingen) was used to phenotype the lymphocyte subsets.

Using the criterion of little or no overlap in confidence intervals, the data reveal the following effects. (1) CD4, CD8, and total lymphocytes have clearly higher values in the two indoor groups than in the two outdoor groups. In contrast, animals from the outdoor groups had more neutrophils than did animals from the indoor groups. This effect is especially strong for females. This may suggest that outdoor groups are exposed to more infectious sources than indoor groups. (2) Values for total red cells, mean corpuscular hemoglobin (MCH), and MCH concentration (MCHC) tend to be higher for indoor groups than for outdoor groups. This may indicate that iron levels are higher for indoor infants, regardless of whether they are nursery or mother reared. (3) Male corn crib monkeys had especially high variability in monocytes and eosinophils, suggesting that some animals had active infections or high stress levels.

CHAPTER TWENTY-SEVEN

Hematology and Serum Chemistry Reference Values for Pigtailed Macaque (*Macaca nemestrina*) Infants

Erika Rainwater

The pigtailed macaque infants described in these tables were assessed for various hematology (Table 27–1) and serum chemistry (Table 27–2) measures during their first postnatal year. Nursery infants were raised under standard conditions in the Infant Primate Research Laboratory (IPRL), as described by Ruppenthal and Sackett (1992; see also Chapter 18 in this volume). Mother-raised infants were housed at the Primate Field Station (PFS) of the Washington National Primate Research Center (WaNPRC). Until they were weaned at an average of 6 months, mother-raised animals lived in harem groups of one male and four to eight females or groups containing only four to eight mother–infant pairs. After weaning, infants lived in mixed-sex peer groups containing 8–16 animals.

The data reported here are from samples taken under normal medical conditions, defined as no treatments for medical purposes within 4 weeks

Table 27–1. Hematology Values[a]

Age	Nursery				Mother			
	Mean	Up	Low	*N*	Mean	Up	Low	*N*
Hemoglobin (g/dl)								
Female								
1–3	13.4	13.7	13.1	294	10.9	12.1	9.7	3
4–6	12.1	12.3	11.9	152	11.5	12.1	10.8	10
7–9	11.7	11.9	11.5	138	11.2	11.5	10.8	30
10–12	11.5	11.7	11.2	129	11.4	11.7	11.1	26
Male								
1–3	13.5	13.7	13.3	382	10.6	11.8	9.4	6
4–6	12.1	12.3	12.0	161	12.0	12.6	11.3	7
7–9	11.7	11.9	11.5	133	11.7	12.0	11.4	28
10–12	11.3	11.5	11.1	151	11.4	11.6	11.2	51
Hematocrit (%)								
Female								
1–3	40.0	40.8	39.1	295	37.1	41.6	32.5	3
4–6	38.0	38.6	37.4	153	36.9	39.0	34.8	10
7–9	38.0	38.8	37.3	138	35.5	36.4	34.7	30
10–12	37.4	38.1	36.6	129	36.4	37.1	35.6	26
Male								
1–3	40.2	40.9	39.5	384	36.5	39.4	33.5	6
4–6	37.4	38.0	36.7	161	38.3	40.6	36.0	7
7–9	37.9	38.6	37.3	133	37.3	38.3	36.4	28
10–12	36.7	37.4	36.1	151	37.6	38.1	37.0	51
Total red cells (10^6/μl)								
Female								
1–3	5.4	5.5	5.4	282	5.7	6.2	5.1	3
4–6	5.9	6.0	5.8	148	5.7	6.0	5.4	10
7–9	5.8	5.9	5.7	136	5.7	5.9	5.5	30
10–12	5.8	5.9	5.6	128	5.8	6.0	5.7	26
Male								
1–3	5.4	5.5	5.3	321	5.7	6.2	5.2	6
4–6	5.9	6.0	5.8	158	6.2	6.7	5.6	7
7–9	5.9	6.0	5.8	130	6.0	6.1	5.8	28
10–12	5.7	5.8	5.6	150	5.8	5.9	5.7	51
Total white cells (10^3/μl)								
Female								
1–3	5.4	5.5	5.4	282	5.7	6.2	5.1	3
4–6	5.9	6.0	5.8	148	5.7	6.0	5.4	10
7–9	5.8	5.9	5.7	136	5.7	5.9	5.5	30
10–12	5.8	5.9	5.6	128	5.8	6.0	5.7	26
Male								
1–3	5.4	5.5	5.3	321	5.7	6.2	5.2	6
4–6	5.9	6.0	5.8	158	6.2	6.7	5.6	7
7–9	5.9	6.0	5.8	130	6.0	6.1	5.8	28
10–12	5.7	5.8	5.6	150	5.8	5.9	5.7	51

Table 27–1. *Continued*

	Nursery				Mother			
Age	Mean	Up	Low	*N*	Mean	Up	Low	*N*
Lymphocytes ($10^3/\mu l$)								
Female								
1–3	5.4	5.5	5.4	282	5.7	6.2	5.1	3
4–6	5.9	6.0	5.8	148	5.7	6.0	5.4	10
7–9	5.8	5.9	5.7	136	5.7	5.9	5.5	30
10–12	5.8	5.9	5.6	128	5.8	6.0	5.7	26
Male								
1–3	5.4	5.5	5.3	321	5.7	6.2	5.2	6
4–6	5.9	6.0	5.8	158	6.2	6.7	5.6	7
7–9	5.9	6.0	5.8	130	6.0	6.1	5.8	28
10–12	5.7	5.8	5.6	150	5.8	5.9	5.7	51
Neutrophils (%)								
Female								
1–3	49.4	51.4	47.5	290	47.1	57.6	36.6	3
4–6	37.3	39.9	34.8	140	55.3	62.3	48.4	10
7–9	43.8	47.7	39.9	101	55.7	60.1	51.4	25
10–12	50.2	53.8	46.7	107	56.1	60.0	52.1	21
Male								
1–3	50.7	52.5	49.0	379	51.7	59.9	43.5	6
4–6	37.3	39.9	34.7	152	36.9	44.5	29.2	6
7–9	43.8	47.2	40.4	106	44.5	48.1	41.0	25
10–12	48.3	51.8	44.8	126	47.5	51.2	43.9	49
Monocytes (%)								
Female								
1–3	4.4	4.8	4.0	222	5.9	10.5	1.3	3
4–6	3.7	4.3	3.1	100	3.9	5.3	2.5	10
7–9	5.4	6.7	4.1	66	6.5	7.3	5.6	24
10–12	4.1	4.9	3.4	83	5.7	6.6	4.8	20
Male								
1–3	4.3	4.7	4.0	256	4.7	6.5	2.9	6
4–6	4.5	5.5	3.4	99	3.0	4.1	1.8	6
7–9	4.2	4.9	3.5	65	5.5	6.5	4.4	25
10–12	4.9	5.7	4.2	104	4.4	5.1	3.8	48
Eosinophils (%)								
Female								
1–3	1.5	1.7	1.2	133	2.3	4.9	—	3
4–6	1.3	1.5	1.0	76	2.1	3.1	1.1	8
7–9	1.3	1.7	0.9	46	1.8	2.7	1.0	8
10–12	1.1	1.4	0.9	62	1.8	2.5	1.0	11
Male								
1–3	1.4	1.5	1.2	164	1.8	2.2	1.4	6
4–6	1.3	1.6	1.1	86	1.4	1.9	0.9	5
7–9	1.2	1.6	0.8	55	1.2	1.7	0.7	19
10–12	1.0	1.3	0.8	74	1.4	1.8	1.0	40

(*Continued*)

Table 27–1. *Continued*

	Nursery				Mother			
Age	Mean	Up	Low	*N*	Mean	Up	Low	*N*
Basophils (%)								
Female								
1–3	0.9	1.0	0.7	120	0.0	—	—	1
4–6	0.8	0.9	0.6	57	0.6	1.0	0.1	6
7–9	0.8	1.0	0.5	43	0.7	1.0	0.4	8
10–12	0.8	1.1	0.5	50	0.7	1.1	0.3	9
Male								
1–3	0.8	1.0	0.7	143	1.5	2.5	0.5	2
4–6	0.6	0.7	0.4	63	1.2	1.4	0.9	3
7–9	0.7	0.9	0.5	50	0.4	0.7	0.2	18
10–12	0.8	1.0	0.7	143	0.5	0.7	0.4	29

[a] Values (mean, upper and lower 95% confidence interval bounds, and sample size) by month of age for female and male pigtailed macaques hand reared in the IPRL nursery or mother reared in the PFS. Ages yielding significant ANOVA sex, rearing, or sex × rearing interaction effects are identified by bold italics and specified in Table 27–3.

Table 27–2. Serum Chemistry Values[a]

	Nursery				Mother			
Age	Mean	Up	Low	*N*	Mean	Up	Low	*N*
Sodium (mEq/liter)								
Female								
4–6	144.0	147.7	140.3	43	139.5	144.1	134.9	3
7–9	142.0	145.2	138.8	25	144.1	145.1	143.1	23
10–12	143.7	146.1	141.3	41	144.0	145.8	142.2	19
Male								
4–6	136.6	140.2	133.0	36	144.6	145.8	143.3	3
7–9	142.2	145.0	139.5	25	144.7	147.2	142.3	13
10–12	136.6	140.2	133.0	36	144.9	146.3	143.6	23
Potassium (mEq/liter)								
Female								
4–6	5.9	8.1	3.7	43	4.0	4.9	3.0	3
7–9	5.6	7.5	3.7	25	4.0	4.1	3.8	23
10–12	4.5	6.1	3.0	41	5.4	8.2	2.6	19
Male								
4–6	8.0	12.0	4.0	35	4.0	4.4	3.6	3
7–9	6.2	8.7	3.7	25	4.4	5.0	3.7	13
10–12	8.3	11.3	5.2	57	4.0	4.2	3.9	23
Chloride (mEq/liter)								
Female								
4–6	109.3	112.5	106.2	39	108.1	114.9	101.3	3
7–9	104.5	107.4	101.5	23	106.6	115.6	97.6	23
10–12	107.3	109.5	105.0	17	109.9	112.8	106.9	19

Table 27–2. *Continued*

	Nursery				Mother			
Age	Mean	Up	Low	*N*	Mean	Up	Low	*N*
Male								
4–6	101.6	104.4	98.7	33	156.1	172.3	154.9	3
7–9	107.6	109.6	105.5	20	110.7	112.8	108.5	13
10–12	106.6	108.8	104.3	28	111.7	113.4	109.9	17
Carbon dioxide (mEq/liter)								
Female								
4–6	16.0	17.7	14.3	30	22.8	27.2	18.5	3
7–9	18.9	21.4	16.3	17	21.9	23.4	20.5	22
10–12	20.0	22.1	18.0	16	24.6	26.7	22.4	19
Male								
4–6	18.2	19.8	16.5	26	22.8	25.6	20.0	3
7–9	19.7	22.2	17.2	11	21.6	23.4	19.8	13
10–12	19.6	21.8	17.5	24	21.9	23.5	20.3	17
Total protein (g/dl)								
Female								
4–6	5.9	6.1	5.7	27	5.6	6.6	4.6	3
7–9	5.9	6.3	5.6	16	6.3	6.5	6.1	23
10–12	5.8	6.2	5.4	15	6.5	6.7	6.3	19
Male								
4–6	6.0	6.2	5.8	18	6.1	6.4	5.9	3
7–9	5.9	6.2	5.6	16	6.2	6.6	5.9	12
10–12	6.2	6.4	6.0	29	6.4	6.7	6.2	18
Albumin (g/dl)								
Female								
4–6	3.9	4.2	3.6	25	2.7	4.3	1.1	3
7–9	3.5	3.9	3.1	16	3.0	3.1	2.8	23
10–12	2.9	3.4	2.4	14	3.0	3.1	2.8	17
Male								
4–6	4.0	4.3	3.7	18	3.5	4.0	2.9	3
7–9	3.4	3.7	3.1	16	3.4	4.2	2.5	12
10–12	3.8	4.2	3.5	23	3.2	3.4	3.1	18
Globulin (g/dl)								
Female								
4–6	1.9	2.4	1.4	9	1.9	—	—	1
7–9	2.7	3.2	2.1	7	3.5	3.8	3.2	8
10–12	2.7	3.2	2.2	7	3.6	4.7	2.4	3
Male								
4–6	1.9	2.2	1.5	13	3.4	4.0	2.7	2
7–9	2.3	2.6	2.0	8	2.7	3.6	1.8	5
10–12	2.1	2.5	1.8	15	2.1	2.1	2.0	2
Calcium (mg/dl)								
Female								
4–6	9.6	9.8	9.3	24	9.0	9.4	8.6	14
7–9	10.2	11.0	9.5	15	9.0	10.4	7.5	3
10–12	9.0	9.4	8.6	14	9.8	9.9	9.6	23

(*Continued*)

Table 27–2. *Continued*

	Nursery				Mother			
Age	Mean	Up	Low	*N*	Mean	Up	Low	*N*
Male								
4–6	9.6	9.9	9.3	17	9.9	10.5	9.4	3
7–9	9.7	10.3	9.1	16	9.7	9.9	9.5	11
10–12	9.5	9.7	9.2	23	9.7	9.9	9.6	18
Total bilirubin (mg/dl)								
Female								
4–6	0.3	0.4	0.2	24	0.5	0.7	0.3	3
7–9	0.4	0.5	0.3	15	0.6	0.8	0.4	23
10–12	0.3	0.4	0.2	14	0.5	0.7	0.3	17
Male								
4–6	0.3	0.4	0.2	17	2.2	5.8	—	3
7–9	0.4	0.5	0.3	15	1.2	2.8	—	12
10–12	0.3	0.3	0.2	20	0.4	0.5	0.4	18
Glucose (mg/dl)								
Female								
4–6	101.5	130.9	72.1	40	86.3	146.1	26.6	3
7–9	138.3	211.2	65.4	25	71.5	85.1	58.0	23
10–12	84.1	107.1	61.2	20	63.6	68.7	58.4	19
Male								
4–6	102.6	135.8	69.4	35	66.1	77.9	54.2	3
7–9	74.8	82.8	66.8	21	69.3	81.3	57.3	12
10–12	78.7	93.1	64.3	37	73.4	79.0	67.8	18
Creatinine (μg/liter)								
Female								
4–6	0.7	0.8	0.5	38	0.8	1.1	0.4	3
7–9	0.7	0.9	0.5	24	0.8	0.9	0.6	23
10–12	0.6	0.8	0.5	19	1.0	1.4	0.5	18
Male								
4–6	0.5	0.6	0.5	29	0.7	0.8	0.6	3
7–9	0.7	0.9	0.5	20	1.6	2.5	0.7	12
10–12	0.6	0.8	0.5	30	1.1	1.7	0.5	18
Alkaline (μg/dl)								
Female								
4–6	830.2	922.4	738.0	16	290.8	—	—	1
7–9	661.2	834.8	487.5	11	635.9	731.2	540.5	19
10–12	749.0	907.9	590.2	11	634.4	757.8	511.0	15
Male								
4–6	950.5	1224.6	676.5	11	991.3	1706.9	275.7	3
7–9	832.8	1037.7	627.8	12	696.8	802.4	591.2	12
10–12	561.5	719.8	403.2	13	686.9	795.7	578.1	15
SGOT (μg/liter)								
Female								
4–6	30.9	39.0	22.8	11	76.5	157.8	—	2
7–9	38.8	44.9	32.7	6	51.8	65.5	38.1	23
10–12	44.2	67.0	21.4	7	50.0	64.1	36.0	17

Table 27–2. *Continued*

	Nursery				Mother			
Age	Mean	Up	Low	*N*	Mean	Up	Low	*N*
Male								
4–6	37.3	51.9	22.7	9	40.1	43.1	37.0	3
7–9	123.9	227.6	20.2	11	66.6	109.6	23.7	12
10–12	40.2	51.4	29.0	10	47.6	52.6	42.7	18
SGPT (μg/liter)								
Female								
4–6	16.4	22.4	10.3	11	54.4	110.0	—	2
7–9	30.4	40.0	20.9	7	28.8	32.5	25.0	22
10–12	34.6	47.1	22.1	8	29.8	34.3	25.3	17
Male								
4–6	22.9	34.8	11.0	8	28.1	33.0	23.2	3
7–9	47.2	79.0	15.3	12	35.0	42.1	27.9	12
10–12	34.9	55.8	14.0	15	30.6	35.7	25.5	17
GGPT (μg/liter)								
Female								
4–6	113.4	142.0	84.7	10	87.1	200.6	—	2
7–9	100.8	123.8	77.8	6	75.7	88.4	62.9	22
10–12	82.4	120.2	44.7	4	80.2	92.8	67.5	17
Male								
4–6	134.5	162.0	107.0	5	107.8	147.9	67.6	3
7–9	131.4	179.3	83.5	7	73.6	92.0	55.1	12
10–12	85.8	118.9	52.7	7	91.3	105.8	76.8	17

[a] Values (mean, upper and lower 95% confidence interval bounds, and sample size) by month of age for female and male pigtailed macaques hand reared in the IPRL nursery or mother reared in the PFS. Ages yielding significant ANOVA sex, rearing, or sex × rearing interaction effects are identified by bold italics and specified in Table 27–3.
SGOT, serum glutamate oxaloacetate; SGPT, serum glutamate transaminase; GGPT, γ-glutamyl transpeptidase.

before or after the sample. Blood was collected between 7:00 am and 4:00 pm for 95% of the samples studied. Blood was drawn at the time of semiannual TB testing, or from a number of specific studies designed to collect samples for normative values. Most preweaning samples from mother-reared infants were taken without anesthesia, while postweaning samples were taken either with or without ketamine anesthesia. Most IPRL samples were taken without anesthesia. A preliminary analysis, adjusted for age and rearing group as covariates, revealed no anesthesia effects on any of the measures summarized here.

Table 27–3. Analysis of Variance Results for Hematology and Serum Chemistry Measures Analyzed for Sex Effects (Age 1–3 Month Hematology) and Sex (S), Rearing (R), and Sex × Rearing (S × R) Effects (Quarters 2, 3, and 4)

	Months 1–3	Months 4–6			Months 7–9			Months 10–12		
	S	S	R	S × R	S	R	S × R	S	R	S × R
Hematology[a]										
Hemoglobin										
Hematocrit						**				
Red cells										
MCH			*			*			*	
MCV	*					*			**	*
MCHC			**			*				*
Platelets										
White cells			*							
Lymphocytes		*		*	**	**	*	*		
Neutrophils										
Monocytes										
Eosinophils									**	
Basophils										
Serum chemistry										
Sodium										
Potassium										
Chloride		**	**	**					**	
Carbon dioxide			**			*			**	
Total protein						*			**	
Albumin			**					**	*	
Globulin					*	*		**		
Calcium								*	**	
Bilirubin		**	**	**					**	
Glucose										
Creatinin		*	**	**						
Alkaline										
SGOT		*								
SGPT		*	**							
GGTP						**				

[a] MCH, mean corpuscular hemoglobin; MCV, mean cell volume; MCHC, MCH concentration; SGOT, serum glutamate oxaloacetate transaminase; SGPT, serum glutamate pyruvate transaminase; GGTP, γ-glutamyl transpeptidase.
* $P \leq 0.05$.
** $P \leq 0.01$.

Most samples consisted of 1–3 ml of blood drawn into unheparinized syringes from a femoral vein, with blood transferred into EDTA tubes and analyzed immediately or centrifuged and frozen for later analysis. Most (98.6%) of the hematology and serum chemistry assays were performed in the clinical laboratory at the PFS or the main facility of the WaNPRC or in the Department of Medicine at the University of Washington Medical Center. Assays used standard automated equipment, often supplemented by manual differential counts. A preliminary analysis, adjusted for age and rearing condition as covariates, revealed no significant laboratory differences on any of the measures reported here.

To maintain a reasonable sample size, data describing nursery- and mother-reared female and male infants were summarized in 3-month age intervals (quarterly). The mean value for each subject in each quarter served as the data. There were insufficient serum chemistry samples for any analyses during months 1–3. Univariate analyses of variance were performed for each hematology and serum chemistry measure with sex, rearing group, and the sex × rearing interaction as factors. Although hematology values for quarter 1 are presented for all groups, they were analyzed only for the nursery-rearing sex effect, as the mother-reared group had too few values for meaningful analysis in quarter 1. Analysis of variance results are presented in Table 27–3.

ACKNOWLEDGMENTS

Collection of these data was supported by NIH NCRR grant RR00166 and NICHHD grants HD08633 and HD02274.

CHAPTER TWENTY-EIGHT

Hematology and Serum Chemistry Reference Values for Mother-Reared Squirrel Monkey (*Saimiri boliviensis boliviensis*) Infants

Lawrence E. Williams

These reference data were obtained from the database maintained at the Squirrel Monkey Breeding and Research Resource (supported by NIH NCRR P40-RR01254). The samples were collected as part of research protocols and clinical observations. Time of day was not controlled across samples, although the majority were collected between 7 am and 10 am (Tables 28–1 and 28–2). All samples were collected from non-fasted, awake (unanesthetized) Bolivian squirrel monkeys (*Saimiri boliviensis boliviensis*) that were manually restrained. A 1- or 3-ml syringe was used to collect 0.5–1.5 ml of whole blood (amount depending on age) from the femoral vein. Following blood collection the animal was returned to its social group and/or placed back on its dam. The whole procedure required less than 2 min.

Table 28–1. Serum Chemistry Reference Values for Bolivian Squirrel Monkeys Aged 1–12 Months

Variable	Obs	Mean	SD	25th percentile	75th percentile
Female					
Alkaline phosphate (μg/liter)	42	1147.55	348.98	695	1798
γ-Glutamyl transferase (μg/liter)	32	15.19	5.58	8	26
Lactic dehydrogenase (μg/liter)	39	146.23	31.46	89	210
Glutamic oxaloacetic transferase (μg/liter)	23	132.26	23.27	98	168
Alanine aminotransferase (μg/liter)	36	75.36	40.23	18	143
Cholesterol (mg/dl)	18	200.72	28.20	148	244
Glucose (mg/dl)	68	105.57	17.21	75	136
Phosphorus (mg/dl)	41	7.04	0.67	6	8
Albumin (g/dl)	49	3.54	0.17	3.2	3.8
Total bilirubin (mg/dl)	19	0.63	0.41	0.2	1.3
Calcium (mg/dl)	39	9.53	0.35	8.9	10.1
Creatinine (μg/dl)	39	0.6	0.08	0.5	0.7
Sodium (mEq/liter)	40	145.29	1.9	142	148
Potassium (mEq/liter)	33	5.13	0.56	4.1	6.1
Globulin (g/dl)	12	2.06	0.52	1.3	3.1
Total protein (g/dl)	94	5.84	0.79	3.6	7.6
Male					
Alkaline phosphate (μg/liter)	38	1390.97	289.47	861	2007
γ-Glutamyl transferase (μg/liter)	17	15.59	3.41	10	22
Lactic dehydrogenase (μg/liter)	39	163.54	55.27	80	289
Glutamic oxaloacetic transferase (μg/liter)	31	121.16	37.77	69	189
Alanine aminotransferase (μg/liter)	51	49.84	42.13	2	132
Cholesterol (mg/dl)	30	162.13	31.89	121	230
Glucose (mg/dl)	64	92.97	15.87	68	120
Phosphorus (mg/dl)	43	7.02	0.84	5.7	8.5
Albumin (g/dl)	39	3.54	0.22	3.2	3.9
Total bilirubin (mg/dl)	20	0.35	0.23	0.1	0.8
Calcium (mg/dl)	32	9.23	0.54	8.2	10.1
Creatinine (μg/dl)	41	0.66	0.11	0.5	0.8
Sodium (mEq/liter)	38	145.20	2.09	141.8	148.4
Potassium (mEq/liter)	42	4.73	0.70	3.8	6.1
Globulin (g/dl)	6	2.43	0.29	2	2.9
Total protein (g/dl)	83	5.81	0.93	2.8	8.0

The data were examined for outliers using a criterion of 1.5 times the interquartile range. Outliers may have been the result of an abnormal medical condition or problems in the sample analysis. All data from a sample containing an outlier were dropped from the dataset before the means, standard deviations, and percentiles were generated.

Table 28–2. Hematology Reference Values for Bolivian Squirrel Monkeys Aged 1–12 Months

Variable	Obs	Mean	SD	25th percentile	75th percentile
Female					
White blood cells ($10^3/\mu l$)	59	9.20	1.16	7.4	11.2
Red blood cells ($10^6/\mu l$)	59	6.55	0.27	6.03	7
Hemoglobin (g/dl)	70	12.94	0.72	11.8	14.3
Mean corpuscular volume (fl)	47	62.19	1.32	60	64.3
Mean corpuscular hemoglobin (MCH—pg)	24	20.06	0.52	19.2	21
MCH concentration (g/dl)	22	32.90	0.85	31.7	34.3
Red cell distribution width	8	13.04	2.70	10.3	17.2
Platelet count	11	418.36	83.78	307	569
Male					
White blood cells ($10^3/\mu l$)	42	9.84	1.73	7.3	12.8
Red blood cells ($10^6/\mu l$)	39	6.42	0.35	5.85	7.05
Hemoglobin (g/dl)	45	13.17	0.84	11.8	14.7
Mean corpuscular volume (fl)	40	62.77	2.06	60	66
Mean corpuscular hemoglobin (MCH—pg)	19	20.03	0.62	19.1	21.11
MCH concentration (g/dl)	22	32.33	0.88	30.9	34.3
Red cell distribution width	4	10.95	0.25	10.7	11.3
Platelet count	4	412.75	64.52	326	476

INDEX